# DEVELOPMENTS IN RELIABLE COMPUTING

Edited by

## TIBOR CSENDES

József Attila University, Szeged, Hungary

**Kluwer Academic Publishers**
Dordrecht / Boston / London

A C.I.P. Catalogue record for this book is available from the Library of Congress.

ISBN 978-90-481-5350-3

Published by Kluwer Academic Publishers,
P.O. Box 17, 3300 AA Dordrecht, The Netherlands.

Sold and distributed in North, Central and South America
by Kluwer Academic Publishers,
101 Philip Drive, Norwell, MA 02061, U.S.A.

In all other countries, sold and distributed
by Kluwer Academic Publishers,
P.O. Box 322, 3300 AH Dordrecht, The Netherlands.

*Printed on acid-free paper*

## Attila József: By the Danube

*Sitting on the steps of the quay*
*I watched a melon rind floating away.*
*Submerged in my fate I barely heard*
*the surface chatter. From the depths: not a word.*
*And just like my heart's high tide*
*the Danube was murky, wise wide.*

*Like muscles at work, muscles that*
*make bricks, dig, hammer and hoe—*
*that's how each movement of each wave*
*kicked into action, tensed and let go.*
*The river, like mother, told tales, lulled me,*
*and washed the city's dirty laundry.*

*A slow drizzle began to fall, but soon*
*gave up, as if it were all the same.*
*Still, I watched the horizon like one*
*inside a cave, watching a steady rain:*
*this gray eternal reain pouring, steadfast,*
*so transparent now, the many-colored past.*

*The Danube flowed on. Small white-*
*crested waves played laughing my way*
*like children in the lap of a fertile*
*young mother who sits dreaming away.*
*Awash in time's flood they chattered,*
*so many headstones, graveyards shattered.*

...

Translated by John Bátki.
Corvina, Budapest, 1997.

# Contents

# Preface

The SCAN conference, the International Symposium on Scientific Computing, Computer Arithmetic and Validated Numerics, takes place biannually under the joint auspices of GAMM (Gesellschaft für Angewandte Mathematik und Mechanik) and IMACS (International Association for Mathematics and Computers in Simulation).

SCAN-98 attracted more than 100 participants from 21 countries all over the world. During the four days from September 22 to 25, nine highlighted, plenary lectures and over 70 contributed talks were given. These figures indicate a large participation, which was partly caused by the attraction of the organizing country, Hungary, but also the effective support system have contributed to the success. The conference was substantially supported by the Hungarian Research Fund OTKA, GAMM, the National Technology Development Board OMFB and by the József Attila University. Due to this funding, it was possible to subsidize the participation of over 20 scientists, mainly from Eastern European countries. It is important that the possibly first participation of 6 young researchers was made possible due to the obtained support. The number of East-European participants was relatively high. These results are especially valuable, since in contrast to the usual 2 years period, the present meeting was organized just one year after the last SCAN-xx conference.

The scientific contents of SCAN meetings traditionally cover all aspects of validation techniques in scientific computing, ranging from hardware requirements, elementary operations, high accuracy function evaluations and interval arithmetic to advanced validating techniques and applications in various fields of practical interest. To emphasize just some of the many topical subjects treated at SCAN-98 I mention the use of validation techniques in the analysis of dynamical systems, parallel validating algorithms, systems of linear and nonlinear equations and global optimization, complexity results for problems with uncertain data, methods for ordinary and partial differential equations, and various applications. The full program and the volume of extended abstracts is available (among other information) on the web site of the conference:

http://www.inf.u-szeged.hu/~scan98

The present proceedings contain 30 contributions presented at SCAN-98, out of which 11 articles appear also in the special issue **5** (3) of the Journal of *Reliable Computing*. We are grateful to all authors for providing us with their interesting and well-written papers, to the more than 60 referees for their speedy and conscientious work. Seeing the quality of the resulting articles, I am sure the flexible settings of the submission and correction deadlines and the page limitation are justified. The two round refereeing procedure and the final editing providing the camera ready versions took some five months only. Due to technical reasons, the articles could not be grouped according to the subfields.

I would like to thank the members of the scientific committee for assisting in preparing the conference and selecting contributions. The advice of Götz Alefeld, Andreas Frommer, Jean-Michel Muller, and Ulrich Kulisch were very valuable, they are kindly acknowledged. I am indebted to George Corliss, who prepared the bibliography and volunteered refereeing many manuscripts in critical phases allowing a timely publication. I wish to express my particular gratitude to the members of the organizing committee involved in preparing and running the conference: András Erik Csallner, Mihály Csaba Markót, and Péter Gábor Szabó. Finally I thank Dmitry Kirshov, Vyacheslav Nesterov and Paul Roos for their kind help in the edition tasks of these volumes.

TIBOR CSENDES
Szeged, March 1999

T. Csendes (ed.), Developments in Reliable Computing 1–16.

1

# Rigorous Global Search: Industrial Applications*

GEORGE F. CORLISS                    georgec@mscs.mu.edu

*Department of Mathematics, Statistics, and Computer Science, Marquette University, P.O. Box 1881, Milwaukee, WI 53201–1881*

and

R. BAKER KEARFOTT                    rbk@usl.edu

*Department of Mathematics, University of Southwestern Louisiana, U.S.L. Box 4–1010, Lafayette, LA 70504–1010*

**Abstract.** We apply interval techniques for global optimization to several industrial applications including Swiss Bank (currency trading), BancOne (portfolio management), MacNeal-Schwendler (finite element), GE Medical Systems (Magnetic resonance imaging), Genome Theraputics (gene prediction), inexact greatest common divisor computations from computer algebra, and signal processing. We describe each of the applications, discuss the solutions computed by Kearfott's GlobSol software (see www.mscs.mu.edu/~globsol), and tell of the lessons learned. In each of these problems, GlobSol's rigorous global optimization provided significant new insights to us and to our industrial partners.

**Keywords:** global optimization, currency trading, portfolio management, finite element analysis, least common denominator.

## 1. Problem domain

For a continuous, differentiable objective function $f : X \subset \mathbb{R}^n \to \mathbb{R}$, consider the global optimization problem,

$$\min_{x \in X} f(x),$$

possibly with linear or nonlinear constraints. We seek validated, tight bounds for the set of minimizers $X^*$ and/or the optimum value $f^*$. Linear systems $Ax = b$ and nonlinear systems $G(x) = 0$ for $G: \mathbb{R}^n \to \mathbb{R}^n$, are also solved by components in our tool box. The rigorous global optimization algorithm[10, 15] uses an exhaustive branch and bound search. It discards regions where the solution is guaranteed *not* to be, it attempts to validate the existence of a unique local minimum in a box, and it makes the enclosing box tight. In contrast, conventional approximate algorithms for global optimization include exhaustive search, simulated annealing [11], and genetic algorithms [1].

---

* This work is supported in part by Sun Microsystems and by the National Science Foundation under various grants.

The validated global optimization algorithm used by the GlobSol software package is related to algorithms described by Hansen [10], Neumaier [18], Ratz [20, 21, 22], Jansson [13, 14], or van Iwaarden [12]. Details of the GlobSol algorithm appear in [15], an overview appears in [5], and details of the applications described here are available from the GlobSol web site: www.mscs.mu.edu/~globsol

A general global optimization problem may be difficult because:

**Feasibility:** Is there a feasible solution? Interval techniques safely discard regions where the solution cannot be and enclose regions guaranteed to be feasible.

**Many local critical points:** Is the local minimum we have found really the global minimum? For an objective function with many local minima, like an egg carton, interval arithmetic can guarantee that a local minimum is indeed a true global minimum.

**Singular or near-singular solutions:** A nearly flat objective function, like the top of an egg carton, exhibits singular (or nearly singular) behavior. Interval techniques can enclose the entire set of points $X^*$ at which the global minimum is attained.

**Uncertain parameters:** Many models contain parameters whose value is not precisely known. Interval techniques automatically handle interval-valued parameters.

**High dimensions:** The algorithm is an exhaustive branch and bound search to guarantee that the global optimum cannot be missed. The code tries to be clever, but the worst-case complexity is $O(2^n)$ for a problem with $n$ independent variables. The complexity is also exponential in the requested tolerance. For some problem classes, the observed complexity is $O(n^3)$ governed by the linear algebra. The largest problem we have successfully handled so far is $n = 128$ (a mildly nonlinear trayed tower model solved by Schnepper, a student of Stadtherr, taking special account of sparsity). Work is in progress with parallel architectures, better heuristics, and sparsity to handle larger problems.

We outline several actual industrial applications we have addressed using the GlobSol software. The intent is to convey the wide domain of applicability of the rigorous global search methods. More details about each application are available from technical reports cited here and may appear in subsequent papers.

## 2. Swiss Bank (Currency Trading)

Swiss Bank's portfolio is managed using a proprietary risk control system. This system aggregates the currency trades and determines an overall value of the portfolio. The total value of the portfolio is evaluated in the light of several risk factors. The objective of the system is to determine the value at risk (VaR), or the maximum overnight loss the bank may face on the portfolio of trades, assuming that the portfolio is illiquid overnight.

The maximum loss is determined for the portfolio closing position by allowing the risk factors to change and calculating the price sensitivities of the trades to these changes in the risk factors. As the risk factors are allowed to change within specified ranges, it becomes difficult to calculate the maximum loss. The theoretical shape of the portfolio frontier becomes more complex as the number of currencies and risk factors increases. Therefore, it becomes important to guarantee that a global maximum is found, as a local maximum may leave a great deal of risk hidden.

Value at Risk (VaR) is a measure of the maximum loss in market value over a given time interval within a given confidence interval. That is, the loss of market value that is exceeded by a given probability. The Bank for International Settlements, in determining adequate bank capital, suggests a 99% confidence interval and a ten day time interval. Given a 99% confidence interval, the VaR approach yields a single estimate of potential loss that would be exceeded in value with only 1 percent probability. As the VaR methodology has gained international acceptance, it has become more important to banks as a risk management tool. Swiss Bank plans to expand its use of the methodology and improve their current approaches. By being able to accurately assess the maximum expected overnight loss exposure, the Bank will be in a much better position to adjust its holdings to reduce that potential loss and satisfy regulatory requirements.

GlobSol was tested using the Garman-Kohlhagen model for valuing foreign exchange options and for determining the VaR for portfolios containing calls and puts on 2, 4, and 7 exchange rates. Initial experiments suggest that the CPU time for these problems grows exponentially with the number of instruments in the portfolio. In these models, market forces work to make the objective functions nearly flat, with many small relative local optima. The guarantee of GlobSol's solutions is that we are certain to have found the best of the local optima.

For the Value at Risk model, we solve the problem of finding the VaR of a portfolio. We claim that not only the maximum loss of the given portfolio containing any finite number of instruments is found, but also guarantee that it is the global maximum that is enclosed in tight bounds.

We used GlobSol to find the minimum variance portfolio for one two week period in October, 1997, of a portfolio containing Swiss Franks, Mexican Pesos, South Korean WONs, Australian Dollars, Netherlands' Guilders, and Singapore Dollars. The minimum variance portfolio had 74.7% Mexican Peso, 8.9% Australian Dollar, 8.4% Netherlands' Guilder, and 8.0% Singapore Dollar. The mini-

mum variance for the two week period was 0.00055% of the portfolio value, while
the variances of the individual currencies ranged from 0.0013% to 0.049%.

The Swiss Bank application is joint work with Shirish Ranjit, Joe Daniels, and
Peter Toumanoff. Details appear in [6, 19] and at
`www.mscs.mu.edu/~globsol/Marquette/apSwiss.html`

## 3. BancOne (Portfolio Management)

We consider the universal portfolio management problem: Given various investment constraints (cash flow, risk element, maturity structure, corporate investment
policies, etc.), what combination of securities will maximize return? This concept
can be most easily related to efficient frontier and security market line solutions of
classic finance. Although quite simple in theory, a solution is difficult to implement
for several reasons:

- Managers may be responsible for several portfolios, each with different investment policies and objectives. This renders the efficient frontier of investments
  subordinate to investment policy.

- Selecting securities from the universe of acceptable investments, pricing options, hedging strategies, even for the most basic fixed income portfolio, can
  require substantial amounts of time.

- Statistics used to judge risk and return are often time consuming to calculate,
  numerous, and of uncertain accuracy. Therefore, these statistics are often not
  considered completely.

For these and other reasons, rule-of-thumb investment management often substi-
tutes for more theoretically sound investment portfolio management and asset techniques. For instance, securities may be selected because they fall within the investment policy of the portfolio, or because the security matches the historical return
or average life of the portfolio.

The required software solution to the problem first involves selecting securities
from a large universe of fixed income securities which meet the required investment
parameters (i.e. the investment policy of the fund). Then, given these securities,
select the optimal mix (or asset allocation) to maximize the expected return of the
portfolio while meeting bank policy objectives regarding the duration to maturity,
cash flow, and risk. In addition, given the selected optimal portfolio(s), the software
solution demonstrates to the money manager the additional opportunities and risks
which exist by relaxing one or several of the policy constraints.

The solution requires flexibility to accommodate a wide range of investment
styles. For instance, risk may be defined in several ways: credit risk, interest rate

risk, variance in return, maturity, etc. Similarly, return may be defined as cash flow yield, coupon yield, total return, capital appreciation, etc. Finally, the final solution also addresses some of the usual portfolio constraints, including the following: tolerance for duration and convexity; required cash flow timing; selected average life of the portfolio; coupon yields or maturity dates; credit ratings; selected responsiveness to changes in partial duration.

A sample fixed-income portfolio is divided into two different models for this analysis. First, we model a portfolio containing only 3-, 6-, and 12-month Treasury bills. To simulate a real-world Treasury bill portfolio, we used actual spot price data for each type of Treasury bill from a recent U.S. Treasury auction. Historical spot price data for each bill was used to calculate the variances of each type of T-bill and covariances between bills. Using a typical set of constraints, GlobSol calculated the optimal allocation to be about 10% in 3-month T-bills, 60% in 6-month T-bills, and 30% in 12-month T-bills, for an average rate of return of 5.363% per year.

Second, we model a portfolio containing only 1-, 5-, 10-, and 30-year Treasury bonds. The objective in each case is to maximize the portfolio's rate of return, given four investment constraints of average duration, cash flow, variance, and total value. For the Treasury bond portfolio model, GlobSol computes the maximum portfolio rate of return subject to four investment constraints. Treasury bond variances and covariances between bonds were calculated using historical data. Hypothetical values were chosen for spot prices, coupon payments, and face values. We computed the rate of return for each bond. For a typical set of constraints, GlobSol determined the optimal allocation to be about 4% in 1-year bonds, 65% in 5-year bonds, 4% in 10-year bonds, and 26% in 30-year bonds. This allocation generated an optimal rate of return of 7.388%.

The GlobSol results guarantee that 1) All constraints are satisfied, and 2) There is no allocation satisfying the investment constraints that offers a higher yield. Guaranteeing that we have determined the optimal allocation is a significant statement to make. This implies that we have assured that all constraints are satisfied. From the perspective of the portfolio manager, this is a guarantee that all portfolio objectives are being met. With financial market volatility constantly changing the value, risk, and return of the portfolio, actual portfolio objectives may not be being met even if the manager is actively trying to meet them. Therefore, the manager may not be able to make this assertion on a consistent basis. The results from GlobSol imply that we can say with 100% certainty that the constraints are being satisfied. More importantly, the generated allocation of securities yields the optimal rate of return given the constraints. The manager can be assured that potential returns are not being missed, and shareholders are benefitting from optimal portfolio management.

The Banc One application is joint work with Paul Thalacker, Joe Daniels, Peter Toumanoff, Kristie Julien, and Ken Myszka. Details appear in [23] and at `www.mscs.mu.edu/~globsol/Marquette/apBancOne.html`

## 4. MacNeal-Schwendler (Finite Element Analysis)

A rocket engine exhaust nozzle's primary function is to direct the outward flow of high-pressure exhaust gas produced by the engine. In this manner, the thrust of the engine is focused, thus achieving the jet propulsion necessary to drive the vehicle forward.

While in operation, a rocket engine nozzle cone experiences an enormous internal material stress due to the massive thrust pressure and temperature created by the ignition of the rocket fuel within the engine. This material stress tends to act on the nozzle cone in a circumferential direction, trying to blow the nozzle apart. This type of material stress is exactly the same stress that causes a toy balloon to rupture from excessive internal air pressure. The task of the rocket nozzle designer is to design a nozzle cone that is strong enough to withstand the extreme forces and temperatures generated within the nozzle during engine operation, while obeying all design specifications.

Design problems like the rocket engine nozzle stress pose incredible challenges to aerospace and materials engineers. To reliably solve these difficult problems in very complex shapes, engineers employ the help of several analysis tools to accurately model the behavior of their designs on a computer before a prototype is actually constructed. One of the most powerful numerical analysis tools utilized by designers is the Finite Element Method.

We decided to pursue the first-principles finite element analysis, or the "all-together" approach, in which the finite element equations are presented as constraints to the optimization problem. Only values satisfying the FEA equations are considered feasible designs over which optimization can be performed. The method takes its name from a technique proposed by John Dennis and Karen Williamson for differential equation parameter identification problems. To gain experience with the all-together finite element approach, we first attacked a VERY simple structural analysis problem, minimizing the maximum stress an axially loaded elastic bar.

We have successfully modeled a 97 node axially loaded bar with the all-together approach. The technique begins with a model consisting of four nodes and three elements. GlobSol solves this problem and returns a solution consisting of four tightly bound node intervals. The results of this model run are then used to build a new model consisting of twice the number of elements. GlobSol now gets half of the model nodes bounded tightly, and once again GlobSol solves the problem. This procedure is repeated until we decide to stop at a model consisting of 96 elements and 97 nodes. The total CPU time required by GlobSol to complete the entire series was about 2.5 hours on a Sun SPARC 10. This model run is significant because it demonstrates that GlobSol can solve a problem based on the all-together technique. Our primary concern is CPU time required.

The MacNeal-Schwendler application is joint work with Frank Fritz, Andy Johnson, Don Prohaska, and Bruce Wade. Details appear in [8, 9] and at `www.mscs.mu.edu/~globsol/Marquette/apMacNeal.html`

## 5. GE Medical Systems (Magnetic Resonance Imaging)

MRI Imaging Instrument is one of the major sophisticated, state-of-the-art, and expensive medical imaging products manufactured by GE Medical Systems. Through a clinical experiment, 3D medical images of a patient are generated by the principle of Nuclear Magnetic Resonance (NMR). One of the key components in the MRI is the MRI coil, which generates the electro-magnetic fields around the patient. The coil is stimulated by current impulses of a certain shape, which are expected to generate an electro-magnetic field of similar shape. However, due to principles of physics, the current impulse is not uniformly distributed in the coil. The current passing through the edge of the coil is greatly distorted. This is called the "Eddy Current" effect, which causes the distortion of the corresponding electro-magnetic fields, resulting a distorted medical image. Thus this eddy current effect must be corrected or compensated using the pre-generated compensation current.

The problem is formulated as a sequence of parameter identification problems:

**Linear Model**
We have a set of inputs (experimental $f(u,t)$) and their associated outputs (experimental $G(t)$). Initially, we assume the model $F$ to be described by a linear, constant coefficient differential equation

$$x'' + a_1 x' + a_2 x = a_3 f(u,t).$$

Then the optimization problem is
**Independents:** $a_1, a_2, a_3$
**Objective:** Sum over all samples $|G(t) - x(t)|_2$

This is an example of a general parameter identification problem. We also consider fitting in a uniform or weighted sense.

**General Model**
We do not expect the linear model to fit sufficiently well, so we explore nonlinear parameterized differential equations

$$x' = F(x,t,f(u,t),a).$$

Then the optimization problem is

**Independents:** $a$

**Objective:** Sum over all samples $|G(t) - x(t)|_2$.

### Discover the Driving Function

Once we have an appropriate model $F$, we determine an optimal driving function $f(u,t)$. For example, we may assume $f$ can be expressed as sum of exponential functions:

$$f(u,t) = \sum \alpha_i \exp(-t/\tau_i),$$

and use optimization to find the best parameters. The optimization problem is

**Independents:** $a_i, \tau_i$

**Objective:** $|G(t) - x(t)|_2$.

In each case of interest, we wish to determine parameters such that the solution to a differential equation fits a target, so we consider the general dynamical systems parameter identification problem. Given a set of data $(t_i, x_i)$, or a target function $G(t)$, and a differential equation

$$x' = f(t, x, a),$$

determine values for the parameters $a$ such that $x(t)$ best fits:

$$\min |G(t) - x(t)|_2.$$

**Independents:**

$a$    parameters in the DE,

$c_j$    coefficients of the basis functions $b_j$, and

$x_i$    approximate solution at the nodes.

**Objective:** Sum over all samples $|G(t) - x(t)|_2$

**Constraints:** $\sum c_j b_j'(t_i) = f(t, \sum c_j b_j(t_i), a)$.

We have solved one example of the second order linear ODE parameter identification problem with 20 nodes using the interval all-together approach, demonstrating the promise of the approach. However, we have not yet been able to scale up our initial success because the differential equation is moderately stiff, the parameters have widely different scales, and the system of constraint equations is poorly conditioned.

The GE Medical Systems application is joint work with Xin Feng, Ruoli Yang, Yunchuan Zhu, Yonghe Yan, and Robert Corless. Details appear in [7] and at www.mscs.mu.edu/~globsol/Marquette/apGEMed.html

## 6. Genome Theraputics (Gene Prediction)

All mammalian genes discovered to date have revealed a pattern to how their information is transcribed and translated. This pattern of information or structure of a gene can form a basis of an intelligent search of a similar DNA sequence or collection of sequences. This premise forms the basis for the identification of a gene by a process of pattern recognition using multiple genomic features from the different regions of a gene (Promoter and Structural regions). Over the last fifteen years, many molecular biologists, mathematicians, and computer scientists have attempted to develop numerical and non-numerical programming methods to analyze, understand, and decode the genetic information within DNA. As methods in molecular genetics expanded, the sequencing of DNA from a variety of species grew dramatically, the structure and function of genes were discovered, and programs and methods have evolved to discover the pattern of information in a gene.

Consider a researcher with a DNA sequence of interest. The sequence is submitted to several gene search engines available on the Internet, and each returns an indication of the genes it recognized and perhaps an indication of its confidence. Usually, the results from different gene search engines are not in perfect agreement. The researcher must reconcile the conflicting answers before beginning expensive laboratory analysis.

Our goal is a meta-engine to

- Accept the researcher's DNA sequence,

- Submit it to multiple Internet gene search engines,

- Gather the results,

- Reconcile the conflicting results, and

- Report a "consensus" answer.

The Meta Gene Prediction Engine will use a neural network trained by the GlobSol global optimization software. The goal is to extract more accurate gene information on stretches of DNA from existing gene prediction systems.

The objective function for training a general neural network, whether we use an $L_2$ or an $L_\infty$ norm, is nearly flat or even singular. Since we do not know in advance the architecture of a net with power sufficient for the problem at hand, we often try a large network. The resulting optimization problem may be over-parameterized. Conventional techniques search down-hill until they find a local minimum, but starting at different points often yields wildly different parameter values for the neural network.

Validated optimization techniques can reliably distinguish local from global minima and can characterize sets of equally good parameter values. However, characterization of sets of solutions is very CPU intensive, so we have only handled very small neural networks. Work is continuing to improve the performance of Glob-Sol for sets of solutions. A prototype of the Meta Gene engine is on the web at `ares.ifrc.mcw.edu/MetaGene`. This prototype does not use GlobSol, but it does use optimization techniques.

The Genome Theraputics application is joint work with Zhitao Wang and Peter Tonellato. Details appear in [25] and at
`www.mscs.mu.edu/~globsol/Marquette/apGeneome.html`

## 7. Inexact Greatest Common Divisor

Given two polynomials $p(x)$ and $q(x)$ with inexact floating-point coefficients, how can we compute their greatest common denominator $d(x)$ (GCD)? The GCD is the polynomial of highest degree that divides both $p(x)$ and $q(x)$ exactly. The GCD is unique, up to multiplication by a constant. If the polynomials are assumed to be known exactly, the GCD can be computed exactly by the Euclidian algorithm, or a variant, as is done by any current computer algebra (CA) system. However, if the coefficients are inexactly known (e.g. floating-point numbers), the polynomials are almost surely relatively prime, and their GCD is 1. This answer is true, but not helpful. Most CA users want a polynomial of as high degree as possible that is the GCD of a pair of 'nearby' polynomials [4].

If $p(x)$ and $q(x)$ are two (or more) univariate polynomials with inexact floating-point coefficients, then an $\varepsilon \geq 0$ GCD of $p$ and $q$ is a polynomial $d(x)$ of highest possible degree such that $d(x)$ exactly divides $p(x) + \Delta p(x)$ and $q(x) + \Delta q(x)$ for some $\Delta p$ and $\Delta q$ with $\|\Delta p(x)\| \leq \varepsilon$ and $\|\Delta q(x)\| \leq \varepsilon$. We use the 2-norm of the vector of coefficients of a polynomial.

We have successfully applied GlobSol to several example problems from the literature, see [3]. For example, let

$$p(x) := x^5 + 5.503x^4 + 9.765x^3 + 7.647x^2 + 2.762x + 0.37725$$
$$q(x) := x^5 - 2.993x^4 - 0.7745x^3 + 2.007x^2 + 0.7605x$$

from [4]. GlobSol found

$$d(x) \in x^2 + [1.\underline{0069986925}, 1.\underline{0069986978}]x + [0.\underline{253494953}, 0.\underline{253494961}]$$

with $\|\Delta p(x)\|^2 + \|\Delta q(x)\|^2 \approx 8 \times 10^{-7}$ in about 160 CPU seconds on a SPARC 10. For this example, giving GlobSol the correct answer as an initial guess made no difference in execution time.

The inexact greatest common divisor application is joint work with Paulina Chin and Robert M. Corless. Details appear in [2, 3] and at
www.mscs.mu.edu/~globsol/Marquette/apGCD.html

## 8. Signal Processing

Parameter estimation plays an important role in many areas of scientific and engineering computation including system identification for the design of control systems, pattern recognition systems, equalization of communications channels, and artificial neural networks. In each of these applications, parametric models are used to either emulate an unknown system, perform a specific task within the system, or mitigate anomalies caused by uncertainty. Conventional estimation methods are often used to compute the parameters so that the mean-squared error between the model output and the desired response is minimized.

Kelnhoffer [17] used interval global optimization techniques very similar to those used by GlobSol to construct bounded response models. That is, he computed intervals of parameter values that are consistent with model data. He applied his technique to problems from system identification, pattern recognition, channel equalization, and artificial neural networks.

One exciting aspect of Kelnhoffer's work is his interpretation of the results as model validation, as first suggested by Walster [24]. If the interval optimization technique returns a solution (a set of parameter values consistent with the data), it has validated the model. Often more helpful, though disappointing, the technique often returns no solution. In that case, we have validated that the model is *inconsistent* with the data. The model builder must construct a new model. Interval techniques have this unique property of helping the modeler stop trying to work with the wrong model.

The details of Kelnhoffer's signal processing application appear in [17].

## 9. On GlobSol's Algorithm

Here, a brief review of the principles of interval branch and bound algorithms is given, as well as an overview of some unique aspects of the GlobSol package. Detailed descriptions appear in [15], [16], or from the World Wide Web page
www.mscs.mu.edu/~globsol/User_Guide/13WhatIs/. The GlobSol working note [5] includes an elementary introduction to the techniques in interval branch and bound algorithms for global optimization.

### 9.1. Branch and Bound

With interval arithmetic to compute rigorous range bounds and automatic differentiation to compute derivatives, the algorithm of GlobSol proceeds by branch and bound. We maintain three lists of boxes:

$\mathcal{L}$    not fully analyzed;

$\mathcal{R}$    small boxes validated to enclose a unique local minimum
(in the unconstrained case), or a feasible point with small objective function value (in the constrained case); and

$\mathcal{U}$    small boxes not validated either to enclose a minimum
or *not* to enclose a minimum.

Find enclosures for $f^*$ and $X^*$ such that $f^* = f(X^*) \le f(X)$, for all $X \in X_0 \subset \mathbb{R}^n$:

Initialize $\mathcal{L} = \{X_0\}$     // List of pending boxes
$\overline{f} = +\infty$     // $f^* \le \overline{f}$
DO WHILE $\mathcal{L}$ is not empty
    Get current box $X^{(c)}$ from $\mathcal{L}$
    IF $\overline{f} < f(X^{(c)})$ THEN
       Discard $X^{(c)}$ // We can guarantee that there is no global minimum in $X^{(c)}$
    ELSE
       Attempt to improve $\overline{f}$
    END IF
    IF we can otherwise guarantee that there is no global minimum in $X^{(c)}$
       THEN Discard $X^{(c)}$
    ELSE IF we can validate the existence of a unique local minimum in $X^{(c)}$
       THEN Shrink $X^{(c)}$, and add it to $\mathcal{R}$
    ELSE IF $X^{(c)}$ is small THEN
       Add $X^{(c)}$ to $\mathcal{U}$
    ELSE
       Split $X^{(c)}$ into sub-boxes, and add them to $\mathcal{L}$
    END IF
END DO

The actual GlobSol code is much more sophisticated than the above suggests, but the concepts are similar. See [15] or [16] for details.

### 9.2. The Unique Features of GlobSol

Some features of GlobSol that distinguish it from other interval global optimization packages are briefly outlined here.

- The objective function and constraints are specified as Fortran 90 programs.

- Globsol can be configured to use <u>constraint propagation</u> (substitution/iteration) on the intermediate quantities in objective and constraint function evaluation. For some objectives and constraints, this can result in significant speedup.

- GlobSol can be configured to use an overestimation-reducing "peeling" process for bound-constraints.

- GlobSol uses a novel and effective point method to find approximate feasible points. When an approximate feasible point is found, GlobSol verifies bounds within which a true feasible point must exist, from which upper bounds for the global optima may be obtained.

- GlobSol has a special augmented system mode for least squares problems.

- Globsol uses epsilon-inflation and set-complementation, with carefully controlled tolerances, to avoid singularity problems, and to facilitate verification.

### 9.3. An Illustrative Example of GlobSol's Use

This simple example (illustrative, not real-world) illustrates use of GlobSol. Suppose the optimization problem is

$$\text{minimize} \quad \phi(X) = -2x_1^2 - x_2^2$$

subject to constraints

$$x_1^2 + x_2^2 - 1 \leq 0$$
$$x_1^2 - x_2 \leq 0$$
$$x_1^2 - x_2^2 = 0.$$

The following Fortran 90 program communicates this objective function and constraints to GlobSol.

```
PROGRAM SIMPLE_MIXED_CONSTRAINTS
  USE CODELIST_CREATION
    PARAMETER (NN=2)
    PARAMETER (NSLACK=0)
    TYPE(CDLVAR), DIMENSION(NN+NSLACK):: X
    TYPE(CDLLHS), DIMENSION(1):: PHI
    TYPE(CDLINEQ), DIMENSION(2):: G
    TYPE(CDLEQ), DIMENSION(1) :: C
    OUTPUT_FILE_NAME='MIXED.CDL'
    CALL INITIALIZE_CODELIST(X)
```

```
        PHI(1) = -2*X(1)**2 - X(2)**2
        G(1) = X(1)**2 + X(2)**2 - 1
        G(2) = X(1)**2 - X(2)
        C(1) = X(1)**2 - X(2)**2
        CALL FINISH_CODELIST
    END PROGRAM SIMPLE_MIXED_CONSTRAINTS
```

A data file of the following form communicates the bounds $[0,1] \times [0,1]$ on the search region to GlobSol.

```
1D-5            ! General domain tolerance
0    1          ! Bounds on the first variable
0    1          ! Bounds on the second variable
```

Separate configuration files supply algorithm options, such as which interval Newton method to use and how to precondition the linear systems.

Running the above program produces an internal representation, or <u>code list</u>. The optimization code then interprets the code list at run time to produce floating point and interval evaluations of the objective function, gradient, and Hessian matrix. An output file of the following form is then produced.

```
Output from FIND_GLOBAL_MIN on 06/28/1998 at 16:28:09.
Version for the system is: June 15, 1998
. . .
LIST OF BOXES CONTAINING VERIFIED FEASIBLE POINTS:
Box no.:1
Box coordinates:
[ 0.707106766730664638E+00,  0.707106795645393915E+00 ]
[ 0.707106766730642544E+00,  0.707106795645415898E+00 ]
PHI:
[-0.150000006134372299E+01, -0.149999993866885117E+01 ]
. . .
Number of bisections: 3
No. dense interval residual evaluations -- gradient
  code list: 83
Number of orig. system inverse midpoint preconditioner
  rows: 3
Number of orig. system C-LP preconditioner rows: 109
Number of Gauss--Seidel steps on the dense system: 112
Number of gradient evaluations from a gradient code
  list: 13
Total number of boxes processed in loop: 7
. . .
Overall CPU time:     0.2277E+00
CPU time in PEEL_BOUNDARY:      0.1365E-03
CPU time in REDUCED_INTERVAL_NEWTON:      0.1458E+00
```

This report says that GlobSol guarantees there is a minimizer satisfying the constraints in the box $[0.\underline{707106}766, 0.\underline{707106}796] \times [0.\underline{707106}766, 0.\underline{707106}796]$. The minimum feasible value of the objective is in the interval $[-1.\underline{500000}062, -1.\underline{499999}937]$. The intervals computed by GlobSol enclose the true values $X^* = (1/\sqrt{2}, 1/\sqrt{2})$ and $f^* = -3/2$. GlobSol required 83 function, and 13 gradient evaluations, and 0.23 CPU seconds on a Sun SPARC Ultra 140.

## 10. Conclusions

We have applied validated global optimization techniques to several practical industrial applications. For each application, we have successfully solved at least modest prototypical problems and have identified modeling and algorithmic advances necessary for success on problems whose size and complexity is of genuine industrial importance. Work is continuing on several of these applications to improve our models and to improve and parallelize the algorithms used by GlobSol.

## References

1. Bill P. Buckles and Fred Petry. *Genetic Algorithms*. IEEE Computer Society Press, 1992.
2. Paulina Chin, Robert M. Corless, and George F. Corliss. Globsol case study: Inexact greatest common denominators. Technical Report, Marquette University Department of Mathematics, Statistics, and Computer Science, Milwaukee, Wisc., 1998.
3. Paulina Chin, Robert M. Corless, and George F. Corliss. Optimization strategies for the floating-point GCD. In *ISSAC Proceedings*, to appear, 1998.
4. Robert M. Corless. Cofactor iteration. *SIGSAM Bulletin: Communications in Computer Algebra*, 30(1):34–38, March 1996.
5. George F. Corliss, Chenyi Hu, R. Baker Kearfott, and G. William Walster. Rigorous global search – Executive summary. Technical Report No. 442, Department of Mathematics, Statistics and Computer Science, Marquette University, Milwaukee, Wisc., April 1997.
6. Joseph Daniels, Shirish Ranjit, R. Baker Kearfott, and George F. Corliss. Globsol case study: Currency trading (Swiss Bank Corp.). Technical Report, Department of Mathematics, Statistics and Computer Science, Marquette University, Milwaukee, Wisc., 1998.
7. Xin Feng, Rudi Yang, Yonghe Yan, Yunchuan Zhu, George F. Corliss, and R. Baker Kearfott. Globsol case study: Parameter optimization for the eddy current compensation of MRI coils (General Electric Medical). Technical Report, Department of Mathematics, Statistics and Computer Science, Marquette University, Milwaukee, Wisc., 1998.
8. Frank Fritz. Validated global optimization and finite element analysis. Master's thesis, Marquette University Department of Mathematics, Statistics, and Computer Science, Milwaukee, Wisc., March 1999.
9. Frank Fritz, George F. Corliss, Andrew Johnson, Donald Prohaska, and Jonathan Hart. Globsol case study: Rocket nozzle design (MacNeal-Schwindler). Technical Report, Department of Mathematics, Statistics and Computer Science, Marquette University, Milwaukee, Wisc., 1998.
10. Elden R. Hansen. *Global Optimization Using Interval Analysis*. Marcel Dekker, New York, 1992.

11. R. Horst and M. Pardalos. *Handbook of Global Optimization*. Kluwer, Dordrecht, Netherlands, 1995.

12. Ron Van Iwaarden. *An Improved Unconstrained Global Optimization Algorithm*. PhD thesis, University of Colorado at Denver, Denver, Colorado, 1996.

13. Christian Jansson. A global optimization method using interval arithmetic. *IMACS Annals of Computing and Applied Mathematics*, 1992.

14. Christian Jansson. On self-validating methods for optimization problems. In J. Herzberger, editor, *Topics in Validated Computations*, pages 381–439, Amsterdam, Netherlands, 1994. North-Holland.

15. R. Baker Kearfott. *Rigorous Global Search: Continuous Problems*. Kluwer Academic Publishers, Dordrecht, Netherlands, 1996.

16. R. Baker Kearfott, M. Dawande, K. S. Du, and Chenyi Hu. Algorithm 737: INTLIB: A portable Fortran 77 interval standard library. *ACM Trans. Math. Software*, 20(4):447–458, 1994.

17. Richard Kelnhoffer. *Applications of Interval Methods to Parametric Set Estimation from Bounded Error Data*. PhD thesis, Marquette University Department of Electrical and Computer Engineering, 1997.

18. Arnold Neumaier. *Interval Methods for Systems of Equations*. Cambridge University Press, Cambridge, 1990.

19. Shirish Ranjit. Risk management of currency portfolios. Master's thesis, Marquette University Department of Economics, Milwaukee, Wisc., April 1998.

20. Dietmar Ratz. An inclusion algorithm for global optimization in a portable PASCAL-XSC implementation. In L. Atanassova and J. Herzberger, editors, *Computer Arithmetic and Enclosure Methods*, pages 329–338, Amsterdam, Netherlands, 1992. North-Holland.

21. Dietmar Ratz. Box-splitting strategies for the interval Gauss–Seidel step in a global optimization method. *Computing*, 53:337–354, 1994.

22. Dietmar Ratz and Tibor Csendes. On the selection of subdivision directions in interval branch-and-bound methods for global optimization. *Journal of Global Optimization*, 7:183–207, 1995.

23. Paul J. Thalacker, Kristie Julien, Peter G. Toumanoff, Joseph P. Daniels, George F. Corliss, and R. Baker Kearfott. Globsol case study: Portfolio management (Banc One). Technical Report, Department of Mathematics, Statistics and Computer Science, Marquette University, Milwaukee, Wisc., 1998.

24. G. William Walster. Philosophies and practicalities of interval arithmetic. *Reliability in Computing*, Ramone E. Moore, ed., pages 309–323. Boston, 1988. Academic Press.

25. Zhitao Wang, Peter Tonellato, George F. Corliss, and R. Baker Kearfott. Globsol case study: Gene prediction (Genome Therapeutics). Technical Report, Department of Mathematics, Statistics and Computer Science, Marquette University, Milwaukee, Wisc., 1998.

T. Csendes (ed.), *Developments in Reliable Computing* 17–30.

# Influences of Rounding Errors in Solving Large Sparse Linear Systems

AXEL FACIUS                                                    axel.facius@math.uni-karlsruhe.de
*Institut für Angewandte Mathematik, Universität Karlsruhe (TH)*

**Abstract.** In many research areas like structural mechanics, economics, meteorology, and fluid dynamics, problems are mapped to large sparse linear systems via discretization. The resulting matrices are often ill-conditioned with condition numbers of about $10^{16}$ and higher. Usually these systems are preconditioned before they are fed to an iterative solver.

Especially for ill-conditioned systems, we show that we have to be careful with these three classical steps — discretization, preconditioning, and (iterative) solving. For Krylov subspace solvers we give some detailed analysis and show possible improvements based on a multiple precision arithmetic. This special arithmetic can be easily implemented using the *exact scalar product* — a technique for computing scalar products of floating point vectors exactly.

**Keywords:** High precision arithmetic, discretization, preconditioning, Krylov subspace methods, large sparse linear systems

## 1. Introduction

This article is a survey of error sources affecting the entire process of solving linear systems, i.e., discretization, preconditioning and finally solving iteratively. In Section 2 we give a short summary of high precision arithmetic and discuss possible hardware supports. In Section 3 we introduce our model problem as well as some verifying solvers and the error types used in this paper. Section 4 demonstrates that once a seemingly tolerable error is introduced there is sometimes no way to obtain an acceptable *solution*. The same problems apply to the preconditioning as shown in Section 5. The only possible way to prevent such problems seems to be to implement preconditioning methods with a more precise arithmetic. In Section 6 we give a short introduction to Krylov subspace solvers and show as the main result the dependence of the Lanczos process on the precision of the arithmetic in use both theoretically and by experiments.

## 2. Short summary of high precision arithmetic

In many numerical algorithms there is a large gap between the theoretical, i.e., mathematical, behavior on the one hand and the finite precision behavior on the other hand. In cases where the accuracy of a result is insufficient or no results can be obtained at all due to poorly conditioned problems, it is desirable to have an arithmetic of higher precision. In many of these cases mathematical software like Mathematica or MAPLE is used because these programs provide arbitrary precision numbers, however at the price of high computational effort. And in many cases it is a prohibitive amount of work or at least impossible to translate an existing code of sometimes thousands of lines to these systems. Thus the question is how to provide high precision arithmetic that is fast and easy to use. In this summary we discuss a solution to this problem that requires very little computational effort and that allows us to increase the precision in critical computations. For this purpose we need a reliable floating point hardware (e.g. according to the IEEE standard 754-1985, [1]) and only one additional hardware operation: this is the *exact scalar product*, as implemented by means of a *long accumulator*, see [9, 10, 11].

Our high precision data type is represented as a sum of $l$ floating point numbers $X = \sum_{i=1}^{l} x_i$ which we call *staggered correction format*. If $|x_1| > \ldots > |x_l|$ and the exponents of two successive summands $x_i, x_{i+1}$ differ at least by the mantissa length $m$ we say that the mantissas in the staggered correction format do not overlap. In this case the number $X$ is represented with a precision of about $m \cdot l$ mantissa digits but with the same exponent range as a basic floating point number. However we increase the machine precision, i.e., we decrease the roundoff unit to $\varepsilon_l = \varepsilon_1^l$ ($\approx 10^{-16l}$ for IEEE double-precision), where $\varepsilon_1$ denotes the machine precision of the underlying floating point system.

The basic operations $+, -, *, /$, and the square-root can be efficiently implemented using sums and scalar products of floating point numbers which can be computed exactly in the long accumulator as shown above. It also shouldn't be too difficult to implement the commonly used elementary functions for this data format.

For example we show an algorithm for multiplication of two staggered precision numbers of length $l_1$ and $l_2$, i.e., to compute $Z = X * Y$ or

$$\sum_{i=1}^{\max\{l_1, l_2\}} z_i := \sum_{i=1}^{l_1} x_i * \sum_{j=1}^{l_2} y_j.$$

1.  $\quad x := (\underbrace{x_1, x_1, \ldots, x_1}_{l_2 \text{elements}}, \ldots \ldots, \underbrace{x_{l_1}, x_{l_1}, \ldots, x_{l_1}}_{l_2 \text{elements}})^T$

2.  $\quad y := (\underbrace{y_1, y_2, \ldots, y_{l_2}, \ldots \ldots, y_1, y_2, \ldots, y_{l_2}}_{l_1 \text{portions}})^T$

3.  $\quad long\_accu := exact\_scalar\_product(x, y)$

4.  $\quad$ **for** $i = 1$ **to** $max(l_1, l_2)$

5.  $\quad\quad z_i = round\_to\_nearest(long\_accu)$

6.                     $long\_accu = long\_accu - z_i$

Because we receive a long accumulator and not just a floating point number as the result of the exact scalar product, we are able to *read out* the staggered precision result to almost any desired precision.

Analyzing Krylov subspace methods for single length matrices and right hand side vectors while computing staggered precision solutions as described below, we found that the most important operations are scalar products of two staggered precision vectors of the same precision and scalar products of a staggered precision vector with an ordinary one (in the matrix vector multiplication). With some extra hardware it is possible to limit the increase of needed clock cycles to be linear in the staggered length $l$. For the latter case we need one register, one multiplier (mul) and one adder (add) that can add a floating point number with a mantissa length of $2m$ digits to a long accumulator. But even the first case — a scalar product of two staggered precision vectors with all components having staggered length $l$ can be computed in $l$-fold time compared to the standard floating point case ($l = 1$), see Table 1.

Therefore we are going to design a hardware with an internal parallel and pipelined structure. The requirements are $2l$ registers, $\lceil l/2 \rceil$ multipliers (mul(i)), $\lceil l/2 \rceil$ adders (add(i))v for double length mantissas, $\lceil l/2 \rceil$ long accumulators (dotaccu(i)), and an adder-tree of adders (dotadd) that can add long accumulators. Figure 1 demonstrates the case $l = 4$.

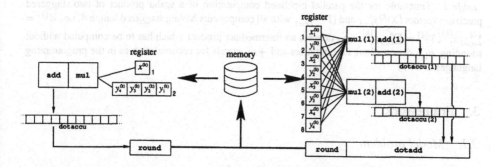

*Figure 1.* A scalar product unit for $l = 4$. The left part computes scalar products of a staggered precision vector with an ordinary vector and the right part performs scalar products of two staggered precision vectors.

In this way it is possible to hide the computation time behind the data movement time which is the maximum speedup that can be reached.

| load | mul(1) | add(1) | mul(2) | add(2) |
|------|--------|--------|--------|--------|
| $\vdots$ | $\vdots$ | $\vdots$ | $\vdots$ | $\vdots$ |
| $x_1^{(k)}$ | | | | |
| $y_1^{(k)}$ | | | | |
| $x_2^{(k)}$ | $p_{11}^{(k)} = x_1^{(k)} y_1^{(k)}$ | | | |
| $y_2^{(k)}$ | $p_{21}^{(k)} = x_2^{(k)} y_1^{(k)}$ | $a_1 += p_{11}$ | | |
| $x_3^{(k)}$ | $p_{12}^{(k)} = x_1^{(k)} y_2^{(k)}$ | $a_1 += p_{21}$ | $p_{22}^{(k)} = x_2^{(k)} y_2^{(k)}$ | |
| $y_3^{(k)}$ | $p_{32}^{(k)} = x_3^{(k)} y_2^{(k)}$ | $a_1 += p_{12}$ | $p_{31}^{(k)} = x_3^{(k)} y_1^{(k)}$ | $a_2 += p_{22}$ |
| $x_4^{(k)}$ | $p_{23}^{(k)} = x_2^{(k)} y_3^{(k)}$ | $a_1 += p_{32}$ | $p_{13}^{(k)} = x_1^{(k)} y_3^{(k)}$ | $a_2 += p_{31}$ |
| $y_4^{(k)}$ | $p_{33}^{(k)} = x_3^{(k)} y_3^{(k)}$ | $a_1 += p_{23}$ | $p_{41}^{(k)} = x_4^{(k)} y_1^{(k)}$ | $a_2 += p_{13}$ |
| $x_1^{(k+1)}$ | $p_{14}^{(k)} = x_1^{(k)} y_4^{(k)}$ | $a_1 += p_{33}$ | $p_{42}^{(k)} = x_4^{(k)} y_2^{(k)}$ | $a_2 += p_{41}$ |
| $y_1^{(k+1)}$ | $p_{24}^{(k)} = x_2^{(k)} y_4^{(k)}$ | $a_1 += p_{14}$ | $p_{43}^{(k)} = x_4^{(k)} y_3^{(k)}$ | $a_2 += p_{42}$ |
| $x_2^{(k+1)}$ | $\vdots$ | $a_1 += p_{24}$ | $p_{34}^{(k)} = x_3^{(k)} y_4^{(k)}$ | $a_2 += p_{43}$ |
| $y_2^{(k+1)}$ | $\vdots$ | $\vdots$ | $p_{44}^{(k)} = x_4^{(k)} y_4^{(k)}$ | $a_2 += p_{34}$ |
| $x_3^{(k+1)}$ | $\vdots$ | $\vdots$ | $\vdots$ | $a_2 += p_{44}$ |
| $\vdots$ | $\vdots$ | $\vdots$ | $\vdots$ | $\vdots$ |

*Table 1.* Timetable for the parallel pipelined computation of a scalar product of two staggered precision vectors $(X^{(k)})_{k=1}^n$ and $(Y^{(k)})_{k=1}^n$ with all components having staggered length 4, i.e., $X^{(k)} = \sum_{i=1}^4 x_i^{(k)}$, $Y^{(k)} = \sum_{i=1}^4 y_i^{(k)}$. ($p_{ij}$ denotes an intermediate product which has to be computed without rounding, $a_{1/2}$ denotes long accumulators and $+=$ stands for accumulation as in the programming languages C/C++)

## 3. Framework

### 3.1. Model problem

Consider a beam of length 1 which is pivoted at $s = 0$ and $s = 1$ and in between it is weighted with a load given by $p(s)$. The bending line $x(s)$ can be described with the ordinary differential equation

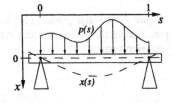

$$\frac{d^2 x}{ds^2}(s) = -\frac{M(s)}{EJ}, \quad \frac{d^2 M}{ds^2}(s) = -p(s),$$

where $M(s)$ is the bending moment, $E$ is the elasticity, and $J$ is the moment of inertia [2]. Differentiating $x(s)$ two times and imposing boundary conditions yields the boundary value problem

$$x^{(iv)}(s) = \frac{1}{EJ}p(s), \quad x(0) = x(1) = 0, \quad x''(0) = \frac{m_0}{EJ}, \quad x''(1) = \frac{m_1}{EJ}, \tag{1}$$

where $m_0 = M(0)$ and $m_1 = M(1)$. In the special case $p(s) \equiv p$ we get

$$x(s) = \frac{s}{24EJ}\left(ps^3 + 2(2m_1 - 2m_0 - p)s^2 + 12m_0 s + p - 8m_0 - 4m_1\right), \quad \text{or}$$

$$x(s) = \frac{s}{24EJ}\left(s^3 - 2s^2 + 1\right), \quad \text{if } m_0 = m_1 = 0, \ p = 1. \tag{2}$$

For discretization we use the grid $s_i = ih$, $i = -1, \dots, n+2$ with $h = 1/(n+1)$. So we have the grid points $s_0, \dots, s_{n+1}$ in $[0, 1]$ and two auxiliary points $s_{-1}$ and $s_{n+2}$ for the boundary conditions. Now approximating the fourth and second order derivative in (1) with the fourth and second order centered differences and by substituting $x_i = x(s_i)$ and $p_i = p(s_i)$ we get the banded system of linear equations

$$\begin{pmatrix} 5 & -4 & 1 & & & & & \\ -4 & 6 & -4 & 1 & & & & \\ 1 & -4 & 6 & -4 & 1 & & & \\ & \ddots & \ddots & \ddots & \ddots & \ddots & & \\ & & \ddots & \ddots & \ddots & \ddots & \ddots & \\ & & 1 & -4 & 6 & -4 & 1 \\ & & & 1 & -4 & 6 & -4 \\ & & & & 1 & -4 & 5 \end{pmatrix} \begin{pmatrix} x_1 \\ x_2 \\ x_3 \\ \vdots \\ \vdots \\ x_{n-2} \\ x_{n-1} \\ x_n \end{pmatrix} = \frac{1}{EJ} \begin{pmatrix} h^4 p_1 - h^2 m_0 \\ h^4 p_2 \\ h^4 p_3 \\ \vdots \\ \vdots \\ h^4 p_{n-2} \\ h^4 p_{n-1} \\ h^4 p_n - h^2 m_1 \end{pmatrix}$$

and the additional equations $x_{-1} = -x_1 + h^2 m_0$, $x_0 = 0$, $x_{n+1} = 0$, and $x_{n+2} = -x_n + h^2 m_1$.

**Note.** If we set $EJ = 1$, $p \equiv 1$, $m_0 = m_1 = 0$ and $n = 2^m - 1$ this system is exactly representable in the IEEE double-precision format. Thus we have no conversion errors in this case.

This coefficient matrix is exactly matrix (4.16) in [5] and its condition number is $\tan^{-4}(\pi/(2n+2))$. Subsequently we will refer this matrix as GK($n$), where $n$ denotes the dimension.

## 3.2. Verifying solvers for linear systems

In order to get reliable solutions of the linear systems and not a mixture of discretization errors, rounding errors etc. we use *verifying solvers*. At this time we use two solvers that are only applicable to relatively small problems. The first one is lss from [6, 7] which doesn't care about any sparsity and therefore needs $O(n^2)$ storage and $O(n^3)$ time. The second one is a solver for banded systems [13]. It is based on a LU-factorization. The subsequent forward and backward substitution is performed with interval arithmetic and there is a tricky algorithm implemented to avoid the wrapping effect. This algorithm needs $O(n(l_1 + l_2))$ storage and $O(n(\max\{l_1, l_2\})^3)$ time if $l_1$ and $l_2$ denotes the upper resp. lower bandwidth.

Since the solutions are computed as interval enclosures we can always determine an upper bound to the absolute and relative error by the diameter resp. the diameter divided by the infimum of the absolute values of the solution.

### 3.3. Some error-types

Suppose $\tilde{x}$ to be an approximation to a possibly unknown exact solution $\hat{x}$ and $[x]$ to be an enclosure of $\hat{x}$. Then we define the relative enclosure error (REE) as

$$\text{REE}([x]) := \frac{\text{diam}([x])}{\langle [x] \rangle}, \quad \text{if } 0 \notin [x]$$

where $\langle [x] \rangle := \inf\{|\xi|, \xi \in [x]\}$ is the mignitude of $[x]$. Further we define the relative approximation error (RAE) as

$$\text{RAE}(\tilde{x}, [x]) := \sup\left\{\left|\frac{\xi - \tilde{x}}{\xi}\right|, \xi \in [x]\right\}, \quad \text{if } 0 \notin [x].$$

This latter error might be enlarged by rough enclosures $[x]$. REE is a measure of the quality of an enclosure $[x]$ and RAE describes the quality of an approximation $\tilde{x}$ if $[x]$ is a *good* enclosure.

## 4.  Discretization problems

In this section we focus on the discretization process which might be sometimes out of our influence but should never be out of our mind. As we will see, an imprecisely discretizised problem may have a solution which is maybe far from the solution of the original problem.

To study the sensitivity of ill-conditioned problems with respect to perturbations in the data we solve the systems $A^{(k)}x^{(k)} = b$, where $A^{(k)} = \text{GK}(1023) + 10^{-k} \cdot E$ with $k \in \{8, 10, 11, 12, 14\}$ and $E$ is a matrix with uniformly distributed random numbers in $[0, 1]$. Let $[x]$ denote the enclosure of the exact solution of the unperturbed GK(1023)-System computed with the 'band'-solver. The relative approximation errors $\text{RAE}(x_j^{(k)}, [x]_j)$ are plotted in Figure 2. The relative enclosure errors are less

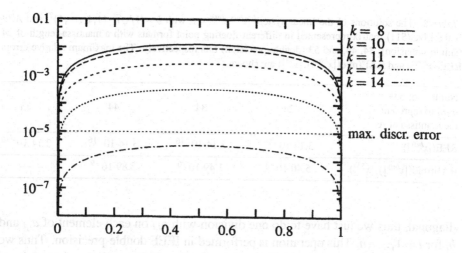

*Figure 2.* The relative errors of the solutions of $A^{(k)}x^{(k)} = b$, where $A^{(k)} = \text{GK}(1023) + 10^{-k} \cdot E$ with $k \in \{8, 10, 11, 12, 14\}$ and a matrix $E$ with $\mathcal{U}[0, 1]$ distributed random numbers as its elements.

than $10^{-15}$ and the discretization errors, i.e., $\text{RAE}(x(s_j), [x]_j)$ are less than $10^{-5}$ in each case ($x(\cdot)$ denotes the exact solution given in (2)).

If we perturb the elements of $A$ only at a level of $10^{-10}$ (which is not unusual if we think of matrices that result from a finite element discretization where each element is obtained by an integration), we get a maximum relative error of about $10^{-2}$. That means we can obtain two correct digits at most, if we don't make any subsequent error in solving the system.

Another example is shown in Table 2. This matrix called LNS(511) results from a fluid flow modeling and its condition number is $6.4 \cdot 10^{15}$ (estimated), see [14]. We solved the system with different numbers of bits used to represent the matrix and the right-hand-side. Again we see, that if we insinuate to get e.g. 24 correct bits (IEEE single-precision) from the discretization-process, we only can hope to obtain at most 3 correct decimal digits from the correct solution.

## 5. Preconditioning

Preconditioning is a very important task in solving a linear system because well conditioned systems are much easier and particularly faster to solve but no one is interested in solutions of problems she didn't ask to solve. Thus the question is, how does preconditioning affect the solution of a linear system.

In this section we investigate the perturbation of the solution of a linear system caused by preconditioning. We used the most trivial preconditioning method: balancing or Jacobi-preconditioning. That means we scale each row of $A$ to get a unit

*Table 2.* The solutions of the linear system $A^{bits}x^{bits} = b$ with $b = 1/\sqrt{n} \cdot (1,1,\dots,1)^T$ and $A^{bits}$ is the LNS(511) matrix represented in different floating point formats with a mantissa-length of 24 (single-precision), 34, 44, and 53 (double-precision) bits respectively. The maximum relative errors $REE([x^{bits}])$ and $RAE(mid([x^{bits}]), [x^{53}])$ are shown.

| Number of bits used to represent the mantissa of A | 24 | 34 | 44 | 53 |
|---|---|---|---|---|
| $REE([x^{bits}])$ | $3.14 \cdot 10^{-16}$ | $3.14 \cdot 10^{-16}$ | $3.14 \cdot 10^{-16}$ | $3.14 \cdot 10^{-16}$ |
| $RAE(mid([x^{bits}]), [x^{53}])$ | $5.40 \cdot 10^{-3}$ | $1.49 \cdot 10^{-6}$ | $3.89 \cdot 10^{-9}$ | — |

diagonal, thus we just have to do one division with $a_{i,i}$ on each element of $a_{.,i}$ and $b_i$ for $i = 1,\dots,n$. This operation is performed in IEEE double-precision. Thus we just introduce an error at roundoff level because the off-diagonal elements of $A$ and the elements of $b$ are not representable in the floating point format. Here we solve our model problem with different dimensions that result in condition numbers of about $10^{15}$. The effect is disastrous as can be seen in Figure 3. The maximum relative errors to the exact solutions are $1.52 \cdot 10^{-2}$ for $n = 2^{13} - 1$, $7.26 \cdot 10^{-2}$ for $n = (2^{13} + 2^{14})/2 - 1$, and $1.98 \cdot 10^{-1}$ for $n = 2^{14} - 1$. Note that there is also a conversion error in the second case because this system is not representable in IEEE double-precision, however this seems not to affect the general behavior.

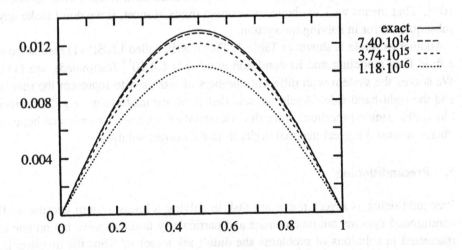

*Figure 3.* The effect of preconditioning to ill-conditioned linear systems. The curves represent the solutions of GK(8191), GK(12287), and GK(16383) (with condition numbers $7.40 \cdot 10^{14}$, $3.74 \cdot 10^{15}$, and $1.18 \cdot 10^{16}$ respectively). The exact solutions of these systems differ so little, that they draw as a single line (solid).

We have to resume that we can't do preconditioning in this traditional way, neither if we are interested in verified solutions, nor if we are only interested in a *good* approximation. It is possible to avoid these problems by doing exact preconditioning but this leads to an interval matrix and therefore to an often unacceptable computational effort to solve these systems. Another, maybe more practical, way might be to use our more accurate arithmetic to avoid the propagation of the various errors up to significant digits. If we apply the preconditioner at each step of an iterative solver (see [3]) rather than precondition the matrix and the right hand side vector in advance then we don't need any additional storage but have a preconditioning with the same accuracy as the iterated solution (see next section).

## 6. Solving via Krylov subspace methods

Aside from classical splitting methods like Jacobi, Gauß-Seidel, or SOR recently Krylov subspace methods like CG, QMR, or BiCGStab push through for solving large sparse linear systems and even for computing eigenvalues and -vectors [4]. Theoretically it can be shown that the linear system solvers converge in at most $n$ steps if $n$ denotes the dimension. Let $A \in \mathbb{R}^{n \times n}$ be a regular matrix and $b \in \mathbb{R}^n$ the right-hand-side vector. The task now is to solve the linear system $Ax = b$. Doing this by Krylov methods, the problem is subdivided into two subproblems

(*i*) Computing an orthonormal basis (ONB) $\{v_1, \ldots, v_m\}$ of the Krylov subspace $\mathcal{K}_m := \mathrm{span}\{b, Ab, \ldots, A^{m-1}b\}$.

(*ii*) Orthogonal projection of the problem from $\mathbb{R}^n$ into $\mathbb{R}^m$, i.e., find a $x_m \in \mathcal{K}_m$ that meets a (Petrov-)Galerkin condition or satisfies a minimal-norm condition.

These two subproblems can also be interpreted as first finding a matrix $H$ which is unitarily similar to $A$ and which is particularly structured, and then solving this simpler system. The parts (*i*) and (*ii*) have to be repeated for $m = 1, 2, \ldots, n$ iteratively.

With the *Arnoldi-algorithm* one iteration step of the first subproblem is performed as follows

$$\tilde{v}_{m+1} = Av_m \qquad \text{(compute a prototype for } v_{m+1})$$

$$\tilde{v}_{m+1} = \tilde{v}_{m+1} - \underbrace{\sum_{i=1}^{m} \langle v_i \mid \tilde{v}_{m+1} \rangle}_{=:h_{im}} v_i \qquad \text{(orthogonalize it against } V_m) \qquad (3)$$

$$v_{m+1} = \frac{\tilde{v}_{m+1}}{\underbrace{\|\tilde{v}_{m+1}\|}_{=:h_{m+1,m}}}. \qquad \text{(and finally normalize it)}$$

This algorithm can be written in matrix form as

$$AV_m = V_m H_m + v_{m+1} h_{m+1,m} e_m^T, \quad (V_m^H V_m = I), \tag{4}$$

where $e_m$ is the $m$th column of an $m$ by $m$ identity matrix and $H$ is an upper Hessenberg matrix with the elements $h_{ij}$.

The (full) GMRES-algorithm is based on this procedure to generate the orthogonal basis $V_m$ but because of its increasing work and storage requirements it may be impractical. The GMRES($j$) algorithm is defined by simply restarting GMRES every $j$ steps, using the latest iterate of the solution as the initial guess for the next cycle. In this way we reduce the storage requirement from $O(n^2)$ for full GMRES to $O(jn)$ for GMRES($j$) but we lose the theoretical property of convergence after at most $n$ iterations [4].

For hermitian $A$ the situation in the Arnoldi-algorithm becomes much more favorable. By multiplying (4) with $V_m^H$ from the left we see

$$H_m = V_m^H A V_m = V_m^H A^H V_m = H_m^H.$$

So $H_m$ is a hermitian upper Hessenberg i.e., a hermitian tridiagonal matrix $T_m$. The *Lanczos-algorithm* [12] computes $T_m = \text{tridiag}(\beta_{j-1}, \alpha_j, \beta_j)$ directly by its two 'new' components per row.

1.       Given $v_1$ with $\|v_1\| = 1$, set $\beta_0 = 0$
2.       **for** $m = 1, 2, \ldots$
3.       $\tilde{v}_{m+1} = A v_m - \beta_{m-1} v_{m-1}$
4.       $\alpha_m = \langle \tilde{v}_{m+1} \mid v_m \rangle$
5.       $\tilde{v}_{m+1} = \tilde{v}_{m+1} - \alpha_m v_m$
6.       $\beta_m = \|\tilde{v}_{m+1}\|$
7.       $v_{m+1} = \tilde{v}_{m+1}/\beta_m$

In this case the recurrence of depth $m$ in (3) reduces to one of depth 2 and again in matrix form we get

$$AV_m = V_m T_m + \beta_m v_{m+1} e_m^T, \quad (V_m^H V_m = I). \tag{5}$$

For a non-hermitian $A$ we get similar results if we iterate a pair of *biorthogonal* bases $V_m$ and $W_m$:

$$AV_m = V_m \hat{T}_m + \gamma_m v_{m+1} e_m^T, \quad A^H W_m = W_m \hat{T}_m^H + \bar{\beta}_m w_{m+1} e_m^T,$$
$$\text{and} \quad V_m^H W_m = I.$$

In this case $\hat{T}_m$ is not hermitian but still tridiagonal ($\hat{T}_m = \text{tridiag}(\gamma_{j-1}, \alpha_j, \beta_j)$) [4]. For simplicity we restrict ourself to the hermitian case.

One well known Krylov subspace method is the conjugate gradient algorithm (CG) of Hestenes and Stiefel [8]. It can be interpreted as a Lanczos procedure and a subsequently $LDL^T$-factorization to solve the tridiagonal system $T\tilde{x} = \tilde{b}$ [3]. In Figure 4 the Euclidean norms of the residual and the error in each step are plotted during solving the GK(255) system. Additionally we show the level of orthogonality of the new basis-vector $v_{m+1}$ to the previous ones: $\max_{k=1}^m \{\langle v_k \mid v_{m+1}\rangle\}$. As we can see there is no convergence at all up to step $m = 1.5n$ and especially no convergence at the theoretically guaranteed step $m = n$. One reason is easy to identify: the basis of the Krylov subspace loses its orthogonality completely at $m \approx 100$ and the *basis*-vectors may even become linearly dependent. So CG can't minimize the residual in the whole $\mathbb{R}^n$ but only in a smaller subspace.

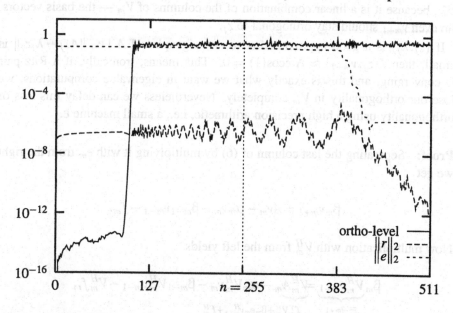

*Figure 4.* The Euclidean norms of the residual and the error, and the level of orthogonality during solving the GK(255) system. (ortho-level = $\max_{k=1}^m \{\langle v_k \mid v_{m+1}\rangle\}$)

An error analysis of the Lanczos procedure [15, 17] shows that in finite precision formula (5) has to be replaced by

$$AV_m = V_m T_m + v_{m+1}\beta_m e_m^T + F_m, \qquad (6)$$

where $F_m$ contains the rounding errors and $\|F_m\|$ is approximately of size $\varepsilon\|A\|$ [16]. $\varepsilon$ denotes the machine precision, e.g. $\approx 10^{-16}$ in IEEE double-precision. Now the question is what happens to the orthogonality condition if $V_m$ is computed in finite precision. The answer is given in lemma 1.

LEMMA 1 *Suppose $(S_m, \Lambda_m)$ with $\Lambda_m := \text{diag}(\lambda_1, \dots, \lambda_m)$ is the exact spectral factorization of the computed $T_m$ (see eqn. (6)) i.e., the $(z_j, \lambda_j)$ are the computed Ritzpairs of $A$ if we define $z_j := V_m s_j$. Then we have*

$$\angle(z_j, v_{m+1}) \approx Arccos\left(\frac{\varepsilon \|A\|}{|\,\|Az_j - \lambda_j z_j\| - \varepsilon\|A\|\,|}\right).$$

**Note.** Notice that $v_{m+1}$ should stay orthogonal to $\mathcal{K}_m = \text{span}(v_1, \dots, v_m)$ and $z_j \in \mathcal{K}_m$, because it is a linear combination of the columns of $V_m$ — the basis vectors. So even $v_{m+1}$ should stay orthogonal to $z_j$.

If $(z_j, \lambda_j)$ is a good approximation to any eigenpair of $A$ i.e., $\|Az_j - \lambda_j z_j\|$ is small, then $\angle(z_j, v_{m+1}) \approx Arccos(1) = 0$. This means, ironically, if a Ritz-pair is converging, and this is exactly what we want in eigenvalue computations, we lose our orthogonality in $V_m$ completely. Nevertheless we can delay this loss of orthogonality using a high precision arithmetic, i.e., a small machine $\varepsilon$.

**Proof:** Separating the last column of (6) by multiplying it with $e_m$ from the right we get

$$\beta_m v_{m+1} = Av_m - \alpha_m v_m - \beta_{m-1} v_{m-1} - f_m.$$

Now multiplication with $V_m^H$ from the left yields

$$\beta_m \underbrace{V_m^H v_{m+1}}_{=: q_{m+1}} = \underbrace{V_m^H A v_m}_{T_m^T V_m^H + \beta_m e_m v_{m+1}^H + F_m^H} - \alpha_m V_m^H v_m - \beta_{m-1} V_m^H v_{m-1} - V_m^H f_m$$

$$\beta_m q_{m+1} = T_m \hat{q}_m - \alpha_m \hat{q}_m - \beta_{m-1} \hat{q}_{m-1} + \beta_m e_m v_{m+1}^H v_m$$
$$+ \underbrace{F_m^H v_m - V_m^H f_m}_{=: g_m}$$

$$\beta_m q_{m+1} e_m^T = T_m \hat{Q}_m - \hat{Q}_m T_m + \hat{G}_m. \tag{7}$$

To facilitate additions of vectors with different lengths and the collection of vectors with different lengths as columns of a matrix, we use the $\hat{\cdot}$-operator which extends a vector to the appropriate size by adding some zero-elements at the bottom.

Let now the $s_j$'s, $z_j$'s, and the $\lambda_j$'s be defined in the same way as in the lemma. Building $s_j^H$ times (7) times $s_j$ for any $j \leq m$ yields

$$\beta_m \underbrace{s_j^H V_m^H v_{m+1}}_{z_j^H} \underbrace{e_m^T s_j}_{s_{mj}} = \underbrace{s_j^H T_m \hat{Q}_m s_j}_{\lambda_j s_j^H} - \underbrace{s_j^H \hat{Q}_m T_m s_j}_{\lambda_j s_j} + s_j^H \hat{G}_m s_j$$

$$\Longleftrightarrow \qquad z_j^H v_{m+1} = \frac{s_j^H \hat{G}_m s_j}{\beta_m s_{mj}}. \tag{8}$$

The numerator of this fraction is approximately of size $\varepsilon \|A\| \|z_j\|$. In order to estimate the denominator we look at the *quality* of the Ritz-pair $(\lambda_j, z_j)$:

$$\|Az_j - \lambda_j z_j\| = \underbrace{\|V_m \tilde{s}_j - \lambda_j V_m s_j\|}_{V_m T_m + \beta_m v_{m+1} e_m^T + F_m}$$

$$\leq \|\beta_m v_{m+1} s_{mj}\| + \|F_m\|$$

$$\approx |\beta_m s_{mj}| + \varepsilon \|A\|. \tag{9}$$

Collecting our results in (8) and (9) we finally obtain

$$\angle(z_j, v_{m+1}) = \text{Arccos}\left(\frac{|z_j^H v_{m+1}|}{\|z_j\| \|v_{m+1}\|}\right) \approx \text{Arccos}\left(\frac{\varepsilon \|A\|}{\|\|Az_j - \lambda_j z_j\| - \varepsilon\|A\|\|}\right) \qquad \blacksquare$$

Finally we show the results of the same problem as illustrated in Figure 4 but now with our staggered-precision arithmetic. In Figure 5 we show the resulting Euclidean norms of the residuals and that of the errors. We achieve a significant reduction of the necessary iterations by increasing the staggered length.

### Acknowledgments

I thank Rudolf Lohner for several valuable discussions on this topic and Ulrike Storck for her helpful remarks on the definition of the error-types used in this paper.

### References

1. American National Standards Institute / Institute of Electrical and Electronic Engineers, New York. *A Standard for Binary Floating-Point Arithmetic*, 1985. ANSI/IEEE Std. 754-1985.
2. L. Collatz. *Differentialgleichungen*. 6. Aufl. Teubner, Stuttgart, 1981.
3. G.H. Golub and C.F. van Loan. *Matrix Computations*. Johns Hopkins, third edition, 1996.
4. A. Greenbaum. *Iterative Methods for Solving Linear Systems*. SIAM, Philadelphia, 1997.
5. R. Gregory and D. Karney. *A Collection of Matrices for Testing Computational Algorithms*. Wiley Interscience, 1969.

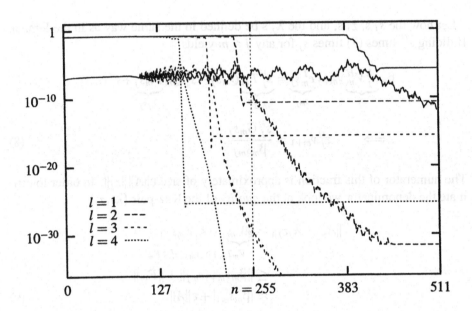

*Figure 5.* The Euclidean norms of the residuals (oscillating) and errors (more or less piecewise constant) during solving the GK(255) system with staggered length $l$ from 1 to 4.

6. R. Hammer, M. Hocks, U. Kulisch, and D. Ratz. *Toolbox for Verified Computing.* Springer, Berlin Heidelberg, 1993.

7. R. Hammer, M. Hocks, U. Kulisch, and D. Ratz. *C++ Toolbox for Verified Computing.* Springer, Berlin Heidelberg, 1995.

8. M.R. Hestenes and E. Stiefel. Methods of conjugate gradients for solving linear systems. *Journal of Research of the National Bureau of Standards*, 49:409–436, 1952.

9. U. Kulisch, editor. *Wissenschaftliches Rechnen mit Ergebnisverifikation – Eine Einführung.* Akademie Verlag, Ost-Berlin, Vieweg, Wiesbaden, 1989.

10. U. Kulisch and W.L. Miranker. *Computer Arithmetic in Theory and Practice.* Academic Press, New York, 1981.

11. U. Kulisch and W.L. Miranker, editors. *A New Approach to Scientific Computation.* Academic Press, New York, 1983.

12. C. Lanczos. An iteration method for the solution of the eigenvalue problem of linear differential and integral operators. *Journal of Research of the National Bureau of Standards*, 45:255–282, 1950.

13. R. Lohner. A verified solver for linear systems with band structure. to be included in: Toolbox for Verified Computing II.

14. The matrix market: a visual web database for numerical matrix data. http://math.nist.gov/MatrixMarket/.

15. C.C. Paige. *The Computation of Eigenvalues and Eigenvectors of Very Large Sparse Matrices.* PhD thesis, University of London, 1971.

16. B.N. Parlett. *The Symmetric Eigenvalue Problem.* Prentice-Hall, Englewood Cliffs, NJ., 1980.

17. H. Simon. *The Lanczos Algorithm for Solving Symmetric Linear Systems.* PhD thesis, University of California, Berkley, 1982.

*T. Csendes (ed.), Developments in Reliable Computing* 31–41.
© 1999 *Kluwer Academic Publishers.*

# A Hardware Approach to Interval Arithmetic for Sine and Cosine Functions

JAVIER HORMIGO, JULIO VILLALBA, AND EMILIO L. ZAPATA

*Dept. Computer Architecture, University of Malaga (Spain)*   {hormigo, julio, ezapata}@ac.uma.es

**Abstract.** The existing software packages for interval arithmetic are often too slow for numerically intensive computation, whereas hardware solutions do not handle elementary functions directly. We propose some minimal modifications to the CORDIC architecture to efficiently support either interval or regular sine and cosine functions. The computational time for an interval operation is close to that of a regular operation for most cases and twice that of a regular operation for the remaining ones. The CORDIC algorithm is also slightly modified to guarantee correct choice of the bounds of the result, and a low worst case error is obtained.

**Keywords:** Interval arithmetic, hardware, sine and cosine functions, CORDIC algorithm

## 1. Introduction

Interval arithmetic [5] provides an efficient method to control errors in numerically intensive computations. Many software packages and libraries have been developed to manage interval arithmetic in order to improve the accuracy and reliability of numerical computation. The main disadvantage of these software tools is their speed. To overcome the speed limitations of existing software tools, specific hardware support is required. There are several works in which special–purpose coprocessors [8] or functional units [12] for interval arithmetic have been designed. Also, there are others in which adding a small amount of hardware to conventional floating point units enables either interval or floating point operations [9]. Nevertheless, the problem of designing specific hardware to perform interval elementary functions has not been studied in detail as yet.

The CORDIC algorithm is a well–known method for computing many elementary functions. Interest in this algorithm has increased in recent years, due to its easy hardware implementation. Due to its iterative nature it could be feasible to extend it to interval arithmetic for a low cost. Specifically, small modifications to the classic CORDIC architecture allow us to compute the interval sine and cosine functions more than 1.5 times faster than the regular one for most cases. Taking into account that previous works needed at least the amount of time of two point function evaluation to perform the interval function [7], this is a very good result.

The rest of the paper is organized as follows: section 2 surveys the CORDIC algorithm itself. The extension of this algorithm to compute interval sine and cosine is presented in section 3, and how a correct choice of the endpoints of the result is guaranteed. In section 4 we see how to reduce the computational time for interval operation. In section 5 we present the architecture which supports the algorithm. Finally, in section 6, we present the conclusions and future works.

## 2. The CORDIC algorithm

The CORDIC algorithm (COordinate Rotation DIgital Computer) was introduced to compute trigonometric functions [11] and generalized to compute linear and hyperbolic functions [13]. This algorithm is attractive from a hardware point of view since it employs only additions and shifts to implement relatively complex functions. Special attention has been paid by different researchers to the improvement of the algorithm in the last few years, as we can see in [6].

The basic task performed by the CORDIC algorithm is to rotate a 2 by 1 vector through an angle using linear, circular or hyperbolic coordinate system. This is accomplished by the CORDIC algorithm by rotating the vector through a sequence of predefined elementary angles whose algebraic sum approximates the desired rotation angle. These elementary angles are stored in a table and have the property that the vector rotation through each of them may be easily computed with a single shift and add operation. Therefore, it is an iterative algorithm, and on each iteration a rotation (called microrotation) through an elementary angle is performed. This formulation leads to a unified procedure to compute a wide range of elementary functions with a fixed number of iterations.

### 2.1. Circular coordinates

Although the CORDIC algorithm can be used in different coordinate systems, we focus on circular coordinates which enables the evaluation of trigonometric functions such as sine and cosine functions.

In circular coordinates, the sequence of elementary angles is given by $\alpha_i = \tan^{-1}(2^{-i})$. Using this set of angles

$$\beta = \sum_{i=0}^{n} \sigma_i \alpha_i + e_i \tag{1}$$

is fulfilled where $e_i \leq \alpha_n$. In this equation $\sigma_i \in \{-1, 1\}$ and specifies the direction of each microrotation. In this case, the basic CORDIC iterations are

$$x(i+1) = x(i) + \sigma_i \cdot 2^{-i} \cdot y(i)$$

$$y(i+1) = y(i) - \sigma_i \cdot 2^{-i} \cdot x(i) \tag{2}$$
$$z(i+1) = z(i) - \sigma_i \cdot \tan^{-1}(2^{-i})$$

where $(x(0), y(0))$ are the initial coordinates of the vector and the $z$ coordinate accumulates the angle. About $n + 1$ iterations are needed to produce $n$ bit precision [3].

The final vector is not only rotated, but is also scaled by the factor

$$K = \prod_{i=0}^{n} \sqrt{1 + \sigma_i^2 \cdot 2^{-2i}} \tag{3}$$

which is a constant since $\mid \sigma_i \mid = 1$. Several techniques can be used to compensate for this scale factor, such as the addition of scaling iteration [2].

## 2.2. Modes of operation

The CORDIC algorithm can be operated in two modes: the rotation mode and the vectoring mode.

- In the rotation mode, the desired rotation angle $\beta$ is given, and the objective is to compute the final coordinates $[x(n + 1), y(n + 1)]$. The $z$ coordinate is initialized to $\beta$ and on each iteration $\sigma_i$ is the sign of $z(i)$, such that after $n + 1$ iterations, the total angle rotated is:

$$\sum_{i=0}^{n} \sigma_i \alpha_i = \beta - z(n+1) \tag{4}$$

where $z(n + 1) \leq \alpha_n$. On each iteration the coordinates go to

$$x \to K(x(0)\cos(\beta) - y(0)\sin(\beta))$$
$$y \to K(y(0)\cos(\beta) + x(0)\sin(\beta)) \tag{5}$$
$$z \to 0$$

- In the vectoring mode, the desired rotation angle $\beta$ is not given, and the objective is to rotate the given initial vector $[x(0), y(0)]$ back to the x-axis so that the angle between them is accumulated in the $z$ coordinate. In order to do this, $z(0)$ is set to 0 and on each iteration $\sigma_i$ is the opposite of the sign of $y(i)$. On each iteration the coordinates go to

$$x \to K\sqrt{x^2(0) + y^2(0)}$$
$$y \to 0 \tag{6}$$
$$z \to z(0) - \tan^{-1}\frac{y(0)}{x(0)}$$

## 2.3. Evaluation of sine and cosine functions

The sine and cosine functions can be simultaneously obtained by means of the CORDIC algorithm using circular coordinates and in the rotation mode. If the initial vector is set to $(\frac{1}{K}, 0)$ and a rotation through $\beta$ is performed [10], we get the sine and cosine of $\beta$ in $y$ and $x$ coordinates respectively (see equation 5). This is represented in Figure 1.a.

It is important to note that, in this case, it is not necessary to compensate for the scale factor, since the initial vector takes this into account.

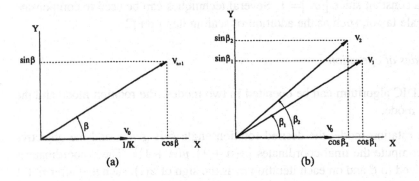

Figure 1. a. Sine and cosine functions evaluation. b. Interval sine and cosine functions.

## 2.4. Basic CORDIC architecture

The basic CORDIC architecture is shown in Figure 2.

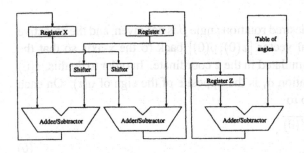

Figure 2. Basic CORDIC architecture.

This performs each CORDIC iteration (see equation 2) in one cycle time: the value $x(i)$, stored in register RX, is used to compute $x(i+1)$ by adding or subtracting (depending on $\sigma_i$) the value $2^{-i}y(i)$ from the barrel shifter. The new value of the $x$ coordinate is stored in RX. A similar operation is performed for the $y$ coordinate. The value of the $z$ coordinate is updated by adding or subtracting the value from the table of angles. The control unit is very simple, since a counter from 0 to $n+1$ can fix the number of bits shifted and index the table of angles on each rotation, while a very simple logic manages the sign of $\sigma_i$ on each iteration.

## 3. Interval sine and cosine functions

In the previous section we have seen how to calculate the sine and cosine functions for a point argument, using the CORDIC algorithm. But, as explained in the introduction, interval arithmetic works with intervals instead of single values, such that the true result is guaranteed to lie within this interval. We call $[\beta_1, \beta_2]$ the input argument.

To compute the sine or cosine functions, and without any loss of generality [7] [1], we first reduce the argument to the range $[\frac{-\pi}{2}, \frac{\pi}{2}]$, since it is the normal working range for the CORDIC algorithm. A method to perform this range reduction can be found in [1]. Basically, the argument is multiplied by $2/\pi$; the integer part gives information about the quadrant, and the fractional part is the argument which we work with. In this bound the sine function is non decreasing, and therefore rounding up the expression $\sin \beta_2$ yields an upper bound, and rounding down the expression $\sin \beta_1$ yields the lower bound. For the cosine function, we have to consider three cases: I) if $\beta_1 < 0$ and $\beta_2 \leq 0$, the function is non decreasing and we use the same solution as the sine function; II) if $\beta_1 \geq 0$ and $\beta_2 \geq 0$, the function is non increasing, and therefore we have to round down the expression $\cos \beta_2$ and round up the expression $\cos \beta_1$ to yield the lower bound and upper bound respectively; III) if $\beta_1 < 0$ and $\beta_2 \geq 0$ the argument interval contains the angle zero, and therefore the upper bound is 1 and the lower bound corresponds to $\cos(max\{|\beta_1|, |\beta_2|\})$.

Hence, to obtain the final result it is generally necessary to perform two CORDIC operations, one rotation through each angle, in such a way that the result lies over the coordinate axes, as shown in Figure 1.b. This takes $2n + 2$ iterations.

Nevertheless, the errors involved in the computation have to be taken into account to guarantee the correct bounds of the final solution. The CORDIC algorithm presents two basic sources of errors [3][4]: the approximation error and the rounding error. The approximation error is due to the quantified representation of a CORDIC rotation angle by finite numbers of elementary angles and is less than the last elementary angle $\alpha_n$. The rounding error is due to the finite precision arithmetic used in practical implementations. The size of these errors can be controlled by adding iterations for the first type of error, and adding some extra bits (guard bits) for the second type of error. In [3] there is a complete guide for selecting the number of iterations and the number of guard bits for the degree of precision required. We now analyze the influence of both sources of error and the necessary modifications introduced to the algorithm in order to ensure a correct result.

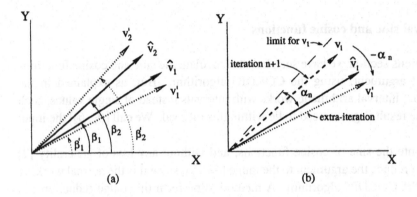

*Figure 3.* a. Correct bounds. b. Addition of one iteration to find the correct bound.

- *The approximation error:* As we saw in section 2, due to this error the real angle rotated is $\beta'$ instead of the ideal value $\beta$ (infinite microrotation). Hence, the vector obtained will be over, or below, the ideal vector depending on the sign of the approximation error (this sign corresponds to the sign of the $z$ coordinate). However, we must obtain a final interval $[\beta'_1, \beta'_2]$ which includes the argument interval $[\beta_1, \beta_2]$, as Figure 3.a shows ($[\beta_1, \beta_2] \subseteq [\beta'_1, \beta'_2]$).

Therefore, control of the approximation error sign is needed to ensure that the final vector is placed in the correct position. If this does not occur, only one extra-iteration through $\alpha_n$ is required to place the vector correctly, since the error is less than $\alpha_n$. In Figure 3.b we have an example of how to find $\beta'_1$ from the vector obtained after the first rotation $v_1$ (n+1 CORDIC microrotations). In this figure, $\hat{v}_1$ is the ideal value (infinite microrotations). In this case $v_1$ is over $\hat{v}_1$ and therefore $z(n + 1) > 0$. To find a lower bound, we perform an extra iteration through a microrotation angle of $-\alpha_n$ in such a way that the vector $v$ is placed beyond the vector $\hat{v}_1$ ($v'_1$ in Figure 3.b ). Hence, up to two iterations are added to the regular CORDIC rotation to guarantee the correct bounds of the result.

- *The accumulated rounding error:* The rounding error accumulated through the successive microrotations (we call $\varepsilon$) has to be taken into account before the final rounding to guarantee correct endpoints. The maximum value of the accumulated rounding error ($\varepsilon_{max}$) is added before rounding up to ensure that the ideal value is less than the value used to round because $\varepsilon$ can be negative [1] [7]. Similarly, it is subtracted before rounding down. The value $\varepsilon_{max}$ depends on the number of guard bits. In [4] [3], it is proved that $\varepsilon_{max} < 1.5(n+1)2^{-p}$, where $p$ is the total number of bits including the guard bits.

## 3.1. Worst case error

An important feature of the algorithms to compute interval functions is how sharp the calculated bounds are. To estimate this, we study the error obtained in the worst case. For the approximation error, it is still $\alpha_n$. If this error is shifted to the $x$ and $y$ coordinates, we have:

$$\cos(\beta - \alpha_n) = \cos(\beta)\cos(\alpha_n) + \sin(\beta)\sin(\alpha_n)$$
$$\sin(\beta - \alpha_n) = \sin(\beta)\cos(\alpha_n) - \cos(\beta)\sin(\alpha_n)$$

Taking into account that for practical implementations $\alpha_n \to 0$, we use the approximation $\sin(\alpha_n) \simeq \alpha_n$ and $\cos(\alpha_n) \simeq 1$, then:

$$\cos(\beta - \alpha_n) \simeq \cos(\beta) + \sin(\beta)\alpha_n$$
$$\sin(\beta - \alpha_n) \simeq \sin(\beta) - \cos(\beta)\alpha_n$$

Since $\| \sin(\beta) \| \leq 1$ and $\| \cos(\beta) \| \leq 1$, then a pessimistic approximation of the maximum error on each coordinate is $\alpha_n$. For the accumulated rounding error, it is positive ($\varepsilon > 0$) in the worst case. Hence, since we add the value $\varepsilon_{max}$ before the final rounding up, the total accumulated error can be $2\varepsilon_{max}$ (similarly, for rounding down). Therefore, taking into account that the final rounding up or rounding down produces a maximum error of one unit in the last place ($ULP$, the minimal distance between two machine numbers), the total worst case error is $\alpha_n + 2\varepsilon_{max} + 1ULP$.

For example, in a typical case, for n bits of precision and n+1 iterations, the value of the $\alpha_n$ is $\alpha_n = \tan^{-1}(2^{-n}) < 2^{-n} = 1ULP$, and taking $\log_2(n+1)+2$ guard bits to ensure a rounding error of $\varepsilon_{max} < 0.5ULPs$ [3][4], the worst case error is $3ULPs$. This result can be improved easily by slightly increasing the number of iterations or/and guard bits. For instance, if two iterations and two bits are added, the worst case error is $1.5ULPs$, which is the same result as in [7]. This result is very low bearing in mind that $1ULP$ is the theoretical limit.

## 4. Reduction of the number of iterations

For any interval input argument, the sine and cosine functions are evaluated in a time which is practically double that of a regular operation. However, this time can be strongly reduced for most cases by taking advantage of the iterative nature of the CORDIC algorithm. If interval arithmetic is focused on bounding the errors in numerical computation, the interval argument is usually narrow. We take advantage of this in such a way that the first vector $(v_1)$ is normally performed by rotating the initial vector $(v_0)$ through $\beta_1$ $(n + 1$ iterations), but the second vector $(v_2)$ is obtained from the first rotated vector $(v_1)$ by rotating it through the incremental angle $\Delta\beta = (\beta_2 - \beta_1)$, as we can see in Figure 4.a. If the incremental angle $\Delta\beta$ is narrow enough, it is possible to perform the second rotation in a few iterations instead of $n + 1$, as we show next.

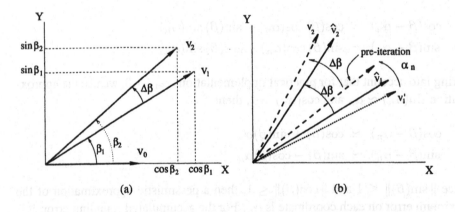

*Figure 4.* a. Rotation using the incremental angle. b. Addition of an extra pre–iteration.

Let us assume that $\Delta\beta = \beta_2 - \beta_1 < 2^{-j}$ and $j > n/3$. In this case the elementary angles fulfill $\alpha_i = \tan^{-1}(2^{-i}) \simeq 2^{-i}$ $(\forall i \geq j)$ [14] and therefore $\Delta\beta < \alpha_j$. From the convergence condition of the CORDIC algorithm $\alpha_i \leq \sum_{p=i+1}^{n} \alpha_p + \alpha_n$. Consequently, to achieve convergence it is not necessary to begin at iteration 0 but at iteration $j + 1$, and rotation through the incremental angle can be carried out in just $n - j$ iterations. Hence, the total number of iterations is $n + 1 + n - j$ instead of $2n + 2$.

Nevertheless, since $v_1$ is rotated to obtain $v_2$, the scale factor is not compensated for. Besides this, since it depends on the first $n/2$ coefficients $(\sigma_i)$, the final values obtained for the incremental rotation can be scaled by different factors depending on the value of $j$. Therefore, the use of scaling iterations or other methods to compensate for the scale factor is not feasible. To prevent this situation, the incremental rotation is only performed if $j > n/2$ (no scaling is introduced by microrotations for the level of precision considered). If this condition is not fulfilled, it is necessary to perform the whole rotation corresponding to the second angle $\beta_2$ instead of the incremental angle ($n + 1$ iterations).

Hence, the computational time depends on the width of the argument. For a width ($w$) less than $2^{n/2}$ $ULPs$ the number of iterations is $n + 1 + \log_2(w)$ (where w is the width of the argument in $ULPs$) and $2n + 2$ in other cases (the iterations used to ensure the correct bounds have not been taken into account).

## 4.1. Correct bounds for incremental rotation

Since the initial vector for the incremental rotation ($v_1$) includes the error due to the first rotation, some additions to the algorithm are necessary in order to ensure a correct choice of the bounds corresponding to this rotation.

First, as we explained in section 3, the final vector of the first rotation ($v_1'$) is always below the ideal vector ($\hat{v}_1$) (since an extra iteration is added if it is over). Nevertheless, before doing the incremental rotation, we need to place the vector ($v_1'$) over the ideal vector to guarantee that with a rotation through $\Delta\beta$ we can reach the location of $v_2$. Hence, an extra pre–iteration through $\alpha_n$ is added before doing the incremental rotation, as Figure 4.b shows.

Consequently, since an extra iteration is added after each rotation when the sign of the approximation error is unsuitable, a total of one extra iteration is required in the best case, and up to three iterations in the worst case in order to obtain correct bounds.

On the other hand, the maximum accumulated rounding error is larger for the incremental rotation, since the initial vector ($v_1$) has the accumulated rounding error produced in the first rotation. The accumulated rounding error produced during the incremental rotation is $0.5\varepsilon_{max}$, because the number of iterations is, at most, $n/2$ for this rotation. Thus, taking into account that the maximum error of $v_1$ is $\varepsilon_{max}$, the maximum error of $v_2$ is $1.5\varepsilon_{max}$. Then, this value is added (or subtracted) instead of $\varepsilon_{max}$ before the final rounding for the incremental rotation.

Moreover, due to the changes to the algorithm, the worst case error is larger for the incremental rotation. It is $2\alpha_n$ for the approximation error, since the error due to the incremental rotation($\alpha_n$) is added to the error in $v_1(\alpha_n)$. This is $3\varepsilon_{max}$ for the accumulated rounding error, since the maximum accumulated error is now $1.5\varepsilon_{max}$. Thus, the total worst case error for the incremental rotation is $2\alpha_n + 3\varepsilon_{max} + 1ULP$ which is $4.5ULPs$ in the typical case, instead of $3ULPs$. Therefore, just a little accuracy is lost for a large improvement in computational time.

## 5. Architecture

In Figure 5 we can see the proposed architecture which is very similar to the regular CORDIC architecture (see Figure 2). The shaded elements have been added to the classical approach to efficiently support interval sine and cosine evaluation. Basically, this is a rotator register where the most significant bit is carried to the control section, to calculate the first iteration on the incremental rotation.

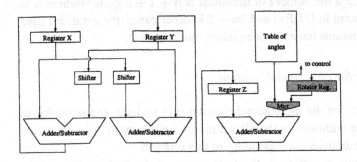

*Figure 5.* Proposed architecture.

Before starting the first rotation, one cycle is employed to calculate the incremental angle $\Delta\beta$, which is loaded in the rotator register (this is performed through the $z$ data path). After this, the first rotation is performed regularly as we explained in subsection 2.4. Besides this, on every iteration of this first rotation, the rotator register is shifted on the left looking for the first most significant bit set to one which gives us the iteration where the incremental rotation must start. After the first rotation, the incremental one is performed starting at the calculated iteration.

It is important to note that the basic structure of the CORDIC architecture has not been changed. Therefore, this architecture is able to support either regular or interval operation. In addition, these proposed minimal modifications can be applied to most of the newest CORDIC architectures yielding higher performance.

## 6. Conclusions

In this paper, we present a hardware solution based on the CORDIC algorithm for computing interval sine and cosine functions. In addition to the improvement gained by using direct hardware support, which is greater than for software implementations, our algorithm computes the interval operation in a time close to one regular operation for most of the interesting cases, whereas the best previous result for elementary functions is twice a regular operation [7]. In order to achieve this, only a very small amount of hardware is added to the classic CORDIC architecture and this allows us to compute either regular or interval operations. Furthermore, correct choice of the bounds of the interval is guaranteed and we provide a study of the worst case error and how to improve it.

This study could be the first step in developing a hardware unit based on CORDIC to evaluate many interval elementary functions.

## References

1. K. Braune. Standard functions for real and complex point and interval arguments with dynamic accuracy. *Computing, Suppl.*, (6):159–184, 1988.
2. G.L. Haviland and A.A. Tuszynski. A CORDIC arithmetic processor chip. *IEEE Trans. on Computers*, C-29(2):68–79, Feb 1980.
3. Y.H. Hu. The quantization effects of the CORDIC algorithm. *IEEE Trans. on Signal Processing*, 40(4):834–844, 1992.
4. K. Kota and J. R. Cavallaro. Numerical accuracy and hardware tradeoffs for CORDIC arithmetic for special purpose processors. *IEEE Trans. on Computers*, 42(7):769–779, July 1993.
5. R. E. Moore. *Reliability in Computing: The Role of Interval Methods in Scientific Computations*. Academic Press, 1988.
6. J. M. Muller. *Elementary Functions: Algorithms and Implementation*. Birkhäuser Boston, 1997.
7. Douglas M. Priest. Fast table-driven algorithms for interval elementary functions. *Proc. 13th Symposium on Computer Arithmetic*, pages 168–174, 1997.
8. M. J. Schulte and E. E. Swartzlander, Jr. Variable-precision, interval arithmetic coprocessors. *Reliable Computing*, 2(1):47–62, 1996.
9. J. E. Stine and M. J. Schulte. A combined interval and floating point multiplier. *Proceeding of 8th Great Lakes Symposium on VLSI*, pages 208–213, February 1998.
10. N. Takagi, T. Asada, and S. Yajima. Redundant CORDIC methods with a constant scale factor for sine and cosine computation. *IEEE Trans. on Computers*, 40(9):989–995, Sept 1991.
11. J.E. Volder. The CORDIC trigonometric computing technique. *IRE Trans. on Electronic Computers*, EC-8(3):330–334, September 1959.
12. J. Wolff von Gudenberg. Hardware support for interval arithmetic. *Scientific Computing and Validated Numerics*, pages 32–37, 1996.
13. J.S. Walther. A unified algorithm for elementary functions. *Proc. Spring Joint Computer Conf.*, pages 379–385, 1971.
14. S. Wang, V. Piuri, and E. E. Swartzlander, Jr. Hybrid cordic algorithms. *IEEE Trans. on Computers*, 46(11):1202–1207, November 1997.

T. Csendes (ed.), Developments in Reliable Computing 43–51.
© 1999 Kluwer Academic Publishers.

# Towards an optimal control of the wrapping effect

WOLFGANG KÜHN                                            kuehn@zib.de
*Konrad-Zuse-Zentrum für Informationstechnik Berlin, Takustraße 7, D-14195 Berlin-Dahlem*

**Abstract.** High order zonotopes, that is, bodies which are the finite sum of parallelepipeds, can be used to compute enclosures for the orbits of discrete dynamical systems. A Flush-When-Full (FWF) strategy on the summands effectively avoids the wrapping effect by wrapping small summands much more often than large ones. In a competitive analysis, different *online* FWF strategies are evaluated according to their performance relative to an optimal *offline* algorithm. Numerical experiments indicate that there is an *optimal* online FWF strategy which significantly outperforms previously employed strategies.

**Keywords:** Wrapping effect, dynamical systems, zonotopes, online algorithms

## 1. Introduction

Given an initial set $\Omega_0$ and maps $f_n : \mathbf{R}^d \to \mathbf{R}^d$, consider the dynamical system

$$\Omega_n = f_n(\Omega_{n-1}), \qquad n = 1, \ldots, N. \tag{1}$$

This system is not suitable for direct execution on a computer. The reasons are that the sets $\Omega_n$ are not representable on a computer, and that it is impossible to evaluate the range of a function exactly because of approximation and rounding errors. We will therefore replace system (1) by the weakened form $\widetilde{\Omega}_0 \supseteq \Omega_0$ and

$$\widetilde{\Omega}_n \supseteq f_n(\widetilde{\Omega}_{n-1}), \qquad n = 1, \ldots, N. \tag{2}$$

In order to evaluate (2) on a computer, it must be feasible to find computer-representable enclosures for the ranges $f_n(\widetilde{\Omega}_{n-1})$.

In [3], a Flush-When-Full (FWF) algorithm is described which computes the sets $\widetilde{\Omega}_n$ in the form of zonotopes [5]. There, it is shown that the overestimation of $\Omega_n$ by $\widetilde{\Omega}_n$, measured in terms of the Hausdorff distance, is sub-exponential: if $m$ is the order (equals number of parallelepiped summands) of the zonotopes, then

$$\mathrm{dist}\,(\Omega_n, \widetilde{\Omega}_n) \leq c_1 \exp(c_2 n^{1/m}) \tag{3}$$

for some constants $c_1$ and $c_2$. The current paper shows how the original FWF zonotope algorithm in [3] can be improved by a trivial modification of an inequality. By comparing the improved algorithm with the best possible zonotope algorithm via a competitive analysis, numerical experiments indicate that the improved version is optimal.

## 2. Zonotopes

A zonotope $Z$ of order $m$ is the finite Minkowski-sum of $m$ parallelepipeds:

$$Z = P^1 + \cdots + P^m. \tag{4}$$

Obviously, the class of all zonotopes is closed under Minkowski-addition and under linear transformations. If the range evaluation of a function is based on *lineariza-tion* plus *added error* term, this closedness property makes zonotopes the natural choice for the enclosure sets $\widetilde{\Omega}_n$. Relevant algebraic and geometric properties of zonotopes are discussed in [3], so only a brief account is given here.

Any parallelepiped $P$ can be written as the linear image $P = AQ$ of the unit cube $Q = [-1, 1]^d$, where the $(d \times d)$ matrix $A$ can be used to represent $P$ on a computer. Note that for intervals, $A$ is diagonal. If $P^j = A^j Q$ in (4) and $B = [A^1, \ldots, A^m]$ is a column augmented $(d \times md)$ matrix which consists of $m$ blocks of square matrices, then $Z = BQ$. Furthermore, if rs$B$ denotes the diagonal (row sum) matrix

$$(\mathrm{rs}\,B)_{ii} = \sum_{j=1}^{md} |B_{ij}|,$$

then the interval hull of $Z$ is given by $\Box Z = (\mathrm{rs}\,B)Q$, and the Euclidean radius of $Z$ is given by

$$\mathrm{rad}\,Z = \left( \sum_i ((\mathrm{rs}\,B)_{ii})^2 \right)^{1/2}.$$

Several examples of zonotopes and their interval hulls are shown in Fig. 2.

## 3. Linearization and Zonotope Extension

Let us assume that, for given $\Omega$, we have a procedure to find an *error* set $\Delta$ and a square matrix $T$ such that $f(\Omega) \subset \Delta + T\Omega$, see Fig. 1.

Often the Mean Value Theorem, in conjunction with interval analysis, will pro-vide such a procedure. Now let us assume that $\Omega = P^1 + \cdots + P^m$ is a zonotope of order $m$, and that $\Delta$ is an interval (this is usually the case). Then $\Omega \to \Delta + T\Omega = \Delta + TP^1 + \cdots + TP^m$ is a zonotope extension for $f$ of order $m + 1$. A successive application of the extension would therefore yield a linear increase of the order in the number of iterations. This is clearly not acceptable.

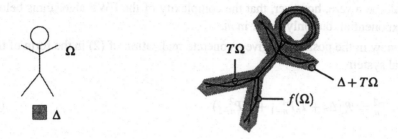

*Figure 1.* "Inflated linear approximation" $\Delta + T\Omega$ for the range $f(\Omega)$

## 4. Reducing the Order of a Zonotope

We can use the interval hull to reduce the order of the zonotope $\Delta + TP^1 + \cdots + TP^m$ by one. For $m = 2$, there are three possibilities:

$$\mathcal{R}(\Delta + TP^1 + TP^2) = \begin{cases} \square(\Delta + TP^1) + & TP^2 & \text{①}; \\ TP^1 & + \quad \square(\Delta + TP^2) & \text{②}; \\ 0 & + \square(\Delta + TP^1 + TP^2) & \text{③}. \end{cases} \quad (5)$$

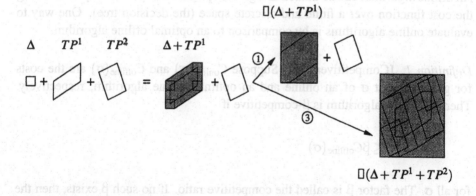

*Figure 2.* Alternatives ① and ③ in (5)

For ③, the order is reduced by two. Note that $\Delta + \square(TP^1 + TP^2)$ is the same as ③ ($\Delta$ is an interval), and does therefore not constitute a separate case. For general $m \geq 2$, one has $2^m - 1$ possibilities to form a nonempty subset of the $m$ terms, a subset to which $\Delta$ is added and of which the interval hull is taken. Because $2^m - 1$ is rapidly increasing, we will only consider $m = 2$ in the subsequent competitive analysis.

You should be aware, however, that the complexity of the FWF algorithms below are not exponential, but only linear in $m$.

We are now in the position to give a concrete realization of (2) in the form of the dynamical system

$$P_n^1 + P_n^2 = \mathcal{R}(\Delta_n + T_n P_{n-1}^1 + T_n P_{n-1}^2) \tag{6}$$

and $\widetilde{\Omega}_n = P_n^1 + P_n^2$.

The purpose of this paper is to study how to best make the *decisions* ①, ②, or ③, and therefore how to best specify $\mathcal{R}$. The objective is to make the decisions such that for given *request* (input sequence) $\{T_n, \Delta_n\}_{n=1}^N$, the *cost* $\mathrm{rad}\,(P_n^1 + P_n^2)$ is minimized uniformly in $n$ (because $\mathrm{dist}\,(\Omega_n, \widetilde{\Omega}_n) \leq \mathrm{rad}\,\widetilde{\Omega}_n$, and the radius is easier to compute than the distance, we only consider radii).

## 5. Competitive Analysis of On-Line Algorithms

An algorithm is *online* if it must make a decision at each time step $n$ based on past requests, and without having information about future requests. An algorithm that can examine all past and future requests before making a decision is *offline*, see [4]. An offline algorithm can be treated as a global optimization problem by minimizing the cost function over a finite and discrete space (the decision tree). One way to evaluate online algorithms is by comparison to an optimal offline algorithm.

*Definition 1.* [Competitiveness]  Suppose $C_{\text{online}}(\sigma)$ and $C_{\text{offline}}(\sigma)$ are the costs for given request $\sigma$ of an online and an optimal offline algorithm, respectively. Then the online algorithm is $\beta$-competitive if

$$C_{\text{online}}(\sigma) \leq \beta C_{\text{offline}}(\sigma)$$

for all $\sigma$. The factor $\beta$ is called the competitive ratio. If no such $\beta$ exists, then the online algorithm is not competitive.

A good online algorithm has a small constant $\beta$. Unfortunately it is often very difficult to compute $C_{\text{online}}$ analytically, and even more difficult to find $C_{\text{offline}}$. For sufficiently small decision space, however, one can find the optimal offline strategy by enumeration, that is, by computing *all* possible constellations. In our case, the cardinality of the decision space is $N^3$ for $m = 2$.

## 6. The Flush-When-Full Online Algorithm

The original FWF online algorithm in [3] (the name Cascade Reduction is used there, but the expression Flush-When-Full is considerably more descriptive; it is borrowed from the Paging Problem [1]) decides as follows: chose ① whenever

$$\mathrm{rad}\, T P_n^1 < \mathrm{rad}\, T P_n^2, \tag{7}$$

otherwise ③, never ②. The heuristic rational for this decision is that for most stages $n$, a small parallelepiped $P_n^1$ is wrapped using ①. If $P_n^1$ is getting bigger than $P_n^2$ (is full) for some stage $n$, it does not make sense to continue choice 1. Instead, $P_n^1$ is flushed to (wrapped with) $P_n^2$, and the process can continue with a small $P_{n+1}^1$. Naturally one will ask whether the heuristic criterion (7) can be improved (flush when how full?), or better yet, optimized.

## 7. A Simplified Dynamical Model

System (6) is too complex for a direct competitive analysis. We will replace it by a much simpler system in $\mathbf{R}^2$ by majorizing the radii.

**LEMMA 1** *Suppose* $\mathrm{rad}\, T_n$ *are orthogonal and* $\mathrm{rad}\, \Delta_n \le \delta$ *for all $n$. If* $\mathrm{rad}\, P_0^j \le r_0^j$ *for $j = 1, 2$ and*

$$(r_n^1, r_n^2) = \begin{cases} (\delta + \alpha r_{n-1}^1, & r_{n-1}^2 & ) & ①; \\ ( r_{n-1}^1, & \delta + \alpha r_{n-1}^2 & ) & ②; \\ ( 0, & \delta + \alpha(r_{n-1}^1 + r_{n-1}^2) ) & ③, \end{cases} \tag{8}$$

*where* $\alpha = \sqrt{d}$, *then* $\mathrm{rad}\, P_n^j \le r_n^j$ *for all $n$.*

**Proof:** Assume $T$ is orthogonal and $P$, $R$ are parallelepipeds. Then $\mathrm{rad}\, TP = \mathrm{rad}\, P$ (Euclidean norm is invariant under orthogonal transformations) and $\mathrm{rad}\, \Box P \le \sqrt{d}\, \mathrm{rad}\, P$. Also the triangle inequality implies $\mathrm{rad}\, (P + R) \le \mathrm{rad}\, P + \mathrm{rad}\, R$. We then have, for example, $\mathrm{rad}\, (\Box(\Delta_n + T_n P_{n-1}^1)) \le \delta + \alpha r_{n-1}^1$, and for the other two terms analogously. ∎

System (8) mimics the zonotope system (6) nicely. But instead of having as request the sequence $\{T_n, \Delta_n\}_{n=1}^N$, the request $\sigma$ now only consists of $N$, the total number of iterates. The following decision strategies for (8) are chosen:

## D-FWF

Corresponding to the FWF strategy (7) for zonotopes, we choose ① whenever $r_n^1 < r_n^2$, otherwise ③, and never ②. 'D' stands for direct, not weighted comparison;

*Table 1.* Decision sequence for $N = 50$. Decision ① in (8) is ·, ② +, and ③ is ∗

| O-FWF | · ∗ · ∗ · · ∗ · · · ∗ · · · · ∗ · · · · · ∗ · · · · · · ∗ · · · · · · · ∗ · · · · · · · · ∗ · · · · |
|-------|---|
| O-OFF | · · + + ∗ · · · · ∗ · · · · · · ∗ · · · · · · · ∗ · · · · · · · · ∗ · · · · · · · · · ∗ · · · · |

## O-FWF

An optimized FWF strategy, where $r_n^1 < r_n^2$ is replaced by

$$r_n^1 \log_2 \left( r_n^1/\delta \right) < r_n^2. \tag{9}$$

The weight $\log_2 \left( r_n^1/\delta \right)$ was found by try and error through numerous numerical experiments;

## O-OFF

The optimal offline, found by minimizing via enumeration. A recursive pseudo C code is listed in the Appendix. Our computational resources put a limit at $N \approx 50$, for which an SGI Onyx computer needs 3 hours to traverse the decision tree.

Fig. 3 shows the cost $r_N^1 + r_N^2$ for $N = 1, \ldots, 50$ of D-FWF, O-FWF, and O-OFF with $\alpha = 2$ (that is $d = 4$), $r_0^1 = r_0^2 = 0$, and $\delta = 1$. The competitive ratios for D-FWF and O-FWF are shown in Fig. 4. O-FWF performs better than the D-FWF, and Fig. 4 suggests that O-FWF is 2-competitive, and that D-FWF is not competitive at all. O-FWF meets optimal offline at the points $N = 1, 3, 6, 10, 15, 21, 28, 36, 45$, and you will note that the gaps increase by 1. We cannot offer an explanation for this surprising fact. The decision sequences for O-FWF and O-OFF are given in Table 1. Clearly, O-OFF also pursues a FWF strategy (many ① decisions, few ③).

## 8.  Improved FWF Zonotope Algorithm and Numerical Results

We will now apply the results of the last sections to the FWF zonotope algorithm. First, (9) is restated in terms of parallelepipeds to get

$$\text{rad}\, T_n P_n^1 \log_2 \left( \text{rad}\, (T_n P_n^1)/\delta_n \right) < \text{rad}\, T_n P_n^2, \tag{10}$$

where $\delta_n = \text{rad}\, \Delta_n$. One can also set $\delta_n$ equal to the sliding average of past errors to guard against too high variations and/or almost vanishing values of $\text{rad}\, \Delta_n$.

The performance of the optimized zonotope algorithm was tested using the Cremona map [2]

$$\begin{pmatrix} x \\ y \end{pmatrix} \mapsto \begin{pmatrix} x\cos\lambda - (y - x^2)\sin\lambda \\ x\sin\lambda + (y - x^2)\cos\lambda \end{pmatrix}.$$

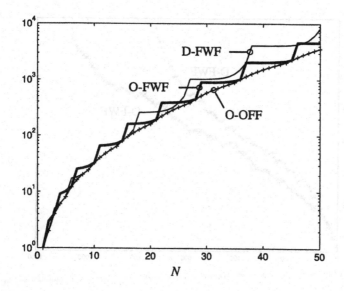

*Figure 3.* Cost (radii) $r_N^1 + r_N^2$ for on- and offline algorithms.

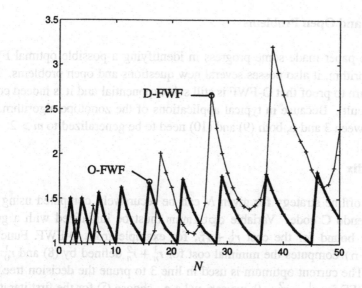

*Figure 4.* Competitive ratios for direct and optimized FWF strategies.

The initial value was $(0, 0.4)$ and $\lambda = 1.3$. The enclosure radii for both optimized and direct FWF are shown in Fig. 5. The optimized algorithm is clearly superior, being able to execute 400 additional iterations before the enclosures blow up.

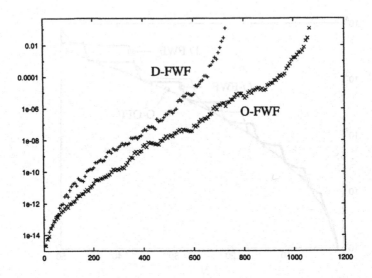

*Figure 5.* Radius of enclosures for order $m = 2$ of original and optimized zonotope algorithm.

## 9. Outlook and Open Problems

Although this paper made some progress in identifying a possible optimal FWF zonotope algorithm, it also posses several new questions and open problems. The most vexing are to proof that O-FWF is still sub-exponential and it is indeed competitive (difficult). Because in typical applications of the zonotope algorithm the order lies between 3 and 6, both (9) and (10) need to be generalized to $m > 2$.

## 10. Appendix

The optimal offline strategy for given $N$ can be recursively computed using the following pseudo C code. Variable optimum must be initialized with a guaranteed upper bound for the cost $r_N^1 + r_N^2$, for example from O-FWF. Function tree(a,b,n) computes the minimal cost for $r_n^1 + r_n^2$ defined by (8) and $r_0^1 = a$ and $r_0^2 = b$. The current optimum is used in line 3 to prune the decision tree. To compute O-OFF for $r_0^1 = r_0^2 = 0$, wa can, w.l.o.g., choose ① for the first iterate in (8), implying $r_1^1 = \delta$ and $r_1^2 = 0$. The remaining decision tree is then descended into by the call tree($\delta$, 0, N-1).

1. double optimum;

2. void tree(double a, double b, int n) {

```
3.          if ( a+b≥optimum ) return;      // Don't descent into this branch.

4.          if ( --n==1 ) {      // Last iterate
5.                cost = δ+α*min(a,b)+max(a,b);
6.                optimum = min(optimum, cost);
7.          } else {
8.                tree(α*a+δ, b, n);
9.                tree(a, α*b+δ, n);
10.               tree(0, α*(a+b)+δ, n);
11.         }
12.   }
```

## References

1. A. Fiat, R. Karp, M. Luby, L. McGeoch, D. Sleator, N. E. Young, Competitive Paging Algorithms, *J. Algorithms*, 12, pp. 685–699, 1991.
2. J. Hale and H. Koçak, Dynamics and Bifurcations, *Springer-Verlag*, 1991.
3. W. Kühn, Rigorously Computed Orbits of Dynamical Systems without the Wrapping Effect, *Computing*, 61, pp. 47–67, 1998.
4. C. C. McGeoch, How to Stay Competitive, *Amer. Math. Monthly*, 101, pp. 897–901, 1994.
5. G. M. Ziegler, Lectures on Polytopes, *Springer-Verlag*, 1995.

T. Csendes (ed.), Developments in Reliable Computing 53–65.
© 1999 Kluwer Academic Publishers.

# On Existence and Uniqueness of Solutions of Linear Algebraic Equations in Kaucher's Interval Arithmetic

ANATOLY V. LAKEYEV                                                                lakeyev@icc.ru

*Institute of Systems Dynamics and Control Theory, Siberian Branch of the Russian Academy of Sciences, 134, Lermontov Str., Irkutsk, 664033, Russia*

## Introduction

Presently there is a growing interest to the investigation of linear algebraic equations in Kaucher's arithmetic [1]. This interest is mainly substantiated by the fact that with the use of solutions of linear algebraic equations in Kaucher's arithmetic allows in some cases to obtain both external and internal estimates of various sets of solutions of linear interval equations. First such results appeared rather long ago and were concerned with the external estimation of the joint set of solutions for the system of linear interval equations by algebraic solving of this system (see, e.g., G. Alefeld and J. Herzberger [2], A. Neumaier [3], and the references in these books). Later, in L.V. Kupriyanova's [4] and S.P. Shary's [5] works, it was shown that with the use of algebraic solutions (but already in an extended Kaucher's interval arithmetic) it is possible to obtain both internal and external estimates for generalized sets of solutions for the systems of linear interval equations.

   This paper investigates the problems of existence and uniqueness of solutions of linear algebraic equations in Kaucher's interval arithmetic. Sufficient conditions, similar to those of nonexpandability introduced by A. Neumaier [3] and J. Rohn [6], of both uniqueness of solutions and - in case of equality of the number of equations to the number of variables - existence of solutions for these equations is obtained. It is also shown that for the matrices of constant signs such conditions are both necessary and sufficient.

## 1.   Linear Algebraic Equations in Kaucher's Arithmetic

Let $\mathbf{IR} = \{[\underline{a}, \overline{a}] \mid \underline{a}, \overline{a} \in \mathbf{R}\}$ be an extended set of intervals (the condition of $\underline{a} \le \overline{a}$ is not needed in this case). Consider the following operations of addition and multiplication of elements of this set introduced by Kaucher [1] (see also [7]).

For the two intervals $\mathbf{a} = [\underline{a}, \overline{a}]$, $\mathbf{b} = [\underline{b}, \overline{b}] \in \mathbf{IR}$ the interval

$$\mathbf{a} + \mathbf{b} = [\underline{a} + \underline{b}, \overline{a} + \overline{b}]$$

is called their sum.

Before defining the operation of multiplication of intervals, introduce the following notations.

Let

$\mathcal{P} = \{[\underline{a}, \overline{a}] \in \mathbf{IR} \mid \underline{a} \geq 0, \overline{a} \geq 0\}$ be a set of nonnegative intervals,

$-\mathcal{P} = \{[\underline{a}, \overline{a}] \in \mathbf{IR} \mid \underline{a} \leq 0, \overline{a} \leq 0\}$ be a set of nonpositive intervals,

$\mathcal{Z} = \{[\underline{a}, \overline{a}] \in \mathbf{IR} \mid \underline{a} < 0 < \overline{a}\}$ be a set of intervals containing zero,

dual $\mathcal{Z} = \{[\underline{a}, \overline{a}] \in \mathbf{IR} \mid \overline{a} < 0 < \underline{a}\}$ be a set of intervals contained in the zero.

Then the multiplication of the two intervals $\mathbf{a} = [\underline{a}, \overline{a}]$, $\mathbf{b} = [\underline{b}, \overline{b}] \in \mathbf{IR}$ is defined by the following table:

| a \ b | $\mathcal{P}$ | $\mathcal{Z}$ | $-\mathcal{P}$ | dual $\mathcal{Z}$ |
|---|---|---|---|---|
| $\mathcal{P}$ | $[\underline{ab}, \overline{ab}]$ | $[\overline{a}\underline{b}, \overline{ab}]$ | $[\overline{a}\underline{b}, \underline{a}\overline{b}]$ | $[\underline{ab}, \underline{a}\overline{b}]$ |
| $\mathcal{Z}$ | $[\underline{a}\overline{b}, \overline{ab}]$ | $[\min\{\overline{a}\underline{b}, \underline{a}\overline{b}\}, \max\{\underline{ab}, \overline{ab}\}]$ | $[\overline{a}\underline{b}, \underline{ab}]$ | $0$ |
| $-\mathcal{P}$ | $[\underline{a}\overline{b}, \overline{a}\underline{b}]$ | $[\underline{a}\overline{b}, \underline{ab}]$ | $[\overline{ab}, \underline{ab}]$ | $[\overline{ab}, \overline{a}\underline{b}]$ |
| dual $\mathcal{Z}$ | $[\underline{ab}, \overline{a}\underline{b}]$ | $0$ | $[\overline{ab}, \underline{a}\overline{b}]$ | $[\max\{\underline{ab}, \overline{ab}\}, \min\{\overline{a}\underline{b}, \underline{a}\overline{b}\}]$ |

From now on, by $\mathbf{IR}^n$ ($\mathbf{R}^n$) and $\mathbf{IR}^{m \times n}$ ($\mathbf{R}^{m \times n}$) we denote the sets on n-vectors and $m \times n$-matrics with the elements from $\mathbf{IR}$ ($\mathbf{R}$), respectively.

Now we can introduce the concept of linear algebraic equation and its solution.

Let $\mathbf{A} = [\underline{A}, \overline{A}] \in \mathbf{IR}^{m \times n}$, $\mathbf{b} = [\underline{b}, \overline{b}] \in \mathbf{IR}^m$. Then in the capacity of the algebraic solution for the equation of the form

$$\mathbf{Ax} = \mathbf{b}, \tag{1}$$

we understand the vector $\mathbf{x} \in \mathbf{IR}^n$ such that multiplication of the matrix $\mathbf{A}$ by this vector (here additions and multiplications of intervals are computed in Kaucher's arithmetic) gives the vector $\mathbf{b}$.

Note, despite the fact that eq.(1) has been called "the linear algebraic equation", it is not linear in the general understanding, since the law of distributivity is not satisfied in interval arithmetic (both in ordinary and in Kaucher's one).

In order to understand to which class eq. (1) really belongs we need some analytic formula for multiplication of intervals in Kaucher's arithmetic.

Henceforth, we use the following notations.

Let $x, y \in \mathbf{R}^n$ be $n$-vectors (in particular - of real numbers from $\mathbf{R}^1$), $A, B \in R^{m \times n}$ be $m \times n$-matrices, then:

- $x \vee y = \max\{x, y\}$, $A \vee B = \max\{A, B\}$ – is the precise upper bound for $x, y$ and $A, B$, respectively;

- $x^+ = \max\{x, 0\}$, $A^+ = \max\{A, 0\}$ – is the positive part of the vector $x$ and the matrix $A$;

- $x^- = \max\{-x, 0\}$, $A^- = \max\{-A, 0\}$ – is the negative part of the vector $x$ and the matrix $A$;

- $|x| = x^+ + x^-$, $|A| = A^+ + A^-$ – is the modulus of the vector and the matrix, respectively.

In this case, the operations max and $|\cdot|$ are understood componentwise for the vectors and elementwise for the matrics.

Using the above notations, for the operation of multiplication of intervals in Kaucher's arithmetic the following statement holds.

LEMMA 1 *[8] If* $\mathbf{a} = [\underline{a}, \overline{a}]$, $\mathbf{b} = [\underline{b}, \overline{b}] \in \mathbf{IR}$ *then*

$$\mathbf{ab} = [(\underline{a}^+ \underline{b}^+) \vee (\overline{a}^- \overline{b}^-) - (\overline{a}^+ \overline{b}^-) \vee (\underline{a}^- \underline{b}^+), (\overline{a}^+ \overline{b}^+) \vee (\underline{a}^- \underline{b}^-) - (\underline{a}^+ \underline{b}^-) \vee (\overline{a}^- \overline{b}^+)]. \tag{2}$$

The proof is obtained by direct choosing of all the variants of sign distribution for $\underline{a}$, $\overline{a}$, $\underline{b}$, $\overline{b}$ and comparison of the obtained result with the above table for multiplication of intervals in Kaucher's arithmetic.

Using this Lemma, an analytical formula for multiplication of the interval matrix by the interval vector can easily be obtained. Furthermore, for the vector $x \in \mathbf{R}^n$ by $D(x) = \operatorname{diag}\{x_1, ..., x_n\}$ we denote the diagonal $n \times n$-matrix with diagonal vector $x$ and by $e^n \in \mathbf{R}^n$ we denote the vector all the components of which are equal to one.

PROPOSITION 1 *[8] If for* $\mathbf{A} \in \mathbf{IR}^{m \times n}$ *and* $\mathbf{x} \in \mathbf{IR}^n$ *we denote* $\mathbf{Ax} = [\underline{Ax}, \overline{Ax}]$ *then*

$$\overline{Ax} = (\overline{A}^+ D(\overline{x}^+) \vee \underline{A}^- D(\underline{x}^-) - \underline{A}^+ D(\overline{x}^-) \vee \overline{A}^- D(\underline{x}^+)) e^n,$$
$$\underline{Ax} = (\underline{A}^+ D(\underline{x}^+) \vee \overline{A}^- D(\overline{x}^-) - \overline{A}^+ D(\underline{x}^-) \vee \underline{A}^- D(\overline{x}^+)) e^n. \tag{3}$$

**Comment 1** It is not difficult to write down also the formulas similar to (3) for the multiplication of two interval matrices. Despite the fact that such formulas are not further used in the present paper, they are of special interest and may be used in both theoretical investigations and in practical applications of interval computations. So, we decided to suggest such formulas.

Let $A \in \mathbf{R}^{m \times n}$ be an $m \times n$-matrix and $B \in \mathbf{R}^{n \times k}$ be an $n \times k$-matrix. Denote by $A \otimes B$ the following $m \times (nk)$-matrix:

if $a_1, ..., a_n \in \mathbf{R}^{m \times 1}$ are vector columns of the matrix $A$, i.e. $A = (a_1 \ ... \ a_n)$ and $b_1, ..., b_n \in \mathbf{R}^{1 \times k}$ are vector-rows of the matrix $B$, i.e. $B = (b_1^T \ ... \ b_n^T)^T$ , then

$$A \otimes B = (a_1 b_1 \ ... \ a_n b_n),$$

(note that each product $a_i b_i, i = \overline{1, n}$ is an m x k-matrix).

Furthermore, denote by $E_k^n = e^n \otimes E_k$ the $(nk) \times k$-matrix, where $e^n \in \mathbf{R}^{n \times 1}$ is a column $n$-vector all the coordinates of which are equal to one, $E_k \in \mathbf{R}^{k \times k}$ is a unit $k \times k$-matrix and "$\otimes$" is the Kronecker's (direct, tensor) product of matrices.

Hence, the following statement holds.

PROPOSITION 2 If $\mathbf{A} = [\underline{A}, \overline{A}] \in \mathbf{IR}^{m \times n}$, $\mathbf{B} = [\underline{B}, \overline{B}] \in \mathbf{IR}^{n \times k}$ and $\mathbf{AB} = [\underline{AB}, \overline{AB}] \in \mathbf{IR}^{m \times k}$ then

$$\overline{AB} = ((\overline{A}^+ \otimes \overline{B}^+) \vee (\underline{A}^- \otimes \underline{B}^-) - (\underline{A}^+ \otimes \overline{B}^-) \vee (\overline{A}^- \otimes \underline{B}^+)) E_k^n,$$

$$\underline{AB} = ((\underline{A}^+ \otimes \underline{B}^+) \vee (\overline{A}^- \otimes \overline{B}^-) - (\overline{A}^+ \otimes \underline{B}^-) \vee (\underline{A}^- \otimes \overline{B}^+)) E_k^n.$$

The proof of Propositions 1 and 2 can be obtained with the use of the formula (2) by simply writing the definitions of multiplication of a matrix by a vector and a matrix by a matrix, respectively, and so we can omit it.

Using the formulas (3), it not difficult to write eq. (1) in its explicit form. In this case, we immediately replace the operation "$\vee$" with its representation containing the modulus (this representation is valid for any vector lattice):

$$A \vee B = \frac{1}{2}(A + B + |A - B|)$$

where $A$ and $B$ are any $m \times n$-matrics. Furthermore, taking account of the relations $D(x)e^n = x$ and $(-x)^+ = x^-$, for $x \in \mathbf{R}^n$, after simple transformations we obtain the following statement.

THEOREM 1 [8] Let $\mathbf{A} = [\underline{A}, \overline{A}] \in \mathbf{IR}^{m \times n}$, $\mathbf{b} = [\underline{b}, \overline{b}] \in \mathbf{IR}^m$. The vector $\mathbf{x} \in \mathbf{IR}^n$ is an algebraic solution of eq. (1) if and only if $\underline{x}$ and $\overline{x}$ satisfy the following system of equations

$$\begin{pmatrix} \overline{A}^+ & \underline{A}^- \\ \underline{A}^- & \overline{A}^+ \end{pmatrix} \begin{pmatrix} \overline{x} \\ -\underline{x} \end{pmatrix}^+ + \left| \begin{pmatrix} \overline{A}^+ & -\underline{A}^- \\ -\underline{A}^- & \overline{A}^+ \end{pmatrix} \begin{pmatrix} D(\overline{x}) \\ D(-\underline{x}) \end{pmatrix}^+ \right| e^n =$$

$$= \begin{pmatrix} \underline{A}^+ & \overline{A}^- \\ \overline{A}^- & \underline{A}^+ \end{pmatrix} \begin{pmatrix} \overline{x} \\ -\underline{x} \end{pmatrix}^- + \left| \begin{pmatrix} \underline{A}^+ & -\overline{A}^- \\ -\overline{A}^- & \underline{A}^+ \end{pmatrix} \begin{pmatrix} D(\overline{x}) \\ D(-\underline{x}) \end{pmatrix}^- \right| e^n + 2 \begin{pmatrix} \overline{b} \\ -\underline{b} \end{pmatrix}. \quad (4)$$

From eq. (4) it is obvious what sort of nonlinearities are contained in eq. (1). In particular, it is obvious that this equation is piecewise linear with respect to $\underline{x}$ and $\bar{x}$. Henceforth we'll need a somewhat different form of eq. (4), by using which it is possible to obtain some sufficient conditions of uniqueness of the solution of eq. (1) which are more close (in some sense) to the necessary ones than the conditions obtained when eq. (4) is used directly.

For this purpose, let us represent the matrix $\mathbf{A}$ as a sum of the 3 matrices:

$$\mathbf{A} = \mathbf{A}_0 + \mathbf{A}_1 + \mathbf{A}_2,$$

where $\mathbf{A}_l = \mathbf{a}_{ij}^l$, $l \in \{0, 1, 2\}$ are such that

$$\mathbf{a}_{ij}^0 = \begin{cases} \mathbf{a}_{ij}, & \text{if } \underline{a}_{ij}\bar{a}_{ij} \geq 0, \\ 0, & \text{else,} \end{cases}$$

$$\mathbf{a}_{ij}^1 = \begin{cases} \mathbf{a}_{ij}, & \text{if } \underline{a}_{ij} < 0 < \bar{a}_{ij}, \\ 0, & \text{else,} \end{cases}$$

$$\mathbf{a}_{ij}^2 = \begin{cases} \mathbf{a}_{ij}, & \text{if } \underline{a}_{ij} > 0 > \bar{a}_{ij}, \\ 0, & \text{else.} \end{cases}$$

It can be readily seen that in this case

$$\bar{A}^+ = \bar{A}_0^+ + \bar{A}_1, \bar{A}^- = \bar{A}_0^- - \bar{A}_2,$$
$$\underline{A}^+ = \underline{A}_0^+ + \underline{A}_2, \underline{A}^- = \underline{A}_0^- - \underline{A}_1.$$

Considering also the fact that for any fixed $i = \overline{1, m}$, $j = \overline{1, n}$ only one of the 3 intervals $\mathbf{a}_{ij}^0$, $\mathbf{a}_{ij}^1$, $\mathbf{a}_{ij}^0$ may be nonzero, eq. (4) may be transformed to the following form

$$\begin{pmatrix} \bar{A}^+ + \bar{A}_0^+ & \underline{A}^- + \underline{A}_0^- \\ \underline{A}^- + \underline{A}_0^- & \bar{A}^+ + \bar{A}_0^+ \end{pmatrix} \begin{pmatrix} \bar{x} \\ -\underline{x} \end{pmatrix}^+ + \left| \begin{pmatrix} \bar{A}_1 & \underline{A}_1 \\ \underline{A}_1 & \bar{A}_1 \end{pmatrix} \begin{pmatrix} D(\bar{x}) \\ D(-\underline{x}) \end{pmatrix}^+ \right| e^n =$$

$$= \begin{pmatrix} \underline{A}^+ + \underline{A}_0^+ & \bar{A}^- + \bar{A}_0^- \\ \bar{A}^- + \bar{A}_0^- & \underline{A}^+ + \underline{A}_0^+ \end{pmatrix} \begin{pmatrix} \bar{x} \\ -\underline{x} \end{pmatrix}^- + \left| \begin{pmatrix} \underline{A}_2 & \bar{A}_2 \\ \underline{A}_2 & \underline{A}_2 \end{pmatrix} \begin{pmatrix} D(\bar{x}) \\ D(-\underline{x}) \end{pmatrix}^- \right| e^n + 2 \begin{pmatrix} \bar{b} \\ -\underline{b} \end{pmatrix}.$$

$$(5)$$

So, the interval vector $\mathbf{x} = [\underline{x}, \bar{x}]$ will be the solution of eq. (1) if and only if $\underline{x}$ and $\bar{x}$ satisfy eq. (5).

## 2. Conditions of Existence and Uniqueness

Consider the following class of equations to which eqs. (4) and (5) belong.

Let $A_1, A_2 \in \mathbf{R}^{s \times r}$ and $L_1, L_2 : \mathbf{R}^r \to \mathbf{R}^{s \times k}$ be linear mappings from $\mathbf{R}^r$ into $\mathbf{R}^{s \times k}$, i.e. $L_1(x)$ and $L_2(x)$ are $s \times k$-matrices whose elements are some linear functions of $x$.

For any $b \in \mathbf{R}^s$ we can write the following equation

$$A_1 x^+ + |L_1(x^+)| e^k = A_2 x^- + |L_2(x^-)| e^k + b, \tag{6}$$

where $x \in \mathbf{R}^r$, $x$ is unknown.

Eqs. (4) and (5) obviously have the form (6) if we put $r = 2n$, $s = 2m$, $k = n$, and, for example for eq. (5)

$$A_1 = \begin{pmatrix} \overline{A}^+ + \overline{A}_0^+ & \underline{A}^- + \underline{A}_0^- \\ \underline{A}^- + \underline{A}_0^- & \overline{A}^+ + \overline{A}_0^+ \end{pmatrix}, \quad A_2 = \begin{pmatrix} \underline{A}^+ + \underline{A}_0^+ & \overline{A}^- + \overline{A}_0^- \\ \overline{A}^- + \overline{A}_0^- & \underline{A}^+ + \underline{A}_0^+ \end{pmatrix},$$

$$x = \begin{pmatrix} \overline{x} \\ -\underline{x} \end{pmatrix}, \quad b = \begin{pmatrix} \overline{b} \\ -\underline{b} \end{pmatrix},$$

$$L_1(x) = \begin{pmatrix} \overline{A}_1 & \underline{A}_1 \\ \underline{A}_1 & \overline{A}_1 \end{pmatrix} \begin{pmatrix} D(\overline{x}) \\ D(-\underline{x}) \end{pmatrix}, \quad L_2(x) = \begin{pmatrix} \underline{A}_2 & \overline{A}_2 \\ \overline{A}_2 & \underline{A}_2 \end{pmatrix} \begin{pmatrix} D(\overline{x}) \\ D(-\underline{x}) \end{pmatrix}.$$

Before giving general conditions of uniqueness, and in the case when $r = s$ also existence, of the solution of eq. (6), consider a particular case of this equation obtained for $L_1(x) = L_2(x) \equiv 0$.

In this case eq. (6) has the form

$$A_1 x^+ = A_2 x^- + b, \tag{7}$$

where $A_1, A_2 \in \mathbf{R}^{s \times r}$, $b \in \mathbf{R}^s$, $x \in \mathbf{R}^r$.

The following statement holds for this equation.

THEOREM 2 *Let eg. (6) have the particular form (7). Then the following two conditions are equivalent:*

*1) eq. (7) has not more than one solution for any $b \in \mathbf{R}^s$;*

*2) the system of equations and inequalities*

$$\begin{cases} (A_1 + A_2)y + (A_1 - A_2)x = 0, \\ |y| \geq |x|, \end{cases} \tag{8}$$

*has the unique solution $x = y = 0$.*

*If, furthermore, $r = s$, these conditions are also equivalent to the following one:*
*3) eq. (7) has a unique solution for any $b \in \mathbf{R}^s$.*

Before proving the theorem let us prove the following lemma. For the vectors $x, y \in \mathbf{R}^n$, by $x \wedge y$ we'll denote the precise lower bound, i.e.

$$x \wedge y = \min\{x, y\},$$

where, as before, the operation min is considered componentwise.

LEMMA 2 *Let $a, b \in \mathbf{R}^n$ and $|a| \leq |b|$. Then*

$$(a+b)^+ \wedge (a-b)^+ = 0, \quad (a+b)^- \wedge (a-b)^- = 0. \tag{9}$$

**Proof:** From the inequality $|a| \leq |b|$ we have that $-|b| \leq a \leq |b|$, and hence, $-|b_i| \leq a_i \leq |b_i|$, $i = \overline{1,n}$. Furthermore, from the inequality $a_i \leq |b_i|$ it follows that when $b_i \geq 0$ then $a_i \leq b_i$, and hence $(a_i - b_i)^+ = 0$ and when $b_i \leq 0$ then $a_i \leq -b_i$, and hence $(a_i + b_i)^+ = 0$.

Consequently, for any $i = \overline{1,n}$ either $(a_i - b_i)^+ = 0$ or $(a_i + b_i)^+ = 0$, and hence, $(a+b)^+ \wedge (a-b)^+ = 0$.

The equality $(a+b)^- \wedge (a-b)^- = 0$ can be obtained similarly from the inequality $-|b| \leq a$. ∎

**Proof of Theorem 2:** Let us first prove the equivalence of conditions 1) and 2).

Let 1) hold, and let $y, x \in \mathbf{R}^r$ satisfy the system (8).

Put

$$x_1 = (x+y)^+ - (x-y)^+, \quad x_2 = (x+y)^- - (x-y)^-.$$

Since $|y| \geq |x|$, from the Lemma 2 it follows that $(x+y)^+ \wedge (x-y)^+ = (x+y)^- \wedge (x-y)^- = 0$ and hence

$$x_1^+ = (x+y)^+, \ x_1^- = (x-y)^+, \ x_2^+ = (x+y)^-, \ x_2^- = (x-y)^-.$$

Next, from the first equation of the system (8) we have $A_1(x+y) = A_2(x-y)$. Hence, from above equalities it follows that

$$A_1(x_1^+ - x_2^+) = A_1((x+y)^+ - (x+y)^-) = A_1(x+y) = A_2(x-y) =$$
$$= A_2((x-y)^+ - (x-y)^-) = A_2(x_1^- - x_2^-),$$

and so $A_1 x_1^+ - A_2 x_1^- = A_1 x_2^+ - A_2 x_2^-$.

Therefore, by putting $b = A_1 x_1^+ - A_2 x_1^- = A_1 x_2^+ - A_2 x_2^-$ we have that $x_1$ and $x_2$ are solutions of eq. (7) with the same $b$. But from 1) we have that $x_1 = x_2$, i.e.

$$(x+y)^+ - (x-y)^+ = (x+y)^- - (x-y)^-, \text{ or}$$
$$x+y = (x+y)^+ - (x+y)^- = (x-y)^+ - (x-y)^- = x - y,$$

and hence, $y = 0$. Since furthermore $|y| \geq |x|$, we have that $x = 0$. Consequently, $x = y = 0$ is a unique solution of (8).

Conversely, let the system (8) have zero as the unique solution, and let $x_1, x_2 \in \mathbf{R}^r$ satisfy eq. (7). Hence $A_1 x_1^+ - A_2 x_1^- = A_1 x_2^+ - A_2 x_2^-$, or $A_1(x_1^+ - x_2^+) = A_2(x_1^- - x_2^-)$.

Now put

$$x = |x_1| - |x_2|, \ y = x_1 - x_2.$$

Hence, since for any $x \in \mathbf{R}^r$ $|x| = x^+ + x^-$, the following equalities obviously hold:

$$x+y = |x_1| + x_1 - |x_2| - x_2 = 2(x_1^+ - x_2^+),$$
$$x-y = |x_1| - x_1 - |x_2| + x_2 = 2(x_1^- - x_2^-).$$

Consequently

$A_1(x+y) = 2A_1(x_1^+ - x_2^+) = 2A_2(x_1^- - x_2^-) = A_2(x-y)$, or
$(A_1 + A_2)y + (A_1 - A_2)x = 0$,

i.e. for $x, y$ the first equation of (8) holds.

Since furthermore it is obvious that

$|x| = ||x_1| - |x_2|| \le |x_1 - x_2| = |y|$,

$x, y$ satisfy the system (8). Consequently, $y = x_1 - x_2 = 0$ and $x_1 = x_2$, i.e. (7) has not more than one solution.

Consider now the case when $r = s$.

Obviously, from 3) we have 1). Therefore, to complete the proof of the theorem it is sufficient to show that from 2) follows the existence of a solution of (7) for any $b \in \mathbf{R}^s$.

Suppose that 2) holds. Now let us rewrite eq. (7) in the following equivalent form, while replacing $x^+$ and $x^-$ with $x^+ = 1/2(|x| + x)$, $x^- = 1/2(|x| - x)$,

$$(A_1 + A_2)x = (A_2 - A_1)|x| + 2b. \tag{10}$$

Note that the matrix $A_1 + A_2$ must be nonsingular, since if $y_0 \ne 0$ such that $(A_1 + A_2)y_0 = 0$ then the pair $(y_0, 0)$ will be a nonzero solution of (8). So, eq. (10) may be rewritten in the form

$$x = (A_1 + A_2)^{-1}(A_2 - A_1)|x| + 2(A_1 + A_2)^{-1}b. \tag{11}$$

Let us introduce the notation $B = (A_1 + A_2)^{-1}(A_2 - A_1)$ and show that the matrix $B$ is nonexpanding [3]. Indeed, if it is not so then there can be found $x_0 \ne 0$ such that $|Bx_0| \ge |x_0|$. Then by putting $y_0 = Bx_0$ we obtain $y_0 = (A_1 + A_2)^{-1}(A_2 - A_1)x_0$ or $(A_1 + A_2)y_0 = (A_2 - A_1)x_0$ and $|y_0| \ge |x_0|$ which contradicts condition 2). But then from Theorem 6.1.5 [3] we have that eq. (11), and consequently eq. (7), have a solution for any $b \in \mathbf{R}^s$.  ∎

**Comment 2** From the last part of the theorem's proof it can be easily seen that for $s = r$ the condition 2) is equivalent to the fact that $\det(A_1 + A_2) \ne 0$ & $(A_1 + A_2)^{-1}(A_2 - A_1)$ is nonexpanding. Therefore, Theorem 2 may be considered as an extension of J. Rohn's [6] and A. Neumaier's [3] results for the systems with moduli to the case when $r \ne s$.

Consider now the general equation (6). In this case the following statement holds.

THEOREM 3 *If the system of inequalities*

$$\begin{cases} (|L_1(x+y)| + |L_2(x-y)|)e^k \ge |(A_1 + A_2)y + (A_1 - A_2)x|, \\ |y| \ge |x|, \end{cases} \tag{12}$$

*has zero as the unique solution, then eq. (6) has not more than one solution for any* $b \in \mathbf{R}^s$. *Moreover, if* $r = s$, *then eq. (6) has a unique solution for any* $b \in \mathbf{R}^s$.

**Proof:** Let us first prove the first statement. Let the system (12) have only the zero solution and let $x_1, x_2 \in \mathbf{R}^r$ be two solutions of eq. (6). Then

$$A_1 x_1^+ - A_2 x_1^- + (|L_1(x_1^+)| - |L_2(x_1^-)|)e^k = A_1 x_2^+ - A_2 x_2^- + (|L_1(x_2^+)| - |L_2(x_2^-)|)e^k,$$

and hence

$$A_1(x_1^+ - x_2^+) + A_2(x_2^- - x_1^-) = (|L_1(x_2^+)| - |L_1(x_1^+)| + |L_2(x_1^-)| - |L_2(x_2^-)|)e^k. \quad (13)$$

Let us put $x = 1/2(|x_1| - |x_2|)$, $y = 1/2(x_1 - x_2)$. Hence, obviously

$$|x| = 1/2||x_1| - |x_2|| \leq 1/2|x_1 - x_2| = |y|$$

and, likewise in proving Theorem 2, it can easily be shown that
$x + y = x_1^+ - x_2^+, x - y = x_1^- - x_2^-$.
Hence from (13) we obtain
$$|A_1(x+y) + A_2(y-x)| = |A_1(x_1^+ - x_2^+) + A_2(x_2^- - x_1^-)| =$$
$$|(|L_1(x_2^+)| - |L_1(x_1^+)| + |L_2(x_1^-)| - |L_2(x_2^-)|)e^k| \leq$$
$$\leq |(|L_1(x_2^+)| - |L_1(x_1^+)|)e^k| + |(|L_2(x_1^-)| - |L_2(x_2^-)|)e^k| \leq$$
$$\leq ||L_1(x_2^+)| - |L_1(x_1^+)||e^k + ||L_2(x_1^-)| - |L_2(x_2^-)||e^k \leq$$
$$\leq |L_1(x_2^+) - L_1(x_1^+)|e^k + |L_2(x_1^-) - L_2(x_2^-)|e^k =$$
$$= (|L_1(x_2^+ - x_1^+)| + |L_2(x_1^- - x_2^-)|)e^k = (|L_1(-(x+y))| + |L_2(x-y)|)e^k =$$
$$= (|L_1(x+y)| + |L_2(x-y)|)e^k.$$
Consequently, the system of inequalities (12) is satisfied for $x, y$, and hence $x = y = 0$. But then $x_1 - x_2 = 2y = 0$ and $x_1 = x_2$, i.e. any two solutions of eq. (6) coincide.

Let now $r = s$ and the system (12) has only a zero solution. We need to show that eq. (6) has a solution for any $b \in \mathbf{R}^s$. For this purpose let us transform eq. (6) to the form (7) and use Theorem 2. Introduce new matrix variables $Z_1, Z_2 \in \mathbf{R}^{s \times k}$ and rewrite eq. (6) in the form of the following system

$$\begin{cases} A_1 x^+ + |Z_1|e^k = A_2 x^- + |Z_2|e^k + b, \\ Z_1 = L_1(x^+), \\ Z_2 = L_2(x^-). \end{cases} \quad (14)$$

Next, by representing $Z_i$ for $i = 1, 2$ in the form $Z_i = Z_i^+ - Z_i^-$ and $|Z_i|$ in the form $|Z_i| = Z_i^+ + Z_i^-$, we transform (14) to the following form

$$\begin{cases} A_1 x^+ + Z_1^+ e^k - Z_2^+ e^k = A_2 x^- - Z_1^- e^k + Z_2^- e^k + b, \\ -L_1(x^+) + Z_1^+ = Z_1^-, \\ Z_2^+ = L_2(x^-) + Z_2^-. \end{cases} \quad (15)$$

Let $\mathcal{E}^k : \mathbf{R}^{s \times k} \to \mathbf{R}^s$ be a linear operator from $\mathbf{R}^{s \times k}$ in $\mathbf{R}^s$, which is defined by multiplication by $e^k$, i.e. if $Z \in \mathbf{R}^{s \times k}$ then $\mathcal{E}^k(Z) = Ze^k$. Then (15) may be represented in the following matrix form

$$
\begin{pmatrix} A_1 & \mathcal{E}^k & -\mathcal{E}^k \\ -L_1 & \mathcal{E} & 0 \\ 0 & 0 & \mathcal{E} \end{pmatrix} \begin{pmatrix} x \\ Z_1 \\ Z_2 \end{pmatrix}^{+} = \begin{pmatrix} A_2 & -\mathcal{E}^k & \mathcal{E}^k \\ 0 & \mathcal{E} & 0 \\ L_2 & 0 & \mathcal{E} \end{pmatrix} \begin{pmatrix} x \\ Z_1 \\ Z_2 \end{pmatrix}^{-} + \begin{pmatrix} b \\ 0 \\ 0 \end{pmatrix}, \quad (16)
$$

where $\mathcal{E}$ is an identical operator, and $0$ is a zero operator.

The system (16) (which is equivalent to eq. (6)) already has the form (7). Hence, according to Theorem 1, it necessitates that the corresponding system of the form (8) has only a zero solution. This system reads

$$
\begin{cases} \begin{pmatrix} A_1 & \mathcal{E}^k & -\mathcal{E}^k \\ -L_1 & \mathcal{E} & 0 \\ 0 & 0 & \mathcal{E} \end{pmatrix} \left( \begin{pmatrix} x \\ V_1 \\ V_2 \end{pmatrix} + \begin{pmatrix} y \\ U_1 \\ U_2 \end{pmatrix} \right) = \begin{pmatrix} A_2 & -\mathcal{E}^k & \mathcal{E}^k \\ 0 & \mathcal{E} & 0 \\ L_2 & 0 & \mathcal{E} \end{pmatrix} \left( \begin{pmatrix} x \\ V_1 \\ V_2 \end{pmatrix} - \begin{pmatrix} y \\ U_1 \\ U_2 \end{pmatrix} \right), \\ \left| \begin{pmatrix} x \\ V_1 \\ V_2 \end{pmatrix} \right| \leq \left| \begin{pmatrix} y \\ U_1 \\ U_2 \end{pmatrix} \right|. \end{cases}
$$

$$(17)$$

When rewriting (17) coordinatewise, after simple transformations we have the following system (18)

$$
\begin{cases} (A_1 + A_2)y + (A_1 - A_2)x = -2V_1 e^k + 2V_2 e^k, \\ U_1 = 1/2 L_1(x+y), \\ U_2 = 1/2 L_2(x-y), \\ |x| \leq |y|, \ |V_1| \leq |U_1|, \ |V_2| \leq |U_2|. \end{cases} \quad (18)
$$

So, we have that if the system (18) has only a zero solution then the system (14) (and hence eq. (6)) has a solution for any $b \in \mathbf{R}^s$.

Let $x, y \in \mathbf{R}^r$, $V_1, V_2, U_1, U_2 \in \mathbf{R}^{s \times k}$ be a solution of (18). Then
$$|(A_1 + A_2)y + (A_1 - A_2)x| = |-2V_1 e^k + 2V_2 e^k| \leq 2|V_1|e^k + 2|V_2|e^k \leq$$
$$\leq 2|U_1|e^k + 2|U_2|e^k = |L_1(x+y)|e^k + |L_2(x-y)|e^k$$
and since $|x| \leq |y|$ the pair $(x,y)$ satisfies (12). But then in accordance with the condition we have $x = y = 0$, and hence $U_1 = 1/2 L_1(x+y) = 0$, $U_2 = 1/2 L_2(x-y) = 0$ and $|V_1| \leq |U_1| = 0$, $|V_2| \leq |U_2| = 0$. Consequently, (18) really has only zero as solution. ∎

**Comment 3** When performing transformations similar to those used in proving the second part of Theorem 2, it is possible to show that the condition of existence of only the zero solution for the system (12) is equivalent to the fact that the following equation

$$A_1 x^+ + |L_1(x^+) + B_1|e^k = A_2 x^- + |L_2(x^-) + B_2|e^k + b, \qquad (19)$$

has not more than one solution for any $b \in \mathbf{R}^s$, $B_1, B_2 \in \mathbf{R}^{s \times k}$ (and in the case when $s = r$ - has a unique solution). Eq. (6) is obtained from (19) in the case when $B_1 = B_2 = 0$. So, the conditions of Theorem 2 are most likely unnecessary in the general case.

Using Theorem 3, it is easy to write down the sufficient conditions of uniqueness (and in the case when $m = n$ also existence) for eqs. (4), (5), and hence for the initial eq. (1). In the general case, these conditions appear to be rather bulky and cannot be practically simplified. Nevertheless, in the given particular case, i.e. when eq. (5) has the form (7), these conditions can be written rather compactly.

Introduce the following definition.

*Definition 1.* The matrix $\mathbf{A} = ([\underline{a}_{ij}, \overline{a}_{ij}]) \in \mathbf{IR}^{m \times n}$, is called a matrix of constant signs if $\underline{a}_{ij} \overline{a}_{ij} \geq 0$ for all $i = \overline{1, m}$, $j = \overline{1, n}$.

Note, in this case, in the decomposition $\mathbf{A} = \mathbf{A}_0 + \mathbf{A}_1 + \mathbf{A}_2$ we have $\mathbf{A} = \mathbf{A}_0$, $\mathbf{A}_1 = \mathbf{A}_2 = 0$. Consequently, eq. (5) (but not eq. (4)) has the form (7), and we can use Theorem 2. After the corresponding transformation of the system (8) we obtain the following corollary from Theorem 2.

For the matrix $\mathbf{A} = [\underline{A}, \overline{A}] \in \mathbf{IR}^{m \times n}$ introduce the denotation $A_c = 1/2(\overline{A} + \underline{A})$, $\Delta = 1/2(\overline{A} - \underline{A})$, $\Delta_0 = 1/2(|\overline{A}| - |\underline{A}|)$.

COROLLARY 1 *Let the matrix* $\mathbf{A} \in \mathbf{IR}^{m \times n}$ *be a matrix of constant signs. The following two conditions are equivalent:*

*1) eq. (1) has not more than one solution for any* $\mathbf{b} \in \mathbf{IR}^m$;
*2) the system of equations and inequalities*

$$\begin{cases} |A_c|y_1 = \Delta x_1, \\ A_c y_2 = \Delta_0 x_2, \\ |x_1 + x_2| \leq |y_1 + y_2|, \\ |x_1 - x_2| \leq |y_1 - y_2|, \end{cases} \qquad (20)$$

*has the unique solution* $x_1 = x_2 = y_1 = y_2 = 0$.

*If, furthermore, $m = n$, these conditions are also equivalent to the following one:*
*3) eq. (1) has a unique solution for any* $\mathbf{b} \in \mathbf{IR}^m$;

**Proof:** It is necessary to show only the fact that the system (8) can be transformed to the form (20). Since the equalities $\mathbf{A} = \mathbf{A}_0$, $\mathbf{A}_1 = \mathbf{A}_2 = 0$ hold for the interval matrix of constant signs, the matrices $A_1$ and $A_2$ in the corresponding equation of the form (7) take the following form

$$A_1 = 2\begin{pmatrix} \overline{A}^+ & \underline{A}^- \\ \underline{A}^- & \overline{A}^+ \end{pmatrix}, \quad A_2 = 2\begin{pmatrix} \underline{A}^+ & \overline{A}^- \\ \overline{A}^- & \underline{A}^+ \end{pmatrix}.$$

Then for $y = \begin{pmatrix} y_1 \\ y_2 \end{pmatrix}$, $x = \begin{pmatrix} x_1 \\ x_2 \end{pmatrix}$, $\in \mathbf{R}^{2n}$ the system (8) writes in the form

$$\begin{cases} \begin{pmatrix} \overline{A}^+ + \underline{A}^+ & \underline{A}^- + \overline{A}^- \\ \underline{A}^- + \overline{A}^- & \overline{A}^+ + \underline{A}^+ \end{pmatrix}\begin{pmatrix} y_1 \\ y_2 \end{pmatrix} + \begin{pmatrix} \overline{A}^+ - \underline{A}^+ & \underline{A}^- - \overline{A}^- \\ \underline{A}^- - \overline{A}^- & \overline{A}^+ - \underline{A}^+ \end{pmatrix}\begin{pmatrix} x_1 \\ x_2 \end{pmatrix} = \begin{pmatrix} 0 \\ 0 \end{pmatrix}, \\ |y_1| \ge |x_1|, \; |y_2| \ge |x_2|. \end{cases}$$

Having written this system out coordinatewise, we obtain

$$\begin{cases} (\overline{A}^+ + \underline{A}^+)y_1 + (\underline{A}^- + \overline{A}^-)y_2 + (\overline{A}^+ - \underline{A}^+)x_1 + (\underline{A}^- - \overline{A}^-)x_2 = 0 \\ (\underline{A}^- + \overline{A}^-)y_1 + (\overline{A}^+ + \underline{A}^+)y_2 + (\underline{A}^- - \overline{A}^-)x_1 + (\overline{A}^+ - \underline{A}^+)x_2 = 0 \\ |y_1| \ge |x_1|, \; |y_2| \ge |x_2|. \end{cases}$$

If we add the first two equations and then subtract the second from the first one, we find out that this system is equivalent to the following one

$$\begin{cases} (\overline{A}^+ + \overline{A}^- + \underline{A}^+ + \underline{A}^-)(y_1 + y_2) + (\overline{A}^+ - \overline{A}^- - \underline{A}^+ + \underline{A}^-)(x_1 + x_2) = 0 \\ (\overline{A}^+ - \overline{A}^- + \underline{A}^+ - \underline{A}^-)(y_1 - y_2) + (\overline{A}^+ + \overline{A}^- - \underline{A}^+ - \underline{A}^-)(x_1 - x_2) = 0 \\ |y_1| \ge |x_1|, \; |y_2| \ge |x_2|, \end{cases}$$

or if we take account of the equalities $\overline{A} = \overline{A}_+ - \overline{A}_-, \underline{A} = \underline{A}^+ - \underline{A}_-, |\overline{A}| = \overline{A}_+ + \overline{A}_-,$ $|\underline{A}| = \underline{A}^+ + \underline{A}_-$, we reveal that the latter system takes the form

$$\begin{cases} (|\overline{A}| + |\underline{A}|)(y_1 + y_2) + (\overline{A} - \underline{A})(x_1 + x_2) = 0, \\ (\overline{A} + \underline{A})(y_1 - y_2) + (|\overline{A}| - |\underline{A}|)(x_1 - x_2) = 0, \\ |y_1| \ge |x_1|, \; |y_2| \ge |x_2|, \end{cases}$$

Next, after the replacement of the variables $u_1 = y_1 + y_2, u_2 = y_1 - y_2, v_1 = -x_1 - x_2, v_2 = -x_1 + x_2$, i.e. $y_1 = \frac{1}{2}(u_1 + u_2), y_2 = \frac{1}{2}(u_1 - u_2), x_1 = -\frac{1}{2}(v_1 + v_2), x_2 = \frac{1}{2}(-v_1 + v_2)$ we obtain

$$\begin{cases} (|\overline{A}| + |\underline{A}|)u_1 = (\overline{A} - \underline{A})v_1, \\ (\overline{A} + \underline{A})u_2 = (|\overline{A}| - |\underline{A}|)v_2, \\ \frac{1}{2}|(u_1 + u_2)| \ge \frac{1}{2}|(v_1 + v_2)|, \\ \frac{1}{2}|(u_1 - u_2)| \ge \frac{1}{2}|(v_1 - v_2)|. \end{cases}$$

Having multiplied the first two equations by $\frac{1}{2}$ and the inequalities by 2, and taking account of above notations, as well as the fact that for the interval matrix of constant signs the equality $|A_c| = \frac{1}{2}(|\overline{A}+\underline{A}|) = \frac{1}{2}(|\overline{A}| + |\underline{A}|)$ holds, we find out that the system takes the following form

$$
\begin{cases}
|A_c|u_1 = \Delta v_1, \\
A_c u_2 = \Delta_0 v_2, \\
|v_1 + v_2| \le |u_1 + u_2|, \\
|v_1 - v_2| \le |u_1 - u_2|,
\end{cases}
$$

i.e. with the precision up to denotations of variables it has the form (20).            ∎

## Acknowledgements

The author thanks the anonymous referees for valuable comments and helpful suggestions on the original version of the paper. This research is conducted within the research grants No 98-01-01137 from RFBR.

## References

1. E. Kaucher, Interval analysis in the extended interval spase **IR**, Computing, Suppl.2 (1980), pp.33-49.
2. G. Alefeld and J. Herzberger, Introduction to Interval Computations, Academic Press, New York, 1983.
3. A. Neumaier, Interval Methods for Systems of Equations, Cambridge University Press, Cambridge, 1990.
4. L.V. Kupriyanova, Inner estimation of the united solution set of interval linear algebraic system, Reliable Computing, 1(1), (1995), pp.15–31.
5. S.P. Shary, Algebraic approach to the interval linear static identification, tolerance and control problems, or One more application of Kaucher arithmetic, Reliable Computing, 2 (1996), pp. 3–33.
6. J. Rohn, Systems of linear interval equations, Lin. Alg. Appl., 126, (1989), pp. 39–78.
7. E. Gardenes and A. Trepat, Fundamentals of SIGLA, an interval computing system over the completed set of intervals, Computing, 24 (1980), pp. 161-179.
8. A.V. Lakeyev, Linear algebraic equations in Kaucher arithmetic, Reliable Computing, 1995, Supplement (Extended Abstracts of APIC'95: International Workshop on Applications of Interval Computations, El Paso, TX, Febr. 23–25, 1995), pp. 130–133.

*T. Csendes (ed.), Developments in Reliable Computing* 67–75.
© 1999 *Kluwer Academic Publishers.*

# A Comparison of Subdivision Strategies for Verified Multi-Dimensional Gaussian Quadrature

BRUNO LANG                                                    lang@rz.rwth-aachen.de
*Aachen University of Technology, Computing Center, Seffenter Weg 23, D-52074 Aachen*

**Abstract.** This paper compares several strategies for subdividing the domain in verified multi-dimensional Gaussian quadrature. Subdivision may be used in two places in the quadrature algorithm. First, subdividing the box can reduce the over-estimation of the partial derivatives' ranges, which are needed to bound the approximation error. Second, if the required error bound cannot be met with the whole domain then the box is split into subboxes, and the quadrature algorithm is recursively applied to these. Both variants of subdivision are considered in this paper.

**Keywords:** Verified Gaussian quadrature, subdivision strategies

## 1. Introduction

In one dimension, the $n$-point Gauss-Legendre quadrature formula is given by

$$I := \int_{-1}^{1} f(x)\, dx \approx \sum_{i=1}^{n} \omega_i^{(n)} \cdot f(x_i^{(n)}) =: A^{(n)}, \tag{1}$$

where the nodes $x_i^{(n)} \in (-1, 1)$ and the weights $\omega_i^{(n)} > 0$ can be determined by solving an appropriate $n \times n$ symmetric tridiagonal eigenproblem [5]. The remainder term,

$$R^{(n)} := I - A^{(n)},$$

is given by

$$R^{(n)} = e^{(n)} \cdot \frac{1}{(2n)!} f^{(2n)}(\xi) \quad \text{for some } \xi \in (-1, 1) \tag{2}$$

with

$$e^{(n)} = \frac{2^{2n+1}}{2n+1} \cdot \binom{2n}{n}^{-2}.$$

(The remainder term has several representations based on the Peano kernel, see [4].)

From (1) and (2) one easily arrives at an enclosure for the integral value,

$$I = \int_{-1}^{1} f(x) \, dx \in A^{(n)} + R^{(n)} ,$$

where the enclosures for the approximation,

$$A^{(n)} = \sum_{i=1}^{n} \omega_i^{(n)} \cdot f(x_i^{(n)}) ,$$

and for the remainder term,

$$R^{(n)} = e^{(n)} \cdot \frac{1}{(2n)!} f^{(2n)}([-1,1]) ,$$

are obtained by substituting enclosing intervals for the nodes and weights (for computing these intervals see [13]) and by enclosing the range of the Taylor coefficients $f^{(2n)}(x)/(2n)!$ over the domain $[-1,1]$ (e.g., using the mitaylor Taylor arithmetic by U. Storck or the techniques described in [3]).

Here and in the remainder of the paper, intervals are denoted by boldface letters, and $f(x)$ is some suitable inclusion of the (real-valued) function $f$'s range over the interval $x$, whereas the actual range is denoted by $f^u(x)$. (This notation is adopted from Kearfott's book [9].)

The idea of using automatic differentiation and interval arithmetic to obtain enclosures for the integral goes back to [6]; Corliss and Rall [2] give a nice overview of techniques for self-validating quadrature in the one-dimensional case.

For $d$-dimensional quadrature, the analogous modifications of a standard ($n_1 \times \ldots \times n_d$)-point product rule [14] yield

$$I = \int_{-1}^{1} \cdots \int_{-1}^{1} f(x_1, \ldots, x_d) \, dx_d \cdots dx_1 \in A^{(n_1 \times \ldots \times n_d)} + R^{(n_1 \times \ldots \times n_d)} ,$$

where

$$A^{(n_1 \times \ldots \times n_d)} = \sum_{i_1=1}^{n_1} \cdots \sum_{i_d=1}^{n_d} \omega_{i_1}^{(n_1)} \cdot \ldots \cdot \omega_{i_d}^{(n_d)} \cdot f(x_{i_1}^{(n_1)}, \ldots, x_{i_d}^{(n_d)}) \qquad (3)$$

and

$$R^{(n_1 \times \ldots \times n_d)} = 2^{d-1} \cdot [-1,1] \cdot \sum_{j=1}^{d} e^{(n_j)} \cdot \max \left| T_{n_j, j}([-1,1]^d) \right| \qquad (4)$$

enclose the approximation and the remainder term, respectively. Here, the $x_{ij}^{(n_j)}$ and $\omega_{ij}^{(n_j)}$, $1 \leq j \leq n_j$, and the constant $e^{(n_j)}$ are just the nodes, weights, and constant of the one-dimensional $n_j$-point formula in direction $x_j$, and

$$T_{n_j,j}([-1,1]^d) = \frac{1}{(2n_j)!} \frac{\partial^{2n_j} f}{\partial x_j^{2n_j}}([-1,1]^d)$$

denotes an enclosure for the range of the $2n_j$-th Taylor coefficient in direction $x_j$ over the domain $[-1,1]^d$.

The above rule applies only to the domain $[-1,1]^d$, but it is easily adapted to more general domains (e.g., of the form $x_1 \in x_1$, $g_1(x_1) \leq x_2 \leq h_1(x_1)$, ..., $g_{d-1}(x_1,\ldots,x_{d-1}) \leq x_d \leq h_{d-1}(x_1,\ldots,x_{d-1})$, where the bounds for $x_i$ depend on the "previous" variables) by a suitable transformation of the variables. As the number of function evaluations in (3) grows heavily with $d$, product rules should be used only for moderate dimensions (up to 12, say).

Typically, the width of $R^{(n_1 \times \ldots \times n_d)}$ by far dominates the width of $A^{(n_1 \times \ldots \times n_d)}$. Therefore the following (recursive) algorithm will in general (see below) determine an enclosure of width $\leq \epsilon_0$ for the integral $I$:

1. Start with the box $x := [-1,1]^d$ (i.e., the whole domain) and $\epsilon := \epsilon_0$.

2. For $1 \leq j \leq d$ and $1 \leq n \leq n_{\max}$ (some maximum order) compute enclosures $T_{n,j}(x)$ of the Taylor coefficients over the current box $x$.

3. If there are combinations $n_1 \times \ldots \times n_d$ of orders $n_j \leq n_{\max}$ such that

$$\text{width}(R^{(n_1 \times \ldots \times n_d)}) \leq \epsilon, \tag{5}$$

where $R^{(n_1 \times \ldots \times n_d)}$ is computed according to (4), and $\epsilon$ denotes the currently required precision,

then determine one of these combinations that minimizes the number $n_1 \times \ldots \times n_d$ of function evaluations in the approximation (a *dynamic programming* problem), and for this combination compute $A^{(n_1 \times \ldots \times n_d)}$ via (3); return $I := A^{(n_1 \times \ldots \times n_d)} + R^{(n_1 \times \ldots \times n_d)}$.

4. **Otherwise** split the current box into $m \geq 2$ subboxes $x^{(1)}, \ldots, x^{(m)}$, apply steps 2. through 4. recursively to these subboxes $x^{(\ell)}$ to obtain enclosures $I^{(\ell)}$ of width $\leq \epsilon^{(\ell)} := \epsilon/m$ for the integrals over each subbox, and return $I := I^{(1)} + \ldots + I^{(m)}$.

If $\text{width}(R^{(n_1 \times \ldots \times n_d)}) \approx \epsilon$ and $\text{width}(A^{(n_1 \times \ldots \times n_d)})$ is not negligible then the enclosure $I$ returned in step 3. may have a width slightly exceeding the prescribed error bound $\epsilon$. If this happens and is not compensated by some enclosures $I$ over other subdomains being narrower than their allowed width $\epsilon$, then the width of the overall enclosure $I$ may exceed $\epsilon_0$. This may be prevented by monitoring the widths of the enclosures $I$ and by recomputing $I$ for a different order $n_1 \times \ldots \times n_d$, if necessary. Additional precaution must be taken to prevent infinite recursion (cf. [2] for one-dimensional quadrature). For the sake of brevity these cases were excluded from the above algorithm.

In the following section we describe several different strategies for effecting the subdivision in step 4. of the above algorithm. The enclosures computed in step 2. (e.g., by standard interval evaluation) tend to severely over-estimate the actual range of the Taylor coefficients over the box $x$. One way to reduce this over-estimation consists in subdividing the box, as discussed in Section 3.

The overall efficiency of the quadrature algorithm depends on the total number of subboxes to be considered. Using suitable subdivision strategies in steps 2. and 4. can help to reduce this number, as shown by our numerical experiments (Section 4).

## 2. Subdivision for the recursive calls

In this section we describe some strategies for generating subboxes (step 4. of the algorithm) when the precision requirement (5) cannot be met with the current box. Subdivision strategies also play an important role in other recursive algorithms, e.g., in branch-and-bound global optimization methods (for investigations in this context see [1, 12, 11]). In our quadrature algorithm we tried the following schemes:

**Cyclic:** Cut the box along *one* direction into $\mu \geq 2$ pieces (resulting in $m = \mu$ subboxes). The direction varies cyclically with the recursion depth.

This strategy is very simple to implement. It produces "almost square" boxes (the ratio between the longest and the shortest edge of a box never exceeds $\mu : 1$).

**Largest:** First determine, for each direction $1 \leq j \leq d$, the "best" Taylor coefficient in this direction, that is,

$$\tau_j := \min_{n=1}^{n_{\max}} \left\{ e^{(n)} \cdot |T_{n,j}(x)| \right\} .$$

Then determine the direction with the largest $\tau_j$ and cut the box along this direction into $\mu$ pieces.

**Full:** Cut the box along *each* direction into $\mu$ pieces of equal size, resulting in $m = \mu^d$ subboxes.

This strategy is motivated by the fact that the precision criterion (5) will be met only if the current box $x$ is "small enough". By generating a larger number of subboxes in each step we will need fewer subdivision steps, that is, fewer intermediate boxes that must be considered. On the other hand, the final subdivision (which accounts for a major portion of the subboxes) may be too fine.

**Largest two:** This may be seen as a compromise between the **Largest** and **Full** schemes:

First determine the $\tau_j$ and then subdivide along the directions $j$ and $j'$ with the largest and second largest $\tau_j$ values. Subdivision in the second direction is done only if we expect that the **Largest** scheme alone would not succeed in sufficiently reducing the width of the remainder terms for the subboxes. Our (purely heuristic) criterion for this decision was

$$\tau_j > \mu \cdot \epsilon / d \,.$$

With the **Largest two** strategy, either $\mu$ or $\mu^2$ subboxes are generated.

In addition to the subdivision strategy, we may choose the subdivision factor $\mu \geq 2$. Similarly to the arguments given with the **Full** strategy, larger values of $\mu$ will help skipping intermediate boxes, at the risk of producing too many (too small) boxes in the last subdivision step.

## 3. Subdivision for enclosing the Taylor coefficients

We investigated two strategies for enclosing the ranges of the Taylor coefficients over the current box:

**Plain:** Use Taylor arithmetic to evaluate the Taylor coefficients on $x$.

**Union:** For larger boxes $x$ and complicated functions with the variables occurring several times, the **Plain** approach will severely over-estimate the actual *range* $T^u_{n,j}(x)$ of the Taylor coefficients. This over-estimation can be reduced by subdividing the box $x$ into subboxes $x^{(\ell)}$ and using that

$$T^u_{n,j}(x) \subseteq \bigcup_\ell T_{n,j}(x^{(\ell)}) \,,$$

where $T_{n,j}(x^{(\ell)})$ denotes the plain interval evaluation of $T_{n,j}$ on the subbox $x^{(\ell)}$.

For the **Union** scheme, we can again choose the subdivision direction(s) and the subdivision factor. In our implementation we tested the **Full** subdivision (along all directions) and the subdivision along the direction(s) that gave the **Largest** and **Largest two**, resp., $\tau_j$ bound(s) in the *father* of the current box.

Note that our subdivision strategies are based on purely uni-directional higher derivatives. For the future we plan to also incorporate mixed partial derivatives in the computations; this will allow us to use centered forms to enclose the Taylor coefficients:

$$T_{n,j}^u(x) \subseteq T_{n,j}(\check{x}) + \sum_{k=1}^d T_{n,j,k}(x) \cdot (x - \check{x}) \qquad \text{for any } \check{x} \in x,$$

where $T_{n,j,k}(x)$ denotes an enclosure of the partial derivative $\partial^{n+1} f(x)/\partial x_j^n \partial x_k$ over the box $x$. The computation of these mixed partial derivatives is more involved and more expensive than the uni-directional derivatives used now (a sophisticated method for determining mixed partials from univariate Taylor series is described in [7]), but they may provide significantly sharper enclosures for the Taylor coefficients.

## 4. Numerical experiments

The quadrature algorithm has been implemented in Pascal-XSC [10] and tested with the following functions (see Figure 1):

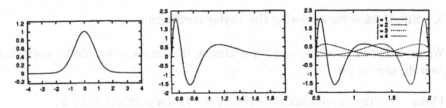

*Figure 1.* Test functions $f_1$, $f_2$, and $f_3$ for $d = 1$ (from left to right).

- $f_1(x) = e^{-\|x\|_2^2}$, $x \in [-4, 4]^d$ for $1 \le d \le 4$.

- $f_2(x) = N(x) \cdot \sin(e^{N(x)})$, $x \in [\alpha, \beta]^d$ for $1 \le d \le 4$, where $\alpha = 0.5$, $\beta = 2$, and

$$N(x) = \frac{1}{d} \cdot \sum_{j=1}^d (x_j - \beta)^2 .$$

This function varies much in the vicinity of $\alpha^d$ and is rather flat elsewhere.

- $f_3(x) = N(x) \cdot \sin(e^{N(x)})$, $x \in [\alpha, \beta]^d$ for $d = 2, 4$, where $\alpha = 0.5$, $\beta = 2$, and

$$N(x) = \frac{1}{d} \cdot \sum_{j=1}^{d} (x_j - \xi_j)^2 \quad \text{with} \quad \xi_j = \alpha + \frac{d-j}{d-1} \cdot (\beta - \alpha).$$

(The partial derivatives in the vicinity of $\alpha^d$ depend heavily on the direction.)

- $f_4$ is defined identically to $f_3$, but before the function is evaluated, the argument is twice rotated to and fro by an angle of $\pi/6$ in the $(x_1, x_2)$ plane (and, for $d = 4$, the same is done in the $(x_3, x_4)$ plane).

  Therefore $f_4$ agrees with $f_3$ for point arguments, whereas non-degenerate interval arguments are blown up, thereby significantly aggravating the over-estimation of the Taylor coefficients' ranges.

- $f_5(r_1, \varphi_1, h_1, r_2, \varphi_2, h_2) = r_1 r_2 \Delta x_3 \|\Delta x\|_2^3$,
  $r_1 \in [\underline{r}_1, \overline{r}_1]$, $\varphi_1 \in [0, 2\pi]$, $h_1 \in [0, \overline{h}_1]$, $r_2 \in [\underline{r}_2, \overline{r}_2]$, $\varphi_2 \in [0, 2\pi]$, $h_2 \in [0, \overline{h}_2]$. Such sextuple integrals arise in determining the gravitational forces between two cylindrical rings with inner (outer) radii $\underline{r}_1$ and $\underline{r}_2$ ($\overline{r}_1$ and $\overline{r}_2$) and heights $\overline{h}_1$ and $\overline{h}_2$, cf. [8]. Here

$$\Delta x = \begin{pmatrix} \Delta x_1 \\ \Delta x_2 \\ \Delta x_3 \end{pmatrix} = G_1 \cdot \begin{pmatrix} r_1 \cos \varphi_1 \\ r_1 \sin \varphi_1 \\ h_1 \end{pmatrix} - G_2 \cdot \begin{pmatrix} r_2 \cos \varphi_2 \\ r_2 \sin \varphi_2 \\ h_2 \end{pmatrix} + \begin{pmatrix} 0 \\ 0 \\ \delta \end{pmatrix}$$

denotes the (directed) distance between two points in the cylindrical rings, $G_1$ and $G_2$ being geometric transformations to account for rotations or displacements of the rings from a common $(x_3\text{-})$axis, and $\delta$ being the distance of the rings in $x_3$ direction.

Due to its higher dimension and to severe over-estimation, this is by far the most demanding of the test functions.

For these functions and various different combinations of subdivision strategies and subdivision factors, the following data were sampled:

- $T$, the total number of Taylor evaluations for the remainder terms (this is also the total number of boxes considered in the whole algorithm)

- $F$, the total number of function evaluations for the approximations, and

- the total time.

The results from over 1000 runs may be summarized as follows.

1. *Subdivision for evaluating the Taylor coefficients did not pay unless the problem was hard.* For the simpler problems $f_1$, $f_2$, $f_3$, and $f_4$ (with $d = 2$), cutting the box into $\tilde{m}$ pieces lead to an almost $\tilde{m}$-fold increase in $T$ without significantly reducing $F$, thus increasing the total time. For $f_4$ and $d = 4$, a subdivision into two pieces along the **Largest** direction reduced $F$ by about 10% while increasing $T$ by a factor of almost 2, resulting in some 5% savings in time. (The reduction of $F$ comes from criterion (5) being achieved with lower-order formulae, due to the tighter enclosures for the $T^u_{n,j}(x)$.) For the hardest problem $f_5$, cutting the box into 16 pieces (4 times along the **Largest two** directions) even halved the total time.

2. **Full** *subdivision never paid.* This is due to the fact that in general the last subdivision steps result in too many (too small) boxes.

3. *The heuristic schemes* **Cyclic**, **Largest**, *and* **Largest two** *performed comparably, with a slight superiority of* **Largest two**. Typically, **Largest** could reduce $T$ by some 10–20%, as compared to **Cyclic**, and **Largest two** gave an additional improvement of roughly the same magnitude. The reductions of $F$ and of the time were much smaller; here the **Largest** and **Largest two** schemes typically gave an improvement of 5–10%. (In some cases, e.g., $f_1$ with $d = 4$, the speedup exceeded 30%.)

4. For almost all runs subdivision factors $\mu = 2$ or $\mu = 3$ were optimal. The times and $T$ and $F$ counts obtained with these two factors differed by up to 40%, but it was impossible to predict which one would perform better for a given problem.

## 5. Conclusions

Given the results from the experiments, the following strategy seems reasonable for a "black-box" integration routine based on Gauss-Legendre quadrature and on purely uni-directional partial derivatives.

- Use the **Largest two** scheme with $\mu = 2$ or $\mu = 3$ for the recursive subdivision of the box.

- Until a certain recursion depth is reached, use **Plain** interval evaluation of Taylor arithmetic to compute the enclosures $T_{n,j}(x)$ of the Taylor coefficients, as this techniques performs best for not-too-difficult problems. Below a certain recursion depth $d_0$ (which grows with the problem dimension $d$) we *assume* that the problem is difficult and switch to the **Largest** or **Largest two** subdivision strategy for evaluating the Taylor coefficients.

As already mentioned in Section 3, using mixed partial derivatives may provide an alternative way to obtain sharper enclosures below some recursion level $d_0$. The feasibility and benefits of this approach will be investigated in the future.

## References

1. S. Berner. *Ein paralleles Verfahren zur verifizierten globalen Optimierung.* PhD thesis, Bergische Universität GH Wuppertal, 1995.

2. G. F. Corliss and L. B. Rall. Adaptive, self-validating quadrature. *SIAM J. Sci. Stat. Comput.,* 8(5):831–847, 1987.

3. G. F. Corliss and L. B. Rall. Computing the range of derivatives. In Edgar Kaucher, S. M. Markov, and Günter Mayer, editors, *Computer Arithmetic, Scientific Computation, and Mathematical Modelling,* volume 12 of *IMACS Annals on Computing and Applied Mathematics,* pages 195–212. Baltzer, Basel, 1991.

4. P. J. Davis and P. Rabinowitz. *Methods of Numerical Integration.* Academic Press, New York, 2nd edition, 1984.

5. G. H. Golub and J. H. Welsch. Calculation of Gauss quadrature rules. *Math. Comp.,* 23:221–230, 1969.

6. J. H. Gray and L. B. Rall. INTE: A UNIVAC 1108/1110 program for numerical integration with rigorous error estimation. MRC Technical Summary Report No. 1428, Mathematics Research Center, University of Wisconsin - Madison, 1975.

7. A. Griwank, J. Utke, and A. Walther. Evaluating higher derivative tensors by forward propagation of univariate Taylor series. Preprint IOKOMO-09-97t, TU Dresden, Inst. of Scientific Computing, 1997. Revised June 1998.

8. O. Holzmann, B. Lang, and H. Schütt. Newton's constant of gravitation and verified numerical quadrature. *Reliable Computing,* 2(3):229–239, 1996.

9. R. Baker Kearfott. *Rigorous Global Search: Continuous Problems.* Kluwer, Dordrecht, 1996.

10. R. Klatte, U. Kulisch, M. Neaga, D. Ratz, and Ch. Ullrich. *Pascal-XSC Language Reference with Examples.* Springer, Berlin, 1992.

11. D. Ratz. Box-splitting strategies for the interval Gauss-Seidel step in a global optimization method. *Computing,* 53:337–353, 1994.

12. D. Ratz and T. Csendes. On the selection of subdivision directions in interval branch-and-bound methods for global optimization. *Journal of Global Optimization,* 7:183–207, 1995.

13. U. Storck. Verified calculation of the nodes and weights for Gaussian quadrature formulas. *Interval Computations,* 4:114–124, 1993.

14. A. H. Stroud. *Approximative Calculation of Multiple Integrals.* Prentice-Hall, New York, 1971.

*T. Csendes (ed.), Developments in Reliable Computing* 77–104.
© 1999 *Kluwer Academic Publishers.*

# INTLAB — INTerval LABoratory

SIEGFRIED M. RUMP                                                      rump@tu-harburg.de
*Inst. f. Informatik III, Technical University Hamburg-Harburg, Eißendorfer Str. 38, 21071 Hamburg,*
*Germany*

**Abstract.**   INTLAB is a toolbox for Matlab supporting real and complex intervals, and vectors, full matrices and sparse matrices over those. It is designed to be very fast. In fact, it is not much slower than the fastest pure floating point algorithms using the fastest compilers available (the latter, of course, without verification of the result). Beside the basic arithmetical operations, rigorous input and output, rigorous standard functions, gradients, slopes and multiple precision arithmetic is included in INTLAB. Portability is assured by implementing *all* algorithms in Matlab itself with exception of exactly one routine for switching the rounding downwards, upwards and to nearest. Timing comparisons show that the used concept achieves the anticipated speed with identical code on a variety of computers, ranging from PC's to parallel computers. INTLAB is freeware and may be copied from our home page.

## 1.  Introduction

The INTLAB concept splits into two parts. First, a new concept of a fast interval library is introduced. The main advantage (and difference to existing interval libraries) is that identical code can be used on a variety of computer architectures. Nevertheless, high speed is achieved for interval calculations. The key is extensive (and exclusive) use of BLAS routines [19], [7], [6].

Second, we aim at an interactive programming environment for easy use of interval operations. Our choice is to use Matlab. It allows to write verification algorithms in a way which is very near to pseudo-code used in scientific publications. The code is an *executable specification*. Here, the major difficulty is to overcome slowness of interpretation. This problem is also solved by exclusive use of high-order matrix operations.

The first concept, an interval library based on BLAS, is the basis of the INTLAB approach. The interval library based on BLAS is very simple to implement in any programming language. It is a fast way to start verification on many computers in a two-fold way: fast to implement and fast to execute.

There are a number of public domain and commercial interval libraries, among them [2], [5], [8], [9], [11], [12], [13], [14], [15], [18] and [27]. To our knowledge, INTLAB is the first interval library building upon BLAS.

The basic assumption for INTLAB is that the computer arithmetic satisfies the IEEE 754 arithmetic standard [10] and, that a permanent switch of the rounding mode is possible. Switching of the rounding mode is performed by setround(i) with $i \in \{-1, 0, 1\}$ corresponding to rounding downwards, to nearest, and rounding upwards, respectively. We assume that, for example, after a call `setround(1)` all subsequent arithmetical operations are using rounding upwards, until the next call of `setround`. This is true for almost all PC's, workstations and many mainframes.

Our goal is to design algorithms with result verification the execution time of which is of the same order of magnitude as the fastest known floating point algorithms using the fastest compiler available. To achieve this goal, it is not only necessary to design an appropriate arithmetic, but also to use appropriate verification algorithms being able to utilize the speed of the arithmetic and the computer. We demonstrate by means of examples the well known fact that a bare count of operations is not necessarily proportional to the actual computing time.

Arithmetical operations in INTLAB are rigorously verified to be correct, including input and output and standard functions. By that it is possible to replace every operation of a standard numerical algorithm by the corresponding interval operations. However, this frequently leads to dramatic overestimation and is not recommended. By the principle of definition, interval operations deliver a superset of the result, which is usually a slight overestimation. It is the goal of validated numerics to design verification algorithms in order to diminish this overestimation, and even to (rigorously) estimate the amount of overestimation. Therefore, for verified solution of numerical problems one should only use specifically designed verification algorithms. Some examples are given in Section 5.

Many of the presented ideas are known in one way or the other. However, the combination produces a very fruitful synergism. Development of validation algorithms is as easy as it could be using INTLAB.

The paper addresses the above mentioned two concepts and is organized as follows. In Section 2 the new interval library approach based on BLAS is presented together with timings for an implementation in C on scalar and parallel computers. Corresponding timings for the INTLAB implementation in the presence of interpretation overhead is presented in Section 3. The latter timings are especially compared with pure floating point. In Section 4 some details of programming in INTLAB are given and the question of correctness of programming is addressed. In the following Section 5 we present some algorithms written in INTLAB together with timing comparisons between verification algorithms and the corresponding pure floating point (Matlab) implementation. We finish the paper with concluding remarks.

## 2.  An interval library based on BLAS

Starting in the late 70's an interface for Basic Linear Algebra Subroutines was defined in [19], [7], and [6]. It comprises linear algebra algorithms like scalar products, matrix-vector products, outer products, matrix multiplication and many more. The algorithms are collected in level 1, 2 and 3 BLAS.

The ingenuous idea was to specify the interface, pass it to the manufacturers and leave the implementation to them. Since BLAS became a major part of various benchmarks, it is the manufacturers own interest to provide very fast implementations of BLAS for their specific computer. This includes all kinds of optimization techniques to speed up code such as blocked code, optimal use of cache, taking advantage of the specific architecture and much more.

In that way BLAS serves a mutual interest of the manufacturer and the user. No wonder, modern software packages like Lapack [4] extensively use BLAS. Beside speed, the major advantage is the fact that a code using BLAS is fast on a variety of computers, without change of the code.

This is the motivation to try to design an interval library exclusively using BLAS routines. At first sight this seems impossible because all known implementations use various case distinctions in the inner loops. In the following we will show how to do it. We assume the reader is familiar with basic definitions and facts of interval arithmetic (cf. [3], [21]).

In the following we discuss implementations in a programming language like C or Fortran; interpretation overhead, like in Matlab, is addressed in Section 3.

We start with the problem of matrix multiplication. Having fast algorithms for the multiplication of two matrices at hand will speed up many verification algorithms. We have to distinguish three cases. We first concentrate on real data, where a real interval A is stored by infimum A.inf and supremum A.sup. If A is a vector or matrix, so is A.inf and A.sup.

## 2.1. Real Point matrix times real point matrix

First, consider the multiplication of two point matrices (infimum and supremum coincide) A and B, both exactly representable within floating point. There is a straightforward solution to this by the following algorithm.

```
setround(-1)
C.inf = A * B;
setround(1)
C.sup = A * B;
```

**Algorithm 1** *Point matrix times point matrix*

We use original INTLAB notation for two reasons. First, the sample code is *executable* code in INTLAB, and second it gives the reader a glimpse on readability of INTLAB notation (see the introduction for setround). Note that this is only *notation*, interpretation overhead is addressed in Section 3.

Note that Algorithm 1 assures that

$$\text{C.inf} \leq \text{A*B} \leq \text{C.sup} \tag{1}$$

including underflow and overflow (comparison of vectors and matrices is always to be understood entrywise); estimation (1) is valid under any circumstances. Moreover, an effective implementation of the floating point matrix multiplication A*B, like the one used in BLAS, will change the order of execution. This does not affect validity of (1).

## 2.2.  *Real point matrix times real interval matrix*

Next, consider multiplication of a real point matrix $A \in M_{m,n}(\mathbb{R})$ and a real interval matrix $\text{intB} \in \mathbb{IM}_{n,k}(\mathbb{R})$. The aim is to compute $C = A * \text{intB}$ with bounds. Consider the following three algorithms to solve the problem.

```
C.inf = zeros(size(A,1),size(intB,2));
C.sup = C.inf;
setround(-1)
for i=1:size(A,2)
 C.inf = C.inf + min(A(:,i)*intB.inf(i,:), A(:,i)*intB.sup(i,:) );
end
setround(1)
for i=1:size(A,2)
 C.sup = C.sup + max(A(:,i)*intB.inf(i,:), A(:,i)*intB.sup(i,:) );
end
```

**Algorithm 2**  *Point matrix times interval matrix, first variant*

By Matlab convention, the minimum and maximum in rows 5 and 9 are entrywise, respectively, the result is a matrix. The above algorithm implements an outer product matrix multiplication.

```
Aneg = min( A , 0 );
Apos = max( A , 0 );
setround(-1)
C.inf = Apos * intB.inf + Aneg * intB.sup;
setround(1)
C.sup = Aneg * intB.inf + Apos * intB.sup;
```

**Algorithm 3**  *Point matrix times interval matrix, second variant*

This algorithm was proposed by A. Neumaier at the SCAN 98 conference in Budapest. Again, the minimum and maximum in lines 1 and 2 are entrywise, respectively, such that for example Apos is the matrix A with negative entries replaced by zero. It is easy to see that both algorithms work correctly. The second algorithm is well suited for interpretation.

```
setround(1)
Bmid = intB.inf + 0.5*(intB.sup-intB.inf);
Brad = Bmid - intB.inf;
setround(-1)
C1 = A * Bmid;
setround(1)
C2 = A * Bmid;
Cmid = C1 + 0.5*(C2-C1);
Crad = ( Cmid - C1 ) + abs(A) * Brad;
setround(-1)
C.inf = Cmid - Crad;
setround(1)
C.sup = Cmid + Crad;
```

**Algorithm 4** *Point matrix times interval matrix, third variant*

This algorithm first converts the second factor into midpoint/radius representation and needs three point matrix multiplications. Recall that Algorithm 3 used four point matrix multiplications. This elegant way of converting infimum/supremum representation into midpoint/radius representation was proposed by S. Oishi during a visit of the author at Waseda university in fall 1998 [22].

### 2.3.   Comparison of results of Algorithms 3 and 4

Without rounding errors, the results of all algorithms are identical because over-estimation of midpoint/radius multiplication occurs only for two thick (non-point) intervals. Practically, i.e. in the presence of rounding errors, this remains true except for the interval factor being of very small diameter. Consider the following test. For A being

   I)   the $5 \times 5$ Hilbert matrix,
  II)   the $10 \times 10$ Hilbert matrix,
 III)   a $100 \times 100$ random matrix,
  IV)   a $100 \times 100$ ill-conditioned matrix (cond(A)=1e12)

   we calculated an approximate inverse R=inv(A) followed by a preconditioning R*midrad(A,e) for different values of e. Here, midrad(A,e) is an interval matrix of smallest width such that A.inf $\leq$ A−e $\leq$ A+e $\leq$ A.sup. Let C3 and C4 denote the result achieved by Algorithm 3 and 4, respectively. The following table displays the minimum, average and maximum of rad(C4)./rad(C3) for different values of the radius e.

|              | case I) | | | case II) | | |
| e            | 1e-16 | 1e-15 | 1e-14 | 1e-16 | 1e-15 | 1e-14 |
| ------------ | ----- | ----- | ----- | ----- | ----- | ----- |
| minimum      | 0.74  | 1.02  | 1.00  | 0.78  | 0.99  | 1.00  |
| average      | 0.85  | 1.04  | 1.01  | 0.90  | 1.02  | 1.00  |
| maximum      | 0.94  | 1.10  | 1.02  | 0.98  | 1.06  | 1.01  |

|              | case III) | | | case IV) | | |
| e            | 1e-16 | 1e-15 | 1e-14 | 1e-16 | 1e-15 | 1e-14 |
| ------------ | ----- | ----- | ----- | ----- | ----- | ----- |
| minimum      | 0.48  | 0.96  | 0.99  | 0.83  | 0.86  | 0.83  |
| average      | 1.09  | 1.00  | 1.00  | 1.02  | 1.01  | 1.02  |
| maximum      | 1.87  | 1.05  | 1.01  | 1.21  | 1.23  | 1.19  |

**Table 5** *Ratio of radii of Algorithm 4 vs. Algorithm 3*

For very small radius of A, sometimes the one, sometimes the other algorithm delivers better results. However, in this case the radii of the final result are very small anyway (of the order of few ulps).

We performed the same test for the solution of systems of linear equations with dimensions up to 500. In this case, *all* computed results were identical in all components up to less than a hundredth of a per cent of the radius. Therefore we recommend to use the faster Algorithm 4.

The two cases discussed, point matrix times point matrix and point matrix times interval matrix, are already sufficient for many verification algorithms from linear to nonlinear systems solvers. Here, matrix multiplication usually occurs as preconditioning.

### 2.4. Real interval matrix times real interval matrix

Finally, consider two interval matrices $\mathtt{intA} \in \mathbb{IM}_{m,n}(\mathbb{R})$ and $\mathtt{intB} \in \mathbb{IM}_{n,k}(\mathbb{R})$. The standard top-down algorithm uses a three-fold loop with an addition and a multiplication of two (scalar) intervals in the most inner loop. Multiplication of two scalar intervals, in turn, suffers from various case distinctions and switching of the rounding mode. This slows down computation terribly. An alternative may be the following modification of Algorithm 2.

```
C.inf = zeros(size(intA,1),size(intB,2));
C.sup = C.inf;
setround(-1)
for i=1:size(intA,2)
  C.inf = C.inf + min(intA.inf(:,i)*intB.inf(i,:), ...
                      intA.inf(:,i)*intB.sup(i,:), ...
                      intA.sup(:,i)*intB.inf(i,:), ...
```

```
                            intA.sup(:,i)*intB.sup(i,:)));
setround(1)
for i=1:size(intA,2)
   C.sup = C.sup + max(intA.inf(:,i)*intB.inf(i,:), ...
                       intA.inf(:,i)*intB.sup(i,:), ...
                       intA.sup(:,i)*intB.inf(i,:), ...
                       intA.sup(:,i)*intB.sup(i,:));
end
```

**Algorithm 6** *Interval matrix times interval matrix*

Correctness of the algorithm follows as before. For the multiplication of two $n \times n$ interval matrices, Algorithm 6 needs a total of $8n^3$ multiplications, additions and comparisons. At first sight this seems to be very inefficient. In contrast, the standard top-down algorithm with a three-fold loop and a scalar interval product in the most inner loop can hardly be optimized by the compiler. However, the timings to be given after the next algorithm tell a different story.

A fast algorithm is possible when allowing a certain overestimation of the result. Algorithm 7 computes the interval hull in the sense of infimum/supremum arithmetic. When computing the result in midpoint/radius arithmetic like in Algorithm 4, better use of BLAS is possible. Consider the following algorithm for the multiplication of two interval matrices intA and intB.

```
setround(1)
Amid = intA.inf + 0.5*(intA.sup - intA.inf);
Arad = Amid - intA.inf;
Bmid = intB.inf + 0.5*(intB.sup - intB.inf);
Brad = Bmid - intB.inf;
setround(-1)
C1 = Amid*Bmid;
setround(1)
C2 = Amid*Bmid;
Cmid = C1 + 0.5*(C2 - C1);
Crad = ( Cmid - C1 ) + ...
            Arad * ( abs(Bmid) + Brad ) + abs(Amid) * Brad;
setround(-1)
C.inf = Cmid - Crad;
setround(1)
C.sup = Cmid + Crad;
```

**Algorithm 7** *Interval matrix multiplication by midpoint/radius*

Unlike Algorithm 4, there may be an overestimation of the result. However, we want to stress that the overestimation is globally limited, independent of the dimension of the matrices. For narrow intervals the overestimation is very small (for quantification see [25]), for arbitrary intervals it is always globally limited by a factor 1.5. That means the radii of the intervals computed by Algorithm 7 are entrywise not larger than 1.5 times the radii of the result computed by power set operations. This has already been observed by Krier [17] in his Ph.D. thesis. For details see also [25]. The analysis holds for rounding-free arithmetic.

Note that Amid, Arad, Bmid and Brad are calculated in Algorithm 7 in such a way that Amid − Arad ≤ A.inf ≤ A.sup ≤ Amid + Arad and Bmid − Brad ≤ B.inf ≤ B.sup ≤ Bmid + Brad is assured. It is easily seen that this is indeed always satisfied, also in the presence of underflow.

### 2.5. Timing for real interval matrix multiplication

An advantage of Algorithm 7 is that only $4n^3$ multiplications and additions plus a number of $0(n^2)$ operations are necessary. With respect to memory and cache problems those four real matrix multiplications are very fast. A disadvantage is the $0(n^2)$ part of the algorithm with some memory operations for larger matrices.

Let A, B ∈ $M_n(\mathbb{R})$ and intA, intB ∈ $\mathbb{IM}_n(\mathbb{R})$, where intA and intB are assumed to be thick interval matrices. An operation count of additions and multiplications for the different cases of real matrix multiplication is as follows.

| | | |
|---|---|---|
| A * B | $n^3$ | for floating point multiplication, |
| A * B | $2n^3$ | for verified bounds, using Algorithm 1, |
| A * intB | $4n^3$ | using Algorithm 3, |
| A * intB | $3n^3$ | using Algorithm 4, |
| intA * intB | $8n^3$ | using Algorithm 6, |
| intA * intB | $4n^3$ | using Algorithm 7 |

**Table 8** *Operation count for real matrix multiplication*

The following table shows computing times of C-routines for multiplication of two interval matrices for the standard top-down approach with scalar interval operations in the most inner loop compared to Algorithms 6 and 7. The timing is on our Convex SPP 2000 parallel computer. Computations have been performed by Jens Zemke.

| dimension | 100 | 200 | 500 | 1000 |
|---|---|---|---|---|
| standard | 0.30 | 2.40 | 72.3 | 613.6 |
| Algorithm 6 | 0.42 | 8.9 | 245 | 1734 |
| Alg. 6 parallel | 1.14 | 10.7 | 216 | 1498 |
| Algorithm 7 | 0.02 | 0.17 | 3.30 | 20.2 |
| Alg. 7 parallel | 0.02 | 0.11 | 0.93 | 5.9 |

**Table 9** *Computing times interval matrix times interval matrix (in seconds)*

The data of the interval matrices were generated randomly such that all intervals entries were strictly positive. That means that no case distinction was necessary at all for the standard top-down approach. It follows that in the examples the standard algorithm needed only $4n^3$ multiplications, additions and comparisons, but also additional $4n^3$ switching of the rounding mode.

The timing clearly shows that the bare operation count and the actual computing time need not be proportional. Obviously, inefficient use of cache and memory slow down the outer product approach Algorithm 6 significantly. The lines 3 and 5 in Table 9 display the computing time using 4 processors instead of 1. We stress that the same code was used. The only difference is that first the sequential BLAS, then the parallel BLAS was linked. For the standard top-down algorithm, use of 1 or 4 processors makes no difference.

For the other operations like addition, subtraction, reciprocal and division, may be used standard implementations combined with the above observations. We skip a detailed description and refer to the INTLAB code.

## 2.6. Complex point matrix multiplication

The next problem are complex operations. Again, we are mainly interested in matrix multiplication because this dominates the computing time of a number of verification algorithms.

Our design decision was to use midpoint/radius representation for complex intervals. For various reasons this is more appropriate for complex numbers than infimum/supremum representation using partial ordering of complex numbers. The definition of midpoint/radius arithmetic goes at least back to Sunaga [26].

In the multiplication of two point matrices A and B a new problem arises, namely that the error during midpoint calculation has to be treated separately. The following algorithm solves this problem.

```
setround(-1)
C1 = real(A)*real(B) + (-imag(A))*imag(B) + ...
     ( real(A)*imag(B) + imag(A)*real(B) ) * j ;
setround(1)
C2 = real(A)*real(B) + (-imag(A))*imag(B) + ...
     ( real(A)*imag(B) + imag(A)*real(B) ) * j ;
C.mid = C1 + 0.5 * (C2-C1);
C.rad = C.mid - C1;
```

**Algorithm 10** *Complex point matrix times point matrix*

First, the implementation assures $C1 \leq A * B \leq C2$ in the (entrywise) complex partial ordering. Second, the midpoint and radius are both calculated with rounding upwards. A calculation shows that indeed

```
C.mid−C.rad ≤ A * B ≤ C.mid+C.rad
```

is always satisfied, also in the presence of underflow.

### 2.7.  Complex interval matrix multiplication

The multiplication of two complex interval matrices does not cause additional problems. The algorithm is as follows, again in original INTLAB code.

```
setround(-1)
C1=real(intA.mid)*real(intB.mid)+(-imag(intA.mid))*imag(intB.mid)+
... (real(intA.mid)*imag(intB.mid)+imag(intA.mid)*real(intB.mid))*j;
setround(1)
C2=real(intA.mid)*real(intB.mid)+(-imag(intA.mid))*imag(intB.mid)+
... (real(intA.mid)*imag(intB.mid)+imag(intA.mid)*real(intB.mid))*j;
C.mid=C1 + 0.5 * (C2-C1);
C.rad=abs( C.mid - C1 ) + ...
           intA.rad*(abs(intB.mid)+intB.rad)+abs(intA.mid)*intB.rad;
```

**Algorithm 11** *Complex interval matrix times interval matrix*

Note that the radius is calculated with only two additional point matrix multiplications. Also note that again midpoint and radius of the result are computed with rounding upwards. Correctness of the algorithm is easily checked. The other complex operations are implemented like the above and do not cause additional difficulties.

We want to stress again that Algorithms 10 and 11 exclusively use BLAS routines. This assures fast execution times. Let $A, B \in M_n(\mathbb{C})$ and $\text{intA, intB}$ $\in \text{IIM}_n(\mathbb{C})$, where $\text{intA}$ and $\text{intB}$ are assumed to be thick interval matrices. An operation count of additions and multiplications for the different cases of complex matrix multiplication is as follows.

| A * B | $4n^3$ | for floating point multiplication, |
|---|---|---|
| A * B | $8n^3$ | for verified bounds, |
| A * intB | $9n^3$ | for verified bounds, |
| intA * intB | $10n^3$ | for verified bounds. |

**Table 12** *Operation count for complex matrix multiplication*

The factor between interval and pure floating point multiplication is smaller than in the real case; it is only 2, 2.25 and 2.5, respectively.

Finally we mention that, as in the real case, the overestimation of the result of interval matrix multiplication using the above midpoint/radius arithmetic cannot be worse than a factor 1.5 in radius compared to power set operations ([17], [25]), and is much better for narrow interval operands [25].

## 3. The interval library and timings in INTLAB

In the preceding section we discussed a general interval library - for implementation in a to-compile programming language like Fortran or C. When moving to an implementation in Matlab, the interpretation overhead becomes a big issue: The "interpretation count" becomes much more important than the operation count. That means algorithms should be designed loop-free using few computing intensive operators. This is true for most of the fast algorithms developed in the previous section. All following computing times are generated on a 120 Mhz Pentium I Laptop and using Matlab V5.1.

### 3.1. Real matrix multiplication

The first example is real matrix multiplication. In many applications this is a computing intensive part of the code. We compare the computing time of the built-in pure floating point routines with INTLAB times for verified computations. Both approaches use extensively BLAS routines. In Table 9 we listed the computing times for C-routines on a parallel computer for compiled code; following are INTLAB times including interpretation overhead on the 120 Mhz Laptop. The following code, for the example $n = 200$, is used for testing.

```
n=200; A=2*rand(n)-1; intA=midrad(A,1e-12); k=10;
tic; for i=1:k, A*A;          end, toc/k
tic; for i=1:k, intval(A)*A;  end, toc/k
tic; for i=1:k, intA*A;       end, toc/k
tic; for i=1:k, intA*intA;    end, toc/k
```

The first line generates an $n \times n$ matrix with randomly distributed entries in the interval $[-1,1]$. INTLAB uses Algorithms 1, 4 and 7. The command tic starts the wall clock, toc is the elapsed time since the last call of tic. The computing times are as follows.

| dimension | pure floating point | verified A*A | verified A*intA | verified intA*intA |
|---|---|---|---|---|
| 100 | 0.11 | 0.22 | 0.35 | 0.48 |
| 200 | 0.77 | 1.60 | 2.84 | 3.33 |
| 500 | 15.1 | 29.1 | 44.7 | 60.4 |

**Table 13** *Real matrix multiplication (computing time in seconds)*

Going from dimension 100 to 200 there is a theoretical factor of 8 in computing time. This is achieved in practice. From dimension 200 to 500 the factor should be

$2.5^3 = 15.6$, but in practice the factor is between 18 and 19. This is due to cache misses and limited memory.

The INTLAB matrix multiplication algorithm checks whether input intervals are thin or thick. This explains the difference between the second and third column in Table 13. According to Algorithms 1, 4, 7 and Table 8, the theoretical factors between the first and following columns of computing times are approximately 2, 3 and 4, respectively. This corresponds very good to the practical measurements, also for $n = 500$. Note that the timing is performed by the Matlab wall clock; actual timings may vary by some 10 per cent.

For the third case, point matrix times interval matrix, alternatively Algorithms 3 and 4 may be used. The computing times are as follows.

| dimension | Algorithm 3 | Algorithm 4 |
|---|---|---|
| 100 | 0.45 | 0.35 |
| 200 | 3.79 | 2.85 |
| 500 | 59.2 | 44.7 |

**Table 14** *Algorithms 3 and 4 for point matrix times interval matrix (computing time in seconds)*

The theoretical ratio 0.75 is achieved in the practical implementation, so we use Algorithm 4 in INTLAB.

### 3.2. Complex matrix multiplication

The computing times for complex matrix multiplication using Algorithms 10 and 11 are as follows.

| dimension | pure floating point | verified A*A | verified A*intA | verified intA*intA |
|---|---|---|---|---|
| 100 | 0.35 | 1.00 | 1.11 | 1.25 |
| 200 | 3.08 | 6.92 | 7.80 | 8.71 |
| 500 | 79.7 | 119.3 | 135.4 | 147.9 |

**Table 15** *Complex matrix multiplication (computing time in seconds)*

For $n = 500$, again cache misses and limited memory in our Laptop slow down the computing time. According to Table 12, the theoretical factors between the first and following columns of computing times are approximately 2, 2.25 and 2.5, respectively. This corresponds approximately to the measurements. The timing depends (at least for our Laptop) on the fact that for A, B $\in M_n(\mathbb{C})$ the built-in multiplication A $*$ B is faster than the multiplication used in Algorithm 10 for smaller dimensions, but slower for larger dimensions. Consider

```
A=rand(n)+j*rand(n); B=A; k=10;
tic; for i=1:k, A * B; end, toc/k
tic; for i=1:k,real(A)*real(B)+(-imag(A)*imag(B);
              real(A)*imag(B)+imag(A)*real(B);end,toc/k
```

for different dimensions n. The timing is as follows.

| n | A * B | real/imaginary part separately |
|------|-------|--------------------------------|
| 100 | 0.37 | 0.47 |
| 200 | 3.35 | 3.35 |
| 500 | 81.2 | 68.7 |

**Table 16** *Different ways of complex matrix multiplication (computing time in seconds)*

### 3.3. Interval matrix multiplication with and without overestimation

Up to now we used the fast version of interval matrix times interval matrix. Needless to say that the standard approach with three loops and case distinctions in the inner loop produces a gigantic interpretation overhead. INTLAB allows to switch online between Algorithms 7 and 6. A timing is performed by the following statements.

```
n=100; A=midrad(2*rand(n)-1,1e-12);
intvalinit('FastIVMult'); tic; A*A; toc
intvalinit('SharpIVMult'); tic; A*A; toc
```

Computing times are 0.4 seconds for fast multiplication according to Algorithm 7, and 23.0 seconds for sharp multiplication according to Algorithm 6.

The difference is entirely due to the severe interpretation overhead. It is remarkable that, in contrast to compiled code, for the interpreted system now Algorithm 6 is apparently the fastest way to compute the product of two thick interval matrices without overestimation. However, we will see in Section 5 that the overestimation in Algorithm 7 due to conversion to midpoint/radius representation is negligible in many applications.

## 4. The INTLAB toolbox

Version 5 of Matlab allows the definition of new classes of variables together with overloading of operators for such variables. The use and implementation is as easy as one would expect from Matlab. However, quite some interpretation overhead has to be compensated. Remember that the goal is to implement everything in Matlab.

## 4.1. The operator concept

For example, a new type `intval` is introduced by defining a subdirectory `@intval` under a directory in the search path of Matlab. This subdirectory contains a constructor `intval`. The latter is a file `intval.m` containing a statement like

```
c = class(c,'intval');
```

This tells the Matlab interpreter that the variable c is of type `intval`, although no explicit type declaration is necessary. Henceforth, a standard operator concept is available with a long list of overloadable operators including $+$, $-$, $*$, $/$, $\setminus$, $'$, $\tilde{}$, [ ], { } and many others.

It has been mentioned before that our design decision was to use infimum/supremum representation for real intervals, and midpoint/radius representation for complex intervals. From the mathematical point of view this seems reasonable. Therefore, our data type `intval` for intervals is a structure with five components:
x.complex boolean, true if interval is complex,
x.inf, x.sup infimum and supremum for real intervals (empty for complex intervals),
x.mid, x.rad midpoint and radius for complex intervals (empty for real intervals).

The internal representation is not visible for the user. For example,

```
inf(x)  or  x.inf  or  get(x,'inf')
```

accesses the infimum of the interval variable x independent of x being real or complex. Operations check the type of inputs to be real or complex.

INTLAB offers three different routines to define an interval:

- by the constructor `intval`, e.g. `x = intval(3);`

- by infimum and supremum, e.g. `x = infsup(4.3-2i,4.5-.1i);`

- by midpoint and radius, e.g. `x = midrad(3.14159,1e-5);`

The first is a bare constructor, the following define an interval by infimum and supremum or by midpoint and radius, respectively. The second example defines a complex interval using partial ordering.

## 4.2. Rigorous input and output

If the input data is not exactly representable in floating point, the (decimal) input data and the internally stored (binary) data do not coincide. This is a problem to all libraries written in a language like Fortran 90, C++ or Matlab. The well known example

$$x = \text{intval}(0.1) \tag{2}$$

converts 0.1 into the nearest floating number, and *then* assigns that value to x.inf and x.sup. However, 0.1 is not exactly representable in binary finite precision so that 1/10 will *not* be included in x.

This is the I/O problem of every interval library. Sometimes, the internal I/O routines do not even satisfy a quality criterion, or at least this is not specified and/or not guaranteed by the manufacturer (during implementation of INTLAB we found plenty of such examples). We had to solve this problem for rigorous input and to guarantee validity of displayed results.

We did *not* want to execute the assignment (2) in a way that a small interval is put around 0.1. One of the reasons not to do this is that intval(-3) would produce an interval of nonzero radius. Another reason is that the routine intval is also used just to change the type of a double variable, and the user would not want to get back an enlarged interval. The better way is to use

$$x = \text{intval}('0.1')$$

instead. In this statement the character string '0.1' is converted into an enclosing interval by means of the INTLAB conversion routine str2intval *with result verification*. For exactly representable input like intval('1') this does not cause a widening of the interval; this is also true, for example, for intval ('0.1e1-1.25i'). Constants with tolerances may be entered directly by replacing uncertainties by '_'. For example,

$$x = \text{intval}('3.14159\_')$$

produces the interval [3.14158, 3.14160]. The underscore _ in input and output is to be interpreted as follows: A correct inclusion is produced by subtracting 1 from and adding 1 to the last displayed figure.

We want to stay with our philosophy to use only Matlab code and not any C-code or assembly language code except the routine for switching the rounding mode. There are two problems in writing a conversion routine in INTLAB. First, the result should be as narrow as possible and second, there should not be too much interpretation overhead. Both problems are solved by a simple trick. We computed 2 two-dimensional arrays power10inf and power10sup of double floating point numbers with the property

$$\text{power10inf}(m, e) \leq m * 10^e \leq \text{power10sup}(m, e), \tag{3}$$

for all $m \in \{1, 2, \ldots, 9\}$ and $e \in \{-340, -339, \ldots, 308\}$. These are one-digit mantissa floating point numbers with exponents corresponding to IEEE 754 double format. Those 11682 numbers were computed off-line and stored in a file. The numbers are sharp. The computation used a rudimentary long arithmetic, all written in INTLAB, and takes less than half a minute on a 120 Mhz Pentium I Laptop.

With these numbers available, conversion is straightforward. For example,

$$x = 0.m_1 m_2 \ldots m_k \cdot 10^e \Rightarrow$$

$$\sum_{i=1}^{k} \texttt{power10inf}(m_i, e - i) \leq x \leq \sum_{i=1}^{k} \texttt{power10sup}(m_i, e - i). \qquad (4)$$

For sharp results, summation should be performed starting with smallest terms. This method using (4) together with (3) can also be vectorized. This allows fast conversion of long vectors of input strings. The sign is treated separately.

The output of intervals is also accomplished using (3) and (4). We convert a double number into a string, convert the string back into double with directed rounding and check whether the result is correct. If not, the string has to be corrected. This would be terribly slow if performed for every single component of an output vector or matrix. However, the code is vectorized such that the interpretation overhead is about the same for scalar or vector argument. The user hardly recognizes a difference in time of internal input/output and rigorous input/output. Summarizing, input and output is rigorously verified to be correct in INTLAB.

### 4.3. Automatic differentiation, pure floating point and rigorous

Most of the other features in INTLAB are more or less straightforward implementations of an interval library using the algorithms described in Section 2 and by using the operator concept in Matlab. It also allows, for example, the implementation of automatic differentiation. For some scalar, vector or matrix x,

```
gx = initvar(x)
```

initializes gx to be the dependent variables with values as stored in x. The corresponding data type is gradient. An expression involving gx is computed in forward differentiation mode with access to value and gradient by substructures .x and .dx. Consider, for example, the evaluation of the function $y = f(x) = \sin(0.1 \cdot \pi \cdot x)$. The following m-file

```
function y = f(x)
    y = sin( 0.1*pi*x );
```
(5)

may be used to evaluate $f(x)$ by y=f(x) for $x$ being a real or complex scalar, vector or matrix, or a sparse matrix. By Matlab convention, evaluation of standard functions with vector or matrix argument is performed entrywise. Similarly,

```
gy = f(gx)
```

gives access to gy.x, the same as $f(x)$, and to the partial derivatives gy.dx. For [m,n] being the size of gx, [m,n,k] is the size of gy.dx, where k is the number of dependent variables. This is true unless n=1, in which case the size of gy.dx is [m,k]. That means, the gradient information is put into an extra dimension with length equal to the number of dependent variables. For example, the input

```
X = initvar([3;-4]);   Y = f(X)
```

produces the output

```
gradient value Y.x =
    0.8090
   -0.9511

gradient derivative(s) Y.dx =
    0.1847         0
         0    0.0971
```

The operator concept follows the usual rules how to choose the appropriate operator. For example, if one of the operands is of type interval, the computation will be carried out in interval arithmetic with verified result. If none of the operands are of type interval, the result is calculated in ordinary floating point arithmetic.

### 4.4.  A problem and its solution

For example,

```
a = 0.1; x = intval(4); (a-a)*x
```

will produce a point interval consisting only of zero, although the multiplication is actually an interval multiplication. However, the left factor is exactly zero. In contrast,

```
a = intval('0.1'); x = intval(4); (a-a)*x
```

produces a small interval around zero. For a function defined in an m-file, like the one defined in (5), this might cause problems. The statement

```
    X = intval([3;-4]);   Y = f(X)
```

evaluates the first product `0.1*pi` in floating point, and only the following multiplication and the `sin` is performed with directed roundings. In order to avoid this, additional knowledge about the type of the input variable x is needed: In case x is floating point, all operations are floating point, in case x is of type `intval`, only interval operations should be used. In fact, the problem is more complicated. In definition (5) we used the built-in variable `pi`, an approximation of the transcendental number $\pi$. For pure floating point calculations, this is legitimate, but not for interval calculations. Alternatively, we may use an interval constant like in the following algorithm:

```
    function y = f(x)
        cPi = midrad(pi,1e-15);                                    (6)
        y = sin( 0.1*cPi*x );
```

But this causes problems in the pure floating point case because the result would *always* be of type `intval`. Our solution to this is a type adjustment by the following algorithm.

```
function  y = f(x)
    cPi = typeadj( midrad(pi,1e-15) , typeof(x) );   (7)
    y = sin( 0.1*cPi*x );
```

The function `typeadj` adjusts the type of the left input argument, in this case `cPi`, to the right argument. If x is of type `intval`, nothing happens, `cPi` is equal to `midrad(pi,1e-15)`. If x is floating point, so will be `cPi` by taking the midpoint of `midrad(pi,1e-15)`, which is `pi` again. In the same way, the statements

```
    X = initvar([3;-4]);   Y = f(X)
    X = initvar(intval([3;-4]));   Y = f(X)
```

will work with definition (7). In the first case, a pure floating point gradient calculation is performed, and in the second case an interval gradient calculation. That means in the second case, `Y.x` and `Y.dx` are of type `intval` being inclusions of the function value and the true gradient.

Similar to the gradient toolbox, there is a slope toolbox in INTLAB for expanding a function with respect to an expansion "point" xs (which may, in fact, be an interval) within an expansion interval x. The above remarks for `typeof` and `typeadj` apply accordingly.

INTLAB makes use of the data types already offered in Matlab, like sparse matrices or complex data. All the functionality of Matlab concerning for example sparse matrix calculations are immediately available in INTLAB. For example, the algorithms listed in Section 3 are suitable for sparse matrices without change. This is the one of the big advantages of the extensive use of BLAS routines.

## 4.5. Correctness of programming

There are various sources of errors in numerical algorithms. To achieve rigorously verified results not only the verification theorem need to be correct but also the implementation itself. Needless to say that all algorithms in INTLAB are examined by a large number of test routines. This improves confidence in the library but is no proof of correctness. However, we will give two arguments applying to INTLAB but not to other numerical algorithms which may imply some additional safety.

The first argument applies to every verification algorithm, namely the necessity of *correctness* of results. For example, $p = 3.141592653589792$ is an excellent approximation for the transcendental number $\pi$; however, the interval $p \pm 10^{-15}$ is a *wrong* answer for the smallest positive zero of $\sin(x)$. It turns out, as a practical experience, that errors sometimes decover by such "inclusions" being only slightly off the true answer. Other examples are inclusions like $[5, 5 + 10^{-16}]$ if the integer 5 is the true result. For many verification algorithms this is much likely pointing to an implementation error.

The second argument applies specifically to INTLAB. Due to the interpretative language algorithms are formulated in a very high level language. Therefore, the algorithms are more like *executable specifications*, and therefore very short, and correspondingly easy to check for correctness.

# 5. Programming and applications in INTLAB

In the following we give some sample programs to demonstrate the use of INT-LAB. All examples are given in original and executable code. Also, we give execution times with comparisons to pure floating point algorithms (those are frequently near the fastest achievable for the computer in use). All of the following computing times are again generated on our 120 Mhz Pentium I Laptop using Matlab V5.1.

## 5.1. Dense linear system

The first example is the solution of a dense system of linear equations. The following code is used in INTLAB.

```
function X = denselss(A,b)          % linear system solver
  R = inv( mid(A) ) ;               % for dense matrices
  xs = R * mid(b) ;
  Z = R * (b-intval(A)*xs) ;        % residual correction
  C=speye(size(A))-R*intval(A);     % residual of preconditioner
  Y = Z;
  E = 0.1*rad(Y)*hull(-1,1) + midrad(0,10*realmin);
  k = 0; kmax = 15; ready = 0;
  while ~ready & k<kmax             % interval iteration
```

```
    k = k+1;
    X = Y + E;
    Y = Z + C * X;
    ready = in0(Y,X);                % check inclusion in interior
  end
  if ready                          % verified inclusion
    X = xs + Y;
  else
    disp('no inclusion achieved for \');
    X = NaN;
  end
```

**Algorithm 17** *Solution of dense linear systems*

This is a well known algorithm for solving systems of linear equations based on the Krawczyk operator [16], [20]. The algorithm computes an inclusion of the difference of the true solution and the approximate solution xs. It is identical to the one presented in [23].

The 4th line Z = R*(b-intval(A)*xs) is a typical statement to ensure interval calculation of the residual. Note that a statement like, for example, Z = R*intval(b-A*xs) would not work correctly in general. The timing in seconds of Algorithm 17 for randomly generated real and complex linear systems of equations is as follows. The second column refers to both point matrix and right hand side, whereas for the third column both matrix and right hand side are thick.

| dimension | pure floating point | verified point | verified interval |
|-----------|---------------------|----------------|-------------------|
| 100       | 0.09                | 0.53           | 0.70              |
| 200       | 0.56                | 3.35           | 4.23              |
| 500       | 8.2                 | 50.9           | 67.6              |

**Table 18** *Solution of real linear systems (computing time in seconds)*

Algorithm 17 requires computation of an approximate inverse and subsequent multiplication of a point matrix times the system matrix. All other computing times are dominated by $O(n^2)$. Hence, the computing time in the real case is $3n^3$ for point and $4n^3$ for interval linear systems, respectively. Hence, the theoretical factor between the first column of computing times and following columns is 9 and 12, respectively.

The actual computing times are much better than this. The reason is that the $n^3$ operations for matrix inversion and matrix multiplication take less than three times the computing time of Gaussian elimination. This is because code for matrix products can be better blocked and optimized. Consider

```
n=200; A=2*rand(n)-1; b=A*ones(n,1); k=10;
tic; for i=1:k, inv(A); end, toc/k
tic; for i=1:k, A*A; end, toc/k
tic; for i=1:k, A\b; end, toc/k
```

taking 1.24 seconds          for matrix inversion,
       0.78 seconds          for matrix multiplication, and
       0.56 seconds          for Gaussian elimination
on the 120 Mhz Laptop. The practical ratio of computing time for Gaussian elimination to inversion is 2.2 instead of a theoretical 3, and the ratio to matrix multiplication only 1.4 instead of 3.

The computing times for the solution of complex systems of linear equations are as follows. As before, point refers to both matrix and right hand side being thin, and interval refers to both being thick. For $n - 500$, again the effect of cache misses is visible.

| dimension | pure floating point | verified point | verified interval |
|-----------|---------------------|----------------|-------------------|
| 100       | 0.22                | 1.92           | 2.05              |
| 200       | 1.65                | 13.1           | 15.1              |
| 500       | 34.0                | 234            | 246               |

**Table 19** *Solution of complex linear systems (computing time in seconds)*

As an example of ease of use of programming in INTLAB, we display the code for the verified solution of over- and underdetermined systems. It is done by solving a larger linear system with square matrix. Brackets are overloaded interval operators to define matrices by means of blocks. The statement isnan is true if verification of the constructed square linear system failed. Otherwise, the code is hopefully self-explaining. Note that the right hand side may be a matrix.

```
[m k] = size(A); [m1 n] = size(b);
if k~=m1
  error('linear system solver: inner matrix dimensions must agree')
end
if m==k                         % linear system with square matrix
  X = denselss(A,b);
else
  if m>k                        % least squares problem
    Y=denselss([A -eye(m);zeros(k) A' ],[ b ; zeros(k,n)]);
    if ~isnan(Y)
      X = Y(1:k,:);
    end
```

```
else                           % minimal norm solution
  Y=denselss([A' -eye(k);zeros(m) A ],[ zeros(k,n) ; b]);
  if ~isnan(Y)
    X = Y(m+1:m+k,:);
  end
end
end
```

**Algorithm 20** *Over- and underdetermined linear systems*

## 5.2. Sparse linear systems

As has been mentioned before, INTLAB supports interval sparse matrices. We use
the algorithm for symmetric positive definite systems as presented in [24]. The
INTLAB code is as follows.

```
function X=sparselss(A,b) % linear system solver for sparse matrices
  [s,xs] = singmin(mid(A),mid(b));
  As = mid(A) - s*speye(size(A));
  C = chol(As);
  setround(-1)
  p1 = C'*C - As;
  setround(1)
  p2 = C'*C - As;
  r = norm(p2-p1,1);
  setround(-1)
  minsvd = s - r;
  if minsvd<0
    disp('matrix too ill-conditioned')
    X = NaN;
  else
    setround(1)
    X = midrad( xs , norm(A*xs-b)/minsvd ) ;
  end
  setround(0)
```

**Algorithm 21** *Sparse linear systems*

The algorithm uses the routine `singmin` for a guess of a lower bound of the
smallest singular value of A. This is a straightforward implementation using an
approximate Cholesky factorization of A and inverse power iteration.

For timing we use the Matlab functions `sprand` to generate a sparse matrix
of dimension 1000 with random entries and random pattern. We add the identity matrix to ensure nonsingularity, and use A*A' to ensure symmetric positive
definiteness. The generation statement is as follows, the density is 0.9% on the
average.

```
n=1000; A=sprand(n,n,2/n)+speye(n); A=A*A'; b=A*ones(n,1);
```

Algorithm 21 performs a factorization of the matrix. Therefore it is important to
reduce fill-in, for example by minimum degree reordering. The Matlab code

```
p = symmmd(A); C = chol(A(p,p));
```

produces a Cholesky factor of the reordered matrix with density of about 11%, without reordering it is more than 50%. We solve the linear system in pure floating point and with verification using the following commands.

```
tic; A(p,p)\b(p); toc
tic; verifylss(A(p,p),b(p)); toc
```

The timing is 2.8 seconds for pure floating point solution, and 18.9 seconds with verification.

For larger dimensions the Matlab routine symmmd becomes very slow; the built-in solver and the verified solver work well for larger dimensions. The latter will be demonstrated for band matrices. The following statements produce a random band matrix with 21 nonzero elements per row and test the built-in linear system solver against our verification algorithm.

```
k=5; A=spdiags(2*rand(n,2*k+1)-1,-k:k,n,n)+speye(n);
A=A*A'; b=A*ones(n,1);
tic, A\b; toc
tic, verifylss(A,b); toc
```

The computing time is slowed down by the fact that the matrix is *not* stored as a band matrix but as a sparse matrix. All matrix operations are general sparse matrix operations. The timing on the 120 Mhz Laptop is as follows.

| n | floating point | verified |
|---|---|---|
| 1000 | 0.12 | 1.10 |
| 2000 | 0.24 | 2.26 |
| 5000 | 0.60 | 5.71 |
| 10000 | 1.21 | 15.0 |
| 20000 | 4.1 | 183 |

**Table 22** *Timing for banded linear systems (in seconds)*

Starting with dimension 10000 the timing is dominated by swapping, displaying mainly lack of memory on our Laptop and a slow disc. Note that due to the sparse matrix storage scheme, Matlab needs for dimension $n = 20000$ already 14.6 MByte of storage for the matrix.

### 5.3. Nonlinear systems

Finally, consider the solution of systems of nonlinear equations. In the formulation of the function it is important to reduce interpretation overhead. That means, any possibility to use high order vector and matrix operations should be used. This is the drawback of interactive environments. This is in particular true when using overloaded operators. Consider, for example,

```
tic; for i=1:1000, y=sqrt(exp(x)-x); end; toc        (8)
```

The computing time is 0.1 seconds for x=0.1, and 9.5 seconds for x=intval(0.1). In contrast, the following does the same in vector notation.

```
X=x*ones(1000,1); tic; Y=sqrt(exp(X)-X); toc
```

Now the computing time is 0.006 seconds for x=0.1, and 0.024 seconds for x=intval(0.1). We also mention that interpretation is particularly slow on the 120 Mhz Pentium I Laptop. The statement (8) executed on a 266 Mhz Pentium II PC needs 0.017 seconds for x=0.1, but 1.51 seconds for x=intval(0.1), a speed-up factor 6 compared to the 120 Mhz Laptop. On the other hand, interpretation is at least partially compensated by the ease of notation. Consider the INTLAB code for the solution of a system of nonlinear equations. It is assumed that a call f(x) calculates the function value at x. An example for a function f is given below. The algorithm uses a modified Krawczyk operator [24].

```
function [ X , xs ] = verifynlss(f,xs)

% floating point Newton iteration
  n = length(xs);
  xsold = xs;
  k = 0;
  while ( norm(xs-xsold)>1e-10*norm(xs) & k<10 ) | k<1
    k = k+1;                % at most 10, at least 1 iteration performed
    xsold = xs;
    x = initvar(xs);        % initialization of gradient calculation
    y = feval(f,x);         % f(x) using gradient variable x
    xs = xs - y.dx\y.x;     % Newton correction
  end

% interval iteration
  R = inv(y.dx);            % approximate inverse of Jacobian
  Z = -R*feval(f,intval(xs)); %initialization of interval iteration
  X = Z;
  E = 0.1*rad(X)*hull(-1,1) + midrad(0,realmin);
  ready = 0; k = 0;
  while ~ready & k<10
    k = k+1;
    Y = hull( X + E , 0 );  % epsilon inflation
    Yold = Y;
    x = initvar(xs+Y);      % interval gradient initialization
    y = feval(f,x);         % f(x) and Jacobian by
    C = eye(n) - R * y.dx;  % automatic differentiation
    i=0;
    while ~ready & i<2      % improved interval iteration
      i = i+1;
```

```
      X = Z + C * Y;
      ready = in0(X,Y);
      Y = intersect(X,Yold);
   end
end
if ready
   X = xs+Y;                % verified inclusion
else
   X = NaN;                 % inclusion failed
end
```

**Algorithm 23** *Verified solution of nonlinear systems*

The first part is a pure floating point iteration, the second part is to calculate an inclusion. Note that the floating point evaluation of f at x with gradient is performed by

```
x = initvar(xs);
y = feval(f,x);
```

in the first part, and the verified evaluation at xs+Y, again with gradient, is performed by

```
x = initvar(xs+Y);
y = feval(f,x);
```

As an example consider a problem given by Abbott and Brent in [1], the discretization of

$$3y''y + y'^2 = 0 \quad \text{with} \quad y(0) = 0 \quad \text{and} \quad y(1) = 20.$$

We solve the discretized system, not the continuous equation. In the paper, the initial approximation is a vector all entries of which are equal to 10, the true solution is $20x^{0.75}$. The discretized problem is specified by the following INTLAB function.

```
function y = f(x)
  y = x;
  n = length(x); v=2:n-1;
  y(1)  = 3*x(1)*(x(2)-2*x(1)) + x(2)*x(2)/4;
  y(v)  =3*x(v).*(x(v+1)-2*x(v)+x(v-1))+(x(v+1)-x(v-1)).^2/4;
  y(n)  = 3*x(n).*(20-2*x(n)+x(n-1)) + (20-x(n-1)).^2/4;
```

Note the vectorized formulation of the function. The dimension of the nonlinear system is determined by the number of elements of the specified approximation. The timing for the solution of the nonlinear system by the following statement

```
tic; X = verifynlss('f',10*ones(n,1)); toc
```

is 5.4 seconds for dimension $n = 50$, 8.5 seconds for dimension $n = 100$, and 20.3 seconds for dimension $n = 200$. The first and last components of the inclusion for $n = 200$ are

```
» X(1:4)'
intval ans =
  0.346256418326_   0.6045521734322   0.8305219234696   1.0376691412984
» X(197:200)'
intval ans =
 19.7005694833674 19.775568557350_  19.8504729393822 19.9252832242374
```

Finally, it may be interesting to analyze the computing intensive parts of the algorithm. This is possible by the Matlab `profile` function. The statements

```
profile verifynlss; n=200; verifynlss('f',10*ones(n,1));
profile report; profile done
```

produce the following output:

```
Total time in "c:\matlab\toolbox\intlab\intval\verifynlss.m":
21.41 seconds

     100% of the total time was spent on lines:
               [18 33 34 19 24 38 23 17 26 32]

                   16:     xsold = xs;
 0.10s,   0%      17:     x = initvar(xs);
10.47s,  49%      18:     y = feval(f,x);
 1.26s,   6%      19:     xs = xs - y.dx\y.x;
                   20:   end

                   22: % interval iteration
 0.15s,   1%      23:   R = inv(y.dx);
 0.42s,   2%      24:   Z = - R * feval(f,intval(xs));
                   25:   X = Z;
 0.03s,   0%      26:   E = 0.1*rad(X)*hull(-1,1) + midrad(0,realmin);
                   27:   ready = 0; k = 0;

                   31:     Yold = Y;
 0.02s,   0%      32:     x = initvar(xs+Y);
 6.07s,  28%      33:     y = feval(f,x);         % f(x) and Jacobian by
 2.68s,  13%      34:     C = eye(n) - R * y.dx; % automatic diff.
                   35:     i=0;

                   37:       i = i+1;
 0.16s,   1%      38:       X = Z + C * Y;
                   39:       ready = in0(X,Y);
```

**Table 24** *Profiling nonlinear system solver*

From the profile follows that more than half of the computing time is spent for the improvement of the poor initial approximation and, that most of the time is spent for the function evaluations in lines 18 and 33. Note that the matrix inversion of the $200 \times 200$ Jacobian in line 23 takes only 1 % of the computing time or about 0.2 seconds. The reason seems that the Jacobian is tridiagonal and Matlab matrix inversion takes advantage of that.

## 6. Future work and conclusions

INTLAB is an interval toolbox for the interactive programming environment Matlab. It allows fast implementation of prototypes of verification algorithms. The goal has been met that the computing time for matrix operations or solution of linear systems is of the order of comparable pure floating point algorithms. For nonlinear systems, the computing time is usually dominated by interpretation overhead for the evaluation of the nonlinear function in use. This overhead is diminished proportional to the possibility of vectorization of the code for the nonlinear function.

Up to now we stay with our philosophy to have everything written in INTLAB, that is Matlab code, except for the routine for switching the rounding mode. This strategy implies greatest portability and high speed on a variety of architectures. Wherever Matlab and IEEE 754 arithmetic with the possibility of switching the rounding mode is available, INTLAB is easy to install and to use.

The (freeware) INTLAB toolbox may be freely copied from the INTLAB home page

http://www.ti3.tu-harburg.de/$\sim$rump/intlab/index.html .

## Acknowledgement

The author is indebted to two anonymous referees for their thorough reading and for their fair and constructive remarks. Moreover, the author wishes to thank Shin'ishi Oishi and his group for many fruitful discussions during his stay at Waseda university, and Arnold Neumaier and Jens Zemke who contributed with their suggestions to the current status of INTLAB. The latter performed all test runs on our parallel computer.

## References

1. J.P. Abbott and R.P. Brent. Fast Local Convergence with Single and Multistep Methods for Nonlinear Equations. *Austr. Math. Soc. 19 (Series B)*, pages 173–199, 1975.
2. ACRITH High-Accuracy Arithmetic Subroutine Library, Program Description and User's Guide. IBM Publications, No. SC 33-6164-3, 1986.
3. G. Alefeld and J. Herzberger. *Introduction to Interval Computations*. Academic Press, New York, 1983.
4. E. Anderson, Z. Bai, C. Bischof, J. Demmel, J. Dongarra, J. Du Croz, A. Greenbaum, S. Hammarling, A. McKenney, S. Ostrouchov, and Sorensen D.C. *LAPACK User's Guide, Resease 2.0*. SIAM Publications, Philadelphia, second edition edition, 1995.
5. *ARITHMOS, Benutzerhandbuch*, Siemens AG, Bibl.-Nr. U 2900-I-Z87-1 edition, 1986.
6. J.J. Dongarra, J.J. Du Croz, I.S. Duff, and S.J. Hammarling. A set of level 3 basic linear algebra subprograms. *ACM Trans. Math. Software*, 16:1–17, 1990.
7. J.J. Dongarra, J.J. Du Croz, S.J. Hammarling, and R.J. Hanson. An extended set of Fortran basic linear algebra subprograms. *ACM Trans. Math. Software*, 14(1):1–17, 1988.

8.  D. Husung. ABACUS — Programmierwerkzeug mit hochgenauer Arithmetik für Algorithmen mit verifizierten Ergebnissen. Diplomarbeit, Universität Karlsruhe, 1988.
9.  D. Husung. Precompiler for Scientific Computation (TPX). Technical Report 91.1, Inst. f. Informatik III, TU Hamburg-Harburg, 1989.
10. *ANSI/IEEE 754-1985, Standard for Binary Floating-Point Arithmetic*, 1985.
11. R.B. Kearfott, M. Dawande, K. Du, and C. Hu. INTLIB: A portable Fortran-77 elementary function library. *Interval Comput.*, 3(5):96–105, 1992.
12. R.B. Kearfott, M. Dawande, and C. Hu. INTLIB: A portable Fortran-77 interval standard function library. *ACM Trans. Math. Software*, 20:447–459, 1994.
13. R. Klatte, U. Kulisch, M. Neaga, D. Ratz, and Ch. Ullrich. *PASCAL-XSC — Sprachbeschreibung mit Beispielen.* Springer, 1991.
14. O. Knüppel. PROFIL / BIAS — A Fast Interval Library. *Computing*, 53:277–287, 1994.
15. O. Knüppel. PROFIL/BIAS and extensions, Version 2.0. Technical report, Inst. f. Informatik III, Technische Universität Hamburg-Harburg, 1998.
16. R. Krawczyk. Newton-Algorithmen zur Bestimmung von Nullstellen mit Fehlerschranken. *Computing*, 4:187–201, 1969.
17. R. Krier. *Komplexe Kreisarithmetik.* PhD thesis, Universität Karlsruhe, 1973.
18. C. Lawo. C-XSC, a programming environment for verified scientific computing and numerical data processing. In E. Adams and U. Kulisch, editors, *Scientific computing with automatic result verification*, pages 71–86. Academic Press, Orlando, Fla., 1992.
19. C.L. Lawson, R.J. Hanson, D. Kincaid, and F.T. Krogh. Basic Linear Algebra Subprograms for FORTRAN usage. *ACM Trans. Math. Soft.*, 5:308–323, 1979.
20. R.E. Moore. A Test for Existence of Solutions for Non-Linear Systems. *SIAM J. Numer. Anal.* 4, pages 611–615, 1977.
21. A. Neumaier. *Interval Methods for Systems of Equations.* Encyclopedia of Mathematics and its Applications. Cambridge University Press, 1990.
22. S. Oishi. private communication, 1998.
23. S.M. Rump. *Kleine Fehlerschranken bei Matrixproblemen.* PhD thesis, Universität Karlsruhe, 1980.
24. S.M. Rump. Validated Solution of Large Linear Systems. In R. Albrecht, G. Alefeld, and H.J. Stetter, editors, *Computing Supplementum*, volume 9, pages 191–212. Springer, 1993.
25. S.M. Rump. Fast and parallel interval arithmetic. *BIT*, 39(3):539–560, 1999.
26. T. Sunaga. Theory of an Interval Algebra and its Application to Numerical Analysis. *RAAG Memoirs*, 2:29–46, 1958.
27. J. Zemke. b4m - BIAS for Matlab. Technical report, Inst. f. Informatik III, Technische Universität Hamburg-Harburg, 1998.

T. Csendes (ed.), *Developments in Reliable Computing* 105–118.
© 1999 *Kluwer Academic Publishers.*

# Verified calculation of the solution of algebraic Riccati equation

WOLFRAM LUTHER AND WERNER OTTEN

{luther, otten}@informatik.uni-duisburg.de
*Gerhard–Mercator–Universität – GH Duisburg, Informatik II, D–47048 Duisburg, Germany*

**Abstract.** In this note we describe a new method to calculate verified solutions of the matrix Riccati equation (ARE) with interval coefficients. Such an equation has to be solved when we want to find the steady state solutions of matrix Riccati differential equation with constant coefficients which arises in the theory of automatic control and linear filtering.

Given the Riccati polynomial $P(X)$ we use the Fréchet–derivative at $X$ to derive a linear equation of type $CX + XD = P$. Applying Brouwer's fixed point theorem, we find an interval matrix $[X]$ that includes a positive definite solution of the equation $P(X) = \Omega$.

First we want to give an outline of linear–quadratic control theory. Then we present results concerning the geometric structures of all solutions and enumerate linearly and quadratically convergent algorithms to find a solution used to construct the optimal feedback control for linear–quadratic optimal control problems.

**Keywords:** Algebraic matrix Riccati equation, interval arithmetic, verified solution

## 1. Basic linear–quadratic control theory

First we will briefly give some basics of linear–quadratic control theory which leads to the problem of verifying solutions of a matrix Riccati equation. A detailed description can be found in [13].

The time–invariant continuous–time linear dynamical system is described by

$$\dot{x}(t) = Ax(t) + Bu(t), \tag{1}$$

with time $t \in [t_0, T]$, the state vector $x(t) \in \mathbf{R}^n$, the control input vector $u(t) \in \mathbf{R}^m$, $A$ and $B$ real matrices of dimension $n \times n$ and $n \times m$, respectively.

Let the cost functional be the quadratic form

$$J(x_0, u, t_0) = \int_{t_0}^{\infty} \left[ x(t)^T Q x(t) + u(t)^T R u(t) \right] dt, \tag{2}$$

where $x_0 := x(t_0)$ is known, $Q$ is a symmetric positive semi–definite $n \times n$ weighting matrix, $R$ is a symmetric positive definite $m \times m$–matrix.

The time–invariant continuous–time linear–quadratic control problem consists in finding a solution $x(t)$ of (1) and a control vector $u(t)$ minimizing (2).

Under certain assumptions it can be shown that there is a unique control vector

$$u(t) = Hx(t), \quad H := -R^{-1} \cdot B^T \cdot X$$

that solves the problem, where $X \in \mathbf{R}^{n \times n}$ is the unique symmetric positive semi–definite stabilizing solution of the algebraic matrix Riccati equation (ARE)

$$P(X) := A^T X + XA - XBR^{-1}B^T X + Q = \Omega \tag{3}$$

with symmetric matrices $Q$ and $R$, $Q$ positive semidefinite and $R$ positive definite.

A complete discussion of the generalized non–symmetric matrix Riccati differential equation is given in Freiling and Jank [4]. They extend results given in Potter [10], Shayman [12], and Willems [14] and derive a fundamental representation formula, a parametrization, and a geometric description of the set of all solutions of the ARE and investigate the non–autonomous case.

We solve the stabilization problem to find a positive semi–definite symmetric stabilizing solution $X$ and the optimal regulator $u(t) = -R^{-1}B^T Xx(t)$. The matrix $F := BR^{-1}B^T$ leads to the stable closed–loop system $\dot{x}(t) = (A - FX)x(t)$. We form the $2n \times 2n$–matrix

$$G = \begin{bmatrix} A & -F \\ -Q & -A^T \end{bmatrix}. \tag{4}$$

The ARE has a stabilizing solution $X \geq 0$ if and only if the pair $(A, B)$ is stabilizable, and the matrix $G$ is a dichotomic matrix, i.e. $\mathrm{Re}(\lambda_i) \neq 0$ for $G$, where $\lambda_i \in \lambda(G)$, $i = 1, ..., 2n$.

A pair $(A, B)$ is stabilizable, if the unstable subspace $N$ is included in the controllability subspace generated by the columns of a certain matrix $\Gamma$ (for a complete definition see [13] pp. 13). Here $N$ denotes the direct sum of linear subspaces $N_i$ corresponding to all unstable eigenvalues $\lambda_i$ of $A$ (with multiplicity $m_i$), where $N_i$ is the null space of the matrix $(A - \lambda_i I)^{m_i}$. By the way, if all eigenvalues $\lambda$ of the matrix $A$ have negative real parts, the pair $(A, B)$ is stabilizable.

The ARE has a unique stabilizing solution $X \geq 0$ if and only if the pairs $(A, B)$ and $(A^T, D^T)$ are stabilizable, where $Q = D^T D$ and $\mathrm{rank}(D) = \mathrm{rank}(Q)$. If the matrix $\begin{bmatrix} D^T & A^T D^T & ....(A^T)^{n-1}D^T \end{bmatrix}^T$ has rank $n$, then the solution $X$ is positive definite.

Potter [10] considers the matrix quadratic equation

$$\Omega = Q + A^T X + XA - XFX,$$

where $Q, A, F$ and $X$ are matrices of dimension $n \times n$ with complex coefficients and where the small superposed letter $T$ denotes transposition and complex conjugation. Potter reduces the problem to an eigenvalue problem for the matrix $G$. For a $2n$–dimensional eigenvector $a$ of $G$ he introduces vectors $e$ and $d$ with $a = \begin{bmatrix} e \\ d \end{bmatrix}$, and $e$ consists of the first $n$ components of $a$ and $d$ of the last $n$ components. Then he proves the following results, using the notations introduced above:

Every solution of the matrix quadratic equation has the form

$$X = [d_1, ..., d_n] \cdot [e_1, ..., e_n]^{-1} = DE^{-1},$$

where the column vectors $e_i$ and $d_i$ form eigenvectors $a_i$ of $G$, $i = 1, ..., n$. Conversely, if $a_i$, $i = 1, ..., n$, are eigenvectors of $G$, and $(e_1, ..., e_n)$ is nonsingular, then

$$X = [d_1, ..., d_n] \cdot [e_1, ..., e_n]^{-1}$$

solves the matrix quadratic equation.

Assume that $a_i, i = 1, ..., n$, are the eigenvectors of $G$, and $\lambda_1, ..., \lambda_n$ are the corresponding eigenvalues. If $Q$ and $F$ are Hermitian matrices and the eigenvalues $\lambda_1, ..., \lambda_n$ satisfy $\lambda_i \neq -\lambda_j^T$, $1 \leq i, j \leq n$, and if $E$ is nonsingular, then $DE^{-1}$ is a Hermitian matrix.

Assume that $Q$ and $F$ are positive semi–definite and Hermitian. If $Q$ or $F$ is nonsingular and $X = DE^{-1}$ is positive definite, then the eigenvalues $\lambda_1, ..., \lambda_n$ have negative real parts. Otherwise, if these eigenvalues of $G$ have negative real parts and $E$ is nonsingular, then $X = DE^{-1}$ is positive semi–definite.

The proof shows that in the case of simple eigenvalues of $G$, the ARE has at most $\binom{2n}{n}$ different solutions.

If $Q$ and $F$ are Hermitian matrices and $\begin{bmatrix} e \\ d \end{bmatrix}$ is an eigenvector corresponding to the eigenvalue $\lambda$, then $\begin{bmatrix} -d \\ e \end{bmatrix}$ is an eigenvector of $G^T$ corresponding to $-\lambda$. Hence $-\lambda^T$ is an eigenvalue of $G$, and $G$ has at most $n$ eigenvalues with negative real parts which can be used to choose eigenvectors in order to construct the positive definite solution $X$.

EXAMPLE 1: ([13], p. 189)

Sima gives as an example a model of a continuous–time chemical reactor, where an ARE of type (3), given by the following matrices, has to be solved:

$$A = \begin{pmatrix} 1.400 & -0.208 & 6.715 & -5.676 \\ -0.581 & -4.290 & 0 & 0.675 \\ 1.067 & 4.273 & -6.654 & 5.893 \\ 0.048 & 4.273 & 1.343 & -2.104 \end{pmatrix}, \quad B = \begin{pmatrix} 0.000 & 0 \\ 5.679 & 0 \\ 1.136 & -3.146 \\ 1.136 & 0 \end{pmatrix},$$

$$R = \begin{pmatrix} 100 & 0 \\ 0 & 100 \end{pmatrix}, \qquad Q = \begin{pmatrix} 1 & 0 & 0 & 0 \\ 0 & 1 & 0 & 0 \\ 0 & 0 & 0 & 0 \\ 0 & 0 & 0 & 0 \end{pmatrix}.$$

To use Potter's method described above, we need the eigenvalues and eigenvectors of the matrix $G$ (cp (4)). With the computer algebra system Maple we have calculated them with a precision Digits := 15, and rounded the results to five significant digits.

$$\lambda_1 = -8.6610, \ \lambda_2 = 8.6610, \ \lambda_3 = -5.0872, \ \lambda_4 = 5.0872,$$
$$\lambda_5 = -2.0075, \ \lambda_6 = 2.0075, \ \lambda_7 = -0.29967, \ \lambda_8 = 0.29967.$$

From the four eigenvectors $a_i$ with negative eigenvalues we form the 4–dimensional vectors $e_i$ and $d_i$ as described above. This leads to the matrices

$$E = \begin{pmatrix} -0.10254 & 0.0053241 & -0.44877 & -0.18987 \\ -0.72320 & 0.018827 & -0.046452 & 0.072939 \\ 0.68966 & -0.083562 & 0.59606 & 0.47738 \\ 0.72522 & -0.096352 & -0.088592 & 0.50523 \end{pmatrix},$$

$$D = \begin{pmatrix} -0.017867 & -0.98498 & -0.065227 & 0.17471 \\ -0.082513 & -0.25797 & -0.0050313 & 0.41054 \\ -0.010473 & -0.71918 & -0.026903 & 0.21139 \\ -0.0022249 & 0.28663 & 0.019349 & 0.22101 \end{pmatrix}$$

and to the solution of the ARE

$$X = D \cdot E^{-1} = \begin{pmatrix} 51.562 & 11.111 & 37.065 & -16.902 \\ 11.111 & 3.1366 & 8.1332 & -3.1494 \\ 37.065 & 8.1332 & 26.702 & -12.056 \\ -16.902 & -3.1494 & -12.056 & 5.9319 \end{pmatrix}. \tag{5}$$

Then the residual is

$$A^T X + XA - XFX + Q =$$
$$= \begin{pmatrix} 0.68 \cdot 10^{-10} & 0.139 \cdot 10^{-10} & 0.53 \cdot 10^{-10} & -0.292 \cdot 10^{-10} \\ 0.191 \cdot 10^{-10} & 0.39 \cdot 10^{-11} & 0.155 \cdot 10^{-10} & -0.78 \cdot 10^{-11} \\ 0.30 \cdot 10^{-10} & 0.59 \cdot 10^{-11} & 0.23 \cdot 10^{-10} & -0.145 \cdot 10^{-10} \\ -0.85 \cdot 10^{-11} & -0.8 \cdot 10^{-12} & -0.81 \cdot 10^{-11} & 0.65 \cdot 10^{-11} \end{pmatrix}.$$

The result of this example will be verified in example 2 in section 4.

## 2. Newton–type algorithm

In this section, we discuss a numerical algorithm for the solution of an ARE. It can be used to derive a starting solution for the verification process.

Kratz and Stickel [7] propose the following algorithm:

Let the matrix polynomial $P(X) := A^T X + XA - XFX + Q$ be given. Then the Fréchet derivative at $X$ applied to the matrix $U$ is:

$$P'(X)[U] = -UFX - XFU + A^T U + UA.$$

If the mapping $P'(X)$ is injective, we call $P'(X)$ regular and can formulate Newton's algorithm:

a)  Start with an appropriate approximation $\widehat{X}$ for the solution of (3).

b)  Solve the linear equation in $U$:

$$P(X) = U(FX - A) + (XF - A^T)U,$$

for example by using the Kronecker product with classical arithmetic as described in the verification step (Section 3) in conjunction with interval arithmetic.

c)  Update: $\widehat{X} := \widehat{X} + U$.

An equivalent Newton–type algorithm given by Sima [13] is the following, where the $k$–th iteration of the Newton process covers the following steps:

a)  Compute $A_k := A + BH_k$.

b)  Solve $A_k^T X_k + X_k A_k + Q_k = \Omega$, $Q_k := H_k^T R H_k + Q$.

c)  Compute $H_{k+1} := -R^{-1} B^T X_k$.

If $Q$ is positive definite and $H_1$ is stabilizing, the sequence $\{X_k\}$ of solutions of Lyapunov equation in (b) is decreasing and tends to a positive semi–definite solution $X$ of the ARE. The convergence rate is quadratic. A stabilizing matrix $H_1$ exists if the pair $(A, B)$ is stabilizable.

In order to calculate such a matrix we can use Armstrong's theorem [2]:

Let the ARE–system defined by the stabilizable pair $(A, B)$. Then $H := -B^T Z^+$, is a stabilizing matrix, $Z^+$ denotes the pseudoinverse to $Z$ and $Z$ is symmetric positive semi–definite and satisfies

$$(A + \delta I)Z + Z(A + \delta I)^T = 2BB^T, \ \delta > \|A\|.$$

## 3.  The verification

In [3] there is given an estimation of the relative error $\kappa(A,B,Q,R)$ of the solution $X$ of an ARE produced by relative errors of order $\varepsilon$ in the data matrices $A,B,Q,R$. However the given formula is very complicated.

Therefore we extend the validation procedure for the calculation of the square root of an interval matrix given in [8] to the ARE problem for interval matrices.

Our algorithm is similar to the methods for computing validated solutions of nonlinear systems proposed by Rump [11]. However we prefer for improved stability a direct reduction to a linear equation system based on the special structure of the problem and the Fréchet derivative of the map $P(X)$ of the given ARE

$$P(X) := A^T X + XA - XFX + Q = \Omega$$

together with suitable criteria concerning the matrices $A$, $F$ and $Q$, which can be point or interval matrices. Our aim is to transform the ARE to an equation of the form $CX + XD = P$ and to calculate a fixed point of this equation. Therefore we assume that $\widehat{X}$ is an approximate solution of the ARE and we set

$$\begin{aligned} A^T X + XA - \widehat{X}FX - XF\widehat{X} &= -P(\widehat{X}) + \Delta F\Delta \\ \Leftrightarrow (A^T - \widehat{X}F)X + X(A - F\widehat{X}) &= -P(\widehat{X}) + \Delta F\Delta. \end{aligned}$$

If $U$ is a fixed point of the map $T : \Delta \to X$ defined by the above equation, we find by replacing $X$ and $\Delta$ by $U$:

$$\begin{aligned} (A^T - \widehat{X}F)U + U(A - F\widehat{X}) + P(\widehat{X}) &= UFU \\ \Leftrightarrow A^T(\widehat{X}+U) + (\widehat{X}+U)A - (\widehat{X}+U)F(\widehat{X}+U) + Q &= \Omega. \end{aligned}$$

which means that $\widehat{X} + U$ is a solution of the ARE.

This leads to the following algorithm for the computation of a verified solution of the ARE. To avoid rounding errors, we use the high precision scalar product for matrix calculations whenever it is possible.

ALGORITHM 1:

a)  Use a classical solution scheme to compute a first approximation $\widehat{X}$ of the ARE together with the matrices mid$(A)$, mid$(Q)$ and mid$(F)$.

b)  Inflate the matrix $\widehat{X}$ to $\widehat{X} + \Delta$ by an error matrix $\Delta$: For example, choose $\Delta := [-\delta,\delta]\widehat{X}$ or $\Delta := [-\delta,\delta](1,1,\ldots,1)^T(1,1,\ldots,1)$.

c)  Transform the interval equation

$$(A^T - \widehat{X}F)X + X(A - F\widehat{X}) = -P(\widehat{X}) + \Delta F\Delta \tag{6}$$

into a linear equation system for which several efficient solution methods are known. With the Kronecker product $A \otimes B$ for a $p \times p$-matrix $A$ and a $q \times q$-matrix $B$ we get a $pq \times pq$-matrix defined by

$$A \otimes B := \begin{pmatrix} a_{11}B & a_{12}B & \cdots & a_{1p}B \\ a_{21}B & a_{22}B & \cdots & a_{2p}B \\ \vdots & \vdots & & \vdots \\ a_{p1}B & a_{p2}B & \cdots & a_{pp}B \end{pmatrix}.$$

Using this and the vec function of a matrix $C$ as a compound column of all columns of $C$ as in Kratz–Stickel [7], an equation system $CX + XD = P$ becomes

$$(I \otimes C + D^T \otimes I) \operatorname{vec} X = \operatorname{vec} P. \tag{7}$$

d)  Use an appropriate interval equation solver to solve the linear equation system (7) for a solution $[X] \in \mathrm{IR}^{n \times n}$.

e)  If the solution interval matrix $[X]$ satisfies the relation $[X] \subset \Delta$, an application of Brouwer's fixed point theorem to the map $T : \Delta \to [X]$, defined by our linear interval equation, ensures that the interval matrix $\widehat{X} + [X]$ encloses $\widehat{X} + U$ as a solution of the ARE.

f)  If the verification is only possible when using a large $\delta$, we can restart in b) with the approximate solution $\widehat{X} := \operatorname{mid}(\widehat{X} + [X])$ and a smaller $\delta$. This procedure can be applied several times.

The method also works if $A$ and $Q$ are replaced by interval matrices $[A]$ and $[Q]$. Then the map $T$ and its fixed point $U$ depend on the choice of matrices $A \in [A]$ and $Q \in [Q]$. However, we must guarantee that for $D \in \Delta$, the linear equation

$$\Omega = (A^T - \widehat{X}F)X + X(A - F\widehat{X}) + A^T\widehat{X} + \widehat{X}A - \widehat{X}F\widehat{X} + Q - DFD$$

has a unique solution $X \in [X] \subset \Delta$. For this purpose, we assume that $\lambda_i + \lambda_j \neq 0$ for the eigenvalues of the matrix $A - F\widehat{X}$.

Our Algorithm 1 can be used to verify all solutions of an ARE regardless whether the solution is positive definite. In control theory, one needs a symmetric positive semi–definite solution. In this case, we have to check whether each symmetric matrix in the inclusion $\widehat{X} + [X]$ is positive semi–definite. In the case of real solutions, we can perform an LU–decomposition of the interval matrix $\widehat{X} + [X]$. If the diagonal elements of the upper triangle matrix $U$ contain only positive values, then every symmetric matrix included in $\widehat{X} + [X]$ is positive definite [15].

It is also possible to perform a Cholesky decomposition of the interval matrix $\widehat{X} + [X] \subseteq [L] \cdot [L]^T$ as given in [1]. If the decomposition is feasible, which means that all diagonal elements of the interval matrix $[L]$ contain only positive intervals, then each symmetric matrix in the inclusion $\widehat{X} + [X]$ is positive definite.

Now we want to complete the defect correction method for solutions of the ARE given by Mehrmann and Tan [9] by a verification step:

If the approximate solution $\widehat{X}$ satisfies the ARE, then the error $E := X - \widehat{X}$ satisfies a Riccati equation of the same type. Let $Y = Y^T$ be the residual of the computed solution $\widehat{X}$. That is, $Y := A^T \widehat{X} + \widehat{X} A - \widehat{X} B R^{-1} B^T \widehat{X} + Q$. Then we have to solve the ARE

$$A_c^T E + E A_c - E B R^{-1} B^T E + Y = \Omega, \quad A_c := A - B R^{-1} B^T \widehat{X}.$$

ALGORITHM 2:

a) For the ARE $A^T X + X A - X F X + Q = \Omega$, calculate an approximate solution $\widehat{X}$ satisfying the ARE.

b) Compute the interval matrices

$$Y = (A^T \widehat{X} + \widehat{X} A - \widehat{X} F \widehat{X} + Q), \quad A_c := (A - F \widehat{X}).$$

c) Use standard methods to solve the problem

$$\mathrm{mid}(A_c^T) E + E \, \mathrm{mid}(A_c) - E F E + \mathrm{mid}(Y) = \Omega, \quad E = X - \widehat{X}.$$

If $\widehat{E}$ is an approximate solution of the above ARE, then the last two steps can be iterated with $\widehat{X} := \widehat{X} + \widehat{E}$.

d) Compute a verified inclusion matrix solution $[E]$ of $A_c^T E + E A_c - E F E + Y = \Omega$ using Algorithm 1. Then $\widehat{X} + [E]$ contains a solution of the ARE $A^T X + X A - X F X + Q = \Omega$.

## 4.  Numerical examples

In this section, we give some numerical examples of the verification of solutions of algebraic Riccati equations. All calculations have been executed using a GNU C++ compiler and the interval arithmetic package Profil/BIAS by Knüppel [6], based on double precision variables. We used the algorithm ILSS from Profil as the interval system equation solver.

In all following examples, the matrices are taken from literature as given there. All components given as integer values are handled as exact and are not inflated to intervals. The other components are first treated as exact to the given decimal digits and rounded to the next machine interval enclosing that number. In a second step the values are treated as uncertain data inflated with the values given in the examples.

EXAMPLE 2: First we verify the solution for the ARE of the continuous–chemical reactor model which we have given in example 1 applying Algorithm 1. As input we have first chosen the matrices $A$ and $B$, where all components are given with four significant decimal digits and have enclosed them in machine intervals with minimal diameter. As an approximative solution $\widehat{X}$, we use the matrix $X$ given in formula (5) with 5 significant digits in each component. The verification step succeeds using $\widehat{X} + \Delta$ with $\Delta = [-\delta, \delta](1,\ldots,1)^T(1,\ldots,1)$, $\delta = 2^{-11}$ and gives the solution $\widehat{X} + [X]$. The verification fails with this starting approximation and $\delta = 2^{-12}$. Then we perform a new verification step using the midpoint matrix of $\widehat{X} + [X]$ as a new approximative solution for the next step. In a second step, the minimal $\delta$ is equal to $2^{-25}$. In a third step, we finally reach a verified solution with $\delta = 2^{-35}$. The maximum of the interval–diameters in $\widehat{X} + [X]$ is $4.4 \cdot 10^{-11}$. Then we also inflate the matrices $A$ and $B$ with a diameter of $10^{-8}$ in each non–zero component, so that the values given in example 1 are approximately the midpoints of the inflated intervals. In this case we can verify the solution and we have a maximum diameter in $\widehat{X} + [X]$ of $6 \cdot 10^{-4}$. In the case of an inflation to intervals with diameters $10^{-5}$ the verification fails.

EXAMPLE 3: The following example is given in Sima ([13], ex. 4.5) and describes an interconnected power system with an ARE. Here the matrices $A$, $B$, $Q$ and $R$ are the following and we get the approximative solution $\widehat{X}$, rounded to 6 digits by using the algorithm from Kratz and Stickel [7] as described in section 2 with the starting matrix $X = 0.5 \cdot I_7$.

$$A := \begin{pmatrix} -0.04165 & 0 & 4.92 & -4.92 & 0 & 0 & 0 \\ -5.21 & -12.5 & 0 & 0 & 0 & 0 & 0 \\ 0 & 3.33 & -3.33 & 0 & 0 & 0 & 0 \\ 0.545 & 0 & 0 & 0 & -0.545 & 0 & 0 \\ 0 & 0 & 0 & 4.92 & -0.04165 & 0 & 4.92 \\ 0 & 0 & 0 & 0 & -5.21 & -12.5 & 0 \\ 0 & 0 & 0 & 0 & 0 & 3.33 & -3.33 \end{pmatrix},$$

$$B := \begin{pmatrix} 0 & 0 \\ 12.5 & 0 \\ 0 & 0 \\ 0 & 0 \\ 0 & 0 \\ 0 & 12.5 \\ 0 & 0 \end{pmatrix}, \quad Q := \begin{pmatrix} 1 & 0 & 0 & 0 & 0 & 0 & 0 \\ 0 & 0 & 0 & 0 & 0 & 0 & 0 \\ 0 & 0 & 0 & 0 & 0 & 0 & 0 \\ 0 & 0 & 0 & 1 & 0 & 0 & 0 \\ 0 & 0 & 0 & 0 & 1 & 0 & 0 \\ 0 & 0 & 0 & 0 & 0 & 0 & 0 \\ 0 & 0 & 0 & 0 & 0 & 0 & 0 \end{pmatrix}, \quad R := \begin{pmatrix} 100 & 0 \\ 0 & 100 \end{pmatrix}.$$

$$\widehat{X} = \begin{pmatrix}
0.581494 & 0.0944149 & 0.464910 & 0.0424524 & -0.134555 & -0.00338797 & -0.0565113 \\
0.0944149 & 0.0455802 & 0.171590 & -0.220239 & -0.00328288 & -0.00478942 & -0.0180646 \\
0.464910 & 0.171590 & 0.679910 & -0.687583 & -0.0561854 & -0.0180971 & -0.0817966 \\
0.0424524 & -0.220239 & -0.687583 & 4.30814 & -0.0437717 & 0.220204 & 0.687100 \\
-0.134555 & -0.00328288 & -0.0561854 & -0.0437717 & 0.582322 & 0.0945507 & 0.465590 \\
-0.00338797 & -0.00478942 & -0.0180971 & 0.220204 & 0.0945507 & 0.0456457 & 0.171837 \\
-0.0565113 & -0.0180646 & -0.0817966 & 0.687100 & 0.465590 & 0.171837 & 0.680895
\end{pmatrix}.$$

To verify the solution we have first chosen the matrices $A$ and $B$ and enclosed all components in machine intervals with minimal diameter. With $\Delta = [-\delta, \delta](1,\ldots,1)^T(1,\ldots,1)$ and $\delta = 2^{-44}$, we inflate the approximation $\widehat{X}$ to $\widehat{X} + \Delta$ and start our verification algorithm. This leads to the following inclusion for a solution of the ARE:

$$\widehat{X} + [X] = \begin{pmatrix}
0.58149430136219^{81}_{20} & 0.09441491013180^{92}_{73} & 0.4649103384670^{309}_{238} \\
0.09441491013180^{92}_{73} & 0.04558023400939^{44}_{37} & 0.1715897711302^{903}_{881} \\
0.4649103384670^{309}_{238} & 0.1715897711302^{903}_{881} & 0.679910096086^{4028}_{3949} \\
0.0424523645718^{324}_{135} & -0.22023924638945^{46}_{96} & -0.6875827628005^{494}_{692} \\
-0.1345548042610^{266}_{305} & -0.00328287715930^{52}_{66} & -0.05618539237706^{45}_{91} \\
-0.0033879731743^{596}_{607} & -0.0047894240874\,07^{0}_{4} & -0.01809708304507^{16}_{28} \\
-0.0565113231467^{794}_{835} & -0.01806456536657^{63}_{77} & -0.0817966347886^{660}_{707}
\end{pmatrix}$$

$$\begin{array}{llll}
0.0424523645718^{324}_{135} & -0.1345548042610^{266}_{305} & -0.0033879731743^{596}_{607} & -0.0565113231467^{794}_{835} \\
-0.22023924638945^{46}_{96} & -0.00328287715930^{52}_{66} & -0.0047894240874\,07^{0}_{4} & -0.01806456536657^{63}_{77} \\
-0.6875827628005^{494}_{692} & -0.05618539237706^{45}_{91} & -0.01809708304507^{16}_{28} & -0.0817966347886^{660}_{707} \\
4.3081381484483^{714}_{198} & -0.0437717490340^{021}_{187} & 0.2202037613341^{535}_{493} & 0.6870997241914^{870}_{703} \\
-0.0437717490340^{021}_{187} & 0.5823223137250^{922}_{870} & 0.0945507123396^{184}_{67} & 0.4655901373428^{523}_{460} \\
0.2202037613341^{535}_{493} & 0.0945507123396^{184}_{67} & 0.04564570222385^{25}_{19} & 0.17183692386829^{89}_{69} \\
0.6870997241914^{870}_{703} & 0.4655901373428^{523}_{460} & 0.17183692386829^{89}_{69} & 0.6808948460970^{204}_{131}
\end{array}$$

In a second test we inflate the input matrices $A$ and $B$ (all components of $Q$ and $R$ are integer values) in all non-zero components with a diameter of $10^{-4}$ and get an inclusion with maximum relative error of 0.02 and at least one correct digit in each component. If we choose greater diameters, the verification fails.

To show that all symmetric matrices inside the solution $\widehat{X} + [X]$ are positive definite, we compute an interval LU–decomposition of the inclusion and verify that all diagonal elements of the upper triangular matrix $U$ consist only of positive elements. In the case of interval input matrices with diameters $10^{-4}$ this test fails.

## 5. Optimization of the numerical algorithm

In Algorithm 1, we have chosen an inflation matrix $\Delta$ with the same interval element in each component of the matrix. If we start with the described Newton method to get an approximative solution of the ARE, we can also introduce an adaptive matrix $\Delta = (\delta_{ij})$ which we get from the Newton iterates in the following way. If $\widehat{X}^{(i)}$ denotes the sequence of iterates of the Newton process and if the approximative solution we use for our algorithm is denoted by $\widehat{X}^{(n)}$, then we set

$$
\delta_{ij} = \begin{cases} \kappa \cdot [-1,1] \cdot |\widehat{X}_{ij}^{(n)} - \widehat{X}_{ij}^{(n-1)}| & \text{if } \widehat{X}_{ij}^{(n)} - \widehat{X}_{ij}^{(n-1)} \neq 0 \\ \kappa \cdot [-2^{-52}, 2^{-52}] \cdot |\widehat{X}_{ij}^{(n-1)}| & \text{if } \widehat{X}_{ij}^{(n)} - \widehat{X}_{ij}^{(n-1)} = 0 \end{cases}
$$

The constant $\kappa$ depends on the condition of the problem and is determined experimentally.

Another way to speed up the algorithm is to use inner dependencies in the linear system, whose verified solution we must calculate. This means, that not for all components the values can vary independently between their upper and lower bounds but the components of the matrix or the right hand side may depend in some sense. Here we present techniques similar to those in [5].

Because the right hand side matrix in the equation of the form $CX + XD = P$ in (6) is symmetric, we are only interested in symmetric solutions $X$, and because of the special structure of the matrix $(I \otimes C + D^T \otimes I)$ vec $X =$ vec $P$, we find that the equations with right hand side $p_{ij}$ and $p_{ji}$ for $i \neq j$ are equal. Hence, we can reduce the dimension of our linear system from $N^2 \times N^2$ (where $N \times N$ is the dimension of $A$) to $\tilde{N} \times \tilde{N}$ with $\tilde{N} = 0.5 \cdot N \cdot (N+1)$.

Using the above described optimizations, we get better performance and we get one more correct digit for the inclusions of our examples because of the reduced dimension.

## 6. Prospects

An algorithm similar to Algorithm 1 can be used to verify the solution of the discrete–time Riccati equation. This system arises from discrete–time systems of the form

$$
x(t+1) = Ax(t) + Bu(t)
$$

with the cost functional

$$
J(x_0, u) = \sum_{t=0}^{\infty} (x(t)^T Q x(t) + u(t)^T R u(t))
$$

for which a control $u(t)$ minimizing the functional has to be calculated. In the time-invariant case, the control is given by

$$u(t) = Hx(t), \qquad H := -(R + B^T X B)^{-1} B^T X A,$$

where $X$ is the unique symmetric positive semi–definite solution of the algebraic matrix discrete–time Riccati equation

$$P(X) := A^T X A - X - A^T X B (R + B^T X B)^{-1} B^T X A + Q = \Omega.$$

In this case, the determination of the fixed point equation, used for the verification is more complicated because of the matrix $X$ inside the inverse $(R + B^T X B)^{-1}$. We put $\Omega = P(\widehat{X} + X)$ with the approximate solution $\widehat{X}$ and construct the fixed point equation. We develop the inverse in terms with quadratic factors of $X$ and replace terms $X$ by the inflation matrix $\Delta$. This leads to the following equation, which we have substituted for the equation (6) in step c) of Algorithm 1.

$$
\begin{aligned}
\Omega \; = \; & (A + BH)^T X (A + BH) - X \\
+ \; & (A + BH)^T \widehat{X} (A + BH) - \widehat{X} + H^T R H + Q \\
- \; & A^T \Delta Z_\Delta \Delta A + A^T \Delta Z_\Delta \Delta Z \widehat{X} A \\
+ \; & A^T \widehat{X} Z_\Delta \Delta Z \Delta A - A^T \widehat{X} Z_\Delta \Delta Z \Delta Z \widehat{X} A \\
=: \; & (A + BH)^T X (A + BH) - X + G(\widehat{X}, \Delta) \\
N \; := \; & (R + B^T \widehat{X} B)^{-1}, \; N_\Delta := (R + B^T (\widehat{X} + \Delta) B)^{-1} \\
H \; := \; & -(R + B^T \widehat{X} B)^{-1} B^T \widehat{X} A = -N B^T \widehat{X} A \\
BH \; = \; & -Z \widehat{X} A \\
Z_\Delta \; := \; & B N_\Delta B^T, \; Z := B N B^T
\end{aligned}
$$

Then we find the interval equation $X = C X C^T + G$, and applying the Kronecker-product we get vec $X - (C \otimes C)$vec $X =$ vec $G$ and can carry out step d) of the algorithm.

We want to discuss these problems in a more detailed way in a further note.

EXAMPLE 4: As an example, we will consider the discrete–time tubular ammonia reactor model from Sima ([13], p. 258) defined by the matrices:

$$
B = \begin{pmatrix}
0.0004760 & -0.00005701 & -0.0083680 \\
0.0000879 & -0.0004773 & -0.0002730 \\
0.0001482 & -0.001312 & 0.0008876 \\
0.0003892 & -0.003513 & 0.0024800 \\
0.0010340 & -0.009275 & 0.0066800 \\
0.0007203 & -0.006159 & 0.0038340 \\
0.0004454 & -0.003683 & 0.0020290 \\
0.0001971 & -0.001554 & 0.0006937 \\
0.0003773 & -0.003028 & 0.0014690
\end{pmatrix}
$$

$$A = \begin{pmatrix} 0.8701 & 0.1350 & 0.01159 & 5.014\cdot10^{-4} & -0.03722 \\ 0.07665 & 0.8974 & 0.01272 & 5.504\cdot10^{-4} & -0.04016 \\ -0.1272 & 0.3575 & 0.81700 & 0.001455 & -0.10280 \\ -0.3635 & 0.6339 & 0.07491 & 0.796600 & -0.27350 \\ -0.9600 & 1.6459 & -0.12890 & -0.005597 & 0.07142 \\ -0.6644 & 1.1296 & -0.08889 & -0.003854 & 0.08447 \\ -0.4102 & 0.6930 & -0.05471 & -0.002371 & 0.06649 \\ -0.1799 & 0.3017 & -0.02393 & -0.001035 & 0.06059 \\ -0.3451 & 0.5804 & -0.04596 & -0.001989 & 0.10560 \end{pmatrix}$$

$$\begin{pmatrix} 0.0003484 & 0 & 0.004242 & 0.007249 \\ 0.0003743 & 0 & 0.004530 & 0.007499 \\ 0.0009870 & 0 & 0.01185 & 0.01872 \\ 0.002653 & 0 & 0.03172 & 0.04882 \\ 0.007108 & 0 & 0.08452 & 0.1259 \\ 0.01360 & 0 & 0.1443 & 0.1016 \\ 0.01249 & 1.063\cdot10^{-4} & 0.09997 & 0.06967 \\ 0.02216 & 0 & 0.2139 & 0.03554 \\ 0.01986 & 0 & 0.2191 & 0.2152 \end{pmatrix}$$

$Q = 50 I_9,\ R = I_3.$

Rounded to four decimal digits the results obtained are $\widehat{X}/10^3 =$

$$\begin{pmatrix} 0.8698 & 0.0411 & 0.1381 & -0.0192 & -0.0593 & -0.0024 & 0.0000 & -0.0245 & -0.0182 \\ 0.0411 & 1.8658 & 0.2075 & 0.1095 & -0.1125 & 0.0055 & 0.0000 & 0.0583 & 0.0493 \\ 0.1381 & 0.2075 & 0.2281 & 0.0401 & -0.0518 & -0.0001 & 0.0000 & -0.0009 & 0.0005 \\ -0.0192 & 0.1095 & 0.0401 & 0.1380 & -0.0385 & 0.0002 & 0.0000 & 0.0024 & 0.0031 \\ -0.0593 & -0.1125 & -0.0518 & -0.0385 & 0.0783 & 0.0002 & 0.0000 & 0.0017 & 0.0005 \\ -0.0024 & 0.0055 & -0.0001 & 0.0002 & 0.0002 & 0.0501 & 0.0000 & 0.0008 & 0.0005 \\ 0.0000 & 0.0000 & 0.0000 & 0.0000 & 0.0000 & 0.0000 & 0.0500 & 0.0000 & 0.0000 \\ -0.0245 & 0.0583 & -0.0009 & 0.0024 & 0.0017 & 0.0008 & 0.0000 & 0.0581 & 0.0054 \\ -0.0182 & 0.0493 & 0.0005 & 0.0031 & 0.0005 & 0.0005 & 0.0000 & 0.0054 & 0.0547 \end{pmatrix}$$

which we use as an approximative solution for the verification. The smallest $\delta$ which allows a verification with the inflation matrix

$$\Delta = [-\delta, \delta](1, 1, \ldots, 1)^T (1, 1, \ldots, 1)$$

and this starting approximation was $\delta = 0.5$. Iterating the algorithm similar to Algorithm 1 f), we get a final verified solution for this problem with the maximum of interval–diameters in $\widehat{X} + [X]$ of $7.2\cdot10^{-11}$. In the last verification step we used an inflation value of $\delta = 2^{-34}$.

If we assume two more correctly rounded digits in each component of the input (and the following digit uncertain) we can guarantee a relative error of $6\cdot10^{-3}$ in the resulting output.

## 7. Conclusions

We have presented an algorithm for the verification of the solution of an algebraic Riccati equation, arising for example from a time–invariant continuous–time dynamical system or a discrete–time system.

The method uses interval equation solvers and Brouwer's fixed point theorem. So it works with uncertain data as input matrices. But the examples also show that the given significant digits from problems found in literature (most times only 4 significant digits) are not always sufficient to guarantee at least one significant digit for the solution. Our approach allows a statement how accurately input data from technical systems must be measured to guarantee a certain number of places in the output result.

## Acknowledgments

The authors would like to thank the referees for their valuable hints and comments.

## References

1. Alefeld, G. and Meyer, G. The Cholesky method for interval data. *Linear Algebra Appl.*, 194:161–182, 1994.
2. Armstrong, E. S. An extension of Bass' algorithm for stabilizing linear continuous constant systems. *IEEE Trans. Automat. Control*, AC–20(4):153–154, 1975.
3. Bunse–Gerstner, A., Byers, R. and Mehrmann, V. Numerical methods for algebraic Riccati equations. In S. Bittanti, editor, *Lecture Notes of the Workshop on "The Riccati Equation in Control, Systems, and Signals", (Como, Italy)*, pages 107–115. Pitagora Editrice, Bologna, 1989.
4. Freiling, G. and Jank, G. Non–symmetric matrix Riccati equations. *J. Analysis Appl.*, 14:259–284, 1995.
5. Jansson, C. Interval linear systems with symmetric matrices, skew-symmetric matrices and dependencies in right hand side. *Computing*, 46:265–274, 1991.
6. Knüppel, O. Profil/BIAS – A fast interval library. *Computing*, 53:277–287, 1994.
7. Kratz, W. and Stickel, E. Numerical solution of matrix polynomial equations by Newton's method. *IMA J. Numer. Anal.*, 7:355–369, 1987.
8. Luther, W. and Otten, W. The complex arithmetic geometric mean and multiple–precision matrix functions. In Alefeld, G., Frommer, A, and Lang, B., editor, *Scientific Computing and Validated Numerics, Proceedings of SCAN–95*, pages 52–58. Akademie Verlag, Berlin, 1996.
9. Mehrmann, V. and Tan, E. Defect correction methods for the solution of algebraic Riccati equations. *IEEE Trans. Automat. Control*, AC–33(7):695–698, 1988.
10. Potter, J. E. Matrix quadratic solutions. *J. SIAM Appl. Math.*, 14:496–501, 1966.
11. Rump, S. M. Improved iteration schemes for the validation algorithms for dense and sparse nonlinear systems. *Computing*, 57:77–84, 1996.
12. Shayman, M. A. Geometry of the algebraic Riccati equation, part I and part II. *SIAM J. Control and Optimization*, 21:375–394, 395–409, 1983.
13. Sima, V. *Algorithms for linear–quadratic optimization*. Marcel Dekker, New York, 1996.
14. Willems, J. C. Least squares stationary optimal control and the algebraic Riccati equation. *IEEE Trans. Autom. Control*, AC–16:621–634, 1971.
15. Zurmühl, R. and Falk, S. *Matrizen und ihre Anwendungen*. Springer, Berlin, 1984.

T. Csendes (ed.), Developments in Reliable Computing 119–130.
© 1999 Kluwer Academic Publishers.

# Expression Concepts in Scientific Computing

MICHAEL LERCH                                            lerch@informatik.uni-wuerzburg.de
*Lehrstuhl Informatik II, Universität Würzburg, D-97074 Würzburg, Germany*

**Abstract.** Most programming languages apply an eager expression evaluation strategy. In Scientific Computing this is not sufficient in many cases. In this paper we show the design of an expressive, efficient, context sensitive and extendable expression concept which allows the implementation of arbitrary evaluation strategies. We apply this mechanism to the automatic computation of sharp enclosures of the range of an arithmetical expression using the mean value form. This also involves a new implementation of an automatic differentiation facility using C++ and the expression template technique [14]. We show that our solution is significantly more efficient than traditional approaches.

## 1. Introduction

Expressions and assignment are central constructs in procedural programming languages. Unfortunately, for Scientific Computing the semantics of expression evaluation is too weak in most available general purpose languages. Expressions are usually evaluated in an eager manner, i.e., the value of each subexpression is computed immediately when all operand values are known. There are many situations where this simple strategy is not adequate. We mention two examples here.

*Example 1: Dot Product Expressions*

An expression one typically can find in verifying algorithms for linear systems is the computation of a defect interval vector $d = \Diamond(b - A\tilde{x})$ for a Matrix $A \in \mathbb{R}^{n \times n}$ and vectors $b, \tilde{x} \in \mathbb{R}^n$ and $d \in \mathbb{IR}^n$. Assuming that appropriate data types are defined and operators are overloaded correspondingly, one can write, e.g. in C++

```
d = b - A*x;
```

The evaluation of this expression is composed of the following steps. First, the product A*x is computed and assigned to a temporary vector variable t1. Then t1 is subtracted from b and the result is stored in another temporary t2. Finally, t2 is assigned to d.

There are two problems with this procedure. First, the evaluation tends to be inefficient, because it involves the creation of intermediate vector objects and additional loops for assignment. Second — even worse in an verifying context — the occurring dot products are not computed with maximum accuracy, which is crucial for obtaining sharp defect enclosures [8].

Of course, this problem can be solved using a special language construct, e.g. #-expressions in Pascal–XSC [6]:

```
d := ##(b - A*x)
```

In a general purpose language like C++ this behaviour has to be simulated by hand, as the following code using the C–XSC library [4] shows:

```
for (int i=0; i<n; i++) {                                    (1)
    accu = b[i];
    accumulate(accu, -A[i], x);
    d[i] = rnd(accu);
}
```

Unfortunately, the expressiveness of the mathematical notation is completely lost here and a descriptive comment is needed.

### Example 2: Range Computation

Consider the task of computing the range of an arithmetical expression $f$. It is well known that the naive interval evaluation $f(x)$ at $x \in \mathbb{IR}$ can lead to a huge overestimation due to the variable dependence problem. For small intervals this overestimation can be reduced significantly by using a centered form, e.g. the mean value form [11]:

```
DerivType f(DerivType x) {
    return x/(1-x);
}

interval mX = _interval(mid(X));                             (2)
fEval(f, mX, fmX);
dfEval(f, X, fX, dfX);
y = fmX - dfX*(X - mX);
```

The shown code uses the automatic differentiation package of the C–XSC Toolbox [4]. The arithmetical expression to be evaluated must be provided by the user in form of a C++ function. Again, the meaning of the program is not obvious without any comment.

Other applications such as parallel evaluation of expressions or automatic identification of important kernel routines (e.g. BLAS or BIAS) also require some kind of lazy evaluation, which is very popular in functional programming languages.

## 2.  Features for a new Expression Concept

The examples have shown that some global view of expressions is needed together with a opportunity of a delayed evaluation. This means that evaluation takes place only when the complete expression structure is known.

Of course, it is essential to keep the abstract mathematical notation in programs as much as possible. Thus, implementation details should be nearly invisible to the end user. Another important point is the context sensitivity of evaluation semantics.

This means that the evaluation semantics should be controllable dynamically by the user or some program state conditions. The design and the implementation should be extendable by new operator and operand types and user-supplied evaluation strategies.

As usual, efficiency is a very important issue, especially in scientific computing. As a minimal requirement, the overhead of a new expression evaluation mechanism should not reduce efficiency of programs significantly, as compared to conventional evaluation.

Finally, in our opinion, an existing, wide-spread general purpose language should be used as an implementation basis. Ideally, it should be possible, to express the new mechanism completely in this language, without the need for additional key-words, preprocessing or the like. In our approach we will use C++ as the imple-mentation language. As an object–oriented language it meets all requirements to seamlessly embed arithmetic data types in the programming environment and also offers a high level of abstraction and reusability.

## 3.  Expression Data Types and Parse Trees

In this section we describe the general framework for the realization of an expres-sion concept.

Provided that the usual mathematical operator notion should be kept and expres-sions should be evaluated lazily, it is clear that all operators in question need to be overloaded such that they successively build some representation of the expression, instead of actually computing values. This representation can later be used for any evaluation purposes. The natural representation for expressions are parse or syntax trees. The nodes of these trees are operators and the leaves are the operands. For example, the corresponding tree for the expression $x + sin(3y)$ is

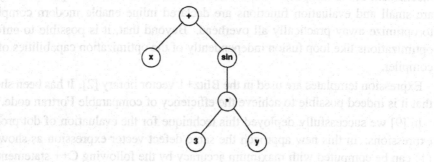

Parse trees are usually implemented as dynamic data structures using pointers or references, respectively. Unfortunately, this approach does not meet the efficiency requirement for two reasons. First, parse trees must be built at runtime before expression evaluation. Even worse, program code using pointers and references is much harder to optimize than code without using aliasing mechanisms. These drawbacks can be eliminated using a relatively new C++ implementation technique called *Expression Templates.*

In C++ a *class template* is a very useful construct to describe parameterized data types. For example, a vector data type which is independent of the element type T can be defined as follows:

```
template <class T>
class Vector {
    T *data;
    // ...
};
```

An *instantiation* provides a class template with actual types to produce concrete classes, i.e.:

```
Vector<Interval> X;
```

Similarly, *function templates* can be defined in C++ with variable function parameter types.

The basic idea with *expression templates* [14, 15] is that expression parse trees can be represented by C++ types in the form of recursively instantiated templates. This means that the tree-like structure of expressions can be mapped to flattened data types. For example, the expression tree in the previous section could be mapped to the following type:

```
Sum<Var<double>, Sin<Prod<Const<double>, Var<double> > > >
```

It is clear, that — in order to produce parse trees — operators need to be overloaded to return template instantiations.

The key to efficiency of this approach is that parse trees are *types* which are constructed at *compile time* using the parsing capabilities and template instantiation mechanism of the compiler. The fact that all introduced runtime expression objects are small and evaluation functions are declared inline enable modern compilers to optimize away practically all overhead. Beyond that, it is possible to enforce optimizations like loop fusion independently of the optimization capabilities of the compiler.

Expression templates are used in the Blitz++ vector library [2]. It has been shown that it is indeed possible to achieve the efficiency of comparable Fortran code.

In [9] we successfully deployed this technique for the evaluation of dot product expressions. In this new approach the same defect vector expression as shown in (1) can be computed with maximum accuracy by the following C++ statements:

```
ExprMode::beginAccurate();
d = b - A*x;
ExprMode::endAccurate();
```

Any legal dot product expression can be computed this way. In fact, this mechanism introduces program parts with accurate evaluation as required e.g. in [13] *without* the need for new keywords. In addition to a high expressiveness, our solution showed a slightly better performance than the C–XSC version and a version in C which was optimized by hand. The accumulator code was identical in all cases, which shows that the achieved improvements were only due to the expression mechanism. Omitting the accurate context statements, a conservative evaluation takes place. In this case our expression template based implementation out-performed the straightforward implementations in C and C++ and showed the same efficiency as a C version which was optimized by hand especially for the used platform.

In the main part of this paper use the expression template technique for a new implementation of an automatic differentiation facility. We the show how to use it to efficiently compute sharp ranges of arithmetical expressions using the well known mean value and introduce a new mechanism for context sensitive expression evaluation.

## 4. Application: Range Computation

Consider the task of computing an enclosure for the range $f^*(x) = \{f(\tilde{x}) \mid \tilde{x} \in x\}$, $x \in I\mathbb{R}^n$ of an arithmetical expression $f : D \subseteq \mathbb{R}^n \longrightarrow \mathbb{R}$. The interval evaluation $f(x)$ of $f$ yields an enclosure $f(x) \subseteq f^*(x)$ with potentially large overestimation. A well-known means of bounding overestimation for small intervals $x$ is the application of centered forms, e.g. the mean value form or slope forms [11]. The mean value form

$$f_m(x) := f(\tilde{z}) + f'(x)(x - \tilde{z}), \qquad \tilde{z} \in x$$

can be computed automatically from an expression for $f$ using means of automatic differentiation [12]. For the first derivatives a differentiation arithmetic is an arithmetic for ordered pairs $U = (u, u')$ where the first component represents the function value and the second the value of the derivative. The rules for this arithmetic can easily be derived from the usual rules of calculus, e.g. for multiplication:

$$U * V = (u * v, u * v' + u' * v)$$

Second derivatives can be computed analogously. The direct implementation of these rules gives the so-called forward mode of automatic differentiation. A typical eager implementation can be found in [4].

The following sections show how to deploy the expression template technique to realize a more expressive and efficient implementation of automatic differentiation. We only consider first derivatives for the one-dimensional case to show the principles. The code can easily be extended to compute gradients and higher order derivatives.

*4.1.* `AutoDiff` *Wrapper Class Template*

The first step is to define a so-called *wrapper class template* [14], which can hold any arbitrary expression. The class is of significant importance to bound the number of needed operator functions (see section 4.3).

```
template <class A>
class ADExpr {
  A a;

public:
  typedef typename A::value_type value_type;

  ADExpr(const A &expr): a(expr) { }

  template <class T>
  value_type eval(const T &x) const {
    return a.eval(x);
  }

  template <class T>
  value_type evalDeriv(const T &x) const {
    return a.evalDeriv(x);
  }
};
```

The class is parameterized by an arbitrary expression type A The constructor accepts an expression object of this type and just stores it by value.

For every expression class template the local type `value_type` gives the type of values the represented expression can take. Obviously, in case of the wrapper class this is the corresponding `value_type` of the wrapped expression type A. Note that in ANSI C++ access to `typedef`'s of template parameters need to be marked with the new keyword `typename` [1].

Because nothing is known about the wrapped expression, the evaluation functions just forward their calls. They return objects of the value type of the expression. The type of the argument at which the expression is to be evaluated is specified as a parameter by means of *member template functions*, a new ANSI C++ construct [1]. This concise formulation allows the arbitrary combination of arithmetical compatible data types. For real applications a method for the simultaneous computation of the function value and the value of the derivatives should be added.

Note that no additional indirection is introduced here since in C++ functions defined in-class are implicitly taken to be inline.

The public part or interface of the wrapper class template also determines the minimal interface of all expression class templates.

## 4.2.   Parse Tree Nodes

The next step is to define a particular class template for every possible kind of parse tree node. In the case of automatic differentiation these are constants, variables, unary and binary operations, standard functions and special nodes for generalized sums, products etc.

For example, the class template for the representation of the identity, i.e. a variable of type T, can be defined as follows:

```
template <class T>
class ADId {
public:
  typedef T value_type;

  ADId(){ }

  template <class T2>
  value_type eval(const T2 &x) const {
    return x;
  }

  template <class T2>
  value_type evalDeriv(const T2 &x) const {
    return 1;
  }
};
```

Provided that a data type Interval is given, interval valued variables can be declared, e.g. using the shortcut

```
typedef ADExpr<ADId<Interval> > ivar;
```

Of course, appropriate preprocessor macros can make such declarations much simpler.

Things get more complicated when dealing with binary operations, for example products. A corresponding class template definition needs to store arbitrary subexpressions in order to allow arbitrary products. Thus, we introduce two type parameters, one for each subexpression:

```
template <class A, class B>
class ADProdExpr {
  A a;
  B b;
public:
  typedef
```

```
    MostCommonType<typename A::value_type,
                    typename B::value_type> traits_type;

  typedef typename traits_type::value_type value_type;

  ADProdExpr(const A &a_expr, const B &b_expr)
    : a(a_expr), b(b_expr) { }

  value_type eval(const value_type &x) const {
    return a.eval(x) * b.eval(x);
  }

  value_type evalDeriv(const value_type &x) const {
    return a.eval(x) * b.evalDeriv(x) +
           a.evalDeriv(x) * b.eval(x);
  }
};
```

A problem rises with the definition of the type value_type. The only sensible solution is to take the mathematically most common type of the value types of A and B, respectively. If the value type of A for example is Interval and the value type of B is double, then Interval should be chosen as the value type of the product expression. Otherwise values of the interval subexpression would be lost. To achieve this we implemented a type propagation mechanism by using a so-called *traits class* [10] MostCommonType. Details are described in [9]. A similar approach can be found in the Blitz++ library [2].

### 4.3.   Operators and Standard functions

Now operator function templates can be defined. Each operator returns an object of the appropriate node type. In principle, for each possible combination of node types a special operator must be defined. The exponentially growing number of operators can be drastically reduced by usage of the wrapper class ADExpr. Consider for example the definition of the multiplication operator for arbitrary expressions:

```
template <class A, class B>
inline
ADExpr<ADProdExpr<ADExpr<A>, ADExpr<B> > >
operator *(const ADExpr<A> &a, const ADExpr<B> &b)
{
    typedef ADProdExpr<ADExpr<A>, ADExpr<B> > ExprT;
    return ADExpr<ExprT>(ExprT(a, b));
}
```

Any two subexpressions can be used as arguments for this operator, as long as they are wrapped in ADExpr. This in turn is ensured by the definition of the result type of the operators. We give an overview of the signatures of all needed operators and standard functions.

*Unary operators* $+$ *and* $-$ *for expression arguments*:

$+ : \texttt{ADExpr<A>} \longrightarrow \texttt{ADExpr<A>}$

$- : \texttt{ADExpr<A>} \longrightarrow \texttt{ADExpr<ADNeg<ADExpr<A>} >$

*Binary operators* $\circ \in \{+, -, *, /\}$ *for expression arguments*:

$\circ : \texttt{ADExpr<A>} \times \texttt{ADExpr<B>} \longrightarrow \texttt{ADExpr<ADOpExpr<ADExpr<A>, ADExpr<B>} > >$

$Op \in \{\texttt{Sum}, \texttt{Diff}, \texttt{Prod}, \texttt{Div}\}$

*Binary operators* $\circ \in \{+, -, *, /\}$ *for mixed scalar/expression arguments*:

$\circ : \texttt{ADExpr<A>} \times \texttt{S} \longrightarrow \texttt{ADExpr<ADOpExpr<ADExpr<A>, ADConst<S>} > >$

$Op \in \{\texttt{Sum}, \texttt{Diff}, \texttt{Prod}, \texttt{Div}\}$

and symmetric versions

*Standard functions*:

$\texttt{sfun} : \texttt{ADExpr<A>} \times \texttt{int} \longrightarrow \texttt{ADExpr<ADSFun<ADExpr<A>} > >$

$SFun \in \{\texttt{Pow}, \texttt{Exp}, \texttt{Sin}, \texttt{Cos}, \ldots\}$

Notice, that only 14 operator functions are needed.

## 4.4. Example: Interval Newton Method

We demonstrate the application of the automatic differentiation facility by computing enclosures for the zero of a function $f$ using the interval Newton method. We define a simple function template `intNewton` which accepts an arithmetical expression for $f$ in form of an expression object of arbitrary type, the search interval X and the relative accuracy eps. The test for applicability of the method is left out here for simplicity.

```
template <class A>
Interval intNewton(const ADExpr<A> &expr,
                   Interval X, double eps) {
  Interval fx, dfx;
  Interval m, fm;
  do {
    m   = Interval(X.mid());
    fm  = expr.eval(m);
    dfx = expr.evalDeriv(X);
    X = X & (m - fm/dfx);
  } while (X.relDiam() > eps);
  return X;
}
```

`intNewton` then can be called by directly passing the function expression as an parameter:

```
Y = intNewton(exp(-3.0*x) - pow(sin(x), 3), X);
```

This procedure is highly efficient because the expression object needs to be constructed only once and the complete evaluation code is expanded inline at the points where `eval` and `evalDeriv` are called.

## 5. Context Sensitive Expression Evaluation

In the last section we have used the methods `eval` and `evalDeriv` to evaluate an expression object and its derivative, respectively. Of course, these computations have exactly the same semantics as traditional eager evaluation. In this section we show how to make evaluation semantics dependent on some context.

We first need a means to manage the current program context state. For this we introduce a class `ExprMode` which provides access to several flags like `accurate`, `useMeanValueForm`, etc.

In order to take advantage of the context state, we define a global function template for the evaluation of expressions:

```
template <class A, class T>
inline
fEval(const ADExpr<A> &expr, T x, T &result)
{
   if (ExprMode::usesMeanValueForm()) {
     T mx = x.mid();
     result = expr.eval(mx) + expr.evalDeriv(x) * (x - mx);
   }
   // ... other forms

   else
     result = expr.eval(x);
}
```

The function takes an expression `expr` and some generic value `t` and computes the value `result` of `expr` at `t` dependent on the flags in `ExprMode`. Function templates for the evaluation of derivatives can be defined analogously. Note that the expression for the computation of the mean value form `expr.eval(mx) + expr.evalDeriv(x) * (x - mx)` corresponds to the more complicated statement sequence shown in code detail (2).

On the user level this setting enables dynamic context switches, e.g.:

```
if (smallEnough(Z))
   ExprMode::beginUseMeanValueForm();
Y = fEval(pow(X, 3) - 3*pow(X, 2) - 2*X + 2, Z)
ExprMode::endUseMeanValueForm();
```

## 6. Experimental Results

All experiments were carried out on a dual Pentium II 333 PC running Linux. We used the commercial C++ compiler KCC [7], currently one of the best optimizing compilers with nearly full ANSI support. The set of test functions to be evaluated contains different kinds of rational functions and polynomials as well as standard functions. The evaluation tests were carried out one million times, the Newton method one thousand times. We show the absolute running times.

|  | C–XSC | | FLLIB | |
|---|---|---|---|---|
| Naive evaluation | 205.7 sec. | | 4.1 sec. | |
| | Forward–AD | ET–AD | Forward–AD | ET–AD |
| Function value | 220.0 sec. | 204.1 sec. | 7.4 sec. | 4.0 sec. |
| Derivative | 296.2 sec. | 276.1 sec. | 12.2 sec. | 8.6 sec. |
| Both | 296.2 sec. | 281.6 sec. | 12.2 sec. | 9.1 sec. |
| Mean value form | 495.7 sec. | 458.7 sec. | 22.4 sec. | 15.6 sec. |
| Newton method | 128.7 sec. | 125.9 sec. | 2.1 sec. | 1.4 sec. |

We compared the automatic differentiation implementation technique in the C–XSC Toolbox (Forward–AD) with our expression template based implementation (ET–AD). The basic definition of interval data types and operations was first taken from the C–XSC library. Since the implementation especially of the standard functions is rather slow in C–XSC, we also show results using FLLIB [5], a fast interval library in ANSI C. The Forward–AD implementation was modified to support the FLLIB interval implementation.

The results show that the Forward–AD technique induces significant overhead when used just for the computation for the function value; in the case of FLLIB it is roughly 80% slower than the naive evaluation. On the other hand the ET–AD based function evaluation is even a little bit faster than naive evaluation. This shows that this technique offers very good optimization possibilities for smart compilers.

With C–XSC the execution speed for the computation of derivatives and mean value forms was only about 5–7% faster using the ET–AD method. The reason for this is that the time for expression administration is almost negligible compared to the time spent in the computational kernels. In contrast, the improvement gained by ET–AD is 25–30% with the faster FLLIB. The same holds for the Newton method.

The C–XSC implementation of the Forward–AD implementation in the C–XSC toolbox offers no possibility to compute the value of the derivative exclusively. This is the reason why the execution times for the derivatives and both the function values and the derivatives are the same. However, it is often the case that only the value of the derivative is needed, e.g. in the Newton method.

The tests were also carried out with the freely available C++ compiler egcs [3]. The overall execution speed was about 10% slower, but the ratio between the Forward–AD and the ET–AD based versions showed to be essentially the same.

A disadvantage of solutions with excessive template usage are long compilations times. A complete rebuild of the ET–AD test suite takes about 6 minutes on our system with KCC, whereas the Forward–AD version compiles in just 11 seconds.

## References

1. ANSI/ISO: *Working Paper for Draft Proposed International Standard for Information Systems– Programming Language C++*, Doc. No. ANSI X3J16/96-0225 ISO WG21/N1043, 1996.
2. Blitz++ Homepage, http://monet.uwaterloo.ca/blitz/
3. egcs Project Homepage, http://egcs.cygnus.com
4. Hammer, R. et al.: C++ *Toolbox for Verified Computing*. Springer, 1995.
5. Hofschuster, W.; Krämer, W.: A Fast Public Domain Interval Library in ANSI C, in: Sydow, A. (ed.), *Proceedings of the 15th IMACS World Congress on Scientific Computation, Modelling and Applied Mathematics*, Vol. 2, 1997, pp. 395–400.
6. Klatte, R. et al.: PASCAL–XSC *Language Reference with Examples*. Springer, 1992.
7. Kuck Associates, Inc., Homepage http://www.kai.com
8. Kulisch, U. and Miranker, W.L.: *Computer Arithmetic in Theory and Practice*, Academic Press, 1981.
9. Lerch, M. and Wolff von Gudenberg, J: Expression Templates for Dot Product Expressions, to appear in *Reliable Computing* 5(1) (1999).
10. Myers, N.C.: Traits: a New and Useful Template Technique. *C++ Report* 7(5) (1995), pp. 32–35.
11. Neumaier, A.: *Interval Methods for Systems of Equations*. Cambridge University Press, 1990.
12. Rall, L.B.: Automatic Differentiation, *LNCS* 120, Springer, 1981.
13. Ullrich, Ch.: Scientific Programming Language Concepts, *ZAMM* 76 S1 (1996), pp. 57–60.
14. Veldhuizen, T.: Expression Templates, *C++ Report* 7(5) (1995), pp. 26–31.
15. Veldhuizen, T. and Ponnambalam, K. : Linear algebra with C++ Template Metaprograms, *Dr. Dobb's Journal of Software Tools* 21(8) (1996), pp. 38–44.

*T. Csendes (ed.), Developments in Reliable Computing* 131–140.
© 1999 *Kluwer Academic Publishers.*

# Performance Evaluation Technique STU and *libavi* Library

RAFAEL SAGULA, TIARAJÚ DIVERIO, AND JOÃO NETTO

{sagula,netto,diverio}@inf.ufrgs.br

*Instututo de Informática and CPGCC/UFRGS, P.O. Box 150640 - 91501-970 Porto Alegre, Brazil*

**Abstract.** In this paper we describe an analysis technique used for performance evaluation of mathematic libraries that manipulate arithmetic intervals. This work is the conclusion of the initial technique presented in SCAN'97 [14]. This Technique called Standard Time Unit (STU) has also been created to solve problems involved in the performance evaluation of libraries that could run in an heterogeneous environment.

**Keywords:** High Accuracy, Performance, Interval, Floating-Point Arithmetic

## 1. Introduction

By 1992, the Mathematics of Computation Group (GMC) of UFRGS started to analyze the quality of numerical results of Calculus. Thus, they developed a study on the limitations of vector processing on Cray Y-MP supercomputers. This study included the identification of not well-conditioned mathematical problems and problems that, when solved on scalar and vector mode, produced different results. They were implemented and executed on both modes. The results were analyzed in order to find and identify the reasons for these differences. This research has been published in the paper: "Errors in vector processing and the *libavi.a* library" [4], by Reliable Computing journal. According to this paper, the reason for these errors was that Cray Y-MP does not use the IEEE 754 standard of floating-point arithmetic. Also, it does not have maximum accuracy operations and it does not carry out the dot product on optimal way.

For this reason, the Mathematics of Computation Group of UFRGS proposed the High Performance with High Accuracy Laboratory [2]. The purpose of this laboratory was to provide an environment where scientific computing problems may be solved with speed, accuracy and reliability, and where the automatic result verification may be carried out by the computer. Therefore, some concepts of Mathematics of Computation and Computational Arithmetic have been incorporated into this system, such as high accuracy arithmetic, interval mathematics, optimal scalar product, automatic verification and methods of inclusion. The Laboratory is composed of the high accuracy arithmetic kernel, the basic interval library and the interval applied modules, for example, interval solution of linear systems of equations [4].

The *libavi.a* is composed of 290 interval routines organized in four modules: the basic module (including the file that contains the real and complex intervals definition and the operations among real intervals); the ci complex intervals module (containing routines for data manipulation of complex interval type); the mvi module (vectors and interval matrices routines); and the aplic module (composite arithmetic operations).

Some tests were developed in order to verify the quality and reliability of the results obtained with the *libavi* /Cray Y-MP. These tests demonstrated the same quality obtained by solving the same problems with Pascal-XSC or C-XSC.

## 2. The need of performance evaluation

For the conclusion of the work developed by GMC/UFRGS with *libavi* (its implementation), it was necessary the evaluation of its performance, as well as the confrontation with other already existent libraries, by means of measuring the performance or analyzing its main features.

However, the performance measure of these libraries requires a previous study of the best technique to be applied in order to prevent common mistakes. One of the difficulties in comparing performances is that in each one the existing libraries use different environments, not only hardware but also software. The *libavi* library, for example, has been developed in the Cray Y-MP Supercomputer using the Cray's Fortran 90 compiler, trying to use to the utmost what the Machine offers. Consequently, programs that use it will probably run faster than those that use libraries such as Profil, Intlib, or Pascal XSC, running on PCs or Workstations, not by the software factor, but by the hardware.

Thus, the development of a standard for measuring the interval software performance requires an impartial analysis and, at the same time, shows the advantages of each software analyzed. Besides this, we must consider the possibility of new tools to be tested in this field, which will fit in the previous method.

### 2.1. Benchmarks

We have studied three types of benchmark, looking for the best way to evaluate the performance of interval libraries. These types are: workload, kernel and application.

Workload - Measures the probability of instruction occurrence. It obtains an average of the instructions' run time . Computers nowadays have a greater number of instructions than the ones used in mixes. In this type of benchmark, we do not use special architectural features. Examples: Flops and Mflops. It is not profitable for this case of study.

Kernel - Evaluates the performance by measuring the time of execution of algorithm that is implemented for this specific goal. Example: Linpack. It has two versions according the size of the matrix.. They may be 100x100 elements or 1000x1000 elements. The second is known as Best Effort, and involves around 10. 000.000 bytes (9766 Kbytes or 9,54 Mbytes - too large to fit in Cache memory). In the case of interval software, the use of benchmarks is presented as the best solution, because the systems are already implemented and usable. However, it must be chosen a method that best represents the items we want to measure. It would be useful to employ a benchmark which is able to measure all the features of the machines in order to perform a numerical analysis. This is a basis for comparing the performance of different interval software.

Application - It is the third way described here for the performance evaluation. First it measures the execution time of all application programs. After that the average run time. is calculated. Example: SPEC. It may be helpful to measure the performance of interval software but, in general, this kind of software is not tax free.

## 3. Performance Evaluation of Interval Libraries

Interval software are numerical libraries that incorporate a high accuracy arithmetic, interval mathematics and optimal dot product. Calculus are carried out using floating-point interval data type and derivatives (matrices, vectors and complex numbers). The computations results are verified automatically by the computers. Studies in the field of the interval arithmetic have been made in order to develop computational tools and interval software that will make possible the use of this arithmetic in computers. Examples of interval software are given in the Table 1.

*Table 1.* Interval software examples.

| Pascal-XSC | PC | Karlsruhe - Germany (Kulisch) | [9] |
|---|---|---|---|
| C-XSC | PC | Karlsruhe - Germany (Kulisch) | [10] |
| Intlib | Sun, PC | Louisiana, USA (Kearfott) | [8] |
| Profil | Sun | Harburg - TUH - Germany (Rump) | [11] |
| libavi.a | Cray Y-MP | Porto Alegre - Brazil (Diverio) UFRGS | [2] |

### 3.1. Desired Benchmark Characteristics

As already mentioned, an interval benchmark must have some special features, such as being relative to system's power on which it is running and showing all the potential of the software tested. Also, it must be portable from one platform to another. The BIAS (Basic Interval Arithmetic Subroutines)'s use is also essential for testing programs that can be written and translated without ambiguities, besides promoting the use of this standard in other applications.

### 3.2. First Evaluation

The tests developed by Prof. Corliss [1] in several packages have been studied for this research. He assumed that the processing speed of the various machines were similar and, consequently, the libraries performance cannot be influenced by this factor. The execution time of the packages was measured and presented with no adjustments. The whole benchmark consists of five little programs (called exercises) that use specific routines of the packages. The goal is to show the behavior of each package upon the various aspects of interval computation. For this, both result accuracy and execution time are measured. The five programs are divided in the following way:

- Test 1: Arithmetic operations +, -, * and /.

- Test 2: Elementary functions

- Test 3: Matrix and vector operations

- Test 4: Interval Newton's methods with one variable.

- Test 5: Global optimization method with one variable

These programs were implemented by using the *libavi.a* library (in Fortran 90) and the execution time of each one was measured. The obvious consequence is that these times were the lowest ones among the other libraries. This good result is due to the high performance of the hardware, which has a peak of 660 MFLOPS as can be seen in Table 2.

*Table 2.* Time values of some interval libraries (in seconds). Corliss work.

| Package | Test 1 | Test 2 | Test 3 | Test 3a | Test 4 | Test 5 |
|---|---|---|---|---|---|---|
| Clemmeson PC, MS Fort 7.0 | 19.99 | - | - | 12.58 | - | - |
| INTLIB PC, MS Fort 7.0 | 89.86 | 89.86 | - | 13.73 | 142.43 | 288.8 |
| INTLIB Sparc 1+, f77 | 35.58 | 74.26 | - | 11.65 | 110.71 | 221.71 |
| C-XSC PC, Borl C++ 3.1 | 52.34 | 56.85 | 78.64 | 25.65 | 27.95 | 50.15 |
| Pascal-XSC PC, Borl C++ 3.1 Win | 40.65 | 74.26 | 106.8 | 32.9 | 50.7 | 81.74 |
| Pascal/BIAS SparcServer 330, gcc | 32 | 341.8 | 61.3 | 24.5 | 122.7 | 178.1 |
| Profil/BIAS SparcServer 330, gcc | 8.5 | 12.2 | 27.8 | 1.4 | 5.5 | 5.4 |
| Libavi, F90, Cray Y-MP2E/232 | 4.57 | 0.16 | 11.72 | - | 0.22 | - |

Corliss' work is completely based on the assumption that the machines used have similar performance. To verify this assumption, he used a PC 486/50MHz and a SUN 4 330 Sparc Server because of their similar execution performance. However, it does not reflect the heterogeneous workstation network, which can be more assorted. Thus, this assumption becomes false, and it is not possible to make any comparison among packages if the benchmark's time unit is in seconds because the elapsed time of execution of a library could be distorted by the environment. The goal became the search of a measure that would be independent of the processing speed by the standardization of a performance's metric.

This problem can be solved if it is possible to establish a relation among the machines used based on their performance. The idea is to use a known and very disseminated benchmark that can establish a ranking between the interval software from their performances on heterogeneous environments. Linpack is a good alternative to the common workload, as it comprehends all existent machines. Besides, it is a scientific (numerical) benchmark. There are two versions of Linpack benchmark. One uses 100x100 matrices and the other ones 1000x1000. The performance results presented by the first one does not consider some architectural aspects, such as the cache. Thus, the use of the other version is more suitable for measuring the whole system performance.

## 4. Technique STU - Standard Time Unit

Nowadays, there are different technologies and architectures used for making processors. Therefore, the performance difference among computers is not due only to the clock speed, but also due to a combination of several factors such as memory, cache and even the interconnection network, in the case of multiprocessor machines. In this way, a performance evaluation of any software that run on these different environments must be done independently of all these factors.

Thus, the comparison of libraries that run on different environments is a difficult task. To solve this problem, the Linpack results (Table 3) can be used to normalize the time values of the libraries (Table 4).

*Table 3.* Linpack result for the three machines.

| Machine | PC 486/50MHz | SUN 4/330 | Cray Y-MP |
|---------|--------------|-----------|-----------|
| MFLOPS  | 2            | 2,5       | 604       |

The technique consists of using a second benchmark in order to normalize the results obtained by an initial benchmark test suite [1]. In this paper the benchmark used for the normalization of the results is Linpack (calculated with best effort for one matrix 1000x1000) [6] but any other relevant benchmark could also be used.

The final results obtained through this technique present more effectively the real performance achieved by the initial benchmark packages. Given the execution time of the five original programs-test and the Mflops measured by Linpack benchmark, it is easy to calculate the Standard Time Unit (STU) for $10^6$ floating-point operations.

$$STU = \frac{10^6 op}{L}$$

Thus, the normalized values ($N[i]$, where $1 \leq i \leq 5$) can be calculated using the following expression:

$N[i] = \frac{T[i]}{STU}$, where $1 \leq i \leq 5$

which is equivalent to

$N[i] = \frac{T[i]}{\left(\frac{10^6 *op}{L}\right)} = \frac{L*T[i]}{10^6 op}$, where $1 \leq i \leq 5$.

STU and T are measured in seconds and L in FLOPS, but if MFLOPS is assumed the final expression would be:

$N[i] = \frac{L*T[i]}{op}$ , where $1 \leq i \leq 5$.

The results are presented in Table 4.

*Table 4.* Benchmark's Normalized Values (using STU)

| Package | Test 1 | Test 2 | Test 3 | Test 3a | Test 4 | Test 5 |
|---|---|---|---|---|---|---|
| Clemmeson PC, MS Fort 7.0 | 39.98 | - | - | 25.16 | - | - |
| INTLIB PC, MS Fort 7.0 | 179.72 | 179.72 | - | 24.46 | 248.96 | 577.6 |
| INTLIB Sparc 1+, f77 | 88.95 | 187.4 | - | 29.125 | 276.775 | 554.275 |
| C-XSC PC, Borl C++ 3.1 | 104.68 | 113.7 | 157.28 | 51.3 | 55.9 | 100.3 |
| Pascal-XSC PC, Borl C++ 3.1 Win | 81.3 | 148.52 | 213.54 | 65.8 | 101.4 | 163.48 |
| Pascal-XSC SparcServer 330, gcc | 80 | 854.5 | 153.25 | 61.25 | 306.75 | 445.25 |
| Profil/BIAS SparcServer 330, gcc | 21.25 | 30.5 | 69.5 | 3.5 | 13.75 | 13.5 |
| Libavi, F90, Cray Y-MP2E/232 | 2760.3 | 96.64 | 6903.7 | - | 132.8 | - |

By using this method, the *libavi* , which was the best library in all tests, becomes the worst one. It is because it does not take advantage of all the power of Cray supercomputer. So, libraries with this characteristic are misjudged.

## 5. Comparison

Several libraries and languages have been tested (including Intlib, Profil, Pascal-XSC, C-XSC and *libavi* ). Table 2 contains the original results of Prof. Corliss' benchmarks showing that *libavi* is the best library in all tests. But the truth is recovered on Table 4, where we have applied the proposed technique to produce normalized values.

At the end of this paper, we present three graphics showing the changes in the ranking order between both benchmarks. We can observe that Profil has the best time and the best performance. In tests 1 and 3, *libavi* changes from the best to the worst place. It happens because interval arithmetic operations and matrix/vector operations were carried out by software and it takes too much time for execution.

In test 4, which implements interval Newton methods, *libavi* changes, but this change is not so significant. We may observe that it has a good performance. On the other hand, we may observe that, in this test, several order inversions happened in relation to Corliss' work.

## 6. Conclusions

This article has presented the final remarks about *libavi* library performance evaluation. It has also described an analysis technique used for performance evaluation of mathematics libraries that manipulate arithmetic intervals. This work is the final conclusion of the initial technique presented in SCAN'97 [14]. This Technique called Standard Time Unit (STU) has also been created to solve problems involved in the performance evaluation of libraries that could run in an heterogeneous environment even with different technologies.

The *libavi* library has been developed with the purpose of exploring the Cray high performance with the use of Interval Mathematics. The results obtained are contained in intervals, which produce a certain reliability. Thus, we will finally have the high accuracy and high performance arithmetic available to the users of Supercomputer Center of UFRGS. Through this work, the interval mathematics has been taken to the field of practice for Cray users and its use has become possible in the solution of numerical problems, especially in the case of problems that are not stable, in which the solution is totally incorrect.

It has been developed in Fortran 90 because, this way it can be ported to other environments. It has not achieved the desired performance, but it made possible to obtain the high accuracy. It was the first interval library written in Fortran 90 (1994/95) and gave us know-how in the development of interval software with high performance. It can be used to solve problems that require high performance, such as the ones from engineering, physics, mathematics and computer science with high accuracy.

## References

1. CORLISS, G. F. Comparing Software Packages for Interval Arithmetic. Preprint presented at SCAN'93, Vienna, 1993.
2. DIVERIO, T. A. Uso efetivo da matemática intervalar em supercomputadores vetoriais. Porto Alegre: CPGCC da UFRGS, 1995. 291p. Tese de doutorado.
3. DIVERIO, T. A. et al. LIBAVI.A Biblioteca de Rotinas intervalares - Manual de utilizacao. Porto Alegre: CPGCC da UFRGS, 1995. 350p
4. DIVERIO, T. A.; FERNANDES, U. A. L; CLAUDIO, D. M. Errors in vector processing and the libavi.a library. Reliable Computing, Moscow, v.2, n.2, p.103-110, 1996 (Proceedings of

SCAN-95, IMACS Annals on Computing and Applied Mathematics, Wuppertal, Set.26-29, 1996).

5. DIVERIO, T. A.; NAVAUX, P. O. A.; CLAUDIO, D. HLBIG, C. A.; SAGULA, R. L; FER-NANDES, U. A. L; High Performance with High Accuracy Laboratory. Revista de Informática Terica e Aplicada, RITA, v.3, n.2, p.35-54. Porto Alegre: Informática da UFRGS 1997.

6. DONGARRA, J. J.: Performance of Various Computers Using Standard Linear Equations Software. http://www.netlib.org/benchmark/performance.ps

7. HAMMER, R. et al.: Numerical Toolbox fox Verified Computing I: basic numerical problems. Berlin, Springer-Verlag, 1993. 337p.

8. KEARFOTT, R. B.; DAWANDE, M.; DU, K.; HU, C. Algorithm 737 - INTLIB: A portable Fortran 77 interval standard-function library. ACM Transactions on Mathematical Software, Vol.20, N.4, December 1994, p.447-459.

9. KLATTE, R; KULISCH, U; NEAGA, M; RATZ, D; ULLRICH, C: - PASCAL-XSC language reference with examples. Berlin, Springer-Verlag, 1993.

10. KLATTE, R; KULISCH, U; WIETHOFF, A; LAWO, C; RAUCH, M: C-XSC a C++ class library for extended scientific computing. Berlin, Springer-Verlag, 1993.

11. KNÜPPEL, O. PROFIL/BIAS - A Fast Interval Library. Computing 53, 1994. P.277-287.

12. KULISCH, U; MIRANKER, K.L. Computer arithmetic in theory and practice, 1981.

13. SAGULA, R. L.; DIVERIO, T. A. Interval Software Performance. In: INTERNATIONAL SYMPOSIUM ON SCIENTIFIC COMPUTING, COMPUTER ARITHMETIC AND VALIDATED NUMERICS, SCAN 97. Lyon, Franca, 10-12, set., 1997. p.III.1-3.

14. SAGULA, R. L. Avaliacao de Desempenho de Bibliotecas Intervalares. Porto Alegre; II da UFRGS, 1997. Trabalho de Diplomacao. 52f.

## Appendix I

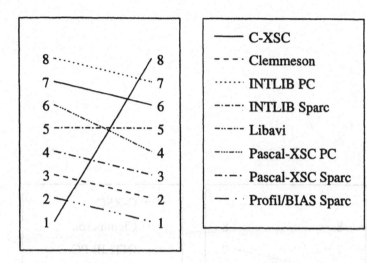

*Figure I.1.* Test 1 ranking crossover

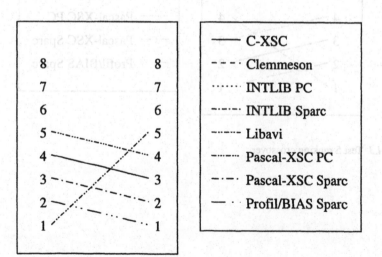

*Figure I.2.* Test 3 ranking crossover

140

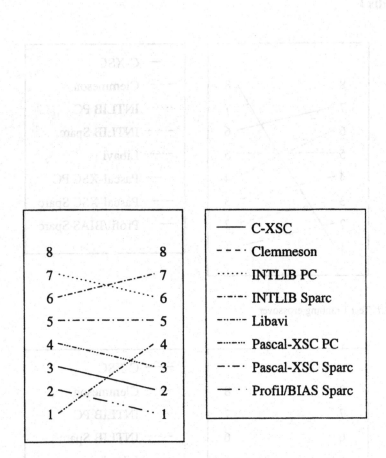

*Figure I.3.* Test 5 ranking crossover

*T. Csendes (ed.), Developments in Reliable Computing* 141–148.
© 1999 *Kluwer Academic Publishers.*

# Single-Number Interval I/O

MICHAEL SCHULTE AND VITALY ZELOV          mschulte@eecs.lehigh.edu
*EECS Dept., Lehigh University, Bethlehem, PA 18015, USA*

G. WILLIAM WALSTER AND DMITRI CHIRIAEV          bill.walster@eng.sun.com
*Sun Microsystems, 901 San Antonio Road, MS UMPK 16-304, Palo Alto, California 94303, USA*

**Abstract.** This paper gives an overview of single-number interval I/O, which provides a convenient method for inputting and outputting intervals. With this technique, each interval is normally represented externally as a single decimal number. Single-number interval I/O has been implemented in interval-enhanced versions of GNU's and Sun's Fortran compilers, an interval calculator, and a Java(TM)[1] application program interface. Complete details on single-number interval I/O in Fortran are given in the Fortran 77 and Fortran 90 Interval Arithmetic Specifications.

**Keywords:** Fortran, interval arithmetic, input, output, conversion, format, compiler.

## 1. Introduction

As noted in [1] and [2], a simple and intuitive method for inputting and outputting interval data is needed to help interval arithmetic gain wide-spread acceptance. It is inconvenient for users to be required to give explicit lower and upper bounds when inputting a number whose exact value is unknown. For example, with conventional interval input, to specify that the number 8.353 is known to four decimal digits requires [8.352, 8.354] to be entered. Providing both interval endpoints takes time and can easily lead to input errors.

With conventional interval output, it is often difficult to determine the relative sharpness of interval results, as is illustrated below.

```
[ 0.14385979,  0.14386012]
[ 3.87564567,  3.87566567]
[12.43123458, 12.43123526]
```

This is especially true when large amounts of interval data need to be examined. Conventional interval I/O is also not convenient for various applications, such as interval spreadsheets or interval calculators, where users would prefer to enter and read each interval as a single decimal number.

Single-number interval I/O helps overcome some of the limitations of conventional interval I/O, by allowing each interval to be input or output using a single decimal number. Because of its usefulness, single-number interval I/O has been included in the Fortran 77 and Fortran 90 Interval Arithmetic Specifications, which provide guidelines for supporting interval data types in Fortran [3], [4]. It has also been implemented in interval-enhanced versions of GNU's [5] and Sun's Fortran compilers, an interval calculator [6], and a Java(TM) application program interface [7]. Rather than replacing conventional interval I/O, single-number interval I/O provides additional flexibility for inputting and outputting interval data.

This paper gives an overview of single-number interval I/O. Section 2 and Section 3 explain the specifications for single-number interval input and output, respectively. Section 4 discusses various implementations of single-number interval I/O. Section 5 presents our conclusions. Detailed specifications for single-number interval I/O in Fortran are given in [3] and [4].

## 2. Single-Number Interval Input

With single-number interval input, the input field can have one of three basic formats: $[x_d]$, $x_d$, or $[x_{d_1}, x_{d_2}]$, where $x_d$, $x_{d_1}$, and $x_{d_2}$ are decimal numbers in either fixed point or scientific notation. These three formats provide flexibility when inputting interval data.

The $[x_d]$ format allows degenerate intervals to be input using a single decimal number. If the input field has the form $[x_d]$, then the internal interval must contain the degenerate interval $[x_d, x_d]$. Examples of inputing intervals using the $[x_d]$ format are given below.

| Input Field | Internal Interval Contains |
|---|---|
| [11.] | [11, 11] |
| [3.24E-3] | [3.24E-3, 3.24E-3] |
| [-22.7] | [-22.7, -22.7] |

The $x_d$ format allows non-degenerate intervals to be input using a single decimal number. The endpoints of the non-degenerate interval are determined by adding and subtracting one unit in the last digit (uld) from $x_d$. Thus, if the input field has the form $x_d$, then the internal interval must contain $x_d + [-1, +1]_{uld}$. Examples of inputing intervals using the $x_d$ format are given below.

| Input Field | Internal Interval Contains |
|---|---|
| 3.1415 | [3.1414, 3.1416] |
| -8.45E-12 | [-8.46E-12, -8.44E-12] |
| 22. | [21, 23] |
| 12.00000 | [11.99999, 12.00001] |

The $[x_{d_1}, x_{d_2}]$ format allows both degenerate and non-degenerate intervals to be input using the conventional format. If the input field has the form $[x_{d_1}, x_{d_2}]$, then the internal interval must contain $[x_{d_1}, x_{d_2}]$. Examples of inputing intervals using the $[x_{d_1}, x_{d_2}]$ format are given below.

| Input Field | Internal Interval Contains |
|---|---|
| [11., 12.] | [11, 12] |
| [3.24E-3, 3.25E-3] | [3.24E-3, 3.25E-3] |
| [22.7, 22.7] | [22.7, 22.7] |

For all three interval formats, the internal interval must contain the interval represented by the input field, regardless of the number of digits in the input field or the precision of the internal representation. If either endpoint is not exactly representable, the left endpoint is rounded downward toward minus infinity and the right endpoint is rounded upward toward positive infinity.

Single-number interval I/O is also designed to handle empty intervals (i.e., intervals that contain no real numbers), and intervals with infinite endpoints. Empty intervals are input and output as [EMPTY], where EMPTY is not case sensitive and may be preceded or followed by blanks. On IEEE 754 compliant processors [8], EMPTY intervals are represented internally by using quiet not-a-numbers (NaNs) for each interval endpoint. Infinite endpoints are input and output using INF, where INF is not case sensitive and is prefixed with a minus or an optional plus sign. Examples of inputting empty intervals and intervals with infinite endpoints are given below.

```
   Input Field              Internal Interval Contains
    [EMPTY]                          [NaN, NaN]
     INF                           [-INF, +INF]
  [-INF, -3]                        [-INF, -3]
  [3, +INF]                         [3, +INF]
```

## 3.  Single-Number Interval Output

With single-number interval output, the output field can have one of three basic formats: $[x_d]$, $x_d$, or $[x_{d_1}, x_{d_2}]$, where $x_d$, $x_{d_1}$, and $x_{d_2}$ are decimal numbers in either fixed point or scientific notation. In the current Fortran implementations of single-number interval I/O, the user has the ability to specify the total output field width ($w$), the number of positions allocated for significant digits ($d$), and (optionally) the number of exponent digits ($e$).

Degenerate internal intervals are typically output using the $[x_d]$ format, where $x_d$ corresponds to the value of both interval endpoints. The decimal point is located in position $p = e + d + 4$ from the right. Examples of the $[x_d]$ output format are given below, assuming $w = 19$, $d = 9$, $e = 3$, and $p = 16$.

```
    Internal Interval              Output Field
                               1234567890123456789
  [0.125, 0.125]               [ 0.125000000E+000]
  [3.126E+12, 3.126E+12]       [ 0.312600000E+013]
  [-33, -33]                   [-0.330000000E+002]
```

Non-degenerate intervals are typical output using the $x_d$ format, where $x_d$ is generated such that $x_d + [-1, 1]_{uld}$ contains the internal interval. The $x_d$ format is also used to output degenerate intervals that have too many significant digits to be output using the $[x_d]$ format. As in the case of the $[x_d]$ format, the decimal point is located in position $p = e + d + 4$ from the right. Examples of the $x_d$ output format are given below, assuming $w = 19$, $d = 9$, $e = 3$, and $p = 16$. As illustrated by these examples, the number of mantissa digits in the output field indicates the relative sharpness of the interval result. In the last example, the number of significant digits is too large to allow the degenerate interval to be output using the $[x_d]$ format.

```
         Internal Interval                Output Field
                                     1234567890123456789
     [2.62890625, 2.6328125]            0.263        E+001
     [-4.298E+12, -4.296E+12]          -0.4297       E+013
     [18., 23.]                         0.2          E+002
     [1234567892., 1234567892.]         0.123456789E+010
```

Sometimes when performing single-number interval output, the $[x_d]$ and $x_d$ single number formats result in unacceptably wide intervals. In cases where the internal interval contains zero or infinity or its single-number representation contains zero, the interval is output using the $[x_{d_1}, x_{d_2}]$ format. In this case, the number of significant digits used for output can either be set to one or selected to maximize the number of significant digits. The number of significant digits output is a quality of implementation opportunity. Examples of the $[x_{d_1}, x_{d_2}]$ output format are given below, assuming $w = 21$ and $d = 11$. In these examples, the maximum number of significant digits are displayed.

```
         Internal Interval                Output Field
                                     123456789012345678901
         [-2, 6]                      [-2.        , 6.        ]
         [1.25, 8.25]                 [ 1.25      , 8.25      ]
         [0.1E+300, INF]              [ 0.1E+300, INF         ]
```

## 4.   Implementations of Single-Number Interval I/O

Single-number interval I/O was first implemented in the University of Minnesota M77 Fortran Compiler [9]. More recently, single number interval I/O has been implemented in interval-enhanced versions of GNU's and Sun's Fortran compilers, an interval calculator, and a JAVA(TM) application program interface. A preliminary version of the interval-enhanced GNU Fortran compiler is available from http://www.eecs.lehigh.edu/~mschulte/compiler/code

### 4.1.   Interval-Enhanced Fortran Compilers

GNU's and Sun's Fortran 77 interval-enhanced compilers support interval I/O according to the Fortran 77 Interval Arithmetic Specification [3]. To support interval I/O, this specification provides four new edit descriptors: VF, VE, VG, and Y. When Fortran's conventional F, E, and G edit descriptors are applied to intervals, they have the same behavior as the Y edit descriptor.

When used to output interval data, the VF, VE, and VG, edit descriptors always output intervals in the $[x_{d_1}, x_{d_2}]$ format, using the Fortran's F, E, and G edit descriptors, respectively, for the interval endpoints. When used to input interval data, the VF, VE, and VG edit descriptors can accept data in any of the three formats supported by single-number interval input.

The Y edit descriptor supports formatted single-number interval I/O. The use of the letter Y was originally suggested by Vladimir Kharchenko, because it serves as a visual indicator of the relationship of the components of an interval value to the single-number used to represent that value, as is illustrate below.

*inf sup*

**Y**

*single-number*
*representation*

When used to output interval data, the Y edit descriptor chooses the appropriate single-number output format (i.e., $[x_d]$, $x_d$, or $[x_{d_1}, x_{d_2}]$) based on the value of the internal interval. When used to input interval data, the Y edit descriptors can accept data in any of the three formats supported by single-number interval input.

The general form of the Y edit descriptor is $Yw.d[Ee]$, where $w$ is the total field width, $d$ is the number of places allocated for displaying significant digits, and $e$ (if present) is the number of digits displayed in the exponent part of the output field. The value for $e$ is optional and defaults to three. When used for output, $w$ should be large enough to account for the opening bracket, the sign for the number, at least one digit to the left of the decimal point, the decimal point, $d$ digits to the right of the decimal point, the exponent indicator, the sign of the exponent, $e$ digits for the exponent, and the closing bracket. Therefore $w$ should be selected such that $w \geq d + e + 7$ if $e$ is present, or $w \geq d + 10$ if $e$ is not present. Examples of Fortran statements that use single-number I/O are shown below.

```
        READ(*, '(Y17.6E4)'), X
        WRITE(*, '(Y22.10)'), X
   10 FORMAT(Y20.10E3)
   20 FORMAT(Y18.7)
```

For single-number interval output, if $e$ is present, then the output field automatically has the form prescribed by Fortran's E edit descriptor. If $e$ is not present, then the compiler must chose whether Fortran's E or F style formatting should be used. The format chosen is the one that can display the greater number of digits in the interval output list. If the number of digits displayed using the E and F formats is the same, the F format is used. Examples of this are shown below for the format Y20.10.

```
        Internal Interval                 Output Field
                                    12345678901234567890
        [2.6289, 2.6328]              2.63
        [1234, 1236]                  0.1235       E+004
        [21, 21]                    [21.0000000000      ]
        [1.2E12, 1.2E12]            [ 0.12           E+013]
```

In addition to supporting VF, VE, VG, and Y interval edit descriptors, Sun's Fortran 90 interval-enhanced compiler provides support for VEN and VES interval edit descriptors [4]. These descriptors are identical to the VE edit descriptor, except that the VES descriptor uses scientific notation (i.e., one significant digit to the left of the decimal point) and the VEN descriptor uses engineering notation (i.e., between one and three significant digits to the left of the decimal point) to output the interval endpoints. The VEN and VES interval edit descriptors are based on the EN and ES real edit descriptors in Fortran 90.

## 4.2. An Interval Calculator

A interval calculator, which is written in Java(TM), has been developed to help familiarize users with single-number I/O and to facilitate simple testing of interval enclosures of mathematical functions [6]. The interval calculator accepts data in any of the three interval input formats described in Section 2. Interval results are displayed in four windows: one window for single-number interval output, and three windows for conventional interval output in decimal, hexadecimal, and octal.

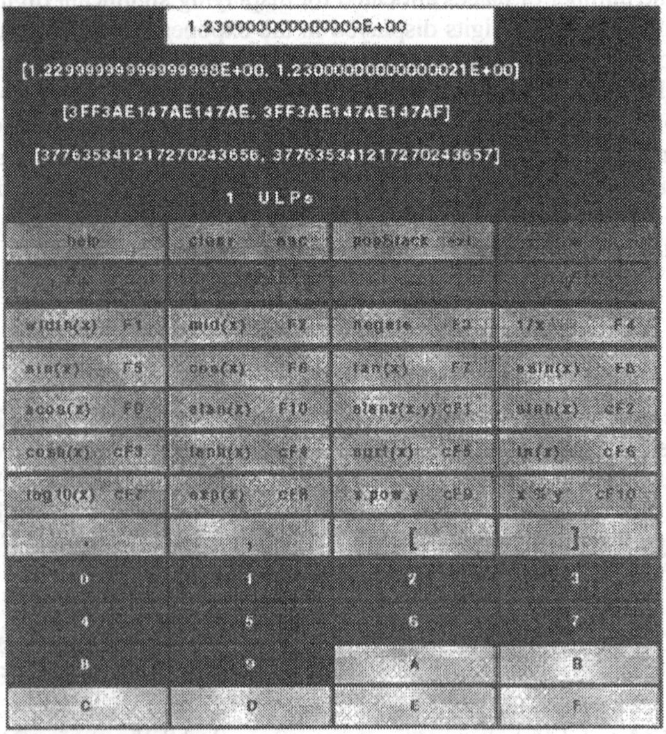

*Figure 1.* Interval Calculator

Figure 1 shows a snapshot of the interval calculator, after the degenerate interval [1.23] has been entered. Since 1.23 is not representable as a floating point number, the interval endpoints are outward rounded. The top window of the calculator indicates that the interval's internal representation is not degenerate (since the number is not enclosed by brackets), and that the lower and upper interval endpoints are very close (since the output is displayed to 16 decimal digits). The next three windows show the interval output in conventional format as decimal, hexadecimal, and octal numbers. As indicated by the fifth window, the width of the internal interval is one unit in the last place (ULP).

In addition to supporting standard and single-number interval I/O the interval calculator also allows the user to perform basic interval arithmetic operations, compute interval enclosures of mathematical functions, and determine the midpoint or width of an interval. An online help menu is provided to facilitate the use of the interval calculator.

### 4.3. Quality of Implementation Opportunities

To allow a wide range of implementations, the only quality requirement for interval I/O in the Interval Arithmetic Specifications is containment. On input the internal interval generated must contain the interval represented in the input field. On output the interval represented in the output field must contain the internal interval that is being output. To provide high-quality interval I/O, the interval-enhanced GNU and Sun Fortran compilers and the interval calculator always produce minimum intervals with containment guaranteed.

When performing single-number interval output, there may be two output representations that are of minimum width and guarantee containment. When this happens, the output interval used is the one whose midpoint is closest to the midpoint of the internal interval. If both output representations' midpoints are the same distance from the midpoint of the internal interval, then by convention the output whose least significant digit is even is used. Examples of this are given below for the Y20.10 edit descriptor.

```
      Internal Interval           Output Field
                                  12345678901234567890
      [2.63, 2.68]                2.7
      [1234, 1235]                0.1234          E+004
      [1.22E12, 1.28E12]          0.12            E+013
```

### 4.4. Internal Base Conversion

Single-number input followed by single-number output can appear to suggest that a decimal digit of accuracy has been lost. This is because radix conversion can cause an increase in the width of the stored input interval. For example, in the interval calculator, the input interval 1.33 is stored internally as approximately

[1.31999999999999984, 1.34000000000000008], due to decimal to binary conversion. When this interval is output in $x_d$ format, the result is 1.3. Although this effect is not desirable, it has to be allowed to prevent a containment violation. To echo single-number input intervals, character I/O and an internal read can be used.

## 5. Conclusions

Single-number interval I/O provides a intuitive method for inputting and outputting intervals. The number of digits and the absence or presence of enclosing brackets indicate the relative sharpness of the interval. Single-number interval I/O has been implemented in interval-enhanced Fortran compilers, an interval calculator, and a Java(TM) application program interface. It can be easily incorporated into compilers for other languages and other interval software tools.

## Acknowledgments

This research is supported by a grant from Sun Microsystems, Inc.

## Notes

1. Sun Microsystems and Java are trademarks or registered trademarks of Sun Microsystems, Inc. in the United States and other countries.

## References

1. G. W. Walster, "Philosophy and Practicalities of Interval Arithmetic," in *Reliability in Computing: The Role of Interval Methods in Scientific Computing*, (Ramon E. Moore, Ed.), Academic Press, pp. 309-323, 1988.
2. G. W. Walster, "Stimulating Support for Interval Arithmetic," in *Applications of Interval Computations*, (R. B. Kearfott and V. Kreinovich Eds.), Kluwer Academic Publishers, pp. 405-416, 1996.
3. D. Chiriaev and G. W. Walster, "Fortran 77 Interval Arithmetic Specification," 1998. Available at http://www.mscs.mu.edu/~globsol/Papers/spec.ps.
4. D. Chiriaev and G. W. Walster, "Fortran 90 Interval Arithmetic Specification," 1999. Available at http://www.mscs.mu.edu/~globsol/Papers/spec90.ps.
5. M. J. Schulte, A. Akkas, V. Zelov, and J. C. Burley, "Adding Interval Support to the GNU Fortran Compiler," accepted for publication in *Proceedings of the International Symposium on Scientific Computing, Computer Arithmetic, and Validated Numerics*, Budapest, Hungary, September, 1998.
6. V. Zelov, G. W. Walster, D. Chiriaev, and M. J. Schulte, "Java(TM) Programming Environment for Accurate Numerical Computing," presented at the *International Workshop on Modern Software Tools for Scientific Computing*, Oslo, Norway, September, 1998.
7. D. Chiriaev, "Framework for Reliable Computing in Java(TM) Programming Language", Manuscript, 1998.
8. ANSI/IEEE 754-1985 Standard for Binary Floating-Point Arithmetic, Institute of Electrical and Electronics Engineers, New York, 1985.
9. "M77 Reference Manual : Minnesota FORTRAN 1977 Standards Version," 1st Edition, University Computer Center, 227 Experimental Engineering, University of Minnesota, 208 Union Street SE, Minneapolis, Minnesota, 55455, 1983.

T. Csendes (ed.), Developments in Reliable Computing 149–158.
149

# Interval Analysis for Embedded Systems

KLAUS MUSCH AND GÜNTER SCHUMACHER

{klaus.musch, guenter.schumacher}@math.uni-karlsruhe.de

*Institut für Angewandte Mathematik, Universität Karlsruhe, Germany*

**Abstract.** An application of interval arithmetic to software testing is described, which has its main importance for embedded systems, in particular safety critical systems. Interval arithmetic allows the full range of input data to be tested and derives predictions about possible variable range violation at runtime. Furthermore, due to a combination with an automatic differentiation-like calculus, it is possible to determine for each variable a sufficient number of digits which shall kept during calculation in order to prevent any serious cancellation.

**Keywords:** interval arithmetic, software testing, ranges of values, condition analysis

## 1. Introduction

When considering software quality and reliability, the numerical behaviour of the used algorithms is an important and currently weakly treated issue since there may be a significant difference between the formal specification (defined for real numbers) and the practical behaviour (when using finite number representations). Failures arising from finite representation would not be detected by any kind of formal verification since these methods operate at the specification level. On the other hand, traditional software test procedures are restricted to code coverage and statistical samples following a specific test strategy. Therefore, software may still fail for a certain set of data not covered by these tests.

During the European Commission funded projects ANTI-CRASH and SAFE a new approach has been established that provides the developer with predictions over the whole range of allowed data. This approach is based on interval arithmetic and has strong relations to algorithms used in global optimisation although the goals are quite different. The approach provides a special splitting strategy for the allowed data ranges which is driven by condition numbers introduced by Richman in 1972 [8]. In this way, the test algorithm successively identifies critical ranges of data sets as well as numerical deficiencies. On the other hand and more welcome, it may also guarantee that such data doesn't exist.

## 2. Mathematical Preliminaries and Notations

Throughout this paper we will consider real functions $f$ over an interval domain $X \in \mathbf{IR}^n$ with values in $\mathbf{R}^m$. $f$ will be assumed to be at least continuous. Moreover, we will assume that for each $x \in X$ the value $f(x)$ can be determined by a finite number of calculation steps using operations from a set of operations $\Omega$. For the moment and for simplicity reasons, $\Omega$ may be just the set of binary operations $\{+, -, \cdot, /\}$ between real numbers (further operations will be discussed later).

In more detail, let $z_i := x_i$, $i = 1, \ldots, n$, then

$$
\begin{aligned}
z_{n+1} &:= \omega_{n+1}(z_{j_{n+1}}, z_{k_{n+1}}), \\
z_{n+2} &:= \omega_{n+2}(z_{j_{n+2}}, z_{k_{n+2}}), \\
&\vdots \\
z_{n+p} &:= \omega_{n+p}(z_{j_{n+p}}, z_{k_{n+p}}),
\end{aligned}
\tag{1}
$$

where $\omega_s \in \Omega$ and $j_s < s$ for $s = n+1, \ldots, n+p$. For any component $f_i(x)$ of $f(x)$ there exists an index $s$ for which $f_i(x) = z_s$ holds.

If finite arithmetic is used, each operation $\omega_s$ is replaced by a corresponding machine operation $\tilde{\omega}_s$, thus leading to a calculation process

$$
\begin{aligned}
\tilde{z}_{n+1} &:= \tilde{\omega}_{n+1}(\tilde{z}_{j_{n+1}}, \tilde{z}_{k_{n+1}}), \\
\tilde{z}_{n+2} &:= \tilde{\omega}_{n+2}(\tilde{z}_{j_{n+2}}, \tilde{z}_{k_{n+2}}), \\
&\vdots \\
\tilde{z}_{n+p} &:= \tilde{\omega}_{n+p}(\tilde{z}_{j_{n+p}}, \tilde{z}_{k_{n+p}}),
\end{aligned}
\tag{2}
$$

for approximations $\tilde{z}_s$.

An *interval extension* $F$ of $f$ is derived from $f$ by simply replacing the operations in $\Omega$ by their corresponding interval operations defined in $\mathbf{IR}$ [1], [2], [6]. With this replacement, (1) becomes a calculation scheme for the interval extension over any interval vector $Y \subset X$:

$$
\begin{aligned}
Z_{n+1} &:= \omega_{n+1}(Z_{j_{n+1}}, Z_{k_{n+1}}), \\
Z_{n+2} &:= \omega_{n+2}(Z_{j_{n+2}}, Z_{k_{n+2}}), \\
&\vdots \\
Z_{n+p} &:= \omega_{n+p}(Z_{j_{n+p}}, Z_{k_{n+p}}),
\end{aligned}
\tag{3}
$$

The values for $Z_i$, $i = 1, \ldots, n$, are initialized by respective subintervals of the component intervals $X_i$.

We will use the notation $d(X)$ to denote the diameter of an interval $X$, $\inf X$ denotes the lower bound and $\sup X$ the upper bound of $X$, $m(x)$ its midpoint. By $|X|$ we will denote the largest absolute value of all $x \in X$.

## 3. Range Analysis

Using the interval extension $F$ to calculate the true range of $f$ over $X$ quickly runs into overestimation due to the worst case reasoning of interval arithmetic. In addition, runtime exceptions such as division by an interval containing zero may arise. Therefore, a splitting of the original interval into smaller subintervals is necessary. This idea was used by Skelboe [12] to introduce a special scheme for calculating the lower and upper bound of a function separately. The well-known scheme uses a sorted list $L$ of pairs $(Y^i, X^i)$ where $Y^i$ is the evaluation result of $F$ over the subinterval $X^i$ of $X$. The sort order is defined as

$$(Y^i, X^i) \leq (Y^j, X^j) \quad :\Longleftrightarrow \quad \inf Y^i \leq \inf Y^j.$$

We will display this algorithm in a general form since we will make use of some of its ideas in the next section. The algorithm determines the lower bound of $f(X)$. The upper bound is achieved by replacing $F$ by $-F$ and the calculated bound by its negation.

1.     **SKELBOE** $(F, X, L)$:
2.        $Y := F(X)$;
3.        $L := \{\}$;
4.        *while* $\mathrm{d}(Y) > \varepsilon$ *loop*
5.           Bisect $X$ into $X^1$ and $X^2$;
6.           $Y^1 := F(X^1)$; $L := L \cup \{(Y^1, X^1)\}$;
7.           $Y^2 := F(X^2)$; $L := L \cup \{(Y^2, X^2)\}$;
8.           $(Y, X) := \mathrm{Pop}(L)$;
9.        *end loop*;
10.  *return* $(Y, X)$;

Here, we have assumed that $n = 1$. The function Pop returns the first pair of list $L$.

The basic assertion of this algorithm is the fact that the minimum of function $f$ is always contained in the $Y$-value of the first pair $(Y, X)$ of list $L$ (cf. [12]). In practice, the following additional features have to be added:

- An upper bound for the number of calls of $F$ will avoid infinite looping. Nevertheless, the concerns of [3] regarding the *cluster problem* has also to be taken into account.

- If $(Y, X)$ is the first pair of $L$, then all pairs $(\tilde{Y}, \tilde{X})$ of $L$ which satisfy the relation $\sup Y < \inf \tilde{Y}$ may be removed from the list. In addition, a midpoint calculation

$F(m(X))$ may be performed which delivers (as its upper bound) a more suitable guess than $\sup Y$.

- If an exception occurs (e.g., overflow, division by zero), the evaluation result is set to $[-\infty, \infty]$. Thus, the corresponding range will be treated first in the next step.

- If $n > 1$, steps 5 to 7 have to be replaced by a cyclic bisection strategy. The bisection is performed only for that component which has the largest diameter.

Of course, this procedure above could be replaced by the "latest" strategy for global optimization [4]. Nevertheless, the chosen strategy was supposed to be as simple as possible without any additional requirement to the function $f$.

## 4. Condition Analysis

In 1972 Richman [8] developed a technique for the automatic calculation of condition numbers for a given function $f$. The basic idea is to calculate an estimation for each $\varepsilon_s := |z_s - \tilde{z}_s|$ which is derived recursively by

$$
\begin{aligned}
\varepsilon_s &= |\omega_s(z_j, z_k) - \tilde{\omega}_s(\tilde{z}_j, \tilde{z}_k)| \\
&\le |\omega_s(z_j, z_k) - \omega_s(\tilde{z}_j, \tilde{z}_k)| + |\omega_s(\tilde{z}_j, \tilde{z}_k) - \tilde{\omega}_s(\tilde{z}_j, \tilde{z}_k)|
\end{aligned}
$$

where we have omitted the subindices of $z_j$ and $z_k$. The first term on the right hand side describes the condition of the specific operation against parameter variation. The second term is the approximation error during calculation which is bounded either by an $\delta_s$ (as an absolute error bound) or by $\delta_s |Z_s|$ (as a relative error bound) for any interval $Z_s$ containing both $z_s$ and $\tilde{z}_s$.

The first term may be estimated by using partial derivatives of $\omega_s$. Let $\partial_1 \omega_s$ be the derivative of $\omega_s$ with respect to the first argument and $\partial_2 \omega_s$ with respect to the second argument. Then,

$$
\begin{aligned}
|\omega_s(z_j, z_k) - \omega_s(\tilde{z}_j, \tilde{z}_k)| &\le |\partial_1 \omega_s(\zeta_j, z_k)||z_j - \tilde{z}_j| + |\partial_2 \omega_s(\tilde{z}_j, \xi_k)||z_k - \tilde{z}_k| \\
&\le |\partial_1 \omega_s(Z_j, Z_k)|\varepsilon_j + |\partial_2 \omega_s(Z_j, Z_k)|\varepsilon_k.
\end{aligned}
$$

In the second step, we have replaced the unknown intermediate values $\zeta_j$ and $\xi_k$, resp., by the enclosing intervals $Z_j$ and $Z_k$. In summary we have the relation

$$
\varepsilon_s \le |\partial_1 \omega_s(Z_j, Z_k)|\varepsilon_j + |\partial_2 \omega_s(Z_j, Z_k)|\varepsilon_k + \delta_s
$$

for absolute errors, or

$$
\varepsilon_s \le |\partial_1 \omega_s(Z_j, Z_k)|\varepsilon_j + |\partial_2 \omega_s(Z_j, Z_k)|\varepsilon_k + \delta_s |Z_s| \tag{4}
$$

for relative errors, respectively. In the special case of a floating-point arithmetic, the $\delta_s$ in (4) are replaced by a global number $\delta$ which is usually the machine precision, if no overflow or underflow occurs. Dividing both sides by $\delta$ results in

$$c_s := \frac{\varepsilon_s}{\delta} \leq |\partial_1 \omega_s(Z_j,Z_k)|\frac{\varepsilon_j}{\delta} + |\partial_2 \omega_s(Z_j,Z_k)|\frac{\varepsilon_k}{\delta} + |Z_s|$$
$$= |\partial_1 \omega_s(Z_j,Z_k)|c_j + |\partial_2 \omega_s(Z_j,Z_k)|c_k + |Z_s|$$

which defines recursively the condition numbers $c_s$ as originally invented by Richman. Nevertheless, as can be seen from the above the calculus can easily be extended to fixed-point calculations as well. This will be reported in more detail in a forthcoming paper [11].

The general approach is also applicable to other operations rather than just the binary $+,-,\cdot,/$ [9]. Table 1 displays the definition of $c_s$ for a number of different operations $\omega_s$.

*Table 1.* Definition Table for Various Operations

| $\omega_s$ | $Z_s$ | $c_s$ |
|---|---|---|
| $+$ | $Z_j+Z_k$ | $|Z_s|+c_j+c_k$ |
| $-$ | $Z_j-Z_k$ | $|Z_s|+c_j+c_k$ |
| $\cdot$ | $Z_j\cdot Z_k$ | $|Z_s|+|Z_k|c_j+|Z_j|c_k$ |
| $/$ | $Z_j/Z_k$ | $|Z_s|+(c_j+|Z_j|c_k)/\text{dist}(Z_k)$ |
| $n$-th power | $Z_j^n$ | $|Z_s|+n|Z_s|c_j/\text{dist}(Z_j)$ |
| exponential | $\exp(Z_j)$ | $|Z_s|+c_j/\exp(\sup(Z_j))$ |
| sine | $\sin(Z_j)$ | $|Z_s|+c_j$ |
| cosine | $\cos(Z_j)$ | $|Z_s|+c_j$ |
| logarithm | $\ln(Z_j)$ | $|Z_s|+1/\text{dist}(Z_j)$ |

The function dist is used for determining the distance of an interval to zero. The power function has the restriction that the argument has to be positive if $n$ is not an integer.

These formulas can be used for a simultaneous calculation of the condition numbers during normal interval calculation. If operator definition is possible for a given programming language the whole calculus can easily be implemented by such newly operators [10].

The interpretation of the final condition number is quite simple. The intermediate result $z_s$ has a possible loss of accuracy of the order of $1+\lceil \log c_s \rceil$ decimal figures. Overestimation of the correct loss of information can be tackled by a similar splitting scheme as Skelboe's method for ranges of values.

Again, the scheme uses a sorted list $L$ of triples $(C^i, X^i, Y^i)$ where $Y^i$ is the evaluation result of $F$ over the subinterval $X^i$ of $X$. The condition interval $C^i$ is generated by the condition number of $F$ over $X^i$ defined above as the upper bound. The lower bound is determined by the condition number of $F$ over the midpoint of $X$. We will denote this process by $(C^i, X^i, Y^i) \leftarrow F(X^i)$. The sort order is defined as

$$(C^i, X^i, Y^i) \leq (C^j, X^j, Y^j) :\Longleftrightarrow \sup C^i \geq \sup C^j.$$

The algorithm reads as follows:

1.   *COND* $(F, X, L)$:
2.       $(C, X, Y) \leftarrow F(X)$;       /* Evaluation of F with corresp. cond. number */
3.       $L := \{\}$;
4.       *while* $d(C) > \varepsilon$ *loop*
5.           Bisect $X$ into $X^1$ and $X^2$;
6.           $(C^1, X^1, Y^1) \leftarrow F(X^1)$; $L := L \cup \{(C^1, X^1, Y^1)\}$;
7.           $(C^2, X^2, Y^2) \leftarrow F(X^2)$; $L := L \cup \{(C^2, X^2, Y^2)\}$;
8.           $(C, X, Y) := \text{Pop}(L)$;
9.       *end loop*;
10.   *return* $(C, X, Y)$

Again, $n = 1$ was assumed. The function Pop returns the first triple of list $L$, $\varepsilon$ is a bound for the resulting diameter.

In practice, additional features are necessary:

- An upper bound for the number of calls of $F$ will avoid infinite looping. Again, the concerns of [3] have to be considered.

- If $(C, X)$ is the first pair of $L$, then all pairs $(\tilde{C}, \tilde{Y})$ of $L$ which satisfy the relation $\inf C > \sup \tilde{C}$ may be removed from the list.

- If an exception occurs (e.g., overflow, division by zero), the evaluation result is set to $[0, \infty]$.

- If $n > 1$, steps 5 to 7 have to replaced by a cyclic bisection strategy. The bisection is performed only for that component which has the largest diameter.

It should be mentioned that the scheme described in this section can be extended to program iterations as well [13], [9], [11]. Nevertheless, in control software iterations are seldomly used since each module normally has a restricted schedule. Despite of that, the scheme has to be extended to program branching. This can be achieved by replacing (real) comparison operations by interval comparisons. If a comparison cannot be decided, a runtime exception is raised which leads to another splitting similar to other exceptions like division by zero.

A more serious problem with branching which deserves more attention is that normal if-then-else statements may lead to a splitting into closed and non-closed definition areas like $[a,b)$ and $[b,c]$, say. Extending the first range to $[a,b]$ would then lead to infinite looping since $b$ would always correspond to the other branch. In fact, the problem of dealing with open sets is outside the scope of this paper and will have a special treatment in [11]. At the moment, and for the purposes of the applications in mind, we have fixed this problem by an additional symmetric reduction of the diameters of the iterated subintervals by a very small percentage, i.e. we continue with an interval $[p,q] \subset [a,b)$ where $p$ and $q$ are close to $a$ and $b$, respectively. The loss of rigorousness was compensated by an independent second (or even further) run with alternate splitting which makes it very unlikely that significant information was lost, at least from a practical point of view.

## 5. Implementation Issues and Examples

Apart from the mathematical problems to be solved, there remains a good deal of implementation work to be done in order to make use of interval analysis for software testing. It turned out that if the source language of the software in question provides certain language features like suitable data types and operators, then the automation of the test procedure can be done by simple textual modification. Otherwise, real compiler techniques have to be developed which would increase the implementation work significantly.

In the following, we specify our tool, called the "Numerical Analysis Tool (NAT)", which summarises the ideas above. This tool is the basic analysis tool which automatically generates a full test bench for a given piece of software. The first prototype runs for Ada and has the following black box characteristics [10]:

- The tool inputs an Ada procedure plus all relevant context information (used packages, etc.). In addition, the following profiles have to be provided:

  - A Numerics Description File (NDF) which contains information about the nature of the used operations, e.g. "ADD means addition of two numbers of type X". All standard mathematical operations of Ada are allowed.

  - An Input Description File (IDF) which contains information about all data necessary to run the procedure. At a first level, this means the specification of ranges for these data. At a further level, additional dependencies among these data can be specified. If, e.g., $x$ and $y$ are input data then a dependency between them could be $x^2 + y^2 \leq 5$.

  - An Output Description File (ODF) which contains information about all (relevant) data changed by the procedure. Similar to the IDF, this could be either simple ranges or additional dependencies. In addition, indications about the accuracy of these data can be provided, e.g., "Absolute tolerable error for variable $x$: 0.0001".

- The tool outputs a source file containing the instrumented original Ada procedure embedded into a driver Ada program (to be passed to an Ada compiler)

In the following we will give a typical example for an Ada subroutine arising in control software:

```
procedure PRESSION is
    INT1, INT2 : CONTROL;
    begin
        INT1   := PACK1.SUB(PRESS1,PRESS2);
        INT1   := PACK2.MULT(INT1,K1);
        INT2   := PACK1.SUB(PRESS2,PRESS3);
```

Very often, arithmetic is used in terms of procedure calls rather than operators due to the direct access of assembly routines for the operations. This demonstrates the necessity of the NDF defined above. In addition, due to storage limitations, results are transported via global variables. This makes it necessary to specify explicitly input and output data as it is done in the IDF and ODF. The transformation of this example would deliver the following:

```
procedure PRESSION is
    INT1, INT2 : r_e_a_1;
    begin
        INT1   := r_e_a_1_sub(PRESS1,PRESS2);
        INT1   := r_e_a_1_mul(INT1,K1);
        INT2   := r_e_a_1_sub(PRESS2,PRESS3);
```

The model arithmetic which provides both interval arithmetic and condition calculus uses the same interface as the original routine. This makes it (at least in principle) possible to restrict transformation to simple textual replacement. The transformed code then is put into a driver program which provides all necessary external declarations and the attachment of the model arithmetic. A sample output result of this driver program would consist of the following information:

*Range checking:*

```
Range check of subroutine PRESSION:

Checking output values:

Analysis for PRESS4        ...completed
Range is:   [-1.32558398437500E+04, 1.34658398437500E+04]
should be: [ 0.00000000000000E+00, 2.10000000000000E+02]

Lower bound violation at the following sample data:
PRESS1              0.000
PRESS2            165.703
K1                 62.754
PRESS3            209.594
K2                 62.754
.   .   .   .   .
```

*Numerical Analysis:*

```
Error check of subroutine PRESSION:

Checking output values:

Analysis for PRESS4        ...completed
Error is:          1.01562500000000E+00
Type precision is: 7.81250000000000E-03
Decimal loss is:        2.1
Binary loss is:         7.0

Maximum error gained at the following sample data:
PRESS1            105.000
PRESS2            144.375
K1                 47.250
PRESS3            105.000
K2                 47.250
```

In this example, a range violation occured which is reported along with a sample data for which violation will happen. On the other hand, the numerical analysis delivers a possible loss of only 2 decimal digits which was tolerable.

## 6.  Conclusions

The practical experiments clearly demonstrate the feasibility of our approach for software testing. It provides a new kind of prediction about the code behaviour and it provides the prediction over the whole range of allowed input data without a formal proof on a symbolic level.

Due to the exponentially growing complexity for increasing number of input data, the approach has its clear practical limits although comprehensive benchmarks are only available for the applications mentioned above for which performance was not an issue. Nevertheless, from this experience it can be stated that the technique is extremely helpful for developers of software for embedded systems. Here, the number of variables is comparibly small while the predictions on the behaviour for all possible data are of high importance, at least for safety critical applications to which the methodology has been applied first.

Readers interested in NAT may contact G. Schumacher for the latest information about the availability.

## Acknowledgments

We would like to thank the European Commission for funding our work during the ESPRIT projects OMI/ANTI-CRASH (EP 20899) and OMI/SAFE (EP 23920).

## References

1. ALEFELD, G.; HERZBERGER, J.: Introduction to Interval Computation. Academic Press, New York (1983)
2. BAUCH, H. ET AL.: Intervallmathematik. BSB B.G. Teubner Verlagsgesellschaft, Leibzig (1987)
3. KEARFOTT, R.B.; DU, K.: The Cluster Problem in Multivariate Global Optimization. J. of Global Optimization 5, pp 253-265 (1994)
4. KEARFOTT, R.B.: A review of techniques in the verified solution of constrained global optimization problems, in: Applications of Interval Computations, ed. by R.B. Kearfott and V. Kreinovich, Kluwer Academic Publishers, Dordrecht, pp 23-59 (1996)
5. KEDEM, G.: Automatic Differentiation of Computer Programs. ACM Trans. Math. Software 6, Nr. 2, pp 150-165 (1980)
6. MOORE,R.: Interval Analysis. Prentice Hall, Englewood Cliffs, New York (1966)
7. RATSCHEK, H.; ROKNE, J.: Computer Methods for the Range of Functions. Ellis Horwood, Chichester (1984)
8. RICHMAN, P.L.: Automatic Error Analysis for Determining Precision. Comm. ACM 15, Nr. 9, pp 813-817 (1972)
9. SCHUMACHER, G.: Genauigkeitsfragen bei algebraisch-numerischen Algorithmen auf Skalar- und Vektorrechnern. Ph.D. Thesis, Universität Karlsruhe (1989)
10. SCHUMACHER, G.; MUSCH, K.: Specification of a Test Case Generator for Ranges of Data. Deliverable T3/3/2, ESPRIT Project No 23920, OMI/SAFE (1998)
11. SCHUMACHER, G.: Computer Aided Numerical Analysis. To appear 2000.
12. SKELBOE, S.: Computation of rational functions. BIT 14, pp 87-95 (1974)
13. TIENARI, M.: On the Control of Floating-Point Mantissa Length in Iterative Computations, in: Proc. of the Intern. Computing Symposium, ed. by A. Günter et al., North Holland, pp 315-322 (1974)

*T. Csendes (ed.), Developments in Reliable Computing* 159–166.
© 1999 *Kluwer Academic Publishers.*

# Prediction by extrapolation for interval tightening methods

YAHIA LEBBAH AND OLIVIER LHOMME    {yahia.lebbah,olivier.lhomme}@emn.fr
*Department of Computer Science, Ecole des Mines de Nantes, 44307 Nantes, France*

**Abstract.** Extrapolation methods are used in numerical analysis to accelerate the convergence of real number sequences. Interval tightening algorithms produce interval vector sequences. Extrapolation can be applied directly on some of these sequences. Nevertheless, bounds are no longer guaranteed. This paper investigates how to use extrapolation methods without losing solutions and reports some experimental results.

**Keywords:** Interval analysis, Constraint satisfaction, Extrapolation methods.

## 1. Introduction

A general strategy for solving non–linear equations with interval methods is "tightening + splitting". The idea is to associate to each variable an interval containing all possible values for the variables. Tightening means tightening the interval vector (the box) containing all possible solutions. Splitting means splitting the box in sub-boxes. It is needed when tightening fails to reduce the box. Then the general strategy is applied over each sub-box recursively.

Interval Newton methods and interval propagation methods are tightening algorithms. The interval Newton methods have been used in interval analysis [11, 1], and have been also well exploited in constraint logic programming [2]. The interval propagation methods [5, 10] have been used in artificial intelligence to solve non–standard constraints (e.g., mixed (discrete and continuous) constraints, non-square systems, etc.).

Interval Newton methods and interval propagation methods may fail to reduce the box, for example due to wide ranges of the derivatives. The usual techniques for overcoming this drawback — splitting the box — leads to a huge number of boxes, especially for high dimensional problems. [12] have successfully used a special propagation method, the $3B$ algorithm [10], combined with interval Newton methods for efficiently solving some global optimization problems. The $kB$ algorithm is the extension to continuous constraints of the $k$–consistency algorithm [8] which is defined on discrete constraints. The $kB$ algorithm is a kind of a limited exploration on the box. Thus the $kB$ algorithm can drastically reduce the need for

$$\vec{I}_{h+1} = \begin{cases} \vec{I} & \text{if } h = 0 \\ Op(\vec{I}_h) & \text{if } h > 0 \end{cases}$$

*Figure 1.* Tightening algorithm sequence schema

splitting. Unfortunately, due to its linear convergence behavior, the $kB$ algorithm is usually slow. To overcome this problem, [9] introduced a general way to accelerate its convergence. [9] focuses on the $3B$ algorithm, whereas in this paper we will be interested in the $kB$ algorithm.

The paper is organized as follows. In Section 2, some definitions concerning non–linear equations are summarized, the $kB$ algorithm will be defined by using the sequence formalism. Section 3 presents extrapolation methods. Section 4 will show our method to accelerate the convergence of the $kB$ algorithm. Section 5 contains experimental results and a discussion.

## 2. Interval methods for solving non-linear systems

Non-linear constraints can be formulated as following:

- $X$ is a set of $v$ variables $x_1, \ldots, x_v$.

- $\vec{I} = (I_1, \ldots, I_v)$ denotes a vector of intervals. The $i^{th}$ component of $\vec{I}$, $I_i$, is the interval containing all acceptable values for $x_i$. $\underline{I_i}$ (resp. $\overline{I_i}$, $wid(I_i)$) denotes the lower bound (resp. the upper bound, the width) of $I_i$.

- $(\mathcal{F} = 0) = \{f_1(x_1, \ldots, x_v) = 0, \ldots f_m(x_1, \ldots, x_v) = 0\}$ denotes a set of constraints.

Interval tightening algorithms can be seen (Figure 1) as a sequence of interval vectors $\{\vec{I}_n\}$. At each iteration $i + 1$, the algorithm computes the terme $\vec{I}_{i+1}$ by applying an operator $Op$ on the previous term $\vec{I}_i$. The fixed point is denoted by $\mathcal{FP}(Op, \vec{I}_1)$. In practice, the algorithm stops before reaching the real fixed point. For example, the algorithm may stop as soon as $Op(\vec{I}) \simeq \vec{I}$, that is a quasi–fixpoint has been reached. In the following, by abuse of notation, $\mathcal{FP}(Op, \vec{I}_1)$ will denote the final result of the algorithm.

$Op$ stands for interval operator. The different interval based algorithms can be instantiated to the algorithmic schema by specializing that operator:

- The interval Newton methods (Hansen and Sengupta, Krawczyk, etc.) can be abstracted as operators by using the sequence formalism. We call these operators 'Newton operators'.

- The standard interval propagation methods [5] (without the $kB$ algorithm which will be presented in the following) are a kind of Gauss–Seidel algorithm, but are applied on non–linear constraints. Propagation works directly on constraints without using derivatives [1]. The propagation algorithms can also be abstracted by a sequence, we call their operators 'propagation operators'.

In the following, the term "standard tightening operators" stands for Newton operators and standard propagation operators. Now, we introduce the $kB$ algorithm.

## 2.1.   The kB algorithm:

The $kB$ algorithm is a tightening algorithm, that is to say that its aim is to reduce the input box to a narrower one. The strategy of this algorithm is the exploration of the extremal sub–boxes of the input box. Such extremal sub–box is in fact a face of the box with a small given thikness. For example, the $3B$ algorithm over a $v$–dimensional box computes the fixed point of a given standard tightening operator on each extremal sub–box (or thick face) of the main box. If one of these computed fixed points is empty, the related sub–box is removed from the main initial box. We get the $kB$ algorithm by applying recursively this process $k - 2$ times on the generated sub–boxes.

The $kB$ operator is parameterized with $k$, and an integer $sp$ which is the splitting parameter.

$$Op2B(\vec{I}) = OpT(\vec{I}), OpT \text{ is a standard tightening operator}$$
$$OpkB(\vec{I}) = \vec{I}', \text{ with } k \geq 3.$$
where $I_i' = [l', r']$,
$$l' = \min\{l | (\exists n_1 \in I\!N, n_1 \geq 0, l = \underline{I_i} + n_1 \times w) \wedge$$
$$\mathcal{FP}(Op|k-1|B, (I_1, \ldots, [l, l+w], \ldots, I_v)) \neq \emptyset\}$$  (1)
$$r' = \max\{r | (\exists n_2 \in I\!N, n_2 \geq 0, r = \overline{I_i} - n_2 \times w) \wedge$$
$$\mathcal{FP}(Op|k-1|B, (I_1, \ldots, [r-w, r], \ldots, I_v)) \neq \emptyset\}$$
where $w = wid(I_i)/sp$

$kB$ algorithm computes the (quasi) fixed point of the operator (1). When $k = 2$, this operator is one of the standard tightening operators. When $k > 2$, this operator explores the extremal sub–boxes.

In the following, the $kB$ algorithm will be taken in its generic sense, and thus will subsume all interval tightening methods.

In pratice, the $kB$ operator is more efficient in term of narrowing than the $|k-1|B$ algorithm, but is more expensive in execution time. In the following sections, we present a way to decrease the execution time of the $kB$ algorithm by using extrapolation.

## 3. Extrapolation methods

Extrapolation methods [4] are used in numerical analysis to accelerate the convergence of real number sequences. Let $\{S_n\}$ be a sequence of real numbers. A sequence $\{S_n\}$ converges if and only if it has a limit $S$: $\lim_{n\to\infty} S_n = S$. Accelerating the convergence of a sequence $\{S_n\}$ amounts of applying a transformation $\mathcal{A}$ which produces a new sequence $\{T_n\}$: $T_n = \mathcal{A}(\{S_n\})$.

For example, the well–known Aitken's $\Delta^2$ process is defined by the transformation $T_n = \frac{S_n S_{n+2} - S_{n+1}^2}{S_{n+2} - 2S_{n+1} + S_n}$.

As given in [4], in order to present some practical interest the new sequence $\{T_n\}$ must exhibit, at least for some particular classes of convergent sequences $\{S_n\}$, the following properties:

1. $\{T_n\}$ converges to the same limit as $\{S_n\}$ : $\lim_{n\to\infty} T_n = \lim_{n\to\infty} S_n$

2. $\{T_n\}$ converges faster than $\{S_n\}$: $\lim_{n\to\infty} \frac{T_n - S}{S_n - S} = 0$

As explained in [4], these properties do not hold for all converging sequences. Particularly, a universal transformation $\mathcal{A}$ accelerating all converging sequences cannot exist [7]. Thus any transformation can accelerate a limited class of sequences. This leads us to a so-called *kernel*[2] of the transformation which is the set of convergent sequences $\{S_n\}$ for which $\exists N, \forall n \geq N, T_n = S$, where $T_n = \mathcal{A}(\{S_n\})$.

Some scalar transformations have been generalized to the vectorial and matricial case. For details about theory and practice of extrapolation methods, see [4].

In the following, results about Aitken's process will be exploited to present an heuristic to detect convergence in $kB$ algorithm.

## 4. Extrapolation for the $kB$ algorithm

In order to optimize the tightening algorithm, it suffices to accelerate the convergence of its associated intervals sequence $\{\vec{I}_n\}$. $\{\vec{I}_n\}$ is a sequence of interval vectors. There does not exist any method to accelerate interval sequences, but an interval can be seen as two reals and $\vec{I}$ can be seen as a 2-columns matrix of reals (the first column is the lower bounds, and the second the upper bounds). Thus we can apply the acceleration methods [4]. By extrapolating the sequence $\{\vec{I}_n\}$ of a particular tightening operator $Op$, with an extrapolation method $\mathcal{A}$, we obtain a new operator $Extrap_{Op}^{\mathcal{A}}$. The operator $Extrap_{Op}^{\mathcal{A}}$ can be applied at each iteration of the operator $Op$ to generate a new sequence. This transformation is correct and complete if the sequences belong to the kernel of the used extrapolation method.

The following theorem on Aitken's process gives us an interesting property to make the prediction when some conditions are verified.

**Theorem 1 (Aitken's kernel [3])** *Let $\{S_n\}$ be a sequence, its limit is $S$. A necessary and sufficient condition to have $S = \Delta^2(\{S_n\})$ for all $n > N$ is $S_n = S + \alpha \times \lambda^n$ with $\lambda \neq 1$ for all $n > N$.*

This theorem means that when we have a *special regularity* $(S_n = S + \alpha \times \lambda^n)$ in the scalar sequence $\{S_n\}$, then the Aitken's process guarantees the acceleration of this sequence.

Nevertheless in the general case, applying directly Aitken's process on the sequences of the $kB$ algorithm gives non guaranteed results because:

- $\{S_n\}$ may be out of the kernel.

- Even if $\{S_n\}$ is in the kernel, it may be difficult to know it.

Now a way of applying extrapolation with guaranteed results is given for $kB$ algorithm.

$kB$ algorithm uses a proof–by–contradiction mechanism: it tries to prove that no solution exists in a sub–box of the main box by $|k - 1|B$ algorithm. If such a proof is found, then the sub–box is removed from the main initial box, else the sub–box is not removed. The point is that we may waste a lot of time trying to find a proof that does not exist. If we could predict with a good probability that such a proof does not exist, we could save time in not trying to find it. The $kB$ operator (1) needs an *empty/non–empty* answer from the $|k - 1|B$ algorithm. We need a method to predict that the sequence generated by the $|k - 1|B$ converges to a non–empty box. This task can be done by extrapolation methods.

When the prediction is *non–empty* then the idea is to avoid a long converging sequence to a non–empty box. When the prediction is *empty* then the algorithm proceeds as usual. The algorithm is well accelerated when many of its scalar sequences are close to the kernel of the acceleration method used.

Aitken's process has been successfuly applied in many fields of applied mathematics. [6] prove that the Aitken's process is optimal for linear convergence sequences. These two reasons enable to take the Aitken's process as an interesting heuristic to make prediction. But it could be interesting to use other extrapolation methods [4]. This is, of course, depending on the problems to solve.

We define a new operator $OpAcckB$ by using this new convergence criterion in the $OpkB$ operator. The $AcckB$ algorithm is the algorithm that computes the fixed point of this operator.

| Problem | kB, in sec | AcckB, in sec |
|---|---|---|
| Moré–Cosnard–60$[-10^{-8},0]$ | 35.29 | 5.02 |
| Moré–Cosnard–80$[-10^{-8},0]$ | ? | 9.63 |
| Moré–Cosnard–160$[-10^{-8},0]$ | ? | 48.25 |
| Moré–Cosnard–320$[-10^{-8},0]$ | ? | 169.14 |
| Economics–4$[-10^8,10^8]$ | 1.70 | 0.49 |
| Economics–5$[-10^8,10^8]$ | 26.10 | 3.92 |
| Economics–6$[-10^8,10^8]$ | 335.30 | 30.46 |
| Broyden–20$[-10^8,10^8]$ | 55.10 | 5.81 |
| Neurophysiology$[-10,10]$ | 2.00 | 0.40 |
| Neurophysiology$[-100,100]$ | 35.70 | 28.47 |
| Combustion | 4.00 | 1.73 |
| Kearfott–1 | 0.10 | 0.05 |
| Kearfott–4 | 10.10 | 17.20 |
| Kearfott–F | 3.90 | 1.82 |
| Wilkinson | 0.10 | 0.17 |

*Figure 2.* Results for standard interval benchmarks

$$OpAcc2B(\vec{I}) = OpT(\vec{I}), \; OpT \text{ is a standard tightening operator}$$
$$OpAcckB(\vec{I}) = \vec{I}', \text{ with } k \geq 3.$$
where $I'_i = [l', r']$,

$l' = \min\{l|(\exists n_1 \in I\!N, n_1 \geq 0, l = \underline{I}_i + n_1 \times w)\wedge$

(a)　　　$\mathcal{FP}(Extrap^{\Delta^2}_{OpAcc|k-1|B}, (I_1, \ldots, [l, l+w], \ldots, I_v)) \neq \emptyset \wedge$　　　　(2)

(b)　　　$\mathcal{FP}(OpAcc|k-1|B, (I_1, \ldots, [l, l+w], \ldots, I_v)) \neq \emptyset\}$

$r' = \max\{r|(\exists n_2 \in I\!N, n_2 \geq 0, r = \overline{I}_i - n_2 \times w)\wedge$

(c)　　　$\mathcal{FP}(Extrap^{\Delta^2}_{OpAcc|k-1|B}((I_1, \ldots, [r-w, r], \ldots, I_v))) \neq \emptyset \wedge$

(d)　　　$\mathcal{FP}(OpAcc|k-1|B, (I_1, \ldots, [r-w, r], \ldots, I_v)) \neq \emptyset\}$

　　where $w = wid(I_i)/sp$

In our implementation, the steps (a) and (b) (resp. (c) and (d)) are interleaved.

## 5. Experimental results

Benchmarks are coming from [13]. Experimentation has been made on a Sun UltraSparc. The precision of the fixed point computations is about $10^{-10}$. The results are given in Figure 2, '?' means that the algorithm generates a big number of boxes that exceeds the computer memory.

　　Those results are obtained for the two algorithms:

1. *kB*: the standard $3B$ algorithm based on [12]. The standard tightening operator is an interval Newton operator.

2. *AcckB*: the *AcckB* algorithm with $k = 3$. The standard tightening operator is an interval propagation operator, because extrapolation is more efficient on this operator than on the Newton operators.

In general the extrapolated algorithm behaves better in execution time. In some cases $3B$ is more efficient, because the extrapolation has made a wrong prediction. For the neurophysiology problem, we have an exponential growth of time when we increase the size of the intervals. Surprisingly, for the Moré–Cosnard nonlinear integral equation, we have approximately obtained a linear growth.

## 6.  Conclusion

Extrapolation methods can be applied as a predictor to prevent slow convergence of tightening algorithms. This prediction is applied on the *kB* algorithm without loss of solutions. It behaves well on standard benchmarks of interval analysis.

We are now working on other ways of extrapolating a sequence of interval vectors, and the improvement of different extrapolation methods on the different tightening algorithms.

## Notes

1. [13] generalised the use of derivatives in propagation algorithms. The obtained algorithms are close to interval Newton methods.
2. The definition of the kernel given here considers only converging sequences.

## References

1. G. Alefeld and J. Hezberger, editors. *Introduction to Interval Computations*. Academic press, 1983.
2. F. Benhamou, D. McAllester, and P. Van Hentenryck. CLP(intervals) revisited. In *Logic Programming - Proceedings of the 1994 International Symposium*, pages 124–138. The MIT Press, 1994.
3. C. Brezinski, editor. *Algorithmes d'accélération de la convergence: étude numériques*. Technip, 1978.
4. C. Brezinski and R. Zaglia, editors. *Extrapolation methods*. Studies in Computational Mathematics. North-Holland, 1990.
5. E. Davis. Constraint propagation with interval labels. *Journal of Artificial Intelligence*, pages 32:281–331, 1987.
6. J. P Delahaye. Optimalité du procédé $\delta^2$ d'aitken pour l'accélération de la convergence linéaire. *RAIRO, Anal. Numér.*, pages 15:321–330, 1981.
7. J. P. Delahaye and B. Germain-Bonne. Résultats négatifs en accélération de la convergence. *Numer. Math*, pages 35:443–457, 1980.
8. E. C. Freuder. Synthesizing constraint expressions. *Communications of the ACM*, pages 21:958–966, 1978.

9. Yahia Lebbah and Olivier Lhomme. Acceleration methods for numeric CSPs. In *AAAI-98*, pages 19–25. MIT Press, July 26–30 1998.
10. O. Lhomme. Consistency techniques for numeric CSPs. In *IJCAI-93*, pages 232–238, 1993.
11. R. Moore, editor. *Interval Analysis*. Prentice Hall, 1966.
12. J.F. Puget and P. Van-Hentenryck. A constraints satisfaction approach to a circuit design problem. *Journal of global optimization*, pages 13(1):75–93, 1998.
13. P. Van-Hentenryck, D. Mc Allester, and D. Kapur. Solving polynomial systems using branch and prune approach. *SIAM Journal on Numerical Analysis*, pages 34(2):797–827, 1997.

*T. Csendes (ed.), Developments in Reliable Computing* 167–188.
© 1999 *Kluwer Academic Publishers.*

# The Contribution of T. Sunaga to Interval Analysis and Reliable Computing

SVETOSLAV MARKOV                                          smarkov@iph.bio.bas.bg
*Institute of Mathematics and Informatics, Bulgarian Academy of Sciences, Sofia, Bulgaria*

KOHSHI OKUMURA                                          kohshi@kuee.kyoto-u.ac.jp
*Department of Electrical Engineering, Kyoto University, Kyoto, Japan*

**Abstract.** The contribution of T. Sunaga to interval analysis and reliable computing is not well-known amongst specialists in the field. We present and comment Sunaga's basic ideas and results related to the properties of intervals and their application.

**Keywords:** History of mathematics, Interval analysis, Reliable computing

> The interval concept is on the borderline linking
> pure mathematics with reality and pure analysis
> with applied analysis.                T. Sunaga

## 1.  Introduction

Interval analysis and reliable computing are interdisciplinary fields, combining both abstract mathematical theories and practical applications related to computer science, numerical analysis and mathematical modelling. At present there exist hundreds of related publications including many monographs, proceedings and collections of papers, more than twenty international meetings have been organized. It is a commonplace view that interval analysis will continue to play a significant role in applied mathematics, especially in mathematical modelling and reliable computing. Undoubtedly the development of interval analysis will become of interest to the history of mathematics. After four decades of intensive and successful development it is now time to look back and give tribute to the pioneering research in this field.

Some early (and rather premature) ideas about an interval calculus can be found in [30], [44]. Two papers can be considered as pioneering works in this field: one by the Japanese mathematician T. Sunaga [38], which summarizes the results of his Master Thesis accomplished in 1956 [36], and another by the Polish mathematician M. Warmus [42]. Both papers have been published almost at the same time and, apparently, completely independent of each other. Several years later a second paper appeared by Warmus [43], also the American mathematician R. Moore accomplished his dissertation in interval analysis [23] and published a monograph [24].

The early works on interval analysis published before the monograph [24] seem not to be well known. In the present article we shall briefly review and comment the work of T. Sunaga [38], which we shall refer to in the sequel as "Sunaga's paper". In some later publications (in Japanese) T. Sunaga uses interval analysis in problems related to mechanics and planning of production [40], [41]. This applied work will not be discussed here; however, in our opinion it also deserves special attention (and, possibly, translation into English). Sunaga's paper contains basic ideas about the algebraic properties of intervals and their application to numerical analysis; however, this work is rarely referenced in the literature.

In the illustrative examples we have done negligible changes in the notations: i) we use the familiar symbol $\{x \mid P(x)\}$ for "the set of elements $x$ such that property $P(x)$ is satisfied" (Sunaga uses verbal expressions); ii) for inclusion we use the familiar notations $\alpha \in B$ or $A \subseteq B$, while Sunaga uses (for both cases) the same notation $\alpha \to B, A \to B$; iii) for the equality of intervals he uses the symbol $A \rightleftharpoons B$, which is associated with the definition: $A \rightleftharpoons B$, iff $A \to B$ and $A \leftarrow B$. It should be mentioned that the use of a single notation "$\to$" instead of both "$\in$" and "$\subseteq$" in numerical calculations is very convenient, because it allows us in the relation $\alpha \to B$ not to specify whether a number $\alpha$ is rounded (and thus is an interval so that we should write $\alpha \subseteq B$) or is exact (in which case we should write $\alpha \in B$); as a typical example see Example 8.2 in [38].

Finally let us mention, that Sunaga's paper is richly illustrated by 17 figures, which contributes to easy reading; and a great number of examples is provided. The final five examples demonstrate the application of interval analysis to reliable computing and present a special interest.

Our article is organized as follows. In Section 2 we briefly review Sunaga's theory of intervals as developed in Sections 1–7 in his work. In Section 3 we present and comment his ideas about the practical implementation of interval analysis to verified solution of numerical problems as given in Sections 8 and 9 of [38]. In Section 4 we make an attempt to briefly trace the further development of the field in relation to Sunaga's work. In the last Section we give a short biography of T. Sunaga.

## 2.  Sunaga's "Theory of an Interval Algebra"

From the introduction of [38] we learn that T. Sunaga arrived to his "Theory of an Interval Algebra" while working in the field of Communication Theory [37]; it is mentioned that this latter work is related to C. Shannon's theory [35] and especially to issues of finiteness and discreteness. The work [26] is mentioned as an example of treating round-off errors by other methods.

**Significance of interval** (Section 1 of [38]). This introductory Section starts with an example of the numerical computation of $\sqrt{2}$ and points out that a real number is actually defined by a sequence of rational intervals, that is intervals with rational endpoints. Therefore "the concept of an interval is more fundamental than that of a real number".

If the rounding mode is known, then an interval may be presented by a single rational number. For example, in the calculation of $\sqrt{2}$, we obtain a sequence of rational numbers, such as 1, 1.4, 1.41, 1.414, ... . The numbers in the sequence can be interpreted as intervals containing the value $\sqrt{2}$, i. e. the number 1.4 as the interval [1.4, 1.5], the number 1.41 as [1.41, 1.42] etc. Thus, in applications, a number is often understood as an interval. "To denote an interval, we need not necessarily use two rational numbers". The shorthand notation $\langle \alpha \rangle$ is introduced and used throughout the paper to denote the tightest interval enclosing a decimal floating point number $\alpha$, assuming that the number has been rounded to the nearest, e. g. $\langle 1.414 \rangle = [1.4135, 1.4145]$. As may be seen from the sequel directed roundings are used systematically in many of the examples, which makes Sunaga's paper an early foregoer of the computer arithmetic approach, cf. e. g. [13].

Further in the introduction a clear distinction is made between the computational approach using intervals and the statistical approach. "The reader may think of such more familiar expressions as '*Statistically* ...' or '*The statistical values are* ... '. But probable or stochastical numerical values are not different from our physical quantities and should also be described by intervals." The author summarizes the discussion of the practical importance of the interval concept by an original thought, which we have chosen as motto of the present article.

In [38] the type of letter used specifies the mathematical object. Small Greek letters $\alpha, \beta, ..., \gamma$, mean real numbers; intervals are denoted by roman letters, thereby upper case roman letters have different meaning than lower case roman letters: the letters $A, B, ..., C$, mean intervals in expressions involving interval arithmetic operations, whereas the letters $a, b, ..., c$, are used to denote variables or parameters of functions and functionals. These variables or parameters may have interval values, then instead of functions we have sets of functions or instead of functionals we have ranges of functions over interval parameters. Boldface roman letters $\mathbf{A}, \mathbf{B}, ..., \mathbf{C}$, denote interval vectors.

**The interval lattice** (Section 2 of [38]). In this second Section the inclusion partial ordering in the set of intervals and the induced lattice operations "join" (joined union, convex hull) $X \vee Y = \inf_{\subseteq}\{Z \mid X \subseteq Z, Y \subseteq Z\}$ and "meet" (intersection) $X \wedge Y = \sup_{\subseteq}\{Z \mid Z \subseteq X, Z \subseteq Y\}$ are discussed in some detail (typically, the definitions of join and meet are given in [38] in verbal form). Why the author starts his exposition with the inclusion relation, and not with the arithmetic operations? He obviously realizes the important role of the partial order in the set of intervals, but also he makes advantage of the fact that order relations belong to a well studied algebraic area — lattice theory [2]. Theorem 1 treats properties of the lattice operations and Theorem 2 states that the set of intervals together with the inclusion relation is a lattice. Another result (not formulated as a theorem) considers the inclusion isotonicity of lattice polynomials.

Let us note that the operation join is extensively used throughout the paper in the special case when the intervals are degenerate, that is numbers. The computational technique using the join of two numbers has been demonstrated in the examples given throughout [38] and in the formulation and proofs of some theorems (cf. e. g. Theorem 3).

**The arithmetic operations** (Section 3 of [38]). In this Section the arithmetic operations for intervals are introduced — to our knowledge for first time. Passing from the authors verbal form to symbolic notations we obtain:

*Definition 6. Sum, difference, product and quotient of X, Y*:

$X + Y = \{\xi + \eta \mid \xi \in X, \eta \in Y\}$;
$X - Y = \{\xi - \eta \mid \xi \in X, \eta \in Y\}$;
$XY = \{\xi\eta \mid \xi \in X, \eta \in Y\}$;
$X/Y = \{\xi/\eta \mid \xi \in X, \eta \in Y\}, 0 \notin Y$.

Definition 7 introduces the operators $-X = 0 - X$ and $X^{-1} = 1/X$. Using these operators subtraction and division are reduced to addition, resp. multiplication by reciprocal, see Theorem 4. Further, Theorem 5 states that commutative and associative laws hold true for addition and multiplication. The last part of Theorem 5 contains the relation $X(Y + Z) \subseteq XY + XZ$, denoted later in the literature on interval analysis as "subdistributive law". It has been noted that the subdistributivity law implies the inclusion $(X - 1)X \subseteq X^2 - X$, and, more generally, that computation of the range of an algebraic polynomial "is better carried out by the Horner's method", symbolically, $(...((A_nX + A_{n-1})X + A_{n-2}X + ... + A_1)X + A_0 \subseteq A_nX^n + A_{n-1}X^{n-1} + ... + A_1X + A_0$. Theorem 6 gives the distributivity of addition, resp., multiplication, with respect to join. The proof supplied demonstrates the simultaneous use of arithmetic and lattice interval operations.

Theorems 7 and 8 contain the cancellation law for addition, resp. for multiplication. The algebraic solution of simple equations in interval arithmetic is discussed, e. g. the solution of $[1,2] + X = [2,5]$ is $X = [1,3]$. To summarize, in this Section it is proved that the set of intervals together with addition (multiplication) is an abelian cancellative monoid (in case of multiplication intervals containing 0 are excluded), moreover the subdistributive relation is established.

The use of join in presenting intervals via endpoints can be seen from the following example:

*Example 3.3.* We have

i) $(\alpha_1 \vee \alpha_2)(\beta_1 \vee \beta_2) = \alpha_1\beta_1 \vee \alpha_1\beta_2 \vee \alpha_2\beta_1 \vee \alpha_2\beta_2$
$$= \min_{i,j}\{\alpha_i\beta_j\} \vee \max_{i,j}\{\alpha_i\beta_j\}.$$

ii) $\langle 1.414 \rangle = 1.414 - 5 \times 10^{-4} \vee 1.414 + 5 \times 10^{-4}$
$$= 1.414 + (-5 \times 10^{-4} \vee 5 \times 10^{-4}) = 1.414 + (-5 \vee 5) \times 10^{-4}.$$

Note that the above notation avoids the necessity to indicate which end-point of a given interval is left and which one is right (in fact this is often unknown in advance — i. e. when the endpoints are represented by some expressions that have to be evaluated).

**Multidimensional intervals** (Section 4 of [38]). In this Section the definitions of the arithmetic operations (addition, subtraction, multiplication by scalar) and the lattice operations are extended to $n$-dimensional intervals of the form $(X_1,...,X_n)$, where $X_1,...,X_n$ are usual (one-dimensional) intervals.. A useful discussion related to multiplication by "interval scalar" is presented. Multiplication of an $n$-dimensional interval by a real scalar as induced by addition of $n$-dimensional intervals is: $x(X_1,...,X_n) = (xX_1,...,xX_n)$. According to Sunaga's general methodology, multiplication by interval scalar should be the extension of the above definition for $x \in X$, that is:

$$X(X_1,...,X_n) = \{(xX_1,...,xX_n) \mid x \in X\}.$$

Sunaga gives the following subdistributive relation (see formula 4.6 in [38]):

$$X(X_1,...,X_n) \subseteq (XX_1,...,XX_n).$$

It is noted that the relation $X(X_1,...,X_n) = (XX_1,...,XX_n)$ does not always hold; and the following simple example for the expressions $X(1,U)$ and $(X,XU)$ for given intervals $X, U$ is discussed:

*Example 16.* $X(1,U) \subseteq (X,XU)$. Let $X = 1 \vee 2$, $U = 1.0 \vee 1.2$, then $X(1,U) = (1 \vee 2)(1, 1.0 \vee 1.2)$; $(X,XU) = (1 \vee 2, 1.0 \vee 2.4)$.

The "wrapping effect" is demonstrated and visualized (cf. Fig. 8 in [38]). A connection is made to similar expressions in the one-dimensional case, e. g. the expressions $X(1+U)$ and $X+XU$. The influence of the errors in the coefficients of a system of linear equations on the solutions of the system is discussed at the end of the Section.

**Interval functions and functionals** (Section 5 of [38]). Note that the value of an expression involving multiplication by interval scalar is not an interval in general. This viewpoint is further extended in the definition of interval function. The value of an interval function is not an interval. According to Sunaga's definition, an interval function is a set of real functions, depending on parameters, taking values from given intervals. For example, the set of functions $f(\,\cdot\,;a) = \{f(\,\cdot\,;\alpha) \mid \alpha \in a\}$ is an interval function. Here $a$ may be a $n$-dimensional interval, e. g. for $n = 2$ we have: $f(\,\cdot\,;a,b) = \{f(\,\cdot\,;\alpha,\beta) \mid \alpha \in a, \beta \in b\}$. An interval functional is a special case of an interval function, when the function depends only on parameter(s). Hence an interval functional is the range of a function over a given ($n$-dimensional) interval. For instance, if $f(\xi)$ is a function defined on some domain, and $a$ be an interval from this domain, then the set $f(a) = \{f(\alpha) \mid \alpha \in a\}$, is an interval functional.

Interval functionals (and functions) involve intervals as arguments; the latter are denoted in the same way as real variables — by lower case Roman letters. For comparison, the intervals involved in an interval-arithmetic expression are denoted by upper case letters. The difference between the two concepts is demonstrated by examples: we have $a(a+b) = a+ab$, but $A(A+B) \neq A+AB$. Another example:

$$x(x-1) = \{\xi(\xi-1) \mid \xi \in x\} \subseteq X(X-1) \subseteq X^2 - X. \tag{1}$$

Such brief notation of an interval functional is very useful for numerical analysis, where a variable $x$ in expressions like $x(x-1)$ may easily turn from a real number to an interval depending on whether $x$ is machine representable or not.

The sharp difference made between ranges of functions and interval arithmetic expressions outlines the subordinate role of interval arithmetic as a tool for numerical computation or enclosure of functional ranges. Thus one of the main objectives of interval analysis is clearly formulated in this Section.

Example (1) expresses the simple rule that the range of a function is contained in the corresponding interval arithmetic expression obtained by formally replacing the real arithmetic operations by interval-arithmetic ones. This rule is not formulated in general. There may be at least two reasons for this: first, the general style of the paper is to present the new ideas by simple examples, second, the author may have considered this rule as a too obvious consequence from the definitions of the interval-arithmetic operations. However, it should be mentioned that this rule has been used at several places in the paper on different expressions of the form $f(x,y,...,z)$. Using Sunaga's notation of an interval functional this rule may be formulated symbolically as:

$$f(x, y, ..., z) \subseteq F(X, Y, ..., Z), \quad x \in X, y \in Y, ..., z \in Z, \tag{2}$$

where $F$ means the interval arithmetic expression obtained by formally replacing the variables of $f$ by the respective intervals and the real arithmetic operations by respective interval-arithmetic ones.

It may be also of some interest to note that in this Section the author uses the concept of functional, as used in the theory of distributions [34], e. g. the Dirac $\delta$ function is a functional, which is extended for interval arguments (and the same for its derivative).

**Differentiation** (Section 6 of [38]). Derivatives are defined in the spirit of the previous Section 5. Thus, if $f$ is a differentiable function on some domain and $a$ is an interval from this domain, then $f'(a)$ means the range of the function $f'$ in the interval $a$, that is, $f'(a) = \{f'(\alpha) \mid \alpha \in a\}$, and is called the *differential coefficient of $f$ on the interval $a$*. The following two propositions are formulated in this section (numbered as formulae 6.4 and 6.5 in Sunaga's paper).

i) If $x, x_1$ and $dx$ are intervals such that $x_1 \subseteq x$, $x_1 + dx \subseteq x$, and $x$ is from the definition domain of $f$, then $f(x_1 + dx) \subseteq f(x_1) + f'(x)dx$.

ii) If $x$, $x_1$, $x_{11}$ and $dx$ are intervals, such that $x_{11} \subseteq x_1 \subseteq x$, $x_{11} + dx \subseteq x$, then $f(x_{11} + dx) \subseteq f(x_{11}) + f'(x_1)dx + (1/2)f''(x)(dx)^2$.

The first proposition is an interval Taylor formula of first order (interval mean-value theorem) and the second proposition is an interval Taylor formula of second order. The interval mean-value theorem has been used several times in the applications, see Section 9. The formula is often used in interval analysis in the special case when $x_1$ and $dx$ are numbers and (only) $x$ is an interval. Sunaga has found a rather general form of the formula. The assumed inclusions of the intervals involved in the second order formula are visualized, see Fig. 12 in [38].

**The topological background** (Section 7 of [38]). In this Section the author looks for already studied algebraic systems, which incorporate the partial order relation inclusion together with arithmetic operations. He thus arrives to analogies between interval systems and topological groups [31]). The idea is to consider intervals as neighbourhoods of numbers and to apply the theory of topological spaces. The inclusion relation plays a basic role in such considerations.

Sunaga gives the definition of a topological group and discusses it's properties in some detail. At the end of the section it is suggested that the set of neighbourhoods, i. e. intervals, is completed into a group, so that inverse elements exist. In this section there are no particular original results, except for the general proposal to use tools from topological algebra for the study of the interval system.

## 3. Sunaga's Proposal for Application of Interval Analysis to Reliable Numerical Computation

In the last two sections of [38] practical application of interval calculus to reliable numerical computation are proposed.

**Practical interval calculus** (Section 8 of [38]). This Section is devoted to the midpoint-radius (center-radius) presentation of intervals. Such presentation is indeed of extraordinary importance in the practical applications. As seen from Example 8.1 several different notations are used for the center-radius form $A = (\alpha, \alpha_0) = \alpha + (-\alpha_0 \vee \alpha_0)$, $\alpha_0 \geq 0$:

*Example 8.1.* $1 \pm 0.01 = (1.00, 10^{-2}) = 1.00; 1$

$$\langle 1.414 \rangle = (1.4140, 5 \times 10^{-4}) = 1.4140; 5.$$

In this example the author introduces another brief notation for intervals in center-radius form: for the interval with center 1.00 and radius 0.01 he writes simply 1.00;1, the interval with center 1.4140 and radius 0.0005 is denoted 1.4140;5 (which was also denoted $\langle 1.414 \rangle$).

The author gives formulae for the arithmetic operations in center-radius form:

**Theorem 11.** Let $A = (\alpha, \alpha_0) = \alpha + (-\alpha_0 \vee \alpha_0)$, $\alpha_0 \geq 0$, $B = (\beta, \beta_0) = \beta + (-\beta_0 \vee \beta_0)$, $\beta_0 \geq 0$, be two intervals. We have:

i) $(\alpha, \alpha_0) + (\beta, \beta_0) = (\alpha + \beta, \alpha_0 + \beta_0)$;

ii) $(\alpha, \alpha_0) - (\beta, \beta_0) = (\alpha - \beta, \alpha_0 + \beta_0)$;

iii) $(\alpha, \alpha_0)(\beta, \beta_0) = (\alpha\beta + \alpha_0\beta_0, \alpha_0\beta + \alpha\beta_0)$, $\alpha \geq \alpha_0 \geq 0$, $\beta \geq \beta_0 \geq 0$;

iv) $\dfrac{(\alpha, \alpha_0)}{(\beta, \beta_0)} = (\dfrac{\alpha\beta + \alpha_0\beta_0}{\beta^2 - \beta_0^2}, \dfrac{\alpha_0\beta + \alpha\beta_0}{\beta^2 - \beta_0^2})$, $\alpha \geq \alpha_0 \geq 0, \beta > \beta_0 \geq 0$.

Sunaga notes that these "operations are similar to those of complex numbers":

$(\xi + i\xi_0) + (\eta + i\eta_0) = (\xi + \eta) + i(\xi_0 + \eta_0)$,
$(\xi + i\xi_0) - (\eta + i\eta_0) = (\xi - \eta) + i(\xi_0 - \eta_0)$,
$(\xi + i\xi_0)(\eta + i\eta_0) = (\xi\eta - \xi_0\eta_0) + i(\xi_0\eta + \xi\eta_0)$,
$(\xi + i\xi_0)/(\eta + i\eta_0) = (\xi\eta + \xi_0\eta_0)/(\eta^2 + \eta_0^2) + i(\xi_0\eta + \xi\eta_0)/(\eta^2 + \eta_0^2)$.

It has been also noticed that the product and the quotient of two intervals $A, B$ do not have as centers the product, resp. the quotient, of the centers of $A$ and $B$. Sunaga gives formulae for the tightest interval with center the quotient of $A, B$, which includes the interval-arithmetic quotient of the intervals $A, B$:

**Theorem 12.** If $\alpha \geq \alpha_0 \geq 0$, and $\beta > \beta_0 \geq 0$, then

$$\frac{(\alpha, \alpha_0)}{(\beta, \beta_0)} \subseteq (\frac{\alpha}{\beta}, \frac{\alpha_0 + (\alpha/\beta)\beta_0}{\beta - \beta_0}). \tag{3}$$

Further, the following important observation is made: "In practical calculations, it is often meaningless to calculate $\alpha\beta$ or $\alpha/\beta$ accurately and numerals should be rounded adequately. In such cases the following theorem is useful:

**Theorem 13.** If $A = (\alpha, \alpha_0)$, $\alpha \in (\alpha', \alpha_0')$, then $A \subseteq (\alpha', \alpha_0 + \alpha_0')$."

*Example 8.2.* We have $1.432 \in 1.43; 14$, hence $1.432; 50 \subseteq 1.43; 6$.

A result for multiplication analogous to Theorem 12 is given using the above result (see Example 8.2, part ii):

If $\alpha \geq \alpha_0 \geq 0$, and $\beta \geq \beta_0 \geq 0$, then

$$(\alpha, \alpha_0)(\beta, \beta_0) \subseteq (\alpha\beta, \alpha_0\beta_0 + \alpha_0\beta + \alpha\beta_0). \tag{4}$$

For practical computations the center-radius form has the advantage that the significant digits are not repeated twice (as in the end-point form, cf. e. g. the interval $[1.4135, 1.4145]$, where the digits 1.41 are repeated twice. The arithmetic rules using center-radius form (Theorem 11) and notation introduced in Example 8.1 (like $1.00; 1$) are demonstrated in some detail in Example 8.3 in [38]:

*Example 8.3.* i) Addition:
$1.689; 4 + 2.745; 1 = 4.434; 5$,
$3.624; 8 + 1.24; 3 = 4.864; 38 \subseteq 4.86; (0.4 + 3.8) \subseteq 4.86; 5$.

ii) Subtraction:
$3.429; 5 - 1.201; 2 = 2.228; 7$,
$6.724; 7 - 2.30; 4 = 4.424; 47 \subseteq 4.42; 6$.

iii) Multiplication:
$(0.4320; 5)(0.3810; 5) \subseteq 0.16459; 43 \subseteq 0.1646; 5$. ...

vi) $\dfrac{7\sqrt{2}-\pi\sqrt{3}}{\pi^2+\sqrt{3}} \in \dfrac{7\times 1.4142140;5-3.1415930;5\times 1.7320510;5}{(3.1415930;5)^2+1.7320510;5}$

$\subseteq \dfrac{9.899498;4-5.441399;3}{9.869607;4+1.732051;1} \subseteq \dfrac{4.458099;7}{11.601658;5} \subseteq 0.384264; 1.$

These examples demonstrate the calculation of guaranteed enclosure of rational expressions — an approach known as "naive interval arithmetic".

**Examples of numerical calculations** (Section 9 of [38]). This Section contains a proposal for an application of interval analysis to reliable numerical computation (on digital computers). The form of the presentation is a detailed discussion of five case studies, which we shall next comment in some detail.

*Example 9.1.* This case study is devoted to the solution of nonlinear equations by means of a proposed Newton-like interval method. The author describes the method on a particular problem, namely the equation $f(x) \equiv x^3 - 3x + 1 = 0$ as follows. It is assumed first that we know an interval $X_0$ containing a solution. The method produces a sequence of intervals $X_1, X_2, \ldots$.

We have $f(0.3) = -f(0.4)$, hence there exists a solution $\alpha \in [0.3, 0.4] = X_0$. The function value at the middle point $\mathrm{mid}(X_0) = 0.35$ is $f(0.35) = -0.007125$. We have $f'(x) = 3(x^2 - 1)$, therefore $f'(X_0) = f'(0.3 \vee 0.4) \subseteq 3([0.3, 0.4]^2 - 1) \subseteq -[2.52, 2.73]$. Denoting $dx = \alpha - \mathrm{mid}(X_0)$ we obtain

$$0 = f(\alpha) = f(\mathrm{mid}(X_0) + dx) = f(\mathrm{mid}(X_0)) + f'(\mathrm{mid}(X_0) + \theta dx)dx,$$

where $0 \leq \theta \leq 1$. Solving for $dx$ we get (using outward rounding!)

$$dx = -\frac{f(\mathrm{mid}(X_0))}{f'(\mathrm{mid}(X_0) + \theta dx)} \in -\frac{f(\mathrm{mid}(X_0))}{f'(X_0)}$$

$$= -\frac{0.007125}{[2.52, 2.73]} \subseteq -[0.0026, 0.0029].$$

Therefore

$$\alpha = \mathrm{mid}(X_0) + dx \in \mathrm{mid}(X_0) - \frac{f(\mathrm{mid}(X_0))}{f'(X_0)}$$

$$= 0.35 - [0.0026, 0.0029] = [0.3471, 0.3474] = X_1.$$

To formulate the above method symbolically, assume that $X_0$ contains a zero of $f(x) = 0$. The next interval $X_1$ containing the solution is computed using outward computer arithmetic by means of the formula:

$$X_1 = \mathrm{mid}(X_0) - f(\mathrm{mid}(X_0))/f'(X_0),$$

where $\text{mid}(X_0)$ is the middle point of $X_0$ and $f'(X_0)$ is the range of the function $f'$ on $X_0$. The procedure can be continued, that is we have

$$X_{n+1} = \text{mid}(X_n) - f(\text{mid}(X_n))/f'(X_n), \quad n = 0, 1, 2, \ldots \tag{5}$$

until the needed accuracy has been achieved. From the derivation of the method it becomes clear, that instead of the midpoint any other point $\gamma$ of the interval $X_0$ can be used; and this is actually done in the next example, where instead of $\gamma = \text{mid}(X_0)$ the left end-point of $X_0$ is taken (due to the fact that a table value is known only for the end-points).

Let us note that during the computations (here and in other examples below) the author makes use of the rule (1) saying that the range of a function is contained in the corresponding interval arithmetic expression obtained by formally replacing the real arithmetic operations by interval-arithmetic ones. For example, the range of the function $f'(x) = 3(x^2 - 1)$ over the interval $X_0 = [0.3, 0.4]$ is computed as follows: $f'(X_0) = 3(x^2 - 1)|_{x \in X_0} \subseteq 3(X_0^2 - 1) = 3([0.3, 0.4]^2 - 1) \subseteq -[2.52, 2.73]$, or in Sunaga's notation $f'(0.3 \vee 0.4) \to 3((0.3 \vee 0.4)^2 - 1) \to -(2.52 \vee 2.73)$.

*Example 9.2.* The author considers one more example for the reliable solution of nonlinear equations. Here the method described in Example 9.1 is implemented to the equation $f(x) = xe^x - 2\sin x$ with initial interval $X_0 = [0.62, 0.63]$.

It is assumed that table values for $e^x$, $\sin x$ and $\cos x$ at the endpoints of the interval $X_0$ are known approximately, namely:

$$e^{0.620} \in \langle 1.8589280 \rangle, \quad e^{0.630} \in \langle 1.8776107 \rangle, \quad \sin 0.620 \in \langle 0.5810352 \rangle,$$
$$\sin 0.630 \in \langle 0.5891447 \rangle, \quad \cos 0.620 \in \langle 0.81387 \rangle, \quad \cos 0.630 \in \langle 0.80815 \rangle.$$

Since there is no table value for $f(0.625)$, that is at the midpoint of the interval $X_0$, the method of Example 9.1. is considered in the form $X_1 = \gamma - f(\gamma)/f'(0.62 \vee 0.63)$, $\gamma \in X_0$, and $\gamma$ is taken to be the left endpoint of $X_0$ (instead of midpoint), $\gamma = 0.62$. Then the first approximation $X_1$ is calculated (using outward roundings) to be $X_1 = [0.6265, 0.6270]$. The computations are performed as follows:

$$f(0.62) = 0.62e^{0.62} - 2\sin 0.62 \in 0.62 \times 1.858928; 1 - 2 \times 0.581035; 1$$
$$\subseteq 1.1525354; 7 - 1.162070; 2 \subseteq -0.009535; 9;$$

$$f'(0.62 \vee 0.63) \subseteq (1 + (0.62 \vee 0.63))e^{0.62 \vee 0.63} - 2\cos(0.62 \vee 0.63)$$
$$\subseteq (1.62 \vee 1.63)(1.858 \vee 1.878) - 2(0.808 \vee 0.814)$$
$$\subseteq (3.009 \vee 3.062) - (1.616 \vee 1.628) \subseteq 1.381 \vee 1.446;$$

$$dx \in -\frac{f(0.62)}{f'(0.62 \vee 0.63)} \subseteq \frac{(9.526 \vee 9.544) \times 10^{-3}}{1.381 \vee 1.446} \subseteq (6.58 \vee 6.91) \times 10^{-3}.$$

Let $\alpha$ be the required root, then $\alpha = 0.62 + dx \in 0.6265 \vee 0.6270 = X_1$. The next step is to proceed in a similar manner, calculating

$$dx \in -f(0.6267)/f'(0.6265 \vee 0.6270).$$

This problem differs from the previous one in the way of computation of the values of the function and its derivative at the end-points of the intervals; here we cannot obtain exact values as in Example 9.1, because we are given inexact table values for the functions $\exp x$, $\sin x$ and $\cos x$, which must be outwardly rounded; for instance: $\exp 0.620 \in \langle 1.18589280 \rangle = 1.858928; 1$. Now, to repeat the process, we need to compute $f(\text{mid}(X_1)) = f(0.6267)$, but such a value is not available in the table used. Hence, this example motivates the need of reliable methods for interpolation, which are discussed in the next example.

*Example 9.3.* In this example the problem of reliable interpolation of functional values with prescribed errors is considered. It is to be noted that in the today's literature of interval analysis there exist many methods devoted to interpolation of points in the plane with given errors in the y-coordinates, without an assumption that the points lie on the graph of a certain smooth function. Here the author considers namely the latter formulation. His method is again based on the interval mean-value theorem and has been demonstrated on the reliable interpolation of nonexact (table) functional values of the function $\exp x$. We give below the arguments of the author.

"Interval relations such as

$e^{0.620} \in 1.85892800; 5$, $e^{0.621} \in 1.86078790; 5$

can be used to evaluate $e^{0.620+v}$, $v \in 0.000 \vee 0.001$, as follows. Using that $f(x_1 + dx) \in f(x_1) + f'(x)dx$, where $x_1 \in x$, $x_1 + dx \in x$, then, by means of the relation

$$e^{0.620+v} \in e^{0.620} + e^{0.620 \vee 0.621}v,$$

we proceed to obtain a sufficiently accurate value. For instance, when $v = 0.0004$, we have:

$$\begin{aligned}
e^{0.6204} &\in 1.85892800; 5 + (1.8589 \vee 1.8608)0.0004 \\
&\subseteq 1.85892800; 5 + (1.8599; 10)4 \times 10^{-4} \\
&\subseteq 1.85892800; 5 + 7.4396; 40 \times 10^{-4} \subseteq 1.85967196; 45.
\end{aligned}$$

Thus, we have seen that making use of differential coefficients, we can increase the accuracy of the interpolation".

*Example 9.4.* This Example is devoted to the reliable computation of definite integrals; as a particular case study the computation of the definite integral $\int_0^1 1/(1+x^2)dx$ by a modification of Simpson's rule is considered. The author's idea is to enclose the remainder term by carefully computing the range of the fourth derivative of the subintegral function (using directed floating-point arithmetic) over each subinterval of the mesh. The idea can be easily extended to a large class of formulae for numerical quadratures. Below we reproduce the calculations from [38]:

"Simpson's rule is as follows:

$$\int_{-h}^{h} f(x)dx \in \tfrac{h}{3}\left(f(h)+f(-h)+4f(0)\right) - \tfrac{h^5}{90}f^{(4)}(-h \vee h),$$

hence it can be written

$$\int_a^b f(x)dx \in \tfrac{h}{3}\left(y_0 + y_{2n} + 2(y_2 + ... + y_{2n-2}) + 4(y_1 + ... + y_{2n-1})\right)$$
$$- \tfrac{h^5}{90}\sum_{i=1}^{n} f^{(4)}\left(x_{2i-1} + (-h \vee h)\right),$$

where $h = (b-a)/2n$. For instance, to integrate numerically

$$\int_0^1 \tfrac{dx}{1+x^2} \left(= \tfrac{\pi}{4}\right)$$

one proceeds as follows

$$f(x) = \tfrac{1}{1+x}, \quad f^{(4)}(x) = \tfrac{4!(1-x^2(10-5x^2))}{(1+x^2)^5}, \quad \tfrac{f^{(4)}(0\vee0.2)}{4!} \subseteq 1 \vee 0.49.$$

Similarly,

$$\tfrac{f^{(4)}(0.2\vee0.4)}{4!} \subseteq (-0.48) \vee 0.54, \quad \tfrac{f^{(4)}(0.4\vee0.6)}{4!} \subseteq -(0.06 \vee 1.15),$$
$$\tfrac{f^{(4)}(0.6\vee0.8)}{4!} \subseteq -(0.11 \vee 0.94), \quad \tfrac{f^{(4)}(0.8\vee1)}{4!} \subseteq -(0.06 \vee 0.49).$$

Then

$$\tfrac{1}{4!}(f^{(4)}(0 \vee 0.2) + f^{(4)}(0.2 \vee 0.4) + f^{(4)}(0.4 \vee 0.6)$$
$$+ f^{(4)}(0.6 \vee 0.8) + f^{(4)}(0.8 \vee 1)) \subseteq -2.57 \vee 1.31.$$

Hence

$$\int_0^1 \tfrac{dx}{1+x^2} \in \tfrac{0.1}{3}(1 + \tfrac{1}{2} + 2(\tfrac{1}{1.04} + \tfrac{1}{1.16} + \tfrac{1}{1.36} + \tfrac{1}{1.64})$$
$$+ 4((\tfrac{1}{1.01} + \tfrac{1}{1.09} + \tfrac{1}{1.25} + \tfrac{1}{1.49} + \tfrac{1}{1.81})) - \tfrac{(0.1)^5 4!}{90}(-2.6 \vee 1.4).$$

Finally one gets $\pi \in 3.141593;2$".

*Example 9.5.* In this Example the IVP for the ODE: $\eta' = f(\xi, \eta)$, $\eta(\xi_0) = y_0$ is considered; the value $y_0$ can be an interval one. Here we slightly change the original notation to make a difference between numerical (real) values $\xi, \eta$ and interval values $x, y$; in [38] all values are denoted using the letters $x, y$, which means that they may be intervals — in consistence with Sunaga's methodology. The author's method can be reformulated as follows. Consider a mesh: $\xi_0, \xi_1, \xi_2, \ldots$ . Denote the range of $\eta$ over a subinterval $x_k = \xi_k \vee \xi_{k+1}$ by $y_{k+1} = \eta(x_k)$, resp. the range of $\eta'$ over $x_k$ by $y'_k = \eta'(x_k)$. Then, using the interval mean-value theorem, we have:

$$\eta'_1 \in y'_1 = f(x_0, y_0), \ \eta_1 \in y_1 = y_0 + f(x_0, y_0)(\xi_1 - \xi_0);$$
$$\eta'_2 \in y'_2 = f(x_1, y_1), \ \eta_1 \in y_2 = y_1 + f(x_1, y_1)(\xi_2 - \xi_1).$$

The author points out that the accuracy can be increased by higher order derivatives. Thus the foundation of a large class of methods for the verified solution of IVP for ODE, using interval mean-value theorem, is given. Using the example: $y' = 2 - x/y$, $y(0) = 1$, he shows the technique of obtaining safe enclosures by special roundings and of increasing the accuracy by using higher order derivatives. His calculations are as follows:

"The x-axis is divided into $x_0 = 0 \vee 0.1$, $x_1 = 0.1 \vee 0.2, \ldots$ . Differential coefficients of high order are calculated from the relations $yy' = 2y - x$, $yy'' + y'^2 = 2y' - 1, \ldots$

Denote $y_0 = y(0 \vee 0.1)$, $y_1 = y(0.1 \vee 0.2), \ldots$ We then have

$$y'_0 \in 2 - \frac{(0 \vee 0.1)}{1} \subseteq 1.9 \vee 2, \ y_1 \in 1 + (1.9 \vee 2)x.$$

To raise its accuracy we must first calculate the differential coefficient of the second order, i. e.

$$y''_0 \in \frac{y_0(2 - y'_0) - 1}{y_0} \subseteq \ldots \subseteq -(0.66 \vee 1).$$

Since

$$y'|_{x=0} = 2, \ y_0 \subseteq 1 + 2x - \frac{0.66 \vee 1}{2}x^2;$$

therefore

$$y(0.1) \in 1 + 0.2 - (0.33 \vee 0.5) \times 10^{-2} \subseteq 1.1957; 9,$$

$y'(0.1) \in 2 - (0.66 \vee 1)0.1 \subseteq 1.917; 17.$

Next we go to the second interval $0.1 \vee 0.2$. In this interval we have

$y_1' \in 2 \vee \left(2 - \frac{0.2}{1.19}\right) \subseteq 2 \vee 1.83,$
$y_1 \in 1.19 \vee 1.20 + (2 \vee 1.83)(x - 0.1),$
$y(0.2) \in 1.19 \vee 1.20 + (2 \vee 1.83)0.1 \subseteq 1.37 \vee 1.40.$

Using this value we can evaluate more accurately, i. e.,

$y_1' \in 2 - \frac{0.1 \vee 0.2}{1.19 \vee 1.40} \subseteq 2 - (0.07 \vee 0.17) \subseteq 1.83 \vee 1.93.$

Hence

$y_1'' \in \frac{y_1'(2 - y_0') - 1}{y_1} \subseteq \frac{(1.83 \vee 1.93)(0.07 \vee 0.17) - 1}{1.19 \vee 1.40} \subseteq -\left(\frac{0.67 \vee 0.88}{1.19 \vee 1.40}\right)$
$\subseteq -(0.47 \vee 0.74);$

$y_1 \in y(0.1) + y'(0.1)(x - 0.1) + \frac{y_1''}{2}(x - 0.1)^2$
$\subseteq 1.1957; 9 + 1.917; 17(x - 0.1) - \frac{(0.47 \vee 0.74)}{2}(x - 0.1)^2;$

$y(0.2) \in 1.1957; 9 + 0.1917; 17 - (30; 7) \times 10^{-4} \subseteq 1.3844; 33.$

As stated above, we can integrate in the interval $0.1 \vee 0.2$ as accurately as we wish and carry it out over a wide region. We cut the $x$-axis into intervals, and reconnect them again".

## 4.  Sunaga's Work and Later Research

Let us mention some later investigations in the field of interval analysis that are related to Sunaga's paper, and some known to us occasions, when his ideas have been exploited and further developed.

*The join notation.* The presentation of an interval as a "join" of its end-points has been used to denote inner arithmetic operations and to perform effective computations with intervals [14]. Recall that the inner difference $A -^- B$ is the algebraic solution of $B + X = A$ or of $A - X = B$ (at least one of these equations has a solution, and if both are solvable, then the solutions coincide). Note that $X = A -^- B$ always exists, therefore it defines an operation, called *inner subtraction*; this algebraically induced operation is useful for the treatment of interval-arithmetic equations and ranges of functions. Therefore it is important to have simple presentation of this operation in terms of interval end-points. The inner difference of the intervals $A = [\alpha_1, \alpha_2]$ and $B = [\beta_1, \beta_2]$ is the interval with end-points $\alpha_1 - \beta_1$ and $\alpha_2 - \beta_2$. We have $\alpha_1 - \beta_1 \leq \alpha_2 - \beta_2$ or $\alpha_1 - \beta_1 > \alpha_2 - \beta_2$ depending on the widths $w$ of the intervals $A$ and $B$, that is we have

$$A -^- B = \begin{cases} [\alpha_1 - \beta_1, \alpha_2 - \beta_2], & \text{if } w(A) \geq w(B), \\ [\alpha_2 - \beta_2, \alpha_1 - \beta_1], & \text{if } w(A) < w(B). \end{cases}$$

Using Sunaga's notation we can simply write $A -^- B = (\alpha_1 - \beta_1) \vee (\alpha_2 - \beta_2)$. Such notation is used for the rest of the inner operations (addition, multiplication and division) and is very useful for performing symbolic manipulations. Similar symbolic notations have been further elaborated in [4], [17] and have contributed much for the algorithmic implementation of the interval arithmetic (especially interval multiplication) in computer algebra systems [32].

*Interval functions, differentiation.* The sharp difference made by Sunaga between ranges of functions and interval arithmetic expressions outlines the subordinate role of interval arithmetic as a formal algebraic tool for the numerical computation or enclosure of ranges. Thus one of the main objectives of interval analysis is clearly formulated in his sections devoted to interval functions and differentiation. Using such methodology theorems for the presentation of ranges of monotone functions have been developed (see, e. g. [3], [16]). Furthermore, Sunaga's derivatives have been expressed in terms of inner interval arithmetic operations, see e. g. [15]; such expressions have been used to formulate efficient Newton-like methods (see, e. g. [5], [14]). The interval mean-value theorem and its application to numerical analysis is probably one of the most important Sunaga's discoveries.

*The center-radius form.* The analogies of the interval arithmetic operations in center-radius form with complex arithmetic noted by Sunaga are quoted in [1]. Such analogies are investigated by E. Kaucher [11] and more recently by V. Zyuzin [45]. A discussion of formulae related to (3) and (4) and their application can be found in [25], where it is noted that these formulae are special case of the complex interval arithmetic using discs, considered by P. Henrici [6]. A study of the algebraic properties of these operations and their relation to stochastic arithmetic has been undertaken in [20].

*The topological background.* In his section about the topological background of interval analysis Sunaga comes to the idea of algebraic embedding of intervals in a system with group properties. It should be noted that M. Warmus [42], [43] also proposes algebraic extensions. As it is well known, such extensions have been investigated in detail about a decade later — the isomorphic algebraic extensions of the order relation and the arithmetic operations are studied by E. Kaucher

[10], [11]. Other important studies of abstract algebraic systems related to intervals, and more generally to convex bodies, have been performed by O. Mayer [21], H. Ratschek and G. Schröder [33] etc. The investigation of the properties of the interval-arithmetic operations by algebraic isomorphic extensions is of considerable interest for the solution of algebraic problems involving intervals. Recent results in this direction (some of them related to the center-radius presentation of intervals) are reported in [18], [19].

*Applications.* Two of the tools found by Sunaga are very important for applications: i) the arithmetic in center-radius form together with formulae for rounding and inclusion, and ii) the interval mean-value theorem and its application to verification algorithms, such as the Newton-like method for nonlinear equations (5), introduced by Sunaga in Example 9.1. In relation to this method it should be noted, that the author does not discuss conditions for convergence and contraction of the sequence of intervals, etc. Similar questions have been later studied by R. Moore, who introduces the notion of finite and global convergence [24] (the Newton-like method of R. Moore differs from (5) by intersecting each new interval by the previous one to assure that the new interval is nested in the previous one). In his chapter on interval arithmetic in [24] R. Moore mentions the work [38] in his references. However, a number of Sunaga's important results are not mentioned there, such as the cancellation law, the center-radius form of the interval arithmetic operations, etc.

Sunaga's method described in Example 9.1 is a foregoer of Newton-like interval methods for solving nonlinear equations, proposed later. Sunaga seems to be the first who uses outwardly directed roundings so that the bounding intervals provide guaranteed enclosures of the solution. In the methods formulated in [5], [14], intersection with previous intervals has been avoided by means of a suitably chosen interval-arithmetic expression for the range of the Newton operator much in the lines of Sunaga's definition of interval function.

The discussion of the relation between the interval approach and the statistical one in Section 1 of [38] deserves attention. It seems to us that this author's view has been later developed in relation to parameter identification, cf. e. g. [22].

Sunaga's idea to use interval analysis for the construction of numerical methods with verification, made interval analysis an important tool in mathematical modelling in applied sciences, such as mechanical engineering [40], [41]. Another research area where interval analysis finds application is electrical engineering (see, e. g. [12], [27], [28], [29]).

A photograph of Teruo Sunaga

## 5.   A Brief Biography of T. Sunaga

- Teruo Sunaga is born April 20, 1929 in Tokyo.

- In March 1954 he graduates from the Department of Applied Mathematics in the Faculty of Engineering of the University of Tokyo and in 1956 he graduates from a master course [36].

- In March 1959 he graduates from the Doctor course of the Graduate School of Mathematical Physics of University of Tokyo.

- On April 1, 1959 he is appointed assistant professor at the Department of Mechanical Engineering for Production of University of Tokyo and on June 16 the same year he becomes associate professor at the same department.

- In March 1961 he becomes Doctor of Engineering from the University of Tokyo [39].

- On April 1, 1974 he becomes Professor at the Department of Mechanical Engineering for Production of University of Tokyo.

- From April 1, 1976 to March 31, 1978 he is Director of the University Computation Center.

- March 31, 1993 retires.

- May 7, 1993 becomes Professor Emeritus at Kyushu University.

- Prof. Teruo Sunaga died February 25, 1995.

## 6. Conclusion

The contribution of T. Sunaga to interval analysis and reliable computing is original and outstanding. He formulates and investigates the basic algebraic properties of the interval arithmetic operations together with the inclusion relation on a firm algebraic foundation. He formulates for the first time an interval form of Taylor's formula (the interval mean-value theorem and a formula of second order) and the center-radius form of the interval-arithmetic operations together with the formulae for "centered" multiplication and division. He demonstrates on a number of case studies how such tools can be applied for reliable numerical computation. He also shows the role of directed roundings for the safe numerical computation. In the Conclusion of his work the author gives a brief, clear and modest summary of his achievements as follows: *"An interval calculus is established algebraically from the lattice theoretical point of view so that it can be applied conveniently to the numerical calculation. Interval functions and functionals, and their differentiation are investigated and used effectively in some examples of applied analysis"*.

T. Sunaga is applied mathematician (by education); he works in diverse applicational areas like communication theory, mechanical engineering and planning of production [37]–[41]. In his applied work he makes a systematic use of interval analysis, which, in our opinion, deserves to be studied.

In [7] (see also [8], [9]) Prof. M. Iri writes: *Sunaga considered all computational procedures, which had been traditionally defined on real numbers, as being too ideal and proposed to replace them by the procedures on real intervals in order to make everything "more realistic". His ideas were certainly influenced by the concepts from topological algebra. ... Sunaga studied many different kinds of numerical procedures including the Taylor-series interval solution of the initial-value problem of ordinary differential equations.*

It is also interesting to read Sunaga's vision about "future problems":

- *To investigate problems of numerical calculation connected with higher dimensional mathematics, for instance, matrix inversion, partial differential equations, etc.,*

- *To investigate direct applications of the interval calculus to physical and engineering problems,*

- *To revise the structure of the automatic digital computer from the standpoint of interval calculus and topology,*

- *To prove the applicability to other fields, of our view that scientific laws should be stated essentially in the language of finite elements and discrete topology.*

It seems that a lot of this ambitious program has been fulfilled, but also a lot remains to be done.

## Acknowledgements

We are grateful to Prof. Eiji Kondo from the Kyushu University and to Prof. Masao Iri from the Chuo University for supplying us with information about Prof. T. Sunaga's scientific work, for his photograph and biographical data. Support from JSPS for a visit of the first author in Kyoto University, where this work has been initiated, and by the Bulgarian National Science Fund under grant No. MM–521/95 is gratefully acknowledged.

## References

1. G. Alefeld, J. Herzberger, Introduction to Interval Computations, Academic Press, New York, 1983.
2. G. Birkhoff, Lattice Theory, AMS Colloquium Public., 25, New York, 1940.
3. N. Dimitrova, S. Markov, On the Interval-arithmetic Presentation of the Range of a Class of Monotone Functions of Many Variables. Computer Arithmetic, Scientific Computation and Mathematical Modelling, E. Kaucher, S. Markov, G. Mayer (Eds.), J. C. Baltzer Publ., IMACS, 12, 1991, 213–228.
4. N. Dimitrova, S. Markov, E. Popova, Extended interval arithmetics: new results and applications. In: Computer arithmetic and enclosure methods (Eds. L. Atanassova, J. Herzberger) North-Holland, Amsterdam, 1992, 225–232.
5. N. Dimitrova, S. Markov, A Validated Newton Type Method for Nonlinear Equations, Interval Computation, 1994, 2, 27-51.
6. P. Henrici, Circular arithmetic and the determination of polynomial zeros. Springer Lecture Notes 228 (1971), 86–92.
7. M. Iri, Guaranteed Accuracy and Fast Automatic Differentiation, KITE Journal of Electronics Engineering, Vol. 4., No. 1A, 34–40, 1993.
8. M. Iri, Automatic Differentiation, Special lecture, Bull. Jap. Soc. Industrial and Applied Mathematics, Vol. 3, 1, March, 1993.

9. M. Iri, The Role of Automatic Differentiation in Nonlinear Analysis and High-Quality Computation, TRISE 96-05, Technical report, Dept. of Information and System Engineering, Faculty of Science and Engineering, Chuo University, Tokyo, Japan, 1–12, 1996.

10. E. Kaucher, Algebraische Erweiterungen der Intervallrechnung unter Erhaltung der Ordnungs- und Verbandstrukturen, Computing Suppl., 1 (1977), 65–79.

11. E. Kaucher, Interval Analysis in the Extended Interval Space IR, Computing Suppl. 2, 33–49 (1980).

12. L. Kolev, Interval Methods for Circuit Analysis, World Scientific Publ., 1993.

13. U. Kulisch, W. L. Miranker, Computer Arithmetic in Theory and Practice, Academic Press, New York, 1981.

14. S. Markov, Some Applications of Extended Interval Arithmetic to Interval Iterations, Computing Suppl. 2, 1980, 69–84.

15. S. Markov, Interval Differential Equations, Interval Mathematics 1980 (Ed. K. Nickel), Acad. Press, 1981, 145-164.

16. S. Markov, On the Presentation of Ranges of Monotone Functions Using Interval Arithmetic; Interval Computations, No. 4(6) (1992), 19–31.

17. S. Markov, On Directed Interval Arithmetic and its Applications, J. UCS, 1, 7 (1995), 510–521.

18. S. Markov, An Iterative Method for Algebraic Solution to Interval Equations, 1998, to appear in Applied Numerical Mathematics.

19. S. Markov, On the Algebraic Properties of Convex Bodies and Some Applications, submitted to J. Convex Analysis.

20. S. Markov, R. Alt, On the Relation Between Stochastic and Interval Arithmetic, preprint.

21. O. Mayer, Algebraische und metrische Strukturen in der Intervallrechnung und einige Anwendungen, Computing 5 (1970), 144–162.

22. M. Milanese, J. Norton, H. P.-Lahanier, E. Walter (Eds.), Bounding Approaches to System Identification. Plenum Press, London, N. Y., 1996.

23. R. Moore, Interval Arithmetic and Automatic Error Analysis in Digital Computing, Applied Math. & Stat. Lab., Stanford University Technical Report No. 25 (1962); also: PhD Dissertation, Stanford University, October 1962.

24. R. Moore, Interval Analysis, Prentice-Hall, Englewood-Cliffs, N. J., 1966.

25. A. Neumaier, A Distributive Interval Arithmetic, Freiburger Intervall-Berichte 82/10, Inst. f. Angewandte Mathematik, U. Freiburg i. Br., 1982, 10, 31–38.

26. J. v. Neuman, H. H. Goldstine, Numerical Inverting of Matrices of High Order, Bull. AMS, 53, 11, 1947, 1021–1099.

27. K. Okumura, S. Saeki and A. Kishima, On an Improvement of an Algorithm Using Interval Analysis for Solution of Nonlinear Circuit Equations, Trans. of IECEJ, J69-A, 4, 489–496, 1986 (in Japanese).

28. K. Okumura, An Application of Interval Operations to Electric Network Analysis, Bull. of the Japan Society for Industrial and Applied Mathematics, 3, 2, 15–27, 1993 (in Japanese).

29. K. Okumura, Recent Topics of Circuit Analysis: an Application of Interval Arithmetic, J. of System Control Information Society of Japan, v. 40, 9, 393–400, 1996 (in Japanese).

30. M. Petrovich, Calculation with Numerical Intervals, Beograd, 1932 (in Serbian, Serbian title: Racunanje sa brojnim razmacima).

31. L. Pontrjagin, Topological Groups, Princeton Univ. Press, Princeton, 1946.

32. E. Popova, C. Ullrich, Embedding Directed Intervals in Mathematica. Revista de Informatica Teorica e Applicada, 3, 2, 1996, pp.99-115.

33. H. Ratschek, G. Schröder, Representation of Semigroups as Systems of Compact Convex Sets, Proc. Amer. Math. Soc. 65 (1977), 24–28.

34. L. Schwartz, Theorie des distributions, I, Hermann, Paris, 1950.

35. C. E. Shannon, The Mathematical Theory of Communication. The University of Illinois Press, Urbana, 1949.
36. T. Sunaga, Geometry of Numerals, Master Thesis, University of Tokio, February 1956.
37. T. Sunaga, A Basic Theory of Communication, Memoirs, 2, G-1 (1958), 426–443.
38. T. Sunaga, Theory of an Interval Algebra and its Application to Numerical Analysis, RAAG Memoirs, 2, Misc. II, 1958, 547–564.
39. T. Sunaga, Algebra of Analysis and Synthesis of Automata (in Japanese) Dr. thesis, University of Tokio, February 1961.
40. T. Sunaga, Differential Decreasing Speed Using Small Number of Differences of Teeths of Gear, Trans. of Japan Society of Mechanical Engineers, v. 39, No. 326, 1973, 3209–3216 (in Japanese).
41. T. Sunaga, Design and Planning for Production, Corona Publ., Tokyo, 1979 (in Japanese).
42. M. Warmus, Calculus of Approximations, Bull. Acad. Polon. Sci., Cl. III, Vol. IV, No. 5 (1956), 253–259.
43. M. Warmus, Approximations and Inequalities in the Calculus of Approximations. Classification of Approximate numbers, Bull. Acad. Polon. Sci., Ser. math. astr. et phys., vol. IX, No. 4, 1961, 241–245.
44. R. Young, The Algebra of Many-valued Quantities, Math. Annalen, 104, 1932, 260–290.
45. V. Zyuzin, On a Way of Representation of the Interval Numbers, SCAN-98 Conference materials (Extended Abstracts), Budapest 1998, 173–174.

*T. Csendes (ed.), Developments in Reliable Computing* 189–202.
© 1999 *Kluwer Academic Publishers.*

# Surface-to-surface intersection with complete and guaranteed results

ERNST HUBER AND WILHELM BARTH{huber,barth}@eiunix.tuwien.ac.at
*Institute of Computer Graphics, Vienna University of Technology, A-1040 Vienna, Karlsplatz 13/186/1, Austria*

**Abstract.** We present an algorithm for the robust computation of the complete intersection curve of two general parametric surfaces based on interval arithmetic. The subdivision algorithm we introduce follows a divide-and-conquer-approach. It avoids loss of any parts of the intersection curve by using safe bounding volumes for all parts (patches) of the surfaces. For each pair of patches, it first checks for intersection of the bounding volumes. If two bounding volumes intersect, it splits one patch, and treats both new pairs recursively until a predefined termination condition is satisfied. We use parallelepipeds as tight bounding volumes. Each parallelepiped considers the shape and orientation of its patch, overestimating it only by second order terms. With the help of interval inclusions for the partial derivatives, we compute parts of the parameter domain, where one patch cannot reach the enclosure of the other one. Cutting off such dispensable regions, and corresponding parts in object space result in a faster convergence of the algorithm. Interval arithmetic is used for all critical operations to achieve robustness.

**Keywords:** surface-to-surface intersection, general parametric surfaces, interval arithmetic

## 1. Introduction

### 1.1. Description of the problem

Geometric modeling is a basic technique for computer aided design and manufacturing processes. Engineers design a product with the help of a CAD program, make some computer aided simulations, finish the test and redesign cycle, and generate input code for NC-machines. Then the product can be manufactured automatically using CAM capabilities.

Performing calculations on an object presumes the existence of a mathematical description of it. In our example, an engineer creates a geometric model during the design cycle of the product. Geometric modeling is not only restricted to typical CAD/CAM applications. Further examples can be found in molecular modeling, in architecture, and in the simulation of environments (virtual reality).

Geometric modeling and operations on these models are summarized in the discipline of computer aided geometric design. An important property of a model is its completeness with respect to the specific application. For example, a wire-frame model is an incomplete geometric description because an object is described only by its edges, thus ambiguous models are possible. To avoid this problem, modern geometric modeling systems keep the complete geometric information (vertices, edges and faces) and topological structure (connection of the edges, vertices and faces) of an object.

Since an edge of an object is defined by the intersection of two surfaces, performing operations on geometric models leads inevitably to the intersection of surfaces. If two surfaces intersect, the result will be either a set of isolated points, a set of curves, a set of overlapping surfaces, or any combination of these three cases [2].

## 1.2.  Previous work

Depending on the properties and the representation of the surfaces, different methods for the computation of the intersection curve have been introduced. Because an exact solution of the surface intersection problem can be given only for some special surface classes, approximation methods are used for the general case. Such methods are qualified according to the attributes efficiency, generality, and robustness [9].

*Marching methods* start from a known common point of both surfaces and march along the intersection curve by successively computing new points and connecting them by line segments. The problem of this approach is its lack of robustness: even simple surfaces possibly intersect in complex curves with several branches, closed loops, and singular points. Such cases are difficult to detect and are a major problem for surface intersection algorithms [2].

*Subdivision methods* divide both surfaces until each patch can be replaced by a simple approximation. For each pair of these approximations (one patch of each surface) the intersection problem can be solved easily. This approach has some weaknesses: it leads to a huge number of, for instance, plane-to-plane intersections and, due to the approximation of the patches, some parts of the solution might be lost. An algorithm using bounding volumes for the surfaces is introduced in [6]. Figure 1a (reduced to 2D) shows that intersection points (equivalent to parts of the intersection curve in 3D) can be lost in algorithms working with approximations. All intersection points are detected if bounding volumes are used (Figure 1b).

The algorithm presented in this paper has following advantages: Due to the usage of interval arithmetic it guarantees that no part of the intersection curve gets lost. It is faster than other safe algorithms because it works with tight enclosures for the patches. It gains further speedup against our algorithm from [7] by cutting off dispensable regions with interval methods.

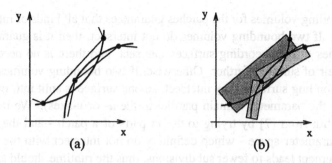

*Figure 1.* 2D-case: Comparison of linear approximations and bounding volumes.

## 2.    A robust surface intersection algorithm

### 2.1.   Basic algorithm

This algorithm computes the intersection curve of general parametric surfaces defined by

$$\vec{s}(u,v) = \left\{ \begin{pmatrix} x(u,v) \\ y(u,v) \\ z(u,v) \end{pmatrix} \middle| \begin{array}{l} u \in U = [\underline{u},\overline{u}] \\ v \in V = [\underline{v},\overline{v}] \end{array} \right\}, \tag{1}$$

where $U$ and $V$ are intervals (Notation: Small letters denote scalar values and functions, capital letters denote intervals and functions returning intervals, vectors are denoted by a $\rightarrow$.) in $\mathbb{R}$. The coordinate functions map a rectangular parameter domain to object space. It is only assumed that these functions are continuous in the whole domain and that the partial derivatives exist and are continuous. The algorithm follows a divide-and-conquer approach, described by the following pseudo code:

**Algorithm 1** *Improved divide-and-conquer SSI algorithm.*

| | |
|---|---|
| 1.    *intersect(Surface $f_i, g_j$)* | |
| 2.          $f_i^* \leftarrow cutDomain(f_i, bounds(g_j))$ | *// see Sections 2.2 and 3.3* |
| 3.          *if boundsIntersect($f_i^*, g_j$) then* | *// see Section 2.3* |
| 4.                *if termCondition($f_i^*, g_j$) then* | *// see Section 4* |
| 5.                      *writeResult()* | |
| 6.          *else* | |
| 7.                *splitSurface($f_i^*, f_{2i+1}, f_{2i+2}$)* | |
| 8.                *intersect($g_j, f_{2i+1}$)* | |
| 9.                *intersect($g_j, f_{2i+2}$)* | |

Using bounding volumes for the patches guarantees that all kinds of intersections are detected. If two bounding volumes do not intersect, then it is guaranteed that no sub-patches of the according surfaces intersect, and there is no need to investigate this pair of surfaces further. Otherwise, if two bounding volumes intersect, their corresponding surfaces may intersect, so one surface is split into two patches (by dividing the parameter domain parallel to the $u$- or $v$-axes). We improve the basic algorithm from [7] by trying to detect parts of a patch - and the according regions in parameter space - which definitely do not intersect with the other one. This improvement leads to fewer subdivisions, thus the runtime should be reduced. After cutting off these parts, we compute tight bounding volumes for the reduced patches and test them against the bounding volume of the other one. This procedure is repeated recursively (the patches are divided alternately), stopping if a predefined termination condition is satisfied.

The algorithm provides two kinds of results. First, it returns in object space a set of intersections of bounding volumes. Secondly it provides for both parameter domains a collection of rectangular pieces supporting the curve. The properties of interval arithmetic guarantee inclusion for all parts of the intersection curve.

## 2.2. Tight bounding volumes

Evaluating the formulas of the coordinate functions of a surface by interval arithmetic provides an axis aligned bounding box (AABB) for the surface. AABBs generally overestimate the enclosed patches (see Figure 2 for the 2D case), thus leading to unnecessary subdivisions and intersection tests. Therefore, we use tight parallelepipeds. A parallelepiped is a better enclosure, taking the shape and orientation of the patch into account [3].

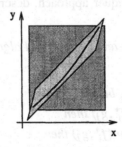

*Figure 2.* 2D-case: Comparison of an axis aligned bounding box and a parallelogram.

We compute interval enclosures for the partial derivatives with respect to $u$ and $v$ during the evaluation of the formulas by a combination of automatic differentiation and interval arithmetic (see [10], Section 2.2), e.g. for $x(u,v)$:

$$\left.\begin{array}{rl} x(u,v) & \in X \\ x_u(u,v) & \in X_u \\ x_v(u,v) & \in X_v \end{array}\right\} \forall u \in U, v \in V.$$

From these intervals we compute an enclosure for the surface. We apply the mean value theorem of differential calculus for two independent variables, e.g. for $x(u,v)$:

$\forall u, u_0 \in U, v, v_0 \in V$ exists a $\tau_u \in [u, u_0] \subseteq U, \tau_v \in [v, v_0] \subseteq V$, so that

$$x(u,v) = x(u_0,v_0) + (u-u_0)x_u(\tau_u,v_0) + (v-v_0)x_v(u,\tau_v), \qquad (2)$$

and analogous for $y(u,v)$ and $z(u,v)$. Replacing the partial derivatives by the corresponding intervals, $u$ by $\underline{u}$ and $v$ by $\underline{v}$ let us rewrite:

$$\vec{s}(\underline{u},\underline{v}) \in \vec{B}(\underline{u},\underline{v}) = \vec{s}(u_0,v_0) + (\underline{u}-u_0)\vec{S}_u + (\underline{v}-v_0)\vec{S}_v \qquad (3)$$

with $\vec{S}_u = (X_u, Y_u, Z_u)$ and $\vec{S}_v = (X_v, Y_v, Z_v)$. This equation provides an enclosure for $\vec{s}(u,v)$ with $u \in [\underline{u}, u_0], v \in [\underline{v}, v_0]$ and a bounding box $\vec{B}(\underline{u},\underline{v})$ containing the corner point $\vec{s}(\underline{u},\underline{v})$. Repeating this step for the other corner points allows us to state the corollary:

COROLLARY 1 *Each convex volume, especially each parallelepiped, enclosing the four corner boxes* $\vec{B}(\underline{u},\underline{v})$, $\vec{B}(\underline{u},\overline{v})$, $\vec{B}(\overline{u},\underline{v})$, *and* $\vec{B}(\overline{u},\overline{v})$ *is a bounding volume for the surface* $\vec{s}(u,v)$ *from Equation (1).*

A parallelepiped can now be computed by finding three pairs of parallel planes, one of them is chosen parallel to a plane approximating the four corner points. Each pair is defined by a normal vector and two constants for the distances from the origin. The distance constants are computed with interval arithmetic in such a way, that the epiped guarantees to enclose completely the four corner boxes.

## 2.3.   *Intersection test*

Another major operation we need for this algorithm is the test whether two bounding volumes intersect (denoted by **boundsIntersect** in Algorithm 1). While this test is trivial for AABBs, it is not for parallelepipeds. Since parallelepipeds are convex polyhedra, we may apply the separating axis theorem [5]:

THEOREM 1 *Two convex polyhedra do not intersect if and only if there exists a separating plane which is either parallel to a face of one polyhedron or which is parallel to at least one edge of each polyhedron.*

An axis normal to a separating plane is called a *separating axis*. Projecting both polyhedra on a separating axis (by an orthogonal projection), we obtain an interval for each polyhedron. If at least one separating axis exists, where the projected polyhedra (the intervals) do not intersect, then the polyhedra do not intersect. Otherwise the polyhedra intersect. For parallelepipeds we have to consider at most 15 separating axes (3 normals to the faces of each epiped and $3 * 3$ normals to planes defined by pairwise combining the edge-directions of both epipeds). With the help of interval arithmetic our algorithm detects cases, where the epipeds surely do not intersect (the contrary is: "they probably intersect").

To take advantage of both bounding volume types, we suggest the combined use of them. This hybrid algorithm first tests whether two AABBs intersect. If this evaluates to true, then in a second stage the parallelepipeds are calculated and tested. The recursion is only entered if the parallelepipeds probably intersect.

## 3. Fast subdivision

In this section we develop the **cutDomain**-step from Algorithm 1. The algorithm relies on enclosing the patches of the surfaces and splitting them by dividing the parameter domain of one patch along the $u$- or $v$-axis. Considering a pair of intersecting patches, the subdivision process can be accelerated by an early detection of parts of a patch which definitely do not intersect with the other one.

Parts of the parameter domain where one patch cannot reach the enclosure of the other one can be determined from the intervals of the partial derivatives (similar to the interval Newton method, see [1], Sections 7 and 19). By cutting off such parts, we reduce the number of surface splits, thus the number of bounding volume computations and intersection tests is reduced.

### 3.1. Condition for dispensable $(u,v)$-parts

Let us consider a patch $\vec{s}$ and a plane $\varepsilon$ which is given by a normal vector $\vec{n}$ and the distance $\lambda$ of the plane from the origin. The problem is stated as follows: determine parts of $\vec{s}$ and the corresponding regions in the parameter domain which definitely lie on the same side of $\varepsilon$ as a corner point $\vec{c} = \vec{s}(u_c, v_c)$ of $\vec{s}$.

For each point of the patch, we can compute an enclosure by an application of the mean value theorem of two independent variables as in Section 2.2:

$$\vec{s}(u,v) \in \vec{c} + \underbrace{(u - u_c)}_{\Delta u}\vec{S}_u + \underbrace{(v - v_c)}_{\Delta v}\vec{S}_v. \tag{4}$$

We start with the computation of the distance $d$ between a corner point $\vec{c}$ and $\varepsilon$. We may state: a surface point $\vec{s}(u, v)$ lies on the same side as $\vec{c}$, if all points of its

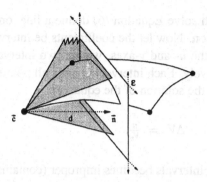

*Figure 3.* Calculating points of surface $\vec{s}$ which definitely lie on the same side of the plane $\varepsilon$ as the corner point $\vec{c}$.

enclosure $(\vec{c} + \Delta u \vec{S}_u + \Delta v \vec{S}_v)$ lie on the same side as $\vec{c}$. The condition for this is:

$$(\Delta u \vec{S}_u + \Delta v \vec{S}_v) \cdot \vec{n} < d, \qquad \text{if } d \geq 0,$$
$$(\Delta u \vec{S}_u + \Delta v \vec{S}_v) \cdot \vec{n} > d, \qquad \text{otherwise.}$$

Only points $\vec{s}(u, v)$, where the according points $(u, v)$ of the parameter domain solve the linear interval equation

$$(\Delta u \vec{S}_u + \Delta v \vec{S}_v) \cdot \vec{n} = d \tag{5}$$

can lie on $\varepsilon$. Thus Equation (5) provides boundaries separating the parameter domain of the patch into three regions:

1. a region defining all points of $\vec{s}$ which definitely lie on the same side of $\varepsilon$ as $\vec{c}_i$,

2. a region defining all points of $\vec{s}$ which possibly lie on $\varepsilon$,

3. a region defining all points of $\vec{s}$ lying definitely on the other side of $\varepsilon$.

Note that region 2 must contain all points that cannot be unambiguously assigned to region 1 or 3 due to the facts that $\vec{S}_u$ and $\vec{S}_v$ are intervals, and rounding errors occur when computing $d$ and solving Equation (5).

## 3.2. The solution of the linear equation

From Equation (5) we get:

$$U_n \Delta u + V_n \Delta v = d, \qquad d \in \mathbb{R}, \tag{6}$$

where $U_n$ and $V_n$ are coefficients defined by the dot product of the vectors for the partial derivatives and the normal vector to the plane. If $U_n$ and $V_n$ were reals, then all pairs $(\Delta u, \Delta v)$ which solve Equation (6) define a line, on the $u$- and $v$-axes we get two points as solution. Now let the coefficients be intervals in $\mathbb{R}$, $U_n = [\underline{u}_n, \overline{u}_n]$ and $V_n = [\underline{v}_n, \overline{v}_n]$. For the $u$- and $v$-axes, we get two intervals $\Delta U = [\Delta \underline{u}, \Delta \overline{u}]$ and $\Delta V = [\Delta \underline{v}, \Delta \overline{v}]$, respectively. Each interval contains all points of the corresponding axis which are parts of the solution of the equation:

$$\Delta U = \tfrac{d}{U_n}, \qquad \Delta V = \tfrac{d}{V_n}.$$

Note that each of these intervals becomes improper (contains $\infty$) if the corresponding denominator interval contains zero (it reaches, e.g. for $u$, from a positive $\Delta \underline{u}$ via $\infty$ to a negative $\Delta \overline{u}$). Each line defined by one point of both intervals $\Delta U$ and $\Delta V$ respectively, is a part of the solution. Depending whether these intervals are proper or improper, we have to distinguish the following cases:

(a)  four cases with proper $\Delta U$ and proper $\Delta V$,

(b)  two cases with proper $\Delta U$ and improper $\Delta V$ and two cases with improper $\Delta U$ and proper $\Delta V$,

(c)  one case with improper $\Delta U$ and improper $\Delta V$.

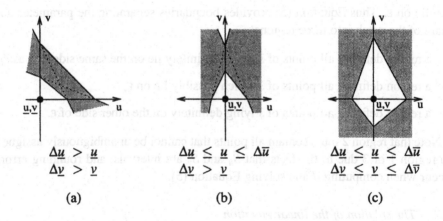

| $\Delta \underline{u} > \underline{u}$ | $\Delta \underline{u} \leq \underline{u} \leq \Delta \overline{u}$ | $\Delta \underline{u} \leq \underline{u} \leq \Delta \overline{u}$ |
| $\Delta \underline{v} > \underline{v}$ | $\Delta \underline{v} > \underline{v}$ | $\Delta \underline{v} \leq \underline{v} \leq \Delta \overline{v}$ |
| (a) | (b) | (c) |

*Figure 4.* Solutions for Equation (6).

Figure 4 shows this for the corner point $c = \vec{s}(\underline{u}, \underline{v})$ of the patch. Figures 4a and 4b show just one of four possible cases. We get the other three cases by appropriate mirroring.

### 3.3. Cutting off parts of the parameter domain

Now we can find the dispensable parts of the parameter domain of a patch $\vec{s}_1$ by applying the results of Section 2.2 to $\vec{s}_1$ and the planes of a parallelepiped $\mathcal{E}_2$ enclosing a patch $\vec{s}_2$. Parallelepiped $\mathcal{E}_2$ is represented by three pairs of parallel planes $(\underline{\varepsilon}_{2j}, \bar{\varepsilon}_{2j})$, $j \in \{1,2,3\}$ containing the faces of $\mathcal{E}_2$. Each pair of parallel planes is given by the normal vector $\vec{n}_{2j}$ and the distances $(\underline{\lambda}_{2j}, \bar{\lambda}_{2j})$ of the planes from the origin.

We determine parts of the parameter domain of $\vec{s}_1$, where $\vec{s}_1$ definitely lies outside the disc defined by one of the three pairs of planes of $\mathcal{E}_2$ (and can not reach $\vec{s}_2$). For each corner point (Note: For enumerating the corner points of a surface (and

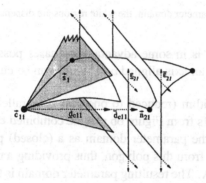

**Figure 5.** Calculating points of surface $\vec{s}_1$ which definitely do not intersect with parallelepiped $\mathcal{E}_2$ defined by 3 pairs of planes $(\underline{\varepsilon}_{2j}, \bar{\varepsilon}_{2j})$.

the according corner points of the parameter domain) we define $(u_{c_1}, v_{c_1}) = (\underline{u}, \underline{v})$, $(u_{c_2}, v_{c_2}) = (\bar{u}, \underline{v})$, $(u_{c_3}, v_{c_3}) = (\underline{u}, \bar{v})$, $(u_{c_4}, v_{c_4}) = (\bar{u}, \bar{v})$.) $\vec{c}_{1i}, i \in \{1,2,3,4\}$ from $\vec{s}_1$ and each pair of parallel planes from $\mathcal{E}_2$, we apply the following steps:

We compute the distances of $\vec{c}_{1i}$ to $\underline{\varepsilon}_{2j}$ and $\bar{\varepsilon}_{2j}$, yielding $\underline{d}_{c_{1i}}$ and $\bar{d}_{c_{1i}}$ with $\underline{d}_{c_{1i}} \leq \bar{d}_{c_{1i}}$. If $\vec{c}_{1i}$ does not lie between $\underline{\varepsilon}_{2j}$ and $\bar{\varepsilon}_{2j}$ (this is true if $\operatorname{sgn}(\underline{d}_{c_{1i}}) = \operatorname{sgn}(\bar{d}_{c_{1i}})$), then we may calculate a region of the parameter domain of $\vec{s}_1$ which is not part of the solution and can be cut off. Next we take the plane from the pair $(\underline{\varepsilon}_{2j}, \bar{\varepsilon}_{2j})$ closer to $\vec{c}_{1i}$. Solving the resulting linear equation as shown in Section 3.2 provides a region that may be cut off. Figure 6 shows several cases of such regions (colored white) for the well behaved case from Figure 4a. If the dispensable region is a triangle in the parameter domain (Figure 6a), then we cannot reduce the domain for this single corner point. In many cases it is possible to reduce the interval of one, or even both parameters (Figure 6b), resulting in an axis aligned box, which can be cut off from the parameter domain. If more than one corner point of $\vec{s}_1$ lies

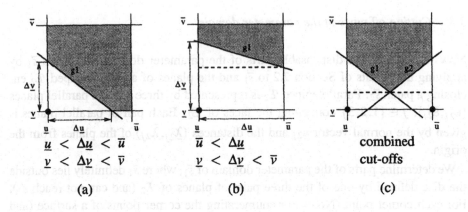

$$u < \Delta\underline{u} < \bar{u}$$
$$\underline{v} < \Delta\underline{v} < \bar{v}$$

(a)

$$\bar{u} < \Delta\underline{u}$$
$$\underline{v} < \Delta\underline{v} < \bar{v}$$

(b)

combined
cut-offs

(c)

*Figure 6.* Reducing the parameter domain, the white regions are dispensable.

exterior of $\mathcal{E}_2$, then it is in some advantageous cases possible to combine these parts and to find again an axis aligned box, which can be cut off (Figure 6c).

The following algorithm (reduced to one pair of parallel planes) takes care of both, the simple cut offs from Figure 6b and the combined cut offs from Figure 6c. For this we consider the parameter domain as a (closed) polygon. Each boundary line cuts off a part from this polygon, thus providing a new (smaller) polygon, which is always convex. The resulting parameter domain is the axis aligned bounding box of this polygon (Figure 7 shows a simple example with only one cut-off per corner).

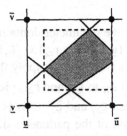

*Figure 7.* Calculating the reduced parameter domain.

This algorithm is summarized by the following pseudo code:

**Algorithm 2** *Reducing the parameter domain of a patch for one pair of parallel planes.*

1.   **cutDomain(Surface $\vec{s}$, Plane $\underline{\varepsilon}, \overline{\varepsilon}$)**
2.       $U_n \leftarrow \vec{S}_u \cdot \vec{n}$
3.       $V_n \leftarrow \vec{S}_v \cdot \vec{n}$
4.       **for** $i \leftarrow 1$ *to* $4$
5.           $[\underline{d},\overline{d}] \leftarrow$ ***Interval***$(dist(c_i,\underline{\varepsilon}), dist(c_i,\overline{\varepsilon}))$
6.           **if** $sgn(\underline{d}) = sgn(\overline{d})$ **then**
7.                $d \leftarrow sgn(\underline{d})\min(abs(\underline{d}), abs(\overline{d}))$
8.                $\Delta U_i \leftarrow \frac{d}{U_n}$
9.                $\Delta V_i \leftarrow \frac{d}{V_n}$
10.       **return cut-off**$(domain, \Delta U_{1...4}, \Delta V_{1...4})$       // *see Figure 7*

## 4. Termination

For termination several conditions can be used: in parameter space simply the size of the parameter domain of a patch relative to the size of the parameter domain of the original surface can be taken. Alternately in object space the volume of the intersection of two bounding volumes can be analysed. For example: if the smallest of the three cross-sections is small, then we have a thin stick as an enclosure for a part of the intersection. Additional restrictions can be made considering the linearity or flatness of the resulting curve or patch.

## 5. Results

The methods of interval arithmetic and automatic differentiation were implemented in an interpretative module for evaluating continuous functions. The whole experimental system was implemented in C++ (see *http://www.apm.tuwien.ac.at/research/ssi/*), for interval arithmetic we used the BIAS-library [8].

Several versions of this algorithm were implemented, one algorithm using AABBs only, one pure epiped algorithm, and a hybrid algorithm. The latter two can be executed with and without cut-offs as explained in Section 3. We demonstrate these algorithms by the following example illustrated in Figure 8:

$$\vec{s}_1(u_1, v_1) = \begin{pmatrix} \cos u_1 * \sin v_1 * (1 - \cos v_1)/2 \\ \sin u_1 * \sin v_1 * (1 - \cos v_1)/2 \\ \cos v_1 \end{pmatrix},$$

("drop") with parameter domain $u_1 \in [-\pi, \pi]$, $v_1 \in [0, \pi]$, and

$$\vec{s}_2(u_2, v_2) = \begin{pmatrix} v_2 \\ 0.4 * \sin(2\pi u_2) \\ 0.4 * \cos(2\pi u_2) \end{pmatrix},$$

("cylinder") with parameter domain $u_2 \in [0, 1]$, $v_2 \in [-1, 1]$.

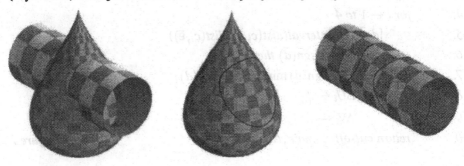

*Figure 8.* Intersecting objects and 3D intersection curve.

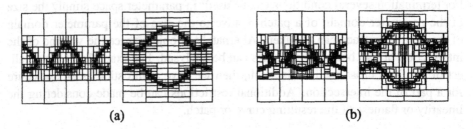

(a)                                                    (b)

*Figure 9.* Parameter domain subdivision, hybrid algorithm without (a) and with (b) cut-offs

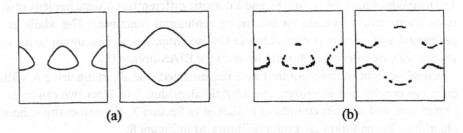

(a)                                                    (b)

*Figure 10.* Intersection curves in the parameter domains, with (a) and without (b) interval arithmetic.

For termination we used the simple condition $A_{D_i} \leq \frac{A_{D_0}}{10000}$, where $A_{D_i}$ is the area of the parameter domain of patch $i$, and $A_{D_0}$ is the area of the parameter domain of the original surface. Table 1 shows the number of iterations (surface splits) required to solve a problem with a predefined precision and the relative consumed time.

*Table 1.* Algorithm comparison.

|  | splits $\vec{s}_1$ | splits $\vec{s}_2$ | rel. time |
|---|---|---|---|
| AABB | 7317 | 8264 | 1.00 |
| Epiped | 3475 | 3734 | 0.93 |
| Hybrid | 2947 | 3264 | 0.72 |
| Epiped$_{+cut-off}$ | 1305 | 1404 | 0.58 |
| Hybrid$_{+cut-off}$ | 1105 | 1234 | 0.48 |

In this example the AABB algorithm shows for both qualification criteria the worst behavior. The epiped and hybrid algorithm are both better. The algorithm using cut-offs shows the best performance. Here the number of subdivisions is decreased to about $\frac{1}{7}$, and runtime is decreased to about $\frac{1}{2}$. This shows also that there is a trade-off between both criteria. Tighter bounding volumes lead to fewer subdivisions but are more expensive to compute.

Figure 9 shows the subdivision of the parameter domains of both surfaces provided by the hybrid algorithm without (Figure 9a) and with (Figure 9b) cut-offs. While the non improved variant bisects the parameter domains, the algorithm using cut-offs decomposes the domains irregularly.

Figure 10b and Figure 11 demonstrate the necessary of not using floating point arithmetic with rounding errors for this problem. Figure 10b shows degenerated intersection curves in the parameter domains of both surfaces. These curves were provided by the hybrid algorithm with cut-offs without using interval arithmetic for critical operations, and many parts of the solution are lost. Figure 11 shows two intersecting parallelepipeds, the resulting intersection volume is approximated by a parallelepiped. These epipeds intersect in a very small volume. Such parts can get lost if floating point arithmetic is used. From some examples where the floating point version returned a similar result as the interval version of the algorithm, we figured out that interval arithmetic increases runtime about 5 to 7 times, depending on the input functions.

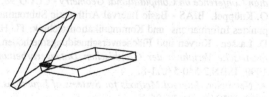

*Figure 11.* Intersecting ("touching") parallelepipeds.

## 6. Conclusion

We have introduced a robust subdivision algorithm for the computation of all parts of the intersection curve of two general parametric surfaces based on interval arithmetic. Surfaces are supplied to the algorithm in a natural way by specifying only the parametric functions. The evaluation of the functions and computation of the partial derivatives is done automatically by the system. Tight parallelepipeds and axis aligned bounding boxes are used as bounding volumes for the patches.

A cut off-step uses the intervals of the partial derivatives, which are already known from the computation of the epipeds. The number of subdivisions of the surfaces can be reduced substantially, resulting also in a remarkable time speedup of the algorithm for almost all investigated examples.

Using interval arithmetic instead of floating point arithmetic for all critical operations guarantees that the intersection curve lies inside the resulting bounding volumes (in object space) or rectangles (in the parameter domains). Using bounding volumes guarantees that all kinds of intersections are detected.

We are investigating further termination conditions and their impact on performance as well as strategies for sorting the resulting hulls. Additionally, efforts will be made to get even tighter inclusions for the patches, e.g. by higher order methods similar to those in [4].

## References

1. G. Alefeld and J. Herzberger. *Introduction to Interval Computations*. Academic Press, Inc., 1983. ISBN 0-12-049820-0.
2. R. Barnhill, G. Farin, M. Jordan, and B. Piper. Surface/Surface Intersection. *Computer Aided Geometric Design*, 4:3–16, 1987.
3. W. Barth, R. Lieger, and M. Schindler. Ray Tracing General Parametric Surfaces Using Interval Arithmetic. *The Visual Computer*, 10(7):363–371, 1994.
4. M. Berz and G. Hofstätter. Computation and Application of Taylor Polynomials with Interval Reminder Bounds. *Reliable Computing*, 4(1):83–97, February 1998.
5. S. Gottschalk. Separating Axis Theorem. Technical Report TR96-024, Department of Computer Science, UNC Chapel Hill, 1996.
6. E.G. Houghton, R.F. Emnett, J.D. Factor, and Ch.L. Sabharwal. Implementation of a Divide-and-Conquer Method for Intersection of Parametric Surfaces. *Computer Aided Geometric Design*, 2:173–183, 1985.
7. E. Huber. Intersecting General Parametric Surfaces Using Bounding Volumes. In *Tenth Canadian Conference on Computational Geometry - CCCG'98*, 1998.
8. O. Knüppel. BIAS - Basic Interval Arithmetic Subroutines. Bericht des Forschungsschwerpunktes Informations- und Kommunikationstechnik, TU Hamburg-Harburg, July 1993.
9. D. Lasser. Kurven und Flächenverschneidungsmethoden. In J. L. Encarnação, editor, *Geometrische Verfahren der Graphischen Datenverarbeitung*, pages 61–87. Springer Verlag, 1990. ISBN 3-540-53011-8.
10. A. Neumaier. *Interval Methods for Systems of Equations*. Cambridge University Press, Inc., 1990. ISBN 0-521-33196-X.

T. Csendes (ed.), *Developments in Reliable Computing* 203–212.
© 1999 *Kluwer Academic Publishers.*

# An Algorithm that Computes a Lower Bound on the Distance Between a Segment and $\mathbb{Z}^2$

VINCENT LEFÈVRE                                                  vincent.lefevre@ens-lyon.fr
*Laboratoire LIP, Project* ARÉNAIRE, *École Normale Supérieure de Lyon*

**Abstract.** We give a fast algorithm for computing a lower bound on the distance between a straight line and the points of a bounded regular grid. This algorithm is used to find worst cases when trying to round the elementary functions correctly in floating-point arithmetic. These worst cases are useful to design algorithms that guarantee the exact rounding of the elementary functions.

**Keywords:** elementary functions, floating-point arithmetic, rounding

## 1. Introduction

Current implementations of the usual math functions are frequently inaccurate for large or special (e.g., close to a multiple of $\pi$ for trigonometric functions) input arguments. Beyond the accuracy problem, this is a serious problem when one is willing to write portable software: calling the same function with the same arguments on different systems may lead to very different results. This is due to the lack of a rigorous specification of these functions in the IEEE-754 floating-point standard. We aim at showing that it is possible to require that these functions should be *exactly rounded* (the IEEE-754 standard requires that addition, subtraction, multiplication, division and square root should be exactly rounded). That is, when computing $f(x)$, where $f$ is exp, log, sin, cos, etc... and $x$ is a "machine number", i.e., a number that can be exactly represented, we want to show that we can always get the machine number that is the closest[1] to $f(x)$ [3]. The basic method for computing an elementary function is first to compute (with a precision that is somewhat higher than the "target" precision) an approximation to $f(x)$. Then we round the approximation. The problem is to know if we get the same result as if we had rounded the exact value $f(x)$. Indeed, if the approximation to $f(x)$ is not accurate enough, we cannot ensure that $f(x)$ is correctly rounded; this problem is known as the *Table Maker's Dilemma*. To solve this problem, we must know with which precision we must carry out the intermediate calculations; that is, we must know the smallest possible non-zero value of $|f(x) - y|$ where $x$ is a machine number and $y$ is either a machine number[2] or the average of two consecutive machine numbers, depending on the rounding mode. The purpose of this paper is to describe a fast algorithm to find this minimum value. We do *not* give here algorithms to compute the elementary functions for the target precision: once the minimum value is found, classical algorithms can be implemented to compute the elementary functions with the required accuracy. We see in the following that the Table Maker's Dilemma is closely related to the problem of computing a lower bound on the distance between a segment and a rectangular grid.

For a given $x$, the distance $|f(x) - y|$ is denoted $d_0$. The tests are performed in two steps [3]: the first step selects the machine numbers $x$ for which $d_0$ is smaller than a given number, thus reducing the number of candidates; the second step will test these few candidates with more accuracy. We will focus on the first step, that needs to be very fast, because the total number of points $x$ is very large: of the order of $10^{20}$ for the double-precision numbers.

We split the considered domain into very small intervals (so small that, in each of these intervals, a degree-1 approximation to $f$ will suffice for the first step), and in each interval $I$, we look for the set $S_I$ of machine numbers $x$ for which $d_0$ is smaller than a given real $\varepsilon_I$. We approximate the function by degree-1 polynomials $\ell_I$ (i.e., segments) with an error less than or equal to $\varepsilon_0$. Approximating the function can be done quickly enough; the part that takes most of the computation time is the tests themselves and we search for a very fast algorithm for them.

On the figure below, the distance $|\ell_I(x) - y|$ is denoted $d$. If $d \geq \varepsilon_I + \varepsilon_0$, then $d_0 \geq \varepsilon_I$. Therefore, in order to find points in $S_I$, we will look for the set $S'_I$ whose points are such that $d < \varepsilon_I + \varepsilon_0$. The intervals $I$ and the numbers $\varepsilon_I$ and $\varepsilon_0$ will be chosen small enough such that $S'_I$ is generally empty (for a reason explained later), but large enough to allow us to compute with small precision numbers: we need to avoid costly multiple-precision calculations.

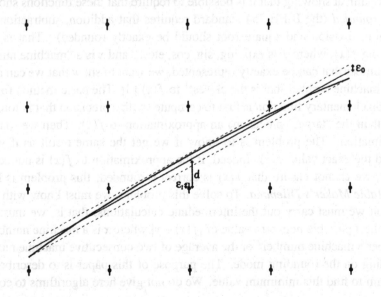

In the chosen domains, the machine numbers are regularly spaced, so that we can multiply the input and output numbers by adequately chosen constant powers of two to consider that they are, in fact, integers; $\varepsilon$ denotes the corresponding (adequately multiplied by a power of 2) value of $\varepsilon_I + \varepsilon_0$. Thus the problem is now: what are the points on the given segment such that the $x$-coordinate is an integer and the distance between the $y$-coordinate and the integers is less than a given $\varepsilon$? We recall that in most cases, there are no such points, from the choice of $\varepsilon$ (the cardinality of $S'_I$ being estimated with the following formula: $N \cdot \varepsilon$, where $N$ is the number of integers in the interval).

The naive approach consists in testing each point whose $x$-coordinate is an integer. If the number of points to test, denoted $N$, is large enough (e.g., 1000 or larger), there exists a faster method, using the fact that the set $S'_I$ is generally empty: we can look for a lower bound on the distance $d$, and if $d$ is larger than $\varepsilon$, then $S'_I$ is empty; otherwise, we can split the interval into subintervals and use this method with different parameters, or use the naive approach.

The segment has an equation of the form $y = ax - b$, where $x$ is restricted to a given interval, e.g., $0 \leq x < N$. In Section 2, we give some mathematical preliminaries and notations. In Section 3, we study the distribution of the points $k.a$ modulo 1, where $k$ is an integer satisfying an inequality $0 \leq k < n$; in particular, we mention a theorem known as the three-distance theorem [1, 4, 5, 6]. In Section 4, we give the algorithm, based on the properties described in Section 3.

## 2. Mathematical Preliminaries – Notations

$\mathbb{R}/\mathbb{Z}$ is the additive group of the real numbers modulo 1. This set can be viewed as a circle, or the segment $[0, 1]$ where both points 0 and 1 are identified (i.e., reals 0 and 1 represent the same point). With this second representation, the point represented by 0 (or 1) can be regarded as an origin. If $a \in \mathbb{R}/\mathbb{Z}$ and $k \in \mathbb{N}$, $k$ is said to be the (group) index of $k.a$ (in the group generated by $a$).

If $a \in \mathbb{R}$, its image in $\mathbb{R}/\mathbb{Z}$ will also be denoted $a$, as there is no possible confusion.

$[x, y]$ represents an interval of real numbers (open, if one has round brackets). $[[x, y]]$ represents an interval of integers. The symbol # denotes the cardinality of a finite set.

## 3. Properties of $k.a$ mod 1

In this section, we study the properties of the points $y = k.a$ modulo 1, where $a$ is a given real number and $k$ is an integer restricted to a given interval, e.g., satisfying $0 \leq k < N$ (where $N$ is a given positive integer). The numbers $a$ and $y$ may be regarded as elements of $\mathbb{R}/\mathbb{Z}$. Let us take for $n \in \mathbb{N}$:

$$E_n = \{k.a \in \mathbb{R}/\mathbb{Z} : k \in \mathbb{N}, \; k < n\}.$$

On examples, we can see that the distribution of the points of $E_n$ has very interesting properties. In particular, we will look for a construction of $E_n$ based on distances between consecutive points on $\mathbb{R}/\mathbb{Z}$.

An example is given on the following figure. We choose a rational number $(17/45)$ for $a$ to make the notations simpler and multiply the rational numbers by 45 to get integers, and instead of dealing with $\mathbb{R}/\mathbb{Z}$, we deal with $\mathbb{Z}/45\mathbb{Z}$.

In this example, we have chosen $n$ small enough to avoid casual equalities and multiple-order points. Since $\mathbb{R}\backslash\mathbb{Q}$ is dense into $\mathbb{R}$ and thanks to topological properties, we can suppose that $a \notin \mathbb{Q}$ for the *mathematical* study, to avoid such problems. Thus the set $E_n$ has exactly $n$ elements (i.e., there is no multiple-order point).

For $0 \leq i < n$, the $e_{n,i}$ denote the values of the points of $E_n$ in $[0,1)$, given in increasing order. We define $e_{n,n} = 1$, which represents the same point as $e_{n,0} = 0$. The distances between two consecutive points on the segment $[0,1]$ (or the circle $\mathbb{R}/\mathbb{Z}$) are the values $e_{n,i+1} - e_{n,i}$ for $0 \leq i < n$.

We now give a new construction of $E_n$ (the equivalence will be proved later), based on distances; this will allow to find our algorithm. For all $n \geq 2$, we define a sign $s_n \in \{-1, +1\}$ and a sequence $S_n$ of $n$ 4-tuples

$$S_n = (d_{n,i}, r_{n,i}, j_{n,i}, k_{n,i})_{0 \leq i < n}$$

where $d_{n,i}$ is a positive real number representing a *distance*, $r_{n,i}$ is a positive integer representing a *rank* (giving an order on the segments $[e_{n,i}, e_{n,i+1}]$ having the same length), and both $j_{n,i}$ and $k_{n,i}$ are elements of $\mathbb{N}$ representing *group indices*. The initial values are:

$$d_{2,0} = a, \quad d_{2,1} = 1 - a, \quad r_{2,0} = r_{2,1} = 1, \quad s_2 = \text{sign}(1 - 2a),$$

$$j_{2,0} = 0, \quad k_{2,0} = 1, \quad j_{2,1} = 1, \quad k_{2,1} = 0.$$

Let us take $D_n = \{d_{n,i} : 0 \leq i < n\}$, which is the set of the distances in $S_n$, $h_n = \max D_n$ and $\ell_n = \min D_n$; we will show that $D_n$ has two or three elements only (this is the three-distance theorem). The sequence $S_n$ and the sign $s_n$ are defined by the initial values and the following transformation. Let $i$ be the unique index such that $d_{n,i} = h_n$ and the rank $r_{n,i}$ is minimal. The 4-tuple $(d_{n,i}, r_{n,i}, j_{n,i}, k_{n,i})$ is replaced by two consecutive 4-tuples defined below; the other terms of the sequence remain unmodified and in the same order. The distances of the two 4-tuples are $\ell_n$ and $h_n - \ell_n$ but the order is determined by $s_n$: $\ell_n$ then $h_n - \ell_n$, if $s_n = +1$; $h_n - \ell_n$ then $\ell_n$, if $s_n = -1$. The new ranks are the smallest positive integers such that all the ranks associated with the distance are different, i.e., all the couples $(d, r)$ in the sequence are different; note that $h_n - \ell_n \neq \ell_n$ since $a$ is irrational. The group indices $(j_{n,i}, k_{n,i})$ are replaced by $(j_{n,i}, n)$ and $(n, k_{n,i})$. Finally, we take $s_{n+1} = s_n \cdot \text{sign}(h_n - 2\ell_n)$, i.e., the sign of $s_n$ changes if and only if $\ell_{n+1} < \ell_n$; this choice ensures that intervals having the same length are split in the same way (see figure).

We can associate a function $f_n : [[0, n-1]] \to \mathbb{R}/\mathbb{Z}$ with each sequence $S_n$, such that each function $f_n$ is a restriction of a function $f : \mathbb{N} \to \mathbb{R}/\mathbb{Z}$ with $f(0) = 0$ and $f(k) - f(j) = d \pmod 1$ for each $(d, r, j, k)$ of a sequence $S_n$.

Let us take the last example. For $n = 2$, we have two points on the circle $\mathbb{Z}/45\mathbb{Z}$, with respective coordinates $f(0) = 0$ and $f(1) = 17$ (modulo 45). These two points form two intervals. The first interval has length 17, the left end is point 0, the right end is point 1, and the rank is 1 (initial interval); thus $(d, r, j, k) = (17, 1, 0, 1)$. The second interval has length $45 - 17 = 28$, the left end is point 1, the right end is point 0, and the rank is also 1; thus $(d, r, j, k) = (28, 1, 1, 0)$. Now, let us consider the $n = 3$ to 4 iteration. For $n = 3$, we have $f(0) = 0$, $f(1) = 17$, and $f(2) = 34$. The interval of length $h_3 = 17$ and the minimal rank is $I = (17, 1, 0, 1)$. This 4-tuple is replaced by $I' = (6,1,0,3)$ and $I'' = (11,2,3,1)$ respectively. Since $d' + d'' = d$, $j' = j$, $k'' = k$, and $k' = j'' = n$, this transformation defines a new point $f(3) = 6$.

We now give the theorem showing that both constructions ($E_n$ and $S_n$) are equivalent. It will be proved later.

THEOREM 1   *For all $n \geq 2$ and $0 \leq i < n$, we have: $d_{n,i} = e_{n,i+1} - e_{n,i}$, $e_{n,i} = j_{n,i} \cdot a$ and $e_{n,i+1} = k_{n,i} \cdot a$, i.e., $\forall k \geq 0$, $f(k) = k.a$.*

Let us take $C_n = \#\{i : d_{n,i} = h_n\}$ and define a sequence $(\gamma_i, \delta_i)$:

$$(\gamma_0, \delta_0) = (a, 1 - a), \quad (\gamma_{i+1}, \delta_{i+1}) = (\min\{\gamma_i, \delta_i\}, |\gamma_i - \delta_i|),$$

i.e., at each iteration, one keeps the smaller element and replaces the larger one by the difference. When we deal with integers, this is the basic iteration of the Euclidean "additive" algorithm for computing gcd's.

The following theorem says that some sequences $S_n$ contain two different distances only, and the next pair of distances is obtained by replacing the larger distance with the difference. Between such two sequences, there is a transient period, where the three distances (both distances of the initial sequence, and the difference) are present.

THEOREM 2   *There exists a strictly increasing function $\varphi : \mathbb{N} \to \mathbb{N}$ such that $\varphi(0) = 2$, and for all $i \geq 0$:*

$$D_{\varphi(i)} = \{\gamma_i, \delta_i\}, \quad \text{and for } \varphi(i) < n < \varphi(i+1), \ D_n = \{\gamma_i, \delta_i, \delta_{i+1}\}.$$

*For all $i$ and $n$ such that $\varphi(i) \leqslant n < \varphi(i+1)$, one has $\varphi(i+1) = n + C_n$. In particular, $\varphi(i+1) - \varphi(i) = C_{\varphi(i)}$.*

**Proof:**   Theorem 2 is a direct consequence of the construction of the sequences $S_n$: we only use the fact that, at each iteration, an interval of length $h_n$ is replaced by two intervals of lengths $\ell_n$ and $h_n - \ell_n$. ∎

Theorem 1 will be deduced from the following lemma.

LEMMA 1   *For all $n$ such that $\#D_n = 2$, i.e., $n \in \varphi(\mathbb{N})$:*

*1. $r_{n,0} = r_{n,n-1} = 1$;*

*2. if $s_n = +1$, then $d_{n,0} = \ell_n$ and $d_{n,n-1} = h_n$;*
   *if $s_n = -1$, then $d_{n,0} = h_n$ and $d_{n,n-1} = \ell_n$;*

*3. $j_{n,0} = k_{n,n-1} = 0$,*
   *$k_{n,0} = \#\{i : d_{n,i} = d_{n,n-1}\}$,*
   *$j_{n,n-1} = \#\{i : d_{n,i} = d_{n,0}\}$;*

*4. for all $(d, r, j, k)$, the values $j - r$ and $k - r$ only depend on the value of $d$.*

This lemma and Theorem 1 can be proved by induction on $n$. The proofs are given in [2].

## 4. Algorithm

We recall that the segment has an equation of the form $y = ax - b$, where $x$ satisfies $0 \leq x < N$.

We will consider the successive $D_{\varphi(i)}$, and memorize the position of point $b$ in the interval that contains this point (the distance from $b$ to the lower bound of the interval) and the way in which the intervals are split, i.e., the values $h_n$, $\ell_n$ and $s_n$, where $n = \varphi(i)$. We recall that, at each iteration, the intervals of length $h_n$ are split into two intervals of lengths $\ell_n$ and $h_n - \ell_n$ (in the order given by $s_n$), and the intervals of length $\ell_n$ remain unchanged. We stop when $n \geq N$, where $N$ is the initial number of values to be tested. Then we can calculate the distance from $b$ to the two ends of the interval.

In fact, we want to know whether the distance between the segment and $\mathbb{Z}^2$ is larger than $\varepsilon$ or not. To avoid calculating the distance from $b$ to the upper end of the interval, we apply the algorithm to $b + \varepsilon$ instead of $b$, i.e., the segment is shifted by $\varepsilon$ downwards, and we only need to know the distance from $b$ to the lower end of the interval, which is directly given by the algorithm.

Note that with this algorithm, we consider more points than wanted. But the number of considered points is bounded from above by twice the initial number of points, i.e., $2N$, which is not too large for our problem, since the value of $N$ can be chosen such that the probability that the test fails is still small.

In order to avoid copying or swapping values and testing "status variables" (such as $s$), we will replace the variables $\ell$ and $h$ by the variables $x = d_{n,0}$ and $y = d_{n,n-1}$ (thus we avoid swapping $\ell$ and $h$ each time $h$ becomes less than $\ell$) and we will remove status variables, like $s$, by duplicating the code: one part for $s = +1$ and one part for $s = -1$. Thus we will know the position of $h$ and $\ell$ without any test: $(x, y) = (\ell, h)$ in the part where $s = +1$, and $(x, y) = (h, \ell)$ in the part where $s = -1$. Instead of comparing $\ell$ and $h$, and updating $s$, we will compare $x$ and $y$ and perform a conditional branch.

We define two new variables: $u$ and $v$, which denote the number of intervals of respective lengths $x$ and $y$; they are only used for calculating $n$. The variable $b$ will be modified in such a way that it always contains the distance from the considered point to the lower end of the interval.

Of course, we will apply the algorithm to rational values, whereas the mathematical study considered irrational values for practical reasons. The algorithm remains the same, but we must be careful concerning the particular cases ($h = \ell$, then $\ell = 0$) and ensure there is no infinite loop.

We have four possible states:

- $h+$:    the interval containing the point has length $h$    and    $s \Rightarrow +1$.
- $h-$:    the interval containing the point has length $h$    and    $s = -1$.
- $\ell-$:    the interval containing the point has length $\ell$    and    $s = -1$.
- $\ell+$:    the interval containing the point has length $\ell$    and    $s = +1$.

In fact, we will group $h-$ and $\ell+$ (point $b$ in the interval of length $x$), as well as $h+$ and $\ell-$ (point $b$ in the interval of length $y$). The algorithm given below can be implemented in different ways; an optimization may require that some instructions are moved, removed or added.

*Initialization:* $x = a$; $y = 1 - a$; $u = v = 1$;
*Infinite loop:*
    **if** $(b < x)$
        **while** $(x < y)$
            **if** $(u + v \geqslant N)$ **exit**
            $y = y - x$; $u = u + v$;
        **if** $(u + v \geqslant N)$ **exit**
        $x = x - y$; $v = v + u$;
    **else**
        $b = b - x$;
        **while** $(y < x)$
            **if** $(u + v \geqslant N)$ **exit**
            $x = x - y$; $v = v + u$;
        **if** $(u + v \geqslant N)$ **exit**
        $y = y - x$; $u = u + v$;

If $b$ is larger than $2\varepsilon$, then the distance between the segment and $\mathbb{Z}^2$ is larger than $\varepsilon$. Otherwise the test fails, and we need a more accurate test (e.g., by splitting the segment or using a slower algorithm).

We notice that this algorithm "contains" Euclid's algorithm, which is used to compute the development of $a$ into a continued fraction.

When the partial quotients of this development are bounded, the number of points generated (i.e. $u + v$) is an exponential function of the number of iterations; therefore, the number of iterations is of complexity $O(\log N)$. In practice, $N$ is bounded (e.g. $N < 100,000$), and one only has to consider the first partial quotients. So, when one of the first partial quotients of the continued fraction of $a$ is very large, the above algorithm is rather slow; for instance, $x$ is much smaller than $y$ (a partial quotient is large) and $u$ is small (the partial quotient is one of the first ones), thus the number $u$ of points added at each iteration is small, and many iterations are needed. It is possible to speed up the algorithm in these cases; however it will slow it down in the general case, which occurs much more often. Different solutions are possible, but they depend on the context in which the algorithm is applied.

## 5. Conclusion

First, the algorithms have been implemented on Sun SparcStations to find the value of $x$, among the $2^{52}$ double-precision floating-point numbers in the interval $[\frac{1}{2}, 1)$, for which the distance between $\exp x$ and a machine number or a number equidistant to two consecutive machine numbers is minimal. The naive algorithm required 3 clock cycles per argument in average. The method described in this paper allowed to deal with 30 arguments per clock cycle in average (with non-optimal parameters). The computations were performed on a few Sparc-4 and Sparc-5 machines and lasted about 10 days. The value $x$ which gives the minimal distance is

$$x = 0.11010110011001111110111110010001101011010011011110000$$
$$= \frac{471483227223279}{562949953421312}.$$

The result $\exp x$ is:

$$10.0100111110000101110010010111100000111101111001110000\ 01^{54}\ 01\ldots$$

(a "0" and 54 "1"s follow the mantissa).

With the above algorithm, we obtained a speed-up of 90 over the naive algorithm. It has been improved a bit to test some other intervals and functions, but can still be improved by choosing better parameters, and by improving the implementation (e.g., using the fact that the slope of the segment increases very slowly for this particular problem). Several functions have been tested: exp from $2^{-1}$ to $2^4$, log from $1 + 2^{-13}$ to 2, sin from $2^{-5}$ to $2^1$, cos from $2^{-2}$ to $2^0$, and we have results for $2^x$ for all $x > 0$ and $\log_2(x)$ for all $x > 1$. The worst non-trivial cases that have currently been found are:

- For

$$x = 0.0111111111001110110011101110011100111010000111101101101$$
$$= \frac{8980155785351021}{18014398509481984},$$

sin $x$ is equal to:

$$0.0111101001100101010000011100110000110001000110100101011\ 1\ 1^{65}\ 0000\ldots$$

- For

$$x = 0.0111101001100101010000011100110000110001000110100101110$$
$$= \frac{4306410053968715}{9007199254740992},$$

arcsin $x$ is equal to:

0.011111111100111011001110111001110011101000011110110110100 $^{64}$ 1000...

Other results can be found on:

```
http://www.ens-lyon.fr/~jmmuller/TMD.html
```

Thanks to the algorithm presented in this paper, we will be able to solve the Table Maker's Dilemma for the double precision in a reasonable time.

## Notes

1. We also consider other "rounding modes", e.g., we may want to get the largest machine number that is less than or equal to $f(x)$.
2. for directed rounding modes

## References

1. P. Alessandri and V. Berthé. Three distance theorems and combinatorics on words. Research report, Institut de Mathématiques de Luminy, Marseille, France, 1997.
2. V. Lefèvre. An algorithm that computes a lower bound on the distance between a segment and $\mathbb{Z}^2$. Research report 97-18, Laboratoire de l'Informatique du Parallélisme, Lyon, France, 1997.
3. V. Lefèvre, J.-M. Muller, and A. Tisserand. Towards correctly rounded transcendentals. In *Proceedings of the 13th IEEE Symposium on Computer Arithmetic*, Asilomar, USA, 1997. IEEE Computer Society Press, Los Alamitos, CA.
4. V. T. Sós. On the distribution mod 1 of the sequence $n\alpha$. *Ann. Univ. Sci. Budapest, Eötvös Sect. Math.*, 1:127–134, 1958.
5. J. Surányi. Über die Anordnung der Vielfachen einer reellen Zahl mod 1. *Ann. Univ. Sci. Budapest, Eötvös Sect. Math.*, 1:107–111, 1958.
6. S. Swierczkowski. On successive settings of an arc on the circumference of a circle. *Fundamenta Math.*, 46:187–189, 1958.

*Reliable Computing* **5:** 213–228, 1999.
© 1999 *Kluwer Academic Publishers.*

213

# Comparing Partial Consistencies*

HÉLÈNE COLLAVIZZA, FRANÇOIS DELOBEL and MICHEL RUEHER
*Université de Nice—Sophia-Antipolis, I3S, ESSI, 930, route des Colles - B.P. 145, 06903
Sophia-Antipolis, France, e-mail: {collavizza,delobel,rueher}@essi.fr*

(Received: 2 November 1998; accepted: 13 January 1999)

**Abstract.** Global search algorithms have been widely used in the constraint programming framework to solve constraint systems over continuous domains. This paper precisely states the relations among the different partial consistencies which are main emphasis of these algorithms.

The capability of these partial consistencies to handle the so-called dependency problem is analysed and some efficiency aspects of the filtering algorithms are mentioned.

## 1. Introduction

Global search algorithms that are based upon partial consistency filtering techniques have proven their efficiency to solve non-trivial constraint systems over the reals. For instance, systems like Newton and Numerica [2], [28] behave better than interval methods on classical benchmarks of numerical analysis and interval analysis (e.g., Moré-Cosnard non-linear integral equation, Broyden banded functions). Moreover, it has been shown recently [9] that combining algorithms based on different partial consistencies can even lead to better performances.

These global search algorithms are actually "branch and prune" algorithms, i.e., algorithms that can be defined as an iteration of two steps:

1. *Pruning the search space* by reducing the intervals associated with the variables.

2. *Generating subproblems* by splitting the domains of a variable (the choice of the variable may be non deterministic or based on some heuristic).

The pruning step achieves a filtering of the domains, in other words, it reduces the intervals associated with the variables until a given partial consistency property is satisfied.

### 1.1. PARTIAL CONSISTENCIES

Informally speaking, a constraint system $C$ satisfies a partial consistency property if a relaxation of $C$ is consistent. For instance, local consistency just requires that, taken individually, the constraints are consistent. The relevance of consistency properties is that whenever a consistency property is violated, there is an associated

---

* This is a revised version of the paper presented at the 4th International Conference on Constraint Programming [6].

recipe for pruning some interval. Most constraint solvers over finite domains [18], [27] are based on a partial consistency named *Arc-Consistency*. Assume $c$ is a $k$-ary constraint over variables $(x_1, ..., x_k)$; $c$ is arc-consistent if, for any value in $x_i$, there exists at least one value in each domain $x_j$ $(j \neq i)$ such that $c$ holds. In the same way, many solvers over continuous domains [1], [25], [28] rely upon relaxations of Arc-Consistency. A relaxation of Arc-Consistency has also been used in the context of global optimization [22].

2B-Consistency (also known as hull consistency) [3], [5], [15], [16] is a relaxation of Arc-Consistency which only requires to check the Arc-Consistency property for each bound of the intervals. The key point is that this relaxation is more easily verifiable than Arc-Consistency itself. Informally speaking, variable $x$ is 2B-Consistent for constraint "$f(x, x_1, ..., x_n) = 0$" if the lower (resp. upper) bound of the domain of $x$ is the smallest (resp. largest) solution of $f(x, x_1, ..., x_n)$. Box-Consistency [2], [11] is a coarser relaxation (i.e., it allows more stringent pruning) of Arc-Consistency than 2B-Consistency. Variable $x$ is Box-Consistent for constraint "$f(x, x_1, ..., x_n) = 0$" if the bounds of the domain of $x$ correspond to the leftmost and the rightmost zero of the optimal interval extension of $f(x, x_1, ..., x_n)$.

*3B-Consistency* and *Bound-Consistency* are higher order extensions of 2B-Consistency and Box-Consistency which have been introduced to limit the effects of a strictly local processing:

- *3B-Consistency* [16] is a relaxation of path consistency [8], a higher order extension of Arc-Consistency. Roughly speaking, 3B-Consistency checks whether 2B-Consistency can be enforced when the domain of a variable is reduced to the value of one of its bounds in the whole system.

- *Bound-Consistency* [24], [28] applies the principle of 3B-Consistency to Box-Consistency: Bound-Consistency checks whether Box-Consistency can be enforced when the domain of a variable is reduced to the value of one of its bounds in the whole system.

## 1.2. AIM OF THE PAPER

This paper investigates the relations between *2B-Consistency*, *Box-Consistency*, *3B-Consistency* and *Bound-Consistency*. More precisely, we prove the following properties:

- 2B-Consistency algorithms actually achieve a weaker filtering (i.e., a filtering that yields bigger intervals) than Box-Consistency, especially when a variable occurs more than once in some constraint (see Proposition 4.2). This is due to the fact that 2B-Consistency algorithms require a decomposition of the constraints with multiple occurrences of the same variable.

- The filtering achieved by Box-Consistency algorithms is weaker than the one computed by 3B-Consistency algorithms (see Proposition 5.2).

This paper also provides an analysis of both the capabilities and the limits of the filtering algorithms which achieve these partial consistencies. We pay special attention to their ability to handle the so-called dependency problem [10].

## LAYOUT OF THE PAPER

Section 2 reviews some basic concepts and introduces the notation used in the rest of the paper. Section 3 is devoted to the analysis of 2B-Consistency. Features and properties of Box-Consistency are the focus of Section 4. 3B-Consistency and Bound-Consistency are introduced in Section 5. Section 6 mentions efficiency issues.

## 2. Interval Constraint Solving

This section recalls some basics of interval analysis [2], [3], [12] and formally defines a constraint system over intervals of real numbers.

### 2.1. NOTATION

We mainly use the notations suggested by Kearfott [13]. Thus, throughout, boldface will denote intervals, lower case will denote scalar quantities, and upper case will denote vectors and sets. Brackets "[ . ]" will delimit intervals while parentheses "( . )" will delimit vectors. Underscores will denote lower bounds of intervals and overscores will denote upper bounds of intervals. $\tilde{x}$ denotes any value in interval $x$ (usually *not* the center of $x$).

We will also use the following notations, which are slightly non-standard:

- $\mathcal{R}^\infty = \mathcal{R} \cup \{-\infty, +\infty\}$ denotes the set of real numbers augmented with the two infinity symbols. $\mathbb{F}$ denotes a finite subset of $\mathcal{R}^\infty$ containing $\{-\infty, +\infty\}$. Practically speaking, $\mathbb{F}$ corresponds to the set of floating-point numbers used in the implementation of non linear constraint solvers;

- if $a$ is a constant in $\mathbb{F}$, $a^+$ (resp. $a^-$) corresponds to the smallest (resp. largest) number of $\mathbb{F}$ strictly greater (resp. lower) than $a$;

- $f, g$ denote functions over the reals; $c : \mathcal{R}^n \to Bool$ denotes a constraint over the reals, $c$ denotes a constraint over the intervals; $Var(c)$ denotes the variables occurring in constraint $c$.

### 2.2. INTERVAL ANALYSIS

DEFINITION 2.1 (Interval). An interval $x = [\underline{x}, \overline{x}]$, with $\underline{x}$ and $\overline{x} \in \mathbb{F}$, is the set of real numbers $\{r \in \mathcal{R} \mid \underline{x} \leq r \leq \overline{x}\}$; if $\underline{x}$ or $\overline{x}$ is the infinity symbol, then $x$ is an opened interval.

$\mathcal{I}$ denotes the set of intervals and is ordered by set inclusion. $\mathcal{U}(\mathcal{I})$ denotes the set of unions of intervals.

DEFINITION 2.2 (Set Extension). Let $S$ be a subset of $\mathcal{R}$. The *Hull* of $S$—denoted $\square S$—is the smallest interval $I$ such that $S \subseteq I$.

The term "smallest subset" (w.r.t. inclusion) must be understood according to the precision of floating-point operations. In the rest of the paper, we consider—as in [15], [2]—that results of floating-point operations are outward-rounded to preserve the correctness of the computation. However, we also assume that the largest computing error when computing a bound of a variable of the initial constraint system is always smaller than one float. This hypothesis may require the use of big floats [4] when computing intermediate results.

DEFINITION 2.3 (Interval Extension [10], [20]).

- $\mathbf{f} : \mathcal{I}^n \to \mathcal{I}$ is an interval extension of $f : \mathcal{R}^n \to \mathcal{R}$ iff $\forall x_1, ..., x_n \in \mathcal{I}$ : $f(\tilde{x}_1, ..., \tilde{x}_n) \in \mathbf{f}(\mathbf{x}_1, ..., \mathbf{x}_n)$.
- $\mathbf{c} : \mathcal{I}^n \to Bool$ is an interval extension of $c : \mathcal{R}^n \to Bool$ iff $\forall \mathbf{x}_1, ..., \mathbf{x}_n \in \mathcal{I}$ : $c(\tilde{x}_1, ..., \tilde{x}_n) \Rightarrow \mathbf{c}(\mathbf{x}_1, ..., \mathbf{x}_n)$.

Similarly, $\mathbf{f}$ is the *natural* interval extension of $f$ (see [20]) if $\mathbf{f}$ is obtained by replacing in $f$ each constant $k$ with the smallest interval containing $k$, each variable $x$ with an interval variable $\mathbf{x}$, and each arithmetic operation with its optimal interval extension [20].

In the rest of this paper, $\mathbf{c}$ denotes the natural interval extension of $c$ and $\oplus$, $\ominus$, $\otimes$, $\oslash$ denote the optimal interval extensions of $+, -, \times, /$.

We now recall a fundamental result of interval analysis with many consequences on efficiency and precision of interval constraint solving methods.

PROPOSITION 2.1 [20]. *Let* $\mathbf{f} : \mathcal{I}^n \to \mathcal{I}$ *be the natural interval extension of* $f : \mathcal{R}^n \to \mathcal{R}$. *If each* $x_i$ *occurs only once in* $f$ *then* $\square\{f(\tilde{x}_1, ..., \tilde{x}_n)\} = \mathbf{f}(\mathbf{x}_1, ..., \mathbf{x}_n)$ *else* $\square\{f(\tilde{x}_1, ..., \tilde{x}_n)\} \subseteq \mathbf{f}(\mathbf{x}_1, ..., \mathbf{x}_n)$.

This result can be trivially extended to relations over $\mathcal{R}^n$:

PROPOSITION 2.2. *Let* $\mathbf{c} : \mathcal{I}^n \to Bool$ *be the natural extension of* $c : \mathcal{R}^n \to Bool$, *if each* $x_i$ *occurs only once in* $c$, *then* $\mathbf{c}(\mathbf{x}_1, ..., \mathbf{x}_n) \Leftrightarrow c(\tilde{x}_1, ..., \tilde{x}_n)$.

## 2.3. INTERVAL CONSTRAINT SYSTEM

DEFINITION 2.4 (CSP). A CSP (Constraint System Problem) [18] is a couple $(X, C)$ where $X = \{x_1, ..., x_n\}$ denotes a set of variables with associated interval domains $\{\mathbf{x}_1, ..., \mathbf{x}_n\}$, and $C = \{c_1, ..., c_m\}$ denotes a set of constraints.

$P_\emptyset$ denotes an empty CSP, i.e., a CSP with at least one empty domain. $\mathbf{X}' \subseteq \mathbf{X}$ means $\mathbf{x}'_i \subseteq \mathbf{x}_i$ for all $i$. We define a CSP $P = (X, C)$ to be smaller than a CSP $P' = (X', C)$ if $\mathbf{X}' \subseteq \mathbf{X}$. We write $P \preceq P'$ for this relation. By convention, $P_\emptyset$ is the smallest CSP.

In the following passage, we define and discuss several kinds of consistency, and the associated filtering of a CSP $P$.

## 3. 2B-Consistency

Most of the CSP systems over intervals (e.g., [1], [3], [23], [25]) compute a relaxation of Arc-Consistency [18] called 2B-Consistency (or Hull consistency). In this section, we give the definition of 2B-Consistency and explain why its computation requires a relaxation of the constraint system.

### 3.1. DEFINITIONS

2B-Consistency [16] states a local property on the bounds of the domains of a variable at a single constraint level. Roughly speaking, a constraint $c$ is 2B-Consistent if, for any variable $x$, there exist values in the domains of all other variables which satisfy $c$ when $x$ is fixed to $\underline{x}$ and $\bar{x}$.

DEFINITION 3.1 (2B-Consistency). Let $(X, C)$ be a CSP and $c \in C$ a $k$-ary constraint over $(x_1, ..., x_k)$. $c$ is 2B-Consistent iff:

$$\forall i, \mathbf{x}_i = \Box\{\tilde{x}_i \mid \exists \tilde{x}_1 \in \mathbf{x}_1, ..., \exists \tilde{x}_{i-1} \in \mathbf{x}_{i-1}, \exists \tilde{x}_{i+1} \in \mathbf{x}_{i+1}, ..., \exists \tilde{x}_k \in \mathbf{x}_k$$
$$\text{such that } c(\tilde{x}_1, ..., \tilde{x}_{i-1}, \tilde{x}_i, \tilde{x}_{i+1}, ..., \tilde{x}_k) \text{ holds}\}.$$

A CSP is 2B-Consistent iff all its constraints are 2B-Consistent.

By definition, 2B-Consistency is weaker than Arc-Consistency. This point is illustrated in Example 3.1.

EXAMPLE 3.1. Let $P_1 = (\{x_1, x_2\}, \{x_1 = x_2 * x_2\})$ be a CSP with $\mathbf{x}_1 = [1, 4], \mathbf{x}_2 = [-2, 2]$. $P_1$ is 2B-Consistent but not arc-Consistent since there is no value in $\mathbf{x}_1$ which satisfies the constraint when $x_2 = 0$.

DEFINITION 3.2 (Closure by 2B-Consistency [16]). The filtering by 2B-Consistency of $P = (X, C)$ is the CSP $P' = (X', C)$ such that:

- $P$ and $P'$ have the same solutions;
- $P'$ is 2B-Consistent;
- $X' \subseteq X$ and the domains in $X'$ are the largest ones for which $P'$ is 2B-Consistent.

We note $\Phi_{2B}(P)$ the filtering by 2B-Consistency of $P$. In the following we will use the term *closure* by 2B-Consistency to emphasize the fact that this filtering always exists and is unique [16].

Proposition 3.1 states a property which is useful when comparing 2B-Consistency and Box-Consistency.

**IN(in** $C$, **inout X)**
Queue $\leftarrow$ $C$;
**while** Queue $\neq \emptyset$
  $c \leftarrow$ POP Queue;
  $\mathbf{X}' \leftarrow$ narrow$(c, \mathbf{X})$;
  **if** $\mathbf{X}' \neq \mathbf{X}$ **then**
    $\mathbf{X} \leftarrow \mathbf{X}'$;
    Queue $\leftarrow$ Queue $\cup$ $\{c' \in C \mid Var(c) \cap Var(c') \neq \emptyset\}$
  **endif**
**endwhile**

*Figure 1.* Algorithm IN.

PROPOSITION 3.1. *Let* $P = (X, C)$ *be a CSP such that no variable occurs more than once in any constraint of* $C$. *Let* $c \in C$ *be a* $k$-*ary constraint over the variables* $(x_1, ..., x_k)$. $P$ *is 2B-Consistent iff* $\forall c \in C$, $\forall i \in 1, ..., k$ *the following relations hold:*

- $c(\mathbf{x}_1, ..., \mathbf{x}_{i-1}, [\underline{\mathbf{x}_i}, \underline{\mathbf{x}_i}^+), \mathbf{x}_{i+1}, ..., \mathbf{x}_k)$, *and*
- $c(\mathbf{x}_1, ..., \mathbf{x}_{i-1}, (\overline{\mathbf{x}_i}^-, \overline{\mathbf{x}_i}], \mathbf{x}_{i+1}, ..., \mathbf{x}_k)$.

*Proof.* Assume that both $c(\mathbf{x}_1, ..., \mathbf{x}_{i-1}, [\underline{\mathbf{x}_i}, \underline{\mathbf{x}_i}^+), \mathbf{x}_{i+1}, ..., \mathbf{x}_k)$ and $c(\mathbf{x}_1, ..., \mathbf{x}_{i-1}, (\overline{\mathbf{x}_i}^-, \overline{\mathbf{x}_i}], \mathbf{x}_{i+1}, ..., \mathbf{x}_k)$ hold. By Proposition 2.2 we have:

1. $\exists \tilde{x}_1 \in \mathbf{x}_1, ..., \exists \tilde{x}_{i-1} \in \mathbf{x}_{i-1}, \exists x_i \in [\underline{x_i}, \underline{x_i}^+), \exists \tilde{x}_{i+1} \in \mathbf{x}_{i+1}, ..., \exists \tilde{x}_k \in \mathbf{x}_k$ such that $c(\tilde{x}_1, ..., \tilde{x}_{i-1}, \tilde{x}_i, \tilde{x}_{i+1}, ..., \tilde{x}_k)$ holds, and
2. $\exists \tilde{x}_1 \in \mathbf{x}_1, ..., \exists \tilde{x}_{i-1} \in \mathbf{x}_{i-1}, \exists x_i \in (\overline{x_i}^-, \overline{x_i}], \exists \tilde{x}_{i+1} \in \mathbf{x}_{i+1}, ..., \exists \tilde{x}_k \in \mathbf{x}_k$ such that $c(\tilde{x}_1, ..., \tilde{x}_{i-1}, \tilde{x}_i, \tilde{x}_{i+1}, ..., \tilde{x}_k)$ holds.

Thus, $\mathbf{x}_i = \square\{\tilde{x}_i \mid \exists \tilde{x}_1 \in \mathbf{x}_1, ..., \exists \tilde{x}_{i-1} \in \mathbf{x}_{i-1}, \exists \tilde{x}_{i+1} \in \mathbf{x}_{i+1}, ..., \exists \tilde{x}_k \in \mathbf{x}_k$ such that $c(\tilde{x}_1, ..., \tilde{x}_i, ..., \tilde{x}_k)$ holds$\}$.
The counterpart results from the definition of 2B-Consistency. $\square$

### 3.2. COMPUTING 2B-CONSISTENCY

2B-Consistency is enforced by narrowing the domains of the variables. Using the above notations, the scheme of the standard interval narrowing algorithm—derived from AC3 [18]—can be written down as in Figure 1. IN implements the computation of the closure by 2B-Consistency of a CSP $P = (X, C)$. narrow$(c, \mathbf{X})$ is a function which prunes the domains of variables $Var(c)$ until $c$ is 2B-Consistent.

The approximation of the projection functions is the basic tool for the narrowing of domains in narrow$(c, \mathbf{X})$. Let $c$ be a $k$-ary constraint over $X = (x_1, ..., x_k)$: for each $i$ in $1, ..., k$, $\pi_i(c, \mathbf{X})$ denotes the projection over $x_i$ of the solutions of $c$ in the space delimited by $\mathbf{X}$.

DEFINITION 3.3 (Projection of a Constraint). $\pi_i(c, \mathbf{X}) : (C, I^k) \rightarrow \mathcal{U}(\mathcal{I})$ is the projection of $c$ on $x_i$ iff $\pi_i(c, \mathbf{X}) = \{\tilde{x}_i \mid \exists(\tilde{x}_1, ..., \tilde{x}_{i-1}, \tilde{x}_{i+1}, ..., \tilde{x}_k) \in \mathbf{x}_1 \times \cdots \times \mathbf{x}_{i-1} \times \mathbf{x}_{i+1}, \cdots \times \mathbf{x}_k$ such that $c(\tilde{x}_1, ..., \tilde{x}_i, ..., \tilde{x}_k)$ holds$\}$.

DEFINITION 3.4 (Approximation of the Projection). $AP_i(c, \mathbf{X}) : (C, \mathcal{I}^k) \rightarrow \mathcal{I}$ is an approximation of $\pi_i(c, \mathbf{X})$ iff $AP_i(c, \mathbf{X}) = \square \ \pi_i(c, \mathbf{X}) = [\min \ \pi_i(c, \mathbf{X}),$ max $\pi_i(c, \mathbf{X})]$. In other words, $AP_i(c, \mathbf{X})$ is the smallest interval encompassing projection $\pi_i(c, \mathbf{X})$.

The following proposition trivially holds:

PROPOSITION 3.2. *Constraint $c$ is 2B-Consistent on* $\mathbf{X}$ *iff for all $i$ in* $\{1, ..., k\}$, $\mathbf{x}_i = AP_i(c, \mathbf{X})$.

In general, $AP_i$ cannot be computed efficiently because it is difficult to define functions min and max, especially when $c$ is not monotonic. For instance, if variable $x$ has multiple occurrences in $c$, defining these functions would require $x$ to be isolated*. Since such a symbolic transformation is not always possible, this problem is usually solved by decomposing the constraint system into a set of primitive constraints for which the $AP_i$ can easily be computed [17]. Primitive constraints are generated syntactically by introducing new variables.

DEFINITION 3.5 (Decomposition of a Constraint System). Let $P = (X, C)$ be a CSP and $c \in C$ a constraint. We define $\mathcal{M}_c \subseteq X$ as the set of variables having multiple occurrences in $c$. $decomp(c)$ is the set of constraints obtained by substituting in $c$ each occurrence of variables $x \in \mathcal{M}_c$ with a new variable $y$ with domain $\mathbf{y} = \mathbf{x}$ and by adding a constraint $x = y$. $New_{(x,c)}$ is the set of new variables introduced to remove multiple occurrences of variable $x$ in $c$, $X_{New} = \bigcup\{New_{(x,c)} \mid x \in X$ and $c \in C\}$. $P_{decomp}$ is the CSP $(X', C')$ where $X' = X \cup X_{New}$, and $C' = \{decomp(c) \mid c \in C\}$.

Decomposition does not change the semantics of the constraint system: $P$ and $P_{decomp}$ have the same solutions since $P_{decomp}$ just results from a rewriting** of $P$. However, a *local* consistency like Arc-Consistency is not preserved by such a rewriting. Indeed, decomposition reduces the scope of local consistency filtering algorithms. Thus, $P_{decomp}$ is a *relaxation* of $P$ when computing a relaxation of Arc-Consistency.

---

* B. Faltings [7] has recently introduced a new method for computing the projection without defining projection function. However, this method requires a complex analysis of constraints in order to find extrema.

** In practice, $c$ is decomposed into binary and ternary constraints for which projection functions are straightforward to compute. Since there are no multiple occurrences in $decomp(c)$ and interval calculus is associative, this binary and ternary constraint system has the same solutions as $P_{decomp}$.

EXAMPLE 3.2 (Decomposition of the Constraint System). Let $c : x_1 + x_2 - x_1 = 0$ be a constraint and $x_1 = [-1, 1]$, $x_2 = [0, 1]$ the domains of $x_1$ and $x_2$. Since $x_1$ appears twice in $c$, its second occurrence will be replaced with a new variable $x_3$: $decomp(c) = \{x_1 + x_2 - x_3 = 0, x_1 = x_3\}$.

In this new constraint system, each projection can easily be computed with interval arithmetic. For instance, $AP_1(x_1 + x_2 - x_3 = 0, (x_1, x_2, x_3))$ is $x_1 \cap (x_3 \ominus x_2)$. However, this decomposition increases the locality problem: the first constraint is checked independently of the second one and so $x_1$ and $x_3$ can take distinct values. More specifically, the initial constraint $c$ is not 2B-Consistent since there is no value of $x_1$ which satisfies $c$ when $x_2 = 1$. On the contrary, $decomp(c)$ is 2B-Consistent since the values $x_1 = -1$ and $x_3 = 0$ satisfy $x_1 + x_2 - x_3 = 0$ when $x_2 = 1$. On the initial constraint, 2B-Consistency reduces $x_2$ to $[0, 0]$ while it yields $x_1 = [-1, 1], x_2 = [0, 1]$ for $decomp(c)$.

*Remark.* Like almost all other examples in this paper, Example 3.2 can be trivially simplified. However, the reader can more easily check partial consistencies on such examples than on non-linear constraints where the same problems occur.

## 4. Box-Consistency

Box-Consistency [2], [11] is a coarser relaxation of Arc-Consistency than 2B-Consistency. It mainly consists of replacing every existentially quantified variable but one with its interval in the definition of 2B-Consistency. Thus, Box-Consistency generates a system of univariate interval functions which can be tackled by numerical methods such as Newton. Contrary to 2B-Consistency, Box-Consistency does not require any constraint decomposition and thus does not amplify the locality problem. Moreover, Box-Consistency can tackle some dependency problems when each constraint of a CSP contains only one variable which has multiple occurrences.

### 4.1. DEFINITION AND PROPERTIES OF BOX-CONSISTENCY

DEFINITION 4.1 (Box-Consistency). Let $(X, C)$ be a CSP and $c \in C$ a $k$-ary constraint over the variables $(x_1, ..., x_k)$. $c$ is Box-Consistent if, for all $x_i$ the following relations hold:

1. $c(x_1, ..., x_{i-1}, [\underline{x_i}, x_i^+), x_{i+1}, ..., x_k)$,
2. $c(x_1, ..., x_{i-1}, (\overline{x_i}^-, \overline{x_i}], x_{i+1}, ..., x_k)$.

Closure by Box-Consistency of $P$ is defined similarly to closure by 2B-Consistency of $P$, and is denoted by $\Phi_{Box}(P)$.

PROPOSITION 4.1. $\Phi_{2B}(P) \preceq \Phi_{Box}(P)$ and $\Phi_{2B}(P) \equiv \Phi_{Box}(P)$ *when no variable occurs more than once in the constraints of* $C$.

*Proof.* From the definitions of 2B-Consistency, Box-Consistency and interval extension of a relation, it results that $\Phi_{2B}(P) \preceq \Phi_{Box}(P)$. By Proposition 2.2 the equivalence holds when no variable occurs more than once in the constraints of $C$. $\square$

It follows that any CSP which is 2B-Consistent is also Box-Consistent. On the contrary a CSP which is Box-Consistent may not be 2B-Consistent (see Example 4.1).

EXAMPLE 4.1. Example 3.2 is not 2B-Consistent for $x_2$ but it is Box-Consistent for $x_2$ since $([-1,1] \oplus [0,0^+] \ominus [-1,1]) \cap [0,0]$ and $([-1,1] \oplus [1^-,1] \ominus [-1,1])$ $\cap [0,0]$ are non-empty.

Of course, the decomposition of a constraint system amplifies the limit due to the local scope of 2B-Consistency. As a consequence, 2B-Consistency on the decomposed system yields a weaker filtering than Box-Consistency on the initial system:

PROPOSITION 4.2. $\Phi_{Box}(P) \preceq \Phi_{2B}(P_{decomp})$.

*Proof.* The different occurrences of the same variable are connected by the existential quantifier as stated in the definition of the 2B-Consistency. However, the decomposition step breaks down the links among these different occurrences and generates a CSP $P_{decomp}$ which is a relaxation of $P$ for the computation of a local consistency. It follows that $\Phi_{Box}(P) \preceq \Phi_{Box}(P_{decomp})$. By Proposition 4.1 we have: $\Phi_{Box}(P_{decomp}) \equiv \Phi_{2B}(P_{decomp})$, and thus $\Phi_{Box}(P) \preceq \Phi_{2B}(P_{decomp})$. $\square$

EXAMPLE 4.2. Let $c$ be the constraint $x_1 + x_2 - x_1 - x_1 = 0$ where $x_1 = [-1,1]$ and $x_2 = [0.5,1]$. $c$ is not Box-Consistent since $[-1,-1^+] \oplus [0.5,1] \ominus [-1,-1^+] \ominus$ $[-1,-1^+] \cap [0,0]$ is empty. But $decomp(c)$ is 2B-Consistent for $x_1$ and $x_2$.

Box-Consistency can tackle some dependency problems in a constraint $c$ which contains only one variable occurring more than once. More precisely, Box-Consistency enables us to reduce the domain $x$ if variable $x$ occurs more than once in $c$ and if $x$ contains inconsistent values. For instance, in Example 4.2, filtering by Box-Consistency reduces $x_1$ because value $-1$ of $x_1$ has no support in $x_2$.

However, Box-Consistency may fail to handle the dependency problem when the inconsistent values of constraint $c$ are in the domain of variable $x_i$ while a variable $x_j$ $(j \neq i)$ occurs more than once in $c$. For instance, in Example 3.2, value 1 of $x_2$ has no support in $x_1$ but Box-Consistency fails to detect the inconsistency because $[-1,1] \oplus [1^-,1] \ominus [-1,1] \cap [0,0]$ is not empty.

## 4.2. COMPUTING BOX-CONSISTENCY

The Box-Consistency filtering algorithm proposed in [2], [28], [29] is based on an iterative narrowing operation using the interval extension of the Newton method. Computing Box-Consistency follows the generic algorithm IN (see Figure 1) used for computing 2B-Consistency. The function $narrow(c, \mathbf{X})$ prunes the domains of

**function LNAR (IN: $f_x$, x, RETURN:** *Interval*)
    $r \leftarrow \overline{x}$
    **if** $0 \notin f_x(x)$ **then return** $\emptyset$
        **else** $i \leftarrow \text{NEWTON}(f_x, x)$
            **if** $0 \in f_x([\underline{i}, \underline{i}^+])$ **then return** $[\underline{i}, r]$
                **else** $\text{SPLIT}(i, i_1, i_2)$
                    $l_1 \leftarrow \text{LNAR}(f_x, i_1)$
                    **if** $l_1 \neq \emptyset$ **then return** $[\underline{l_1}, r]$
                        **else return** $[\underline{\text{LNAR}(f_x, i_2)}, r]$
                **endif**
        **endif**
    **endif**

*Figure 2.* Function LNAR.

the variables of $c$ until $c$ is Box-Consistent. Roughly speaking, for each variable $x$ of constraint $c$, an interval univariate function $f_x$ is generated from $c$ by replacing all variables but $x$ with their intervals. The narrowing process consists of finding the leftmost and rightmost zeros of $f_x$. Figure 2 shows function LNAR which computes the leftmost zero of $f_x$ for initial domain $I_x$ of variable $x$ (this procedure is given in [29]).

Function LNAR first prunes interval x with function NEWTON which is an interval version of the classical Newton method. However, depending on the value of x, Newton may not reduce x enough to make x Box-Consistent. So, a split step is applied in order to ensure that the left bound of x is actually a zero. Function SPLIT divides interval i in two intervals $i_1$ and $i_2$, $i_1$ being the left part of the interval. The splitting process avoids the problem of finding a safe starting box for Newton (see [11]). As mentioned in [29], even if $f_x$ is not differentiable, the function LNAR may find the leftmost zero thanks to the splitting process (in this case, the call to function NEWTON is just ignored). Notice that Box-Consistency can be computed in such a way because it is defined on interval constraints whereas the existential quantifiers in the definition of 2B-Consistency require the use of projection functions.

## 5. 3B-Consistency and Bound-Consistency

2B-Consistency and Box-Consistency are only partial consistencies which are often too weak for computing an relevant superset of solutions of a CSP. In the same way that Arc-Consistency has been generalized to higher consistencies (e.g., path consistency [18]), 2B-Consistency and Box-Consistency can be generalized to higher order consistencies [16].

## 5.1. 3B-CONSISTENCY

DEFINITION 5.1 (3B-Consistency [16]). Let $P = (X, C)$ be a CSP and $x$ a variable of $X$. Let also:

- $P_{x \leftarrow [\underline{x}, \underline{x}^+)}$ be the CSP derived from $P$ by substituting $x$ with $[\underline{x}, \underline{x}^+)$;
- $P_{x \leftarrow (\overline{x}^-, \overline{x}]}$ be the CSP derived from $P$ by substituting $x$ with $(\overline{x}^-, \overline{x}]$.

$x$ is 3B-Consistent iff $\Phi_{2B}(P_{x \leftarrow [\underline{x}, \underline{x}^+)}) \neq P_\emptyset$ and $\Phi_{2B}(P_{x \leftarrow (\overline{x}^-, \overline{x}]}) \neq P_\emptyset$. A CSP is 3B-Consistent iff all its domains are 3B-Consistent.

It results from Definition 5.1 that any CSP which is 3B-Consistent is also 2B-Consistent [16]. The generalization of the 3B-Consistency to $k$B-Consistency is straightforward and is given in [16], [17].

3B-Consistency is less local than 2B-Consistency or Box-Consistency. Proposition 5.1 shows that 3B-Consistency always prunes more strongly than Box-Consistency, even if 3B-Consistency is achieved on the decomposed system and Box-Consistency on the initial system.

PROPOSITION 5.1. *Let $P = (X, C)$ be a CSP. If $P_{decomp}$ is 3B-Consistent then $P$ is Box-Consistent.*

*Proof.* Since Box-Consistency is a local consistency we just need to show that the property holds for a single constraint.

Assume $c$ is a constraint over $(x_1, ..., x_k)$, $x$ is one of the variables occuring more than once in $c$ and $New_{(x,c)} = (x_{k+1}, ..., x_{k+m})$ is the set of variables introduced for replacing the multiple occurrences of $x$ in $c$. Suppose that $P_{decomp}$ is 3B-Consistent for $x$.

Consider $P_1$, the CSP derived from $P_{decomp}$ by reducing domain $x$ to $[\underline{x}, \underline{x}^+)$. $P_1$ is 2B-Consistent for $x$ and thus the domain of all variables in $New_{(x,c)}$ is reduced to $[\underline{x}, \underline{x}^+)$; this is due to the equality constraints added when introducing new variables. From Proposition 3.1, it results that the following relation holds:

$$c'(x_1, ..., x_{i-1}, [\underline{x}, \underline{x}^+), x_{i+1}, ..., x_k, [\underline{x}, \underline{x}^+), ..., [\underline{x}, \underline{x}^+), x_{k+m+1}, ..., x_n)$$

$c'$ is the very same syntactical expression as $c$ (where some variables have been renamed).

$(x_{k+m+1}, ..., x_n)$ are the domains of the variables introduced for replacing the multiple occurrences of $\mathcal{M}_c \setminus \{x\}$. As the natural interval extension of a constraint is defined over the intervals corresponding to the domains of the variables, relation $c(x_1, ..., x_{i-1}, [\underline{x}, \underline{x}^+), x_{i+1}, ..., x_k)$ holds too.

The same reasoning can be applied when $x$ is replaced with its upper bound $(\overline{x}^-, \overline{x}]$. So we conclude that $x$ is also Box-Consistent. $\quad\square$

EXAMPLE 5.1. Let $C = \{x_1 + x_2 = 100, x_1 - x_2 = 0\}$ and $x_1 = [0, 100], x_2 = [0, 100]$ be the constraints and domains of a given CSP $P$.

$\Phi_{3B}(P_{decomp})$ reduces the domains of $x_1$ and $x_2$ to the interval $[50,50]$ whereas $\Phi_{Box}(P)$ does not achieve any pruning ($P$ is Box-Consistent).

The following proposition is a direct consequence of Proposition 5.1:

PROPOSITION 5.2. $\Phi_{3B}(P_{decomp}) \preceq \Phi_{Box}(P)$.

Thus, 3B-Consistency allows us to tackle at least the same dependency problems as Box-Consistency. However, 3B-Consistency is not effective enough to tackle the dependency problem in general (see Example 5.2).

EXAMPLE 5.2. Let $c$ be the constraint $x_1 * x_2 - x_1 + x_3 - x_1 + x_1 = 0$ where $\mathbf{x_1} = [-4, 3], \mathbf{x_2} = [1, 2]$ and $\mathbf{x_3} = [-1, 5]$. $decomp(c) = \{x_1 * x_2 - x_4 + x_3 - x_5 + x_6 = 0, x_1 = x_4 = x_5 = x_6\}$. $c$ is not 2B-Consistent since there are no values in $\mathbf{x_1}$ and $\mathbf{x_2}$ which verify the relation when $x_3 = 5$.

However, $decomp(c)$ is 3B-Consistent. Indeed, the loss of the link between the two occurrences of $x_1$ prevents the pruning of $x_3$.

A question which naturally arises is that of the relation which holds between $\Phi_{2B}(P)$ and $\Phi_{3B}(P_{decomp})$: Example 5.3 shows that $\Phi_{2B}(P) \preceq \Phi_{3B}(P_{decomp})$ does not hold and Example 5.2 shows that $\Phi_{3B}(P_{decomp}) \preceq \Phi_{2B}(P)$ does not hold, even if only one variable occurs more than once in each constraint of $P$. It follows that no order relation between $\Phi_{3B}(P_{decomp})$ and $\Phi_{2B}(P)$ can be exhibited.

EXAMPLE 5.3. Let $P$ be a CSP defined by $C = \{x_1 + x_2 = 10; x_1 + x_1 - 2x_2 = 0\}$ where $\mathbf{x_1} = \mathbf{x_2} = [-10, 10]$. $decomp(x_1 + x_1 - 2x_2 = 0) = \{x_1 + x_3 - 2x_2 = 0, x_3 = x_1\}$. $P$ is 2B-Consistent but $P_{decomp}$ is not 3B-Consistent: Indeed, when $x_1$ is fixed to 10, $\Phi_{2B}(P_{\mathbf{x_1} \leftarrow [10^-, 10]}) = P_\emptyset$ since $\mathbf{x_2}$ is reduced to $\emptyset$. In this case, the link between $x_1$ and $x_3$ is preserved and 3B-Consistency reduces $\mathbf{x_2}$ to $[5, 5]$.

## 5.2. BOUND-CONSISTENCY

Bound-Consistency was suggested in [17] and was formally defined in [28]. Informally speaking, Bound-Consistency applies the principle of 3B-Consistency to Box-Consistency: it checks whether Box-Consistency can be enforced when the domain of a variable is reduced to the value of one of its bounds in the whole system.

DEFINITION 5.2 (Bound-Consistency). Let $(X, C)$ be a CSP and $c \in C$ a $k$-ary constraint over the variables $(x_1, ..., x_k)$. $c$ is Bound-Consistent if for all $x_i$, the following relations hold:

1. $\Phi_{Box}(c(\mathbf{x_1}, ..., \mathbf{x_{i-1}}, [\underline{\mathbf{x}}, \underline{\mathbf{x}}^+), \mathbf{x_{i+1}}, ..., \mathbf{x_k})) \neq P_\emptyset$,
2. $\Phi_{Box}(c(\mathbf{x_1}, ..., \mathbf{x_{i-1}}, (\overline{\mathbf{x}}^-, \overline{\mathbf{x}}], \mathbf{x_{i+1}}, ..., \mathbf{x_k})) \neq P_\emptyset$.

Since $\Phi_{Box}(P) \preceq \Phi_{2B}(P_{decomp})$ it is trivial to show that $\Phi_{Bound}(P) \preceq \Phi_{3B}(P_{decomp})$. Bound-Consistency achieves the same pruning as 3B-Consistency when applied to Examples 5.1 and 3.2.

## 6. Efficiency Issues

The aim of numerical CSP is not to compute partial consistencies but to find accurate solutions; that is, either small intervals containing isolated solutions, or intervals which tightly encompass sets of continuous solutions. Thus, in practical systems (e.g., Numerica [28], PROLOG IV [25]), partial consistencies are combined with several search heuristics and splitting techniques. Experimental results of Numerica and Newton are very impressive. However it is difficult to draw a conclusion from the published benchmarks because these systems differ in several critical points:

- They use different splitting heuristics.
- There are significant variations in the implementation of the filtering algorithms (precision parameters in order to force an early halt to the propagation process, constraint ordering, detection of cycles, ...).
- They use different implementation languages (Prolog, C, ...).

So we limit the discussion to a brief examination of three key points which help to better understand the performances of the different systems:

1. **Cost of the basic narrowing operator**: Performing interval newton method on univariate functions is more expensive than computing projection functions on primitive constraints. For instance, let $C = \{x^2 = 2\}$ and $x = [1, 10]$. Box-Consistency requires 6 narrowing steps with the Newton method (about 100 interval operations) whereas 2B-Consistency only requires the computation of one relational square root operation and one intersection over intervals [9]. Thus, 2B is more efficient than Box on problems where the projection functions compute an accurate result.

   The gain of performance due to accurate projection functions is well illustrated using the pentagon problem. This problem consists of finding the coordinates of five points of a circle such that these points define a convex pentagon. The constraint system consists of five quadratic equations. To avoid an infinite number of solutions, the first point is given, and the five points are ordered to avoid symmetrical solutions. On this example, Bound-Consistency and 3B-Consistency achieve the same pruning. According to [1] Box is about forty times slower than 2B on this example.

2. **Expansion of the constraint system**: Decomposition of the initial constraint system may generate a huge number of primitive constraints when variables occur more than once. For instance, consider the classical "Broyden 160" example (160 initial variables, 160 constraints). Box-Consistency will generate 160 variables, 1184 univariate functions whereas 2B-Consistency will generate 2368 variables, 6944 ternary projection functions.

   From a practical point of view, 2B-Consistency is seriously weakened by the decomposition required for computing the narrowing functions. On the other hand, the univariate functions generated by Box-Consistency can be handled very efficiently using Newton-like methods.

3. **Precision of the computation**: For a fixed final precision, the efficiency of the computation may strongly depend on the accuracy of the partial consistency filtering algorithm. For instance, consider again the resolution of the "Broyden 160" problem by combining Box-Consistency filtering and a domain splitting strategy. If the final intervals have to be computed with a size smaller than or equal to $10^{-8}$, the computation is about 10 times faster with a coarse relaxation of Box-Consistency than with an accurate one [9].

It appears that the following approaches are promising:

- combining different partial consistencies [9];
- pre-processing of the constraints (e.g., symbolic transformations);
- intelligent search strategies (e.g., use of extrapolation techniques before starting a costly filtering process [14], dynamic choice of the filtering precision).

## 7. Conclusion

This paper has investigated the relations among 2B-Consistency, 3B-Consistency, Box-Consistency and Bound-Consistency. The main result is a proof of the following properties:

$$\Phi_{Bound}(\mathbf{P}) \preceq \Phi_{3B}(P_{decomp}) \preceq \Phi_{Box}(P),$$

$$\Phi_{2B}(\mathbf{P}) \preceq \Phi_{Box}(\mathbf{P}) \preceq \Phi_{2B}(\mathbf{P}_{decomp}),$$

$$\Phi_{3B}(\mathbf{P}_{decomp}) \text{ and } \Phi_{2B}(\mathbf{P}) \text{ are not comparable.}$$

The advantage of Box-Consistency is that it generates univariate functions which can be tackled by numerical methods such as Newton, and which do not require any constraint decomposition. On the other hand, 2B-Consistency algorithms require a decomposition of the constraints with multiple occurrences of the same variable. This decomposition increases the limitations due to the local nature of 2B-Consistency. As expected, higher consistencies—e.g., 3B-Consistency and Bound-Consistency—can reduce the drawbacks due to the local scope of the inconsistency detection.

Efficiency of the filtering algorithms is a critical issue, but it is difficult to draw a conclusion from the published benchmarks. Further experimentation combining these different partial consistencies with various search techniques is required to better understand their advantages and drawbacks and to define the class of application in which each of them is most relevant.

## 8. Acknowledgements

Thanks to Olivier Lhomme, Christian Bliek and Bertrand Neveu for their careful reading and helpful comments on earlier drafts of this paper. Thanks also to

Frédéric Goualard, Laurent Granvilliers, and Arnold Neumaier for interesting discussions.

Thanks also go to a referee for his many comments, which hopefully led to significant improvements of this paper.

## References

1. Benhamou, F., Goualard, F., and Granvilliers, L.: Programming with the DecLIC Language, in: *Proceedings of the Second Workshop on Interval Constraints*, Port-Jefferson, NY, 1997.
2. Benhamou, F., Mc Allester, D., and Van Hentenryck, P.: CLP(Intervals) Revisited, in: *Proc. Logic Programming: Proceedings of the 1994 International Symposium*, MIT Press, 1994.
3. Benhamou, F. and Older, W.: Applying Interval Arithmetic to Real, Integer and Boolean Constraints, *Journal of Logic Programming* **32** (1997), pp. 1–24.
4. Brent, R. P.: A FORTRAN Multiple-Precision Arithmetic Package, *ACM Trans. on Math. Software* **4** (1) (1978), pp. 57–70.
5. Cleary, J. C.: Logical Arithmetic, *Future Computing Systems* **2** (2) (1987), pp. 125–149.
6. Collavizza, H., Delobel, F., and Rueher, M.: A Note on Partial Consistencies over Continuous Domains Solving Techniques, in: *Proc. CP98 (Fourth International Conference on Principles and Practice of Constraint Programming)*, LNCS 1520, Springer Verlag, 1998.
7. Faltings, B.: Arc-Consistency for Continuous Variables, *Artificial Intelligence* **65** (1994), pp. 363–376.
8. Freuder, E. C.: Synthesizing Constraint Expressions, *Communications of the ACM* **21** (1978), pp. 958–966.
9. Granvilliers, L.: *On the Combination of Box-Consistency and Hull-Consistency*, Workshop ECAI on Non Binary-Constraints, Brighton, United Kingdom, 1998
10. Hansen, E.: *Global Optimization Using Interval Analysis*, Marcel Dekker, NY, 1992.
11. Hong, H., Stahl, V.: Safe Starting Regions by Fixed Points and Tightening, *Computing* **53** (1994), pp. 323–335.
12. Hyvönen, E.: Constraint Reasoning Based on Interval Arithmetic: The Tolerance Propagation Approach, *Artificial Intelligence* **58** (1992), pp. 71–112.
13. Kearfott, R. B.: *Rigorous Global Search: Continuous Problems*, Kluwer Academic Publishers, Dordrecht, Netherlands, 1996.
14. Lebbah, Y. and Lhomme, O.: Acceleration Methods for Numeric CSPs AAAI, MIT Press, 1998.
15. Lee, J. H. M. and Van Emden, M. H.: Interval Computation as Deduction in CHIP, *Journal of Logic Programming* **16** (1993), pp. 255–276.
16. Lhomme, O.: Consistency Techniques for Numeric CSPs, in: *Proc. IJCAI93*, Chambery, France, 1993, pp. 232–238.
17. Lhomme, O. and Rueher, M.: Application des techniques CSP au raisonnement sur les intervalles, *RIA (Dunod)* **11** (3) (1997), pp. 283–312.
18. Mackworth, A.: Consistency in Networks of Relations, *Artificial Intelligence* **8** (1) (1997), pp. 99–118.
19. Montanari, U.: Networks of Constraints: Fundamental Properties and Applications to Picture Processing, *Information Science* **7** (2) (1974), pp. 95–132.
20. Moore, R.: *Interval Analysis*, Prentice Hall, 1966.
21. Neumaier, A.: *Interval Methods for Systems of Equations*, Cambridge University Press, 1990.
22. Neumaier, A., Dallwig, S., and Schichl, H.: GLOPT—A Program for Constrained Global Optimization, in: Bomze, I. et al. (eds.), *Developments in Global Optimization*, Kluwer Academic Publishers, Dordrecht, 1997, pp. 19–36.
23. Older, W. J. and Vellino, A.: Extending Prolog with Constraint Arithmetic on Real Intervals, in: *Proc. of IEEE Canadian Conference on Electrical and Computer Engineering*, IEEE Computer Society Press, 1990.
24. Puget, J.-F. and Van Hentenryck, P.: A Constraint Satisfaction Approach to a Circuit Design Problem, *Journal of Global Optimization* **13** (1998), pp. 75–93.

25. Prologia: *PrologIV Constraints Inside*, Parc technologique de Luminy—Case 919 13288 Marseille cedex 09, France, 1996.
26. Rueher, M. and Solnon, C.: Concurrent Cooperating Solvers within the Reals, *Reliable Computing* **3** (3) (1997), pp. 325–333.
27. Tsang, E.: *Foundations of Constraint Satisfaction*, Academic Press, 1993.
28. Van Hentenryck, P., Deville, Y., and Michel, L.: *Numerica. A Modeling Language for Global Optimization*, MIT Press, 1997.
29. Van Hentenryck, P., McAllester, D., and Kapur, D.: Solving Polynomial Systems Using a Branch and Prune Approach, *SIAM Journal on Numerical Analysis* **34** (2) (1997).

*Reliable Computing* **5**: 229–240, 1999.
© 1999 *Kluwer Academic Publishers.*

# Verified Computation of Fast Decreasing Polynomials

NELI S. DIMITROVA and SVETOSLAV M. MARKOV
*Institute of Mathematics and Informatics, Bulgarian Academy of Sciences, Acad. G. Bonchev str.,
Bl. 8, BG-1113 Sofia, Bulgaria, e-mail: {nelid, smarkov}@iph.bio.bas.bg*

(Received: 28 October 1998; accepted: 4 February 1999)

**Abstract.** In this paper the problem of verified numerical computation of algebraic fast decreasing polynomials approximating the Dirac delta function is considered. We find the smallest degree of the polynomials and give precise estimates for this degree. It is shown that the computer algebra system Maple does not always graph such polynomials reliably because of evaluating the expressions in usual floating-point arithmetic. We propose a procedure for verified computation of the polynomials and use it to produce their correct graphic presentations in Maple.

## 1. Introduction

The algebraic polynomials $P$ are called fast decreasing on $[-1, 1]$ if they satisfy the properties

$$P(0) = 1; \quad |P(x)| \le e^{-\varphi(x)}, \quad x \in [-1, 1], \tag{1.1}$$

where $\varphi$ is an even, nondecreasing and right continuous on $[0, 1]$ function [4]. For a given $\varphi$ the problem for finding the smallest possible degree (which depends on $\varphi$) of the polynomials $P$ satisfying (1.1) is solved in [4], more precisely, bounds for the smallest possible degree of $P$ are given.

Fast decreasing polynomials play an important role in many applications of mathematical analysis and technical sciences such as approximation theory, orthogonal polynomials, moment problems, syntheses of antennae and electrical circuits etc.

Fast decreasing polynomials $P$ produce an approximation of the interval function

$$\delta(x) = \begin{cases} [0, 1], & x = 0; \\ 0, & 0 < |x| \le 1, \end{cases}$$

which is the complete graph of the Dirac delta function according to the theory of Hausdorff approximations of functions [5]. For this purpose let $\varphi$ in (1.1) be of the form

$$\varphi(x) = \begin{cases} 0, & x \in [0, \alpha); \\ -\ln \beta, & x \in [\alpha, 1], \end{cases}$$

where $0 < \alpha < 1$ and $0 < \beta < 1 / e$ are given numbers [4]. Then conditions (1.1) reduce to

$$P(0) = 1; \quad |P(x)| \leq 1, \ |x| < \alpha; \tag{1.2}$$

$$|P(x)| \leq \beta, \ |x| \in [\alpha, 1]. \tag{1.3}$$

The following bounds for the smallest possible degree $k(\alpha, \beta) = 2n$ of $P(x)$ from (1.2), (1.3) are given in [4]:

$$\frac{1}{9} \left( \frac{\ln \beta}{\ln(1 - \alpha)} + 1 \right) \leq k(\alpha, \beta) \leq 18 \left( \frac{\ln \beta}{\ln(1 - \alpha)} + 1 \right). \tag{1.4}$$

The polynomials $P(x) = P_{2n}(x)$, $n = n(\alpha, \beta)$, are of the form

$$P_{2n}(x) = \frac{T_n \left( \dfrac{1 + \alpha^2 - 2x^2}{1 - \alpha^2} \right)}{T_n \left( \dfrac{1 + \alpha^2}{1 - \alpha^2} \right)}, \tag{1.5}$$

where $T_n$ is the $n$-th Chebyshev polynomial [1]. In fact, $P_{2n}(x)$ are the polynomials of degree $2n$ with constant term 1, which deviate least from the origin on $[-1, -\alpha] \cup [\alpha, 1]$.

The polynomial $P_{2n}(x)$ given by (1.5) satisfies (1.2). Since $|T_n(x)| \leq 1$ for $x \in [-1, 1]$ is valid, (1.3) reduces to finding the smallest integer $n$ such that the inequality

$$1 \leq \beta T_n \left( \frac{1 + \alpha^2}{1 - \alpha^2} \right) \tag{1.6}$$

is fulfilled.

In the next section we shall find the smallest degree $2n$, $n = n(\alpha, \beta)$, of the polynomials $P_{2n}(x)$ and improve the bounds given by (1.4) for this degree. In Section 3 we show graphic presentations of some polynomials $P_{2n}(x)$ in the computer algebra system Maple. Since for higher degrees $2n$ the graphs differ qualitatively from the theoretical assumptions, we propose a procedure written in Maple for validated computation of the values of these polynomials at given points. This procedure is further used to present correct graphs of the polynomials. We then give a number of numerical examples, demonstrating the advantages of our procedure.

## 2. Bounds for the Degree of the Polynomial

We use the presentation

$$T_n \left( \frac{1 + \alpha^2}{1 - \alpha^2} \right) = \frac{1}{2} \left( \left( \frac{1 + \alpha^2}{1 - \alpha^2} + \sqrt{ \left( \frac{1 + \alpha^2}{1 - \alpha^2} \right)^2 - 1 } \right)^n \right.$$

$$\left. + \left( \frac{1 + \alpha^2}{1 - \alpha^2} - \sqrt{ \left( \frac{1 + \alpha^2}{1 - \alpha^2} \right)^2 - 1 } \right)^n \right).$$

Then (1.6) is equivalent to

$$\left(\frac{1+\alpha}{1-\alpha}\right)^n + \left(\frac{1-\alpha}{1+\alpha}\right)^n \geq \frac{2}{\beta}. \tag{2.1}$$

By substituting in (2.1)

$$t := \left(\frac{1+\alpha}{1-\alpha}\right)^n, \quad t > 1,$$

we obtain

$$\beta t^2 - 2t + \beta \geq 0.$$

The solution of the last inequality is easily seen to be

$$t \geq \frac{1 + \sqrt{1 - \beta^2}}{\beta}.$$

Further we obtain consecutively

$$\left(\frac{1+\alpha}{1-\alpha}\right)^n \geq \frac{1 + \sqrt{1 - \beta^2}}{\beta},$$

$$n \ln \frac{1+\alpha}{1-\alpha} \geq \ln \frac{1 + \sqrt{1 - \beta^2}}{\beta},$$

$$n \geq \frac{\ln(1 + \sqrt{1 - \beta^2}) - \ln \beta}{\ln(1 + \alpha) - \ln(1 - \alpha)}.$$

Denote

$$\tilde{n} = \frac{\ln(1 + \sqrt{1 - \beta^2}) - \ln \beta}{\ln(1 + \alpha) - \ln(1 - \alpha)} \tag{2.2}$$

and take

$$n = \text{trunc}(\tilde{n}) + 1, \tag{2.3}$$

where the function $\text{trunc}(\tilde{n})$ presents the integer part of the argument $\tilde{n}$. Then $2n$, $n = n(\alpha, \beta)$, is the smallest degree of the polynomial $P_{2n}(x)$ satisfying (1.6) and therefore (1.3).

PROPOSITION 2.1. *Let $\tilde{n}$ be defined by (2.2) with $0 < \alpha < 1$ and $0 < \beta < 1/e$. Then the following inequalities hold true:*

$$\frac{1}{2} \frac{\ln \beta}{\ln(1 - \alpha)} < \tilde{n} < 2 \frac{\ln \beta}{\ln(1 - \alpha)} + 1. \tag{2.4}$$

*Proof.* Denote

$$a = -\ln(1 - \alpha), \quad b = -\ln \beta, \quad c = \ln(1 + \alpha), \quad d = \ln\left(1 + \sqrt{1 - \beta^2}\right).$$

It is easy to see that

$$a, c, d > 0, \quad b > 1, \quad a > c, \quad b > d \tag{2.5}$$

hold true. The left-hand side inequality in (2.4) is then equivalent to

$$\frac{b}{a} < 2\frac{b+d}{a+c} \quad \text{and} \quad b(c-a) < 2ad,$$

which is always satisfied because of (2.5). The right-hand side inequality in (2.4) is equivalent to

$$\frac{b+d}{a+c} < 2\frac{b}{a} + 1 \quad \text{or} \quad a(b-d) + 2bc + ac + a^2 > 0.$$

The latter is again satisfied because of (2.5). □

COROLLARY 2.1. *Let $P_{2n}(x)$ be the fast decreasing polynomial of smallest degree $2n$ satisfying the properties (1.2) and (1.6) with $n = n(\alpha, \beta)$ from (2.3). The following bounds for $2n$ are then valid:*

$$\max\left\{\frac{1}{9}\left(\frac{\ln \beta}{\ln(1-\alpha)} + 1\right), \frac{\ln \beta}{\ln(1-\alpha)}\right\} < 2n < 4\left(\frac{\ln \beta}{\ln(1-\alpha)} + 1\right).$$

*Proof.* The conclusion follows from the obvious inequalities

$$2\tilde{n} < 2n < 2\tilde{n} + 2. \qquad \Box$$

*Remarks.* Denote

$$L = \frac{\ln \beta}{\ln(1-\alpha)}, \quad L_1 = \frac{1}{9}(L+1), \quad U = 4(L+1).$$

Obviously Corollary 2.1 implies an improved upper bound $U$ for the smallest degree $2n$ of the polynomial $P_{2n}(x)$ when compared with the one given by (1.4). Actually the bounds for $2n$ are given by

$$\texttt{trunc}(\max\{L, L_1\}) + 1 \le 2n \le \texttt{trunc}(U). \tag{2.6}$$

The following relation between $L$ and $L_1$ can be easily seen:

$$L_1 \ge L \iff L \le \frac{1}{8} \iff \alpha + \beta^8 \ge 1.$$

In practice we are interested in small values for $\alpha$ and $\beta$, e.g. such that $\alpha + \beta \le 1$. For these $\alpha$ and $\beta$ we have $L \ge 1$ and $L_1 < L$. In this case (2.6) implies

$$\texttt{trunc}(\max\{L, L_1\}) + 1 = \texttt{trunc}(L) + 1 \ge 2,$$

that is the polynomials which approximate the function $\delta(x)$ are of smallest degree $2n = 2$.

Table 1. Lower and upper bounds for the smallest degree $2n$ w.r.t. $\alpha$ and $\beta$.

| $\alpha$ | $\beta$ | $L_1$ | $L$ | $U$ | $\bar{n}$ | $2n$ |
|---|---|---|---|---|---|---|
| 0.9999 | 0.367 | 0.12321 | 0.10883 | 4.4353 | 0.16762 | 2 |
| 0.999 | 0.36 | 0.12754 | 0.14789 | 4.5916 | 0.22113 | 2 |
| 0.99 | 0.01 | 0.22222 | 1.0 | 8.0 | 1.0009 | 4 |
| 0.8 | 0.1 | 0.27008 | 1.4307 | 9.7227 | 1.3623 | 4 |
| 0.1 | 0.1 | 2.5394 | 21.854 | 91.417 | 14.916 | 30 |
| 0.06 | 0.07 | 4.8864 | 42.978 | 175.91 | 27.893 | 56 |
| 0.05 | 0.08 | 5.5823 | 49.241 | 200.96 | 32.146 | 66 |
| 0.02 | 0.02 | 21.627 | 193.64 | 778.55 | 115.11 | 232 |

Table 1 shows some values for $L_1$, $L$, $U$, $\bar{n}$, and $2n$ with respect to different $\alpha$ and $\beta$. In the first and second lines of the table, $\alpha + \beta^8 = 1.000229\ldots > 1$ and $\alpha + \beta^8 = 0.99928\ldots < 1$ respectively are satisfied, in the last four lines the value of $\alpha + \beta$ is substantially less than one. Note that $1/e \in [0.3678794411, 0.3678794413]$ holds true. The computations are performed in the computer algebra system Maple.

## 3. Graphic Presentation of Fast Decreasing Polynomials

The polynomial $P_{2n}(x)$ can be computed explicitly according to (1.5) using the symbol replacement command subs in Maple:

```
> T:=Chebyshev(x,n):
> T1:=subs(x=(1+alpha**2)/(1-alpha**2),T):
> T2:=subs(x=(1+alpha**2-2*x**2)/(1-alpha**2),T):
> P:=expand(T1/T2);
```

Here Chebyshev(x,n) means a procedure for recursive computation of the Chebyshev polynomial of $n$-th degree $T_n(x)$ in the variable $x$. Setting $\alpha$ to a concrete rational value we obtain the corresponding polynomial P $= P_{2n}(x)$. For example, the command

```
> Px:=expand(subs(alpha=1/10,P)):
```

produces the polynomial Px $= P_{2n}(x)$ with exact (rational) coefficients, that is no rounding errors occur in the presentation of Px. Further the execution of

```
> plot(Px, x = -1..1);
```

delivers the graphic of Px $= P_{2n}(x)$. Figure 1 shows $P_{30}(x)$ for $x \in [-1, 1]$ and $\alpha = \beta = 0.1$. The graphic is placed in an $(\alpha, \beta)$-strip around the Dirac delta function. The graphic presentation of the polynomial $P_{56}(x)$, $\alpha = 0.06$, $\beta = 0.07$ and $x \in [0, 1]$ is given in Figure 2. Unfortunately, for $x \in [0.9, 1]$ (see also Figure 3) the picture differs qualitatively from the actual result given by condition (1.3). Similar effects

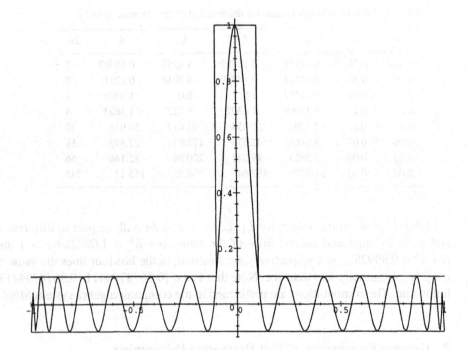

*Figure 1.*   $P_{30}(x)$, $\alpha = 0.1$, $\beta = 0.1$, $x \in [-1, 1]$.

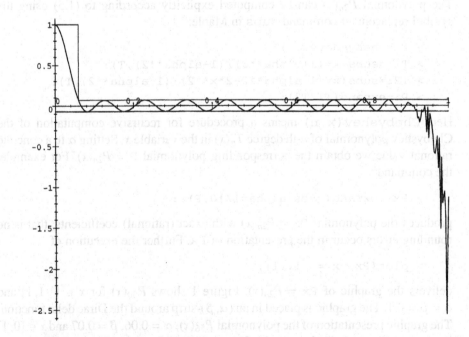

*Figure 2.*   $P_{56}(x)$, $\alpha = 0.06$, $\beta = 0.07$, $x \in [0, 1]$.

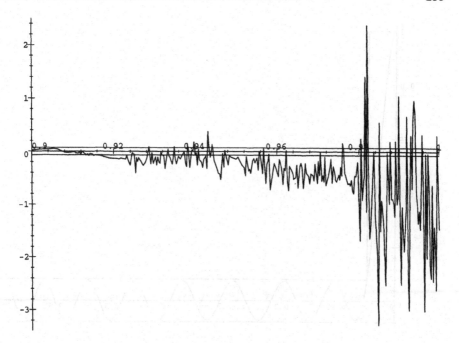

*Figure 3.*   $P_{56}(x)$, $\alpha = 0.06$, $\beta = 0.07$, $x \in [0.9, 1]$.

can be observed by plotting $P_{66}(x)$ with $\alpha = 0.05$, $\beta = 0.08$ on Figures 4, 5 and 6. As mentioned in [2], Maple's `plot` command "computes the value of a given expression at a modest number of equally spaced points in the specified interval. It then selects more points to compute within those subintervals where there is a large amount of fluctuation. This technique is known as adaptive plotting... The adaptive plotting approach does not always lead to 'infallible' pictures."

To demonstrate these plotting effects we use again the `subs` command in Maple to evaluate the polynomial $P_{2n}(x)$ at different points $x_i \in [-1, 1]$ in the usual floating-point arithmetic. Some results are given by Tables 2 and 3 in the columns $P_{56}(x_i)$ and $P_{66}(x_i)$ respectively. Remember that $P_{56}(x_i)$ and $P_{66}(x_i)$ should satisfy condition (1.3) that is $|P_{56}(x_i)| \leq 0.07$ and $|P_{66}(x_i)| \leq 0.08$ have to be fulfilled.

In what follows we propose a procedure written in Maple for evaluation of Chebyshev polynomials in computer interval arithmetic. The computations are based on the well known formula

$$T_n(x) = 2xT_{n-1}(x) - T_{n-2}(x).$$

Thereby we use the interval arithmetic package *INTPAK* implemented in Maple V Release 3 [3].

```
> IntEvalCheb:=proc(xv,n)
>    local ixv;
>    option remember;
```

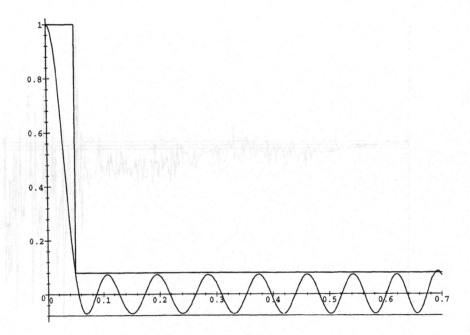

*Figure 4.* $P_{66}(x)$, $\alpha = 0.05$, $\beta = 0.08$, $x \in [0, 0.7]$.

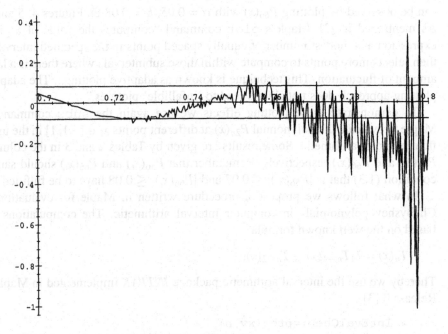

*Figure 5.* $P_{66}(x)$, $\alpha = 0.05$, $\beta = 0.08$, $x \in [0.7, 0.8]$.

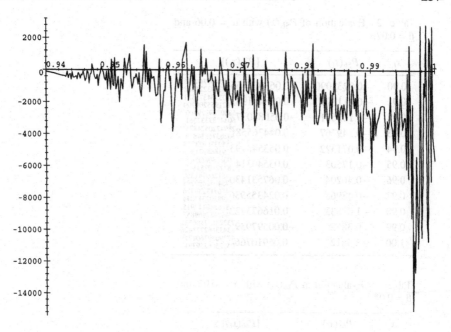

*Figure 6.* $P_{66}(x)$, $\alpha = 0.05$, $\beta = 0.08$, $x \in [0.94, 1]$.

```
>    if type(xv,'interval') then ixv:=xv
>    elif type(xv,'interval_comp') or type(xv,'integer')
>      then ixv:=construct(xv)
>    elif type(xv, 'rational') then
>      ixv:=construct(xv,'rounded')
>    fi;
>    if n=0 then RETURN(eval(construct(1.)))
>    elif n=1 then RETURN(eval(ixv))
>    else
>      RETURN(((construct(2.) &* eval(ixv) ) &*
>      IntEvalCheb(xv,n-1)) &- IntEvalCheb(xv,n-2))
>    fi
> end:
```

The output from IntEvalCheb is an interval enclosing the exact polynomial value $T_n(x)$ at the "point" $x = $ xv. Using this procedure we are able to compute guaranteed bounds for the value of $P_{2n}(x)$ at a given point $x = $ xv by means of the following commands:

```
> ialpha:=construct(alpha,'rounded'):
> iv:=(construct(1.) &+ &sqr(ialpha))
>      &/ (construct(1.)&- &sqr(ialpha)):
> IT1:=IntEvalCheb(iv,n):
```

Table 2. Evaluation of $P_{56}(x)$ with $\alpha = 0.06$ and $\beta = 0.07$.

| $x_i$ | $P_{56}(x_i)$ | $[P_{56}(x_i)]$ |
|-------|---------------|------------------|
| 0.90 | 0.0232090 | $0.068067465^{355469744956}_{162693224448}$ |
| 0.91 | −0.011876 | $0.028481982^{103704371521}_{021955209700}$ |
| 0.92 | −0.102671 | $-0.056880037^{5880769329800}_{6044383332361}$ |
| 0.93 | −0.149767 | $-0.044365558^{2894220424875}_{3026374834818}$ |
| 0.94 | −0.071322 | $0.053543193^{559336923718}_{391664971308}$ |
| 0.95 | −0.17503 | $0.035403144^{2008380868908}_{1871280251162}$ |
| 0.96 | −0.46204 | $-0.067531430^{481260141360}_{754052490298}$ |
| 0.97 | −0.48968 | $0.024355536^{456913239776}_{198302317044}$ |
| 0.98 | −1.07833 | $0.016675422^{916280142915}_{442482469857}$ |
| 0.99 | 0.3852 | $-0.005979939^{73328527186532}_{83433662771426}$ |
| 1.00 | −2.5812 | $0.069107660^{4729039760477}_{0552506259391}$ |

Table 3. Evaluation of $P_{66}(x_i)$ with $\alpha = 0.05$ and $\beta = 0.08$

| $x_i$ | $P_{66}(x_i)$ | $[P_{66}(x_i)]$ |
|-------|---------------|------------------|
| 0.950 | −1790.604 | $0.0393837^{80949717858602}_{70099117465438}$ |
| 0.955 | −810.17 | $-0.0364816^{30178802653642}_{41260843910044}$ |
| 0.960 | 50.46 | $-0.0731325^{45869575411536}_{67302542694798}$ |
| 0.965 | −475.13 | $-0.0186730^{05341846056596}_{16236908073950}$ |
| 0.970 | −1242.93 | $0.063750^{724496051997657}_{699133058886717}$ |
| 0.975 | −2591.00 | $0.0456569^{98274609087359}_{70636613514287}$ |
| 0.980 | −2458.13 | $-0.057467^{868038753268166}_{909061556363026}$ |
| 0.985 | −2096.5 | $-0.0329284^{11646205435696}_{66596643561258}$ |
| 0.990 | −910.2 | $0.073276^{958336635929493}_{863397091564796}$ |
| 0.995 | −1609.6 | $-0.069550^{724129591389172}_{861119077492987}$ |
| 0.998 | −5955.8 | $0.037281^{210618741091688}_{030899752013744}$ |
| 1.000 | −6939.7 | $-0.07346^{3886898985447140}_{4238844879000255}$ |

```
> ivx:=((construct(1.) &+ &sqr(ialpha)) &-
>      (construct(2.) &* &sqr(construct(xv,'rounded'))))
>      &/ (construct(1.)&- &sqr(ialpha)):
> IT2:=IntEvalCheb(ivx,n):
> IP:=IT1 &/ IT2;
```

Tables 2 and 3 in columns $[P_{56}(x_i)]$ and $[P_{66}(x_i)]$ respectively summarize the outputs from the execution of the above commands. All computations are performed with Digits:=20, i.e. with 20 decimal digits in the mantissa. Obviously the absolute values of the produced intervals satisfy condition (1.3), that is $|[P_{56}(x_i)]| \leq 0.07$ and $|[P_{66}(x_i)]| \leq 0.08$ are valid.

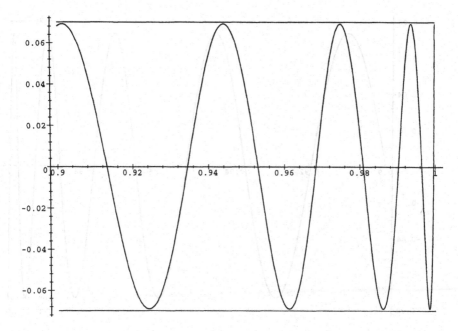

*Figure 7.*   $P_{56}(x)$, $\alpha = 0.06$, $\beta = 0.07$, $x \in [0.9, 1]$.

This approach is further used to plot the polynomials $P_{56}(x)$ and $P_{66}(x)$, $x \in [0.9, 1]$. We construct a regular mesh $x_i = x_0 + (i-1)h$ for $i = 1, 2, ..., N$ with $x_0 = 0.9$, $h = 0.0001$ and $x_N = 1$. The centers of the computed intervals $[P_{56}(x_i)]$ and $[P_{66}(x_i)]$ are taken as approximate values for the polynomials. The produced collections of points connected by lines are then plotted for each polynomial separately. The respective graphics are shown by Figures 7 and 8.

## 4.  Conclusions

In this work we present an algorithm for the computation of the even fast decreasing polynomials $P_{2n}(x)$ approximating the Dirac delta function for given $\alpha$ and $\beta$. The approximation is understood within the theory of Hausdorff approximations of functions. We first compute the smallest degree $2n$ of these polynomials and propose improvements of known upper and lower bounds of the degree $2n$. Since the computer algebra system Maple does not always produce 'infallible' computations and plottings because of evaluating expressions in usual floating-point arithmetic, we propose a procedure written in Maple for verified computation of Chebyshev polynomials and use it to plot correctly the polynomials $P_{2n}(x)$.

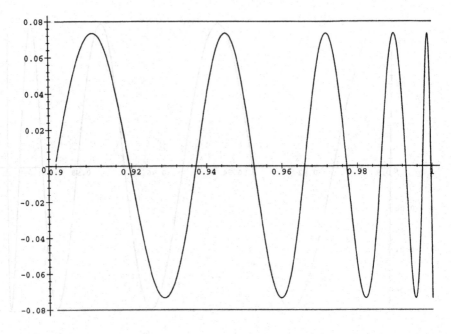

*Figure 8.*   $P_{66}(x)$, $\alpha = 0.05$, $\beta = 0.08$, $x \in [0.9, 1]$.

## Acknowledgement

This work has been partially supported by the Bulgarian National Science Fund under grant No. MM. 521/95.

## References

1. Ahieser, N. I.: *Elements of the Elliptic Functions Theory*, Moscow, Nauka, 1970 (in Russian).
2. Char, B. W., Geddes, K. O., Gonnet, G. H., Leong, B. L., Monagan, M. B., and Watt, S. M.: *First Leaves: A Tutorial Introduction to Maple V*, Springer Verlag, 1992.
3. Connel, A. E. and Corless, R. M.: An Experimental Interval Arithmetic Package in Maple, *Interval Computations* 2 (1993), pp. 120–134.
4. Ivanov, K. G. and Totik, V.: Fast Decreasing Polynomials, *Constructive Approx.* 6 (1990), pp. 1–20.
5. Sendov, B.: *Hausdorff Approximations*, Kluwer Academic Publ., London, 1990.

*Reliable Computing* **5**: 241–253, 1999.
© 1999 *Kluwer Academic Publishers.*

# An Accurate Distance-Calculation Algorithm for Convex Polyhedra *

EVA DYLLONG, WOLFRAM LUTHER and WERNER OTTEN
*Gerhard-Mercator-Universität-GH Duisburg, Informatik II, D-47048 Duisburg, Germany,*
*e-mail:* {*dyllong, luther, otten*}*@informatik.uni-duisburg.de*

(Received: 9 November 1998; accepted: 19 December 1998)

**Abstract.** The paper describes an efficient and accurate algorithm to calculate the distance between convex polyhedra. The closest points between two objects can be calculated by simple projections and can be followed continuously in time. The polyhedra are given by the vertices. Interval data are supported. The accuracy of the calculation is explored.

## 1. Introduction

In the fields of computational geometry, robotics and Computer Aided Design there is an abundance of literature to calculate the distance between convex and non-convex objects. For convex polyhedra there exists a lot of algorithms working in linear time $O(n)$, or nearly linear time $O(n \log n)$ where $n$ denotes the number of vertices [1], [2]. The Gilbert method [6] takes the centroid of each polyhedron, a vector $\vec{v}$ connecting both and constructs candidates for the two nearest points taking the maximum of scalar products between $\vec{v}$ (resp. $-\vec{v}$) and the vertices of the polyhedra. Calculating a sequence of better approximations by a local search, the iteration process finds the distance between both objects. This local search was proposed by Sato et al. [9] to enhance the computational speed of the original Gilbert algorithm. The testing of non-empty intersection is possible in $O(\log n_P \cdot \log n_Q)$-time for convex polyhedra $P$ and $Q$ which have been preprocessed in linear time to a hierarchical representation. Even the intersection depth in any direction can be obtained under this bound. These results were shown by Dobkin et al. [3], [4]. Note that complexity bounds can be reduced when the closest points are followed in an incremental way.

A different approach is based on a preprocessing for representing an object as a binary tree of spheres and a final search on the tree structure in a depth of first order [8]. The algorithm according to Sato et al. [9] allows an efficient collision detection. It is applicable to convex and non-convex objects and utilizes different preprocessing steps—a construction of a convex hull from a general non-convex

---

* This work was supported by the Deutsche Forschungsgemeinschaft within the scope of the project A8 of the Sonderforschungsbereich 291.

polyhedron, a construction of adjacent-vertex lists and a construction of a binary tree of covering spheres for non-convex polyhedra.

However, our scope is different from these results. Using classical path planning and collision-detection strategies, the robot moves in the environment to perform controlling and repairing actions in the neighbourhood of obstacles or the end effector has to track a target object. Obstacles are often modeled or reconstructed from sonar and visual data leading to uncertain information. We are interested in a simple algorithm to calculate the distance between a point and a convex polyhedron or two convex polyhedra with interval vertices, therefore we only use accurate operations in order to calculate it fast and, if possible, avoid accumulation errors. A part of our results is to express the distance $d(\vec{y}(t), P)$ between a moving point $\vec{y}(t) := (y_1(t), y_2(t), y_3(t))$, for instance the location of a sensor, and a component of a target $P$ described by a convex polyhedron as a well-defined sequence of projections onto certain planes and edges of the facets. The part of the algorithm deciding whether a point resides within a polygonal facet or not, can be useful, e.g. for multilegged walking robots. Here the projection of the center of gravity must lie within the polygon defined by the different leg-positions to guarantee stability.

## 2.  Distance Between a Point and a Convex Polyhedron

Given a point $\vec{y} = (y_1, y_2, y_3)$ and a non-degenerate convex polyhedron $P$ bounded by

$$\partial P = \{S_i, \ i = 1, ..., m; \ \vec{v}_k, \ k = 1, ..., n\}$$

with $m$ facets and $n$ vertices. All coordinates are machine numbers belonging to the floating-point screen $S := S(\mathcal{B}, \ell, em, eM)$ with its base $\mathcal{B}$, mantissa length $\ell$ and $[em, eM]$ smallest and largest allowable exponent, respectively. The relative error for all elementary screen operations is assumed to be bounded by $\varepsilon_\ell := \mathcal{B}^{1-\ell}$ if no underflow occurs (another possibility for example would be using machine interval arithmetic, where the bounds are obtained automatically, even if underflow occurs). We assume logically that all vertices of one facet are coplanar—in most cases the facets will be triangles—and address the problem of finding the shortest straight line segment $[\vec{y}, \vec{x}]$ between a point $\vec{y}$ and the polyhedron with $\vec{x} \in \partial P$. We call $\vec{x}$ an opposite distance point if $\vec{y} \in P - \partial P$. In the sequel we denote the vertices belonging to the facet $S_i$ by $\vec{s}_{ik}, k = 1, ..., t_i, t_i > 2$, given in counter-clockwise order when looking from outside of $P$ to $S_i$, and $[\vec{s}_{ik}, \vec{s}_{i(k+1)}]$ are the edges of the facet $S_i$, $k = 1, ..., t_i, \vec{s}_{i(t_i+1)} := \vec{s}_{i1}$. In the following the operator $\cdot_\#$ describes the high-precise dot product resulting in a data type *dotprecision* (this data type accommodates the full floating-point range with double exponents; see [7]). The operator $\#\square(\cdot)$ denotes the correct rounding, i.e. the evaluation of an expression using a data type *dotprecision* followed by a rounding (to nearest).

At the beginning, before starting our algorithm, we calculate the correctly rounded cross product

$$\vec{n}_{i2} = \#\square\left((\vec{s}_{i2} - \vec{s}_{i1}) \times_{\#} (\vec{s}_{i3} - \vec{s}_{i2})\right) = \#\square(\vec{s}_{i2} \times_{\#} \vec{s}_{i3} + \vec{s}_{i1} \times_{\#} \vec{s}_{i2} - \vec{s}_{i1} \times_{\#} \vec{s}_{i3})$$

with $\vec{x} \times_{\#} \vec{y} := (x_2 \cdot_{\#} y_3 - x_3 \cdot_{\#} y_2, x_3 \cdot_{\#} y_1 - x_1 \cdot_{\#} y_3, x_1 \cdot_{\#} y_2 - x_2 \cdot_{\#} y_1)$ for $\vec{x} = (x_1, x_2, x_3)$, $\vec{y} = (y_1, y_2, y_3)$, and the normal vector

$$\vec{n}_i = \frac{\vec{n}_{i2}}{\sqrt{\#\square(\vec{n}_{i2} \cdot_{\#} \vec{n}_{i2})}}$$

directed to the exterior of the polyhedron for all $i = 1, ..., m$. Then we denote by $E_i$ the plane described by

$$E_i : \vec{x} \cdot \vec{n}_i - \vec{s}_{i1} \cdot \vec{n}_i = 0.$$

$E_i$ is not the plane generated by $S_i$, since $E_i$ is defined by a perturbed normal vector, nevertheless in the following we call $E_i$ the plane supporting $S_i$. We start the algorithm by determining the projections $\vec{x}_i$ of the point $\vec{y}$ to all planes $E_i$ and the lengths $l_i$ of the projection-rays. Then, we decide whether point $\vec{y}$ lies within the polyhedron or not. If the point is in the polyhedron a smallest $|l_i|$ defines an opposite distance-point $\vec{x} \in \partial P$, the solution may not be unique in this case. Otherwise, we have to decide whether a point $\vec{x}_i$ resides within a polygonal facet of the polyhedron or not. If no point $\vec{x}_i$ is in the facet $S_i$, we test for all edges $[\vec{s}_{ik}, \vec{s}_{i(k+1)}]$ of $S_i$, $k = 1, ..., t_i$, whether the projection $\vec{p}_{ik}$ of $\vec{y}$ on the supporting line belongs to this edge of $P$. If not, the distance-point $\vec{x}$ is a corner-point. Otherwise, we choose the nearest projection point along an edge. The algorithm works in linear time $O(n)$ and it reads as follows:

**Algorithm 1: Distance Point to Polyhedron**

A: We calculate the distance between point $\vec{y}$ and each plane $E_i$ by the aid of the long scalar product using a data type *dotprecision* [7]

$$l_i := \#\square(\vec{y} \cdot_{\#} \vec{n}_i - \vec{s}_{i1} \cdot_{\#} \vec{n}_i).$$

We memorize the sign of $l_i$, $i = 1, ..., m$. If all $l_i$ are negative, then the point $\vec{y}$ is located in the interior of the polyhedron. For an opposite distance-point $\vec{x} \in \partial P$ (may not be unique) it holds that

$$\vec{x} := \vec{y} + |l_{i_0}| \cdot \vec{n}_{i_0},$$

and the distance is given by $|l_{i_0}| = \min_{1 \le i \le m} |l_i|$ or we set the distance to zero if necessary for application.

B: Assume now that there is at least one $l_i > 0$, then the point $\vec{y}$ is outside. All facets $S_i$ with $l_i > 0$ are visible from $\vec{y}$. Then, we calculate accurately for all $i$ with $l_i > 0$ the projections

$$\vec{x}_i := \vec{y} - l_i \cdot \vec{n}_i$$

onto $E_i$. Now we must decide whether $\vec{x}_i$ is in $S_i$. Here we apply a simple test: $\vec{x}_i$ resides within the facet $S_i$, if and only if $\vec{x}_i$ lies on the left of all lines

$$\vec{u}_{ik}(t) := \vec{s}_{i(k-1)} + t \cdot (\vec{s}_{ik} - \vec{s}_{i(k-1)}), \quad -\infty < t < \infty, \quad k = 2, \dots, t_i + 1,$$

$$\vec{s}_{i(t_i+1)} := \vec{s}_{i1}.$$

For this purpose we calculate the scalar product of the cross product

$$\#\Box\big((\vec{s}_{ik} - \vec{s}_{i(k-1)}) \times_\# (\vec{x}_i - \vec{s}_{ik})\big)$$

and the vector $\vec{n}_{i2}$. We denote this scalar product by $d_{ik}$ and use the formula

$$(a \times b) \cdot (c \times d) = (a \cdot c)(b \cdot d) - (a \cdot d)(b \cdot c),$$

i.e. we have

$$d_{ik} := \#\Box\Big(\big(\#\Box((\vec{s}_{ik} - \vec{s}_{i(k-1)}) \cdot_\# (\vec{s}_{i2} - \vec{s}_{i1}))\big) \cdot_\# \big(\#\Box((\vec{x}_i - \vec{s}_{ik}) \cdot_\# (\vec{s}_{i3} - \vec{s}_{i2}))\big)$$

$$- \big(\#\Box((\vec{s}_{ik} - \vec{s}_{i(k-1)}) \cdot_\# (\vec{s}_{i3} - \vec{s}_{i2}))\big) \cdot_\# \big(\#\Box((\vec{x}_i - \vec{s}_{ik}) \cdot_\# (\vec{s}_{i2} - \vec{s}_{i1}))\big)\Big)$$

$$= \#\Box \begin{pmatrix} \#\Box(\vec{s}_{ik} \cdot_\# \vec{s}_{i2} - \vec{s}_{ik} \cdot_\# \vec{s}_{i1} - \vec{s}_{i(k-1)} \cdot_\# \vec{s}_{i2} + \vec{s}_{i(k-1)} \cdot_\# \vec{s}_{i1}) \\ \cdot_\# \#\Box(\vec{s}_{i3} \cdot_\# \vec{x}_i - \vec{s}_{ik} \cdot_\# \vec{s}_{i3} - \vec{s}_{i2} \cdot_\# \vec{x}_i + \vec{s}_{ik} \cdot_\# \vec{s}_{i2}) \\ -\#\Box(\vec{s}_{ik} \cdot_\# \vec{s}_{i3} + \vec{s}_{i2} \cdot_\# \vec{s}_{i(k-1)} - \vec{s}_{i(k-1)} \cdot_\# \vec{s}_{i3} - \vec{s}_{ik} \cdot_\# \vec{s}_{i2}) \\ \cdot_\# \#\Box(\vec{x}_i \cdot_\# \vec{s}_{i2} - \vec{s}_{ik} \cdot_\# \vec{s}_{i2} + \vec{s}_{i1} \cdot_\# \vec{s}_{ik} - \vec{x}_i \cdot_\# \vec{s}_{i1}) \end{pmatrix}.$$

We memorize the sign of all $d_{ik}$. If all these products are non-negative, then $\vec{x}_i$ belongs to the polygonal surface $S_i$. We set the distance-point $\vec{x} := \vec{x}_i$ and the distance $d := l_i$. The polyhedron is convex, so there exists at most one $S_i$ with $d_{ik} \geq 0$ for all $k$.

C: If there is no $S_i$ with $d_{ik} \geq 0$, $k = 1, \dots, t_i$, then we must decide for all $[\vec{v}_s, \vec{v}_r]$ with $\exists_{i,k}([\vec{v}_s, \vec{v}_r] = [\vec{s}_{ik}, \vec{s}_{i(k+1)}] \wedge l_i > 0)$, if the projection of $\vec{y}$ to the edge

$$\vec{u}(t) := \vec{v}_s + t(\vec{v}_r - \vec{v}_s)$$

meets a point with parameter $0 \leq t \leq 1$.
We form the dot products (i.e. the accurately calculated scalar products)

$$\kappa := (\vec{y} - \vec{v}_s) \cdot_\# (\vec{v}_r - \vec{v}_s) \quad \text{and} \quad \mu := (\vec{v}_r - \vec{v}_s) \cdot_\# (\vec{v}_r - \vec{v}_s).$$

If $\kappa < 0$ or $\kappa > \mu$, then the projection ray does not meet the section between $\vec{v}_s$ and $\vec{v}_r$. Otherwise, the projection point on $[\vec{v}_s, \vec{v}_r]$ is given by

$$\vec{x}_{sr} := \vec{v}_s + \frac{\#\Box(\kappa)}{\#\Box(\mu)}(\vec{v}_r - \vec{v}_s)$$

and the square of the distance is given by

$$d_{sr}^2 := \frac{\#\Box\big((\#\Box((\vec{v}_r - \vec{v}_s) \times_\# (\vec{y} - \vec{v}_r))) \cdot_\#(\#\Box((\vec{v}_r - \vec{v}_s) \times_\# (\vec{y} - \vec{v}_r))))\big)}{\#\Box(\vec{v}_r \cdot_\# \vec{v}_r - \vec{v}_r \cdot_\# \vec{v}_s - \vec{v}_s \cdot_\# \vec{v}_r + \vec{v}_s \cdot_\# \vec{v}_s)}.$$

If several projection points on edges exist we choose the nearest one. We set

$$d := d_{s'r'} = \min_{sr} \sqrt{d_{sr}^2} \quad \text{and} \quad \vec{x} := \vec{x}_{s'r'}.$$

D: In all other cases the shortest path joins point $\vec{y}$ and a vertex-point $\vec{x}$ with $\vec{x} \in V := \{\vec{v}_k \mid \exists_{1 \leq i \leq m} \vec{v}_k \in S_i \wedge l_i > 0\}$.

If the distance-ray finishes on a facet $S_i$ we can introduce a local right-handed coordinate system $\{(\vec{s}_{il} - \vec{s}_{i(l-1)}), (\vec{s}_{i(l+1)} - \vec{s}_{il}), \vec{n}_{il}\}$ and describe the path $\vec{x}(t)$ on $S_i$ whenever the point $\vec{y}(t)$ is moving outside. Along an edge, the curve vector can be represented with respect to adjacent coordinate systems supported by $E_i$ and $E_j$ as long as it does not switch to $S_j$. Finally, at a vertex $\vec{v}$ there are $\deg(\vec{v})$ different local coordinate systems, where $\deg(\vec{v})$ denotes the number of adjacent facets at the vertex $\vec{v}$.

## 3. Error Discussion

For three different points, which are coplanar, three different normalized machine vectors may exist to approximate the normal vector of this plane. Using the dot product we can exactly calculate the direction of the normal vector, but we have to round it to the next machine vector for putting it into the equation of the plane. The rounding to the length $\|\vec{n}\| = 1$ leads to errors, too, since $\|\vec{n}\|^2 = n_1^2 + n_2^2 + n_3^2 = 1$ fulfilling $n_1 \cdot n_2 \cdot n_3 \neq 0$ has no solution in the screen of machine numbers. We denote by $\varepsilon_\ell$ and $\varepsilon'_\ell$ the rounding error of the user-screen and the work-screen with guard digits respectively; $\varepsilon'_\ell \leq \varepsilon_\ell \leq 2^{-52}$. For the evaluation of the square root we use Newton's method. K. Werner gives a complete error analysis for this evaluation in [10] (the IEEE norm 754/854 guarantees that the square root has machine accuracy $\varepsilon'_\ell$ for single and double data format). His result reads as follows:

Starting from an initial value $y_0 = (1 + x) / 2$ where $x \in [0.5, 2] \cap S$, $B = 2^\beta$, $\varepsilon'_\ell \leq 2^{-52}$, the relative error of the square root $\sqrt{x}$ calculated by Newton's method $y_n = y_{n-1} - 0.5(y_{n-1}^2 - x)$, $n \geq \lceil \log_2(\beta(\ell' - 1) - 2) \rceil + 2$ is bounded by $1.50001\varepsilon'_\ell$.

Then with $\vec{N}_{ik} = (\vec{s}_{ik} - \vec{s}_{i(k-1)}) \times (\vec{s}_{i(k+1)} - \vec{s}_{ik})$ it holds

$$\vec{n}_{ik} \in \vec{N}_{ik}(1 \pm 0.5\varepsilon'_\ell), \qquad \vec{n}_{ik} \cdot \# \vec{n}_{ik} \in \|\vec{N}_{ik}\|^2 (1 \pm 1.001\varepsilon'_\ell),$$

$$\#\square(\vec{n}_{ik} \cdot \# \vec{n}_{ik}) \in \|\vec{N}_{ik}\|^2 (1 \pm 1.502\varepsilon'_\ell).$$

Calculating the root $\sqrt{\#\square(\vec{n}_{ik} \cdot \# \vec{n}_{ik})}$ and the quotient $\vec{n}_{ik} / \sqrt{\#\square(\vec{n}_{ik} \cdot \# \vec{n}_{ik})}$ yield the estimation $\vec{n}_i \in \vec{N}_i(1 \pm 3.76\varepsilon'_\ell)$ for the normalized machine vector $\vec{n}_i$ (see [10]); $\vec{N}_i$ denotes the real normal vector of plane $E_i$. Then, if we work with two guard digits and the high-precise scalar product with a final rounding in the user-screen, we finally reach $\vec{n}_i \in \vec{N}_i(1 \pm \varepsilon_\ell)$. It is obvious that an approximation of the normal vector depends on the choice of vertices for its calculation. To make the total error as small as possible in step A of our algorithm we calculate again the scalar product $\vec{n}_i \vec{y} - \vec{n}_i \vec{s}_{i1}$ using data type *dotprecision* to avoid an accumulation of errors during the building of sum. The absolute error of a machine approximation to $l_i$ is bounded by

$$(|(N_i)_1 y_1| + |(N_i)_2 y_2| + |(N_i)_3 y_3| + |(N_i)_1 (s_{i1})_1| + |(N_i)_2 (s_{i1})_2| + |(N_i)_3 (s_{i1})_3|) 4.27\varepsilon'_\ell.$$

Table 1. Absolute resp. relative errors of the end-point and distance.

| | |
|---|---|
| A,B: | $x_v = x_{i,v} = X_v + \delta_{1,v}(\|\vec{y}\| \cdot 11.032 + \sigma_i \cdot 10.032)\varepsilon'_\ell$ |
| | $d = D + \delta'_1(\|\vec{y}\| + \sigma_i) \cdot 4.27\varepsilon'_\ell$ |
| C: | $x_v = X_v + \delta_{2,v}(\|\vec{y}\| \cdot 2.505 + \sigma_i \cdot 14.515)\varepsilon'_\ell$ |
| | $d = D(1 + \delta'_2 \cdot 3.003\varepsilon'_\ell)$ |
| D: | $x_v = X_v$ |
| | $d = D(1 + \delta'_3 \cdot 1.76\varepsilon'_\ell)$ |

$v = 1, 2, 3, \quad \sigma_i := \max_k \|\vec{s}_{ik}\|, \quad |\delta_{j,v}| \leq 1, \quad |\delta'_j| \leq 1, \quad j = 1, 2, 3,$

$D = \|\vec{y} - \vec{X}\|, \quad \vec{X} \in \partial P$ the real end-point

In step B we take $\vec{n}_{i2}$ instead of the normalized vector $\vec{n}_i$ to decide whether $\vec{x}_i$ is inside $S_i$. For that purpose we always use the high-precision dot product to make the test reliable. The square of the distance between $\vec{x}$ and $\vec{y}$ in step D can be calculated exactly. A relative error appears only by rounding to the work-screen and is bounded by $0.5\varepsilon'_\ell$. After extracting the root the error increases to $1.76\varepsilon'_\ell$. Finally, in Table 1 we give the absolute resp. relative errors of the calculated end-point $\vec{x} = (x_1, x_2, x_3)$ and the distance $d = \|\vec{y} - \vec{x}\|$ for the cases discussed in steps A, B, C and D of Algorithm 1. It should be mentioned that the relative errors given above need to be replaced by suitable absolute errors if underflow occurs.

## 4. Speeding Up the Algorithm

There is a simple way to reduce the number of discussed facets $S_i$ in step B and edges in step C of Algorithm 1. If the polyhedron $P$ and the point $\vec{y}$ fulfill the assumptions of the following theorem we only have to inspect facets and edges adjacent to the nearest vertex-point of the set $\mathcal{V} := \{\vec{v}_k \mid \exists_{1 \leq i \leq m} \vec{v}_k \in S_i \wedge (\vec{y} \cdot \vec{n}_i - \vec{v}_k \cdot \vec{n}_i) > 0\}$.

THEOREM 4.1. *Let $P$ be a convex polyhedron bounded by its facets and vertices $\partial P = \{S_i, i = 1, ..., m; \vec{v}_k, k = 1, ..., n\}$ and let $\vec{y}$ be a point outside. In addition, let $\vec{n}_i$ be the normal vector of $S_i$ directed to the exterior of the polyhedron and $\mathcal{V} = \{\vec{v}_k \mid \exists_{1 \leq i \leq m} \vec{v}_k \in S_i \wedge (\vec{y} \cdot \vec{n}_i - \vec{v}_k \cdot \vec{n}_i) > 0\}$. Assume that $P$ has the following property: For each facet $S_i$ the common point of the perpendicular bisectors erected at the midpoint of each two adjacent edges of $S_i$ is inside $S_i$.*

*Then the nearest point $\vec{x}$ on the surface of $P$ with respect to $\vec{y}$ belongs to a facet with the nearest vertex-point $\vec{v} \in \mathcal{V}$.*

*Proof.* Assume that this implication is wrong: Let $\vec{y}$ be a point outside of $P$ for which there is no $S_i$ with $\vec{x} \in S_i$ and $\vec{v} \in S_i$; $\vec{x}$ denotes the nearest point of $P$ and $\vec{v} \in \mathcal{V}$ the nearest vertex-point of $P$ with respect to $\vec{y}$. Then let $S_i$ be a facet of $P$ with $\vec{v} = \vec{s}_{ik_0} \in S_i, k_0 \in \{1, ..., t_i\}$, for which $[\vec{x}_i, \vec{v}) \cap S_i \neq \emptyset$ holds; $\vec{x}_i$ denotes the projection-point of $\vec{y}$ onto the plane supporting $S_i$ and $[\vec{x}_i, \vec{v})$ is the half-open straight line between $\vec{x}_i$ and $\vec{v}$. Such facet $S_i$ exists, since we assume that $\vec{x}$ doesn't belong to a facet with the vertex $\vec{v}$. Then it is valid

$$d^2(\vec{v}, \vec{x}_i) = d^2(\vec{y}, \vec{v}) - d^2(\vec{x}_i, \vec{y}) \le d^2(\vec{y}, \vec{s}_{ik}) - d^2(\vec{x}_i, \vec{y}) = d^2(\vec{s}_{ik}, \vec{x}_i)$$

since $\vec{s}_{ik} \in \mathcal{V}$ if $\vec{s}_{ik_0} \in \mathcal{V}$. Here, $d(\cdot, \cdot)$ denotes the Euclidean distance. Hence $d(\vec{v}, \vec{x}_i) \le d(\vec{s}_{ik}, \vec{x}_i)$ and particularly

$$d(\vec{s}_{ik_0}, \vec{x}_i) \le d(\vec{s}_{i(k_0-1)}, \vec{x}_i) \quad \text{and} \quad d(\vec{s}_{ik_0}, \vec{x}_i) \le d(\vec{s}_{i(k_0+1)}, \vec{x}_i).$$

Then $\vec{x}_i$ belongs to the quadrangle $M_{ik_0}$ with the vertices $\vec{s}_{ik_0}$, $(\vec{s}_{i(k_0-1)} + \vec{s}_{ik_0}) / 2$, $\vec{m}_{ik_0}$, $(\vec{s}_{ik_0} + \vec{s}_{i(k_0+1)}) / 2$, if $[\vec{x}_i, \vec{v}] \cap S_i \ne \emptyset$; $\vec{m}_{ik_0}$ denotes the common point of the perpendicular bisectors erected at the midpoint of $[\vec{s}_{i(k_0-1)}, \vec{s}_{ik_0}]$ and $[\vec{s}_{ik_0}, \vec{s}_{i(k_0+1)}]$. Hence $\vec{x}_i$ lies on the facet $S_i$ because of the properties of $P$. Then $\vec{x}_i$ is the nearest point of $P$ with respect to $\vec{y}$ because the plane supporting $S_i$ separates point $\vec{y}$ and the half-space including the convex polyhedron $P$. That implies

$$\vec{x} = \vec{x}_i \in S_i.$$

Then we have $\vec{x} \in S_i$ and $\vec{v} \in S_i$ which is inconsistent with the assumption. $\qquad \square$

COROLLARY 4.1. *We adopt the definitions and assumptions of Theorem 4.1. Then the following assertion holds:*

*If the nearest point $\vec{x}$ belongs to an edge of $P$ then at least one of the end-points of this edge has the shortest distance to $\vec{y}$ of all $\vec{v}_k \in \mathcal{V}$.*

*Proof.* Let $\vec{x} \in [\vec{s}_{ik_0}, \vec{s}_{i(k_0+1)}]$ be the nearest point of $P$ with respect to $\vec{y}$. Then Theorem 4.1 implies that the nearest visible vertex-point with respect to $\vec{y}$ belongs to the facet $S_i$, i.e.

$$\vec{v} = \vec{s}_{ik} \quad \text{for at least one} \quad k \in \{1, ..., t_i\}.$$

Assume that $d(\vec{y}, \vec{s}_{ik'}) > d(\vec{y}, \vec{v})$ for $k' = k_0, k_0 + 1$. Let $\vec{x}_i$ be the projection-point of $\vec{y}$ onto the plane $E_i$ supporting $S_i$.

Then we obtain with the same reasoning as in the proof of Theorem 4.1 the following inequalities

$$d(\vec{x}_i, \vec{s}_{ik}) \le d(\vec{x}_i, \vec{s}_{i(k-1)}) \quad \text{and} \quad d(\vec{x}_i, \vec{s}_{ik}) \le d(\vec{x}_i, \vec{s}_{i(k+1)}),$$

and even

$$d(\vec{x}_i, \vec{x}) \le d(\vec{x}_i, \vec{x}_{S_i}) \quad \text{for all} \quad \vec{x}_{S_i} \in S_i.$$

The last inequality implies that the line supporting $[\vec{s}_{ik_0}, \vec{s}_{i(k_0+1)}]$ separates $\vec{x}_i$ and the facet $S_i$ with reference to $E_i$. Hence it follows that $[\vec{x}_i, \vec{s}_{ik}) \cap S_i \ne \emptyset$. Using the other inequality we obtain $\vec{x}_i \in M_{ik}$ and finally $\vec{x}_i = \vec{x}$ (see the proof of Theorem 4.1). Since $\vec{x}$ lies on an edge and $\vec{x} \notin [\vec{s}_{i(k-1)}, \vec{s}_{ik}]$ and $\vec{x} \notin [\vec{s}_{ik}, \vec{s}_{i(k+1)}]$ it follows that $\vec{x}_i = \vec{m}_{ik}$. Then it holds

$$d(\vec{x}_i, \vec{v}) = d(\vec{x}_i, \vec{s}_{ik}) = d(\vec{x}_i, \vec{s}_{i(k-1)}) = d(\vec{x}_i, \vec{s}_{i(k+1)}),$$

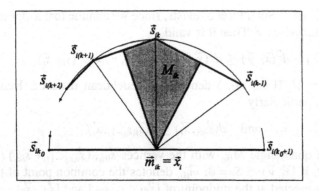

*Figure 1.*    Graphic sketch to the proof of Corollary 4.1.

and using orthogonality we get

$$d(\vec{y}, \vec{v}) = d(\vec{y}, \vec{s}_{ik}) = d(\vec{y}, \vec{s}_{i(k-1)}) = d(\vec{y}, \vec{s}_{i(k+1)}).$$

Replace $\vec{v} = \vec{s}_{ik}$ by $\vec{s}_{i(k+1)}$ and carry out the steps of this proof as above. Repeat it until $k = k_0$. Then we obtain

$$d(\vec{y}, \vec{v}) = d(\vec{y}, \vec{s}_{ik_0}) = d(\vec{y}, \vec{s}_{i(k_0-1)}) = d(\vec{y}, \vec{s}_{i(k_0+1)})$$

in contradiction to the assumption.                                                                     □

Theorem 4.1 is applicable in most cases, because in practice it is not necessary that each facet $S_i$ of $P$ fulfills the assumption but only the facets adjacent to a next vertex-point. In particular, at first we calculate the distance between $\vec{y}$ and all visible vertices (i.e. belonging to visible facets). Then only for a nearest vertex-point $\vec{v}_{k_0}$ we test whether the common point of the perpendicular bisectors erected at the midpoint of each two adjacent edges to $\vec{v}_{k_0}$ of $S_i$ is inside $S_i$. The test is easy to realize and generally it is based on the test "point in polygon" from step B, but often the facets are triangles, so we only need to test whether the angle adjacent to the nearest vertex is acute. In computer graphics most common representations for 3D polygon mesh surfaces use strips, where every consecutive set of three vertices defines an acute triangle, which always fulfills the assumption. If the assumption is fulfilled then in step B and C we only need to discuss the facets resp. edges adjacent to $\vec{v}_{k_0}$ instead of all visible facets or edges. That involves a significant reduction of computations in both steps.

Another possibility to speed up the algorithm to logarithmic order is to represent the bounded convex polyhedron $P$ with vertex set $V(P)$ in a hierarchical way by preprocessing it in linear time [5]. A sequence $P_1, P_2, P_3, ..., P_k$ is a hierarchical representation of $P$ [3] if

- $P_1 = P, P_k$ is a simplex, that is, $P_k$ is the convex hull of four non-coplanar points,
- $P_{i+1} \subset P_i, V(P_{i+1}) \subset V(P_i), 1 \leq i < k,$

- the vertices $V(P_i) - V(P_{i+1})$ form an independent set in $P_i$, $1 \le i < k$, i.e. the vertices it contains are pairwise non-adjacent (two vertices are said to be adjacent if they are incident upon a common edge).

The representation is compact if $k \le c \log m$, $\sum_{i=1}^{k} |P_i| \le c \cdot m$ and the degree of each new vertex in $V(P_i) - V(P_{i+1})$ is uniformly bounded by $c$, for some fixed constant $c$. We denote by $|P|$ the total number of facets of $P$. A compact hierarchical representation can be constructed in time $O(n)$ (see [5]). If we have a compact hierarchical representation of $P$, we can maintain the closest pair $(\vec{y}, \vec{x}_i)$, $i = k, ..., 1$, stepping through this representation from a simplex to $P$ and considering for $\vec{x}_i$ only the new facets in the direction of $\vec{y}$. It suffices to notice that the pair can be updated in constant time per step. A proof is based on the following properties of hierarchical representations:

If $E$ is a plane supporting $P_{i+1}$ (i.e. supporting one facet of $P_{i+1}$), then either $E$ supports $P_i$ or a unique vertex $v \in P_i$ exists so that $E$ separates $v$ and $P_{i+1}$. Each facet $S_p$ of $P_{i+1}$ that is not a facet of $P_i$ has a pointer to the unique vertex of $P_i$ that lies in the half-space opposite to $P_{i+1}$ with respect to the plane supporting $S_p$.

## 5. Distance Between Two Convex Polyhedra

To compute the distance between a polyhedron $P$ and a polyhedron $Q$ bounded by $\partial Q = \{S'_j, j = 1, ..., m'; \vec{v}'_k, k = 1, ..., n'\}$ with the vertices belonging to the facet $S'_j$ denoted by $\vec{s}'_{jl}$, $l = 1, ..., t'_j$, we can use the algorithm explained in the previous paragraphs, but we also need a method to prove whether an edge of one polyhedron intersects a facet of the other one or to calculate the distance between two edges.

At the beginning we calculate both centroid points $\vec{z}_P$ of $P$ and $\vec{z}_Q$ of $Q$ and test with Algorithm 1 if $\vec{z}_P$ lies in $Q$ or $\vec{z}_Q$ in $P$. If it applies to one of these points then the polyhedra have a non-empty intersection and we can stop. Otherwise, we compute, as in the preparation for Algorithm 1, the normal vectors $\vec{n}_i$ of facet $S_i$ and $\vec{n}'_j$ of facet $S'_j$ directed to the exterior of the polyhedra. Then we calculate for all $i$ and $j$ with

$$\#\square(\vec{z}_Q \cdot_{\#} \vec{n}_i - \vec{s}_{i1} \cdot_{\#} \vec{n}_i) \ge 0 \quad \text{and} \quad \#\square(\vec{z}_P \cdot_{\#} \vec{n}'_j - \vec{s}'_{j1} \cdot_{\#} \vec{n}'_j) \ge 0$$

the value $\#\square(\vec{n}_i \cdot_{\#} \vec{n}'_j)$ and store all pairs $(i, j)$ with $\#\square(\vec{n}_i \cdot_{\#} \vec{n}'_j) \le 0$.

These pairs $(i, j)$ form the set $S$, and the facets $S_i$ and $S'_j$ are called relevant pairs of facets. The following steps are implemented only for pairs of facets, vertex-points or edges belonging to relevant pairs of facets.

### Algorithm 2: Distance Polyhedron to Polyhedron

A: By application of Algorithm 1 we calculate the distances

$$d(\vec{v}_i, Q), \ \forall i \text{ with } \exists_j (i, j) \in S \quad \text{and} \quad d(\vec{v}'_j, P), \ \forall j \text{ with } \exists_i (i, j) \in S.$$

If $\vec{v}_i$ is in $Q$ or $\vec{v}'_j$ in $P$ then the polyhedra have a non-empty intersection and we can stop (the distance is equal to zero), otherwise we set

$$d := \min_{i,j}\{d(\vec{v}_i, Q), d(\vec{v}'_j, P)\}.$$

B: If $d > 0$, we test whether an edge of $Q$ intersects a facet of $P$. For this purpose we calculate (for all $(i,j) \in S$)

$$r'_{jli} := \#\square(\vec{s}'_{j(l+1)} \cdot \# \vec{n}_i - \vec{s}'_{jl} \cdot \# \vec{n}_i) \quad \text{and}$$

$$d'_{jli} := \#\square(\vec{s}'_{jl} \cdot \# \vec{n}_i - \vec{s}_{i1} \cdot \# \vec{n}_i) \quad \text{for} \quad l \in \{1, ..., t'_j\}, \ \vec{s}'_{j(t'_j+1)} := \vec{s}'_{j1}.$$

If $[\vec{s}'_{jl}, \vec{s}'_{j(l+1)}]$ and the plane $E_i$ supporting facet $S_i$ are parallel then $r'_{jli}$ is equal to zero, otherwise we set $t := d'_{jli} / r'_{jli}$. If $0 \le t \le 1$, the edge intersects plane $E_i$ and we have to decide whether the intersection point $\vec{s}'_{jl} + t(\vec{s}'_{j(l+1)} - \vec{s}'_{jl})$ lies within facet $S_i$ (see step B of Algorithm 1) or not. If it is in $S_i$ then the polyhedra intersect and their distance is equal to zero. We also test all relevant facets of $Q$ and edges of $P$.

C: Assume now that $P$ and $Q$ do not intersect (else we would get a solution in step B) then we test whether there are edges of $Q$ parallel to a facet of $P$ (we test all relevant edges of $P$ and facets of $Q$ analogously). For that purpose we explore the dot products $r'_{jli}$ (for all $(i,j) \in S$ and $l = 1, ..., t'_j$). If $r'_{jli} = 0$, edge $[\vec{s}'_{jl}, \vec{s}'_{j(l+1)}]$ is parallel to plane $E_i$. In this case we determine the projection of $[\vec{s}'_{jl}, \vec{s}'_{j(l+1)}]$ onto $E_i$ and analyse it for an intersection with $S_i$ (a polygon in $E_i$). Only in the case of a non-empty intersection $d'_{jli}$ indicates the distance between the edge and the facet. Let $S'_{j_0}$ be a facet of $Q$ with $S_{j_0} \cap S_j = [\vec{s}'_{jl}, \vec{s}'_{j(l+1)}]$ and $l_0 \in \{1, ..., t'_{j_0}\}$ an index with $[\vec{s}'_{j_0 l_0}, \vec{s}'_{j_0(l_0-1)}] = [\vec{s}'_{jl}, \vec{s}'_{j(l+1)}]$. Then if $\#\square(\vec{s}'_{j(l-1)} \cdot \# \vec{n}_i - \vec{s}'_{jl} \cdot \# \vec{n}_i) \ge 0$, $\#\square(\vec{s}'_{j_0(l_0+1)} \cdot \# \vec{n}_i - \vec{s}'_{j_0 l_0} \cdot \# \vec{n}_i) \ge 0$ and $0 \le d'_{jli} < d$ we replace $d$ by $d'_{jli}$, determine the end-points of the shortest distance and stop the algorithm.

D: The end-points of the shortest straight line between two separated polyhedra can lie within their edges. For that reason we finally calculate the distance between the lines

$$\vec{s}_{ik} + t(\vec{s}_{i(k+1)} - \vec{s}_{ik}) \quad \text{and} \quad \vec{s}'_{jl} + t'(\vec{s}'_{j(l+1)} - \vec{s}'_{jl}), \quad -\infty < t, \ t' < \infty,$$

by solving a system of two equations with two variables; $(i,j) \in S$. The equations result from first derivatives of the function $f$ (the square of the distance)

$$f(t, t') = \|\vec{s}_{ik} + t(\vec{s}_{i(k+1)} - \vec{s}_{ik}) - \vec{s}'_{jl} - t'(\vec{s}'_{j(l+1)} - \vec{s}'_{jl})\|^2$$

in the variables $t$ and $t'$ by setting both to zero. The determinant of the system vanishes only if the edges are parallel (then they lie in one plane and the determination of their distance or the end-points is simple) otherwise it is positive. The solution is of interest if $0 < t < 1$ and $0 < t' < 1$, i.e. the end-points lie within the edges. In this case we calculate the distance between them and update $d$.

If there is a pair $(i_0, j_0) \in S$ with $\#\Box(\vec{n}_{i_0} \,\text{\textschwa\#} \vec{n}'_{j_0}) \approx -1$ we can speed up Algorithm 2 by testing in step C first the edges of $S_{i_0}$ and facet $S'_{j_0}$ as well as the edges of $S'_{j_0}$ and facet $S_{i_0}$.

A hierarchical representation of both polyhedra is highly recommended. Then we can carry out the test as above only for the simplex of $P$ and $Q$ at first, then we build up both polyhedra and follow the closest points only in the direction in which the polyhedra are facing each other. This is more efficient than testing almost $n_P \cdot n_Q$ vertices and $m_P \cdot m_Q$ facets for it.

For collision-detection or in path planning it is usually important to know if the polyhedra intersect or not, and it is not necessary to know the accurate distance between them. In this case we can stop the algorithm already after step B.

When we are only interested in a lower bound $b$ of the distance between $P$ and $Q$ we can stop the algorithm even after step A if $b^2 := d^2 - (\text{diam}_p / 2)^2 > 0$ where $d :=$ $\min_{i,j}\{d(\vec{v}_i, Q), d(\vec{v}'_j, P)\} = d(\vec{v}_{i_0}, Q)$ and $\text{diam}_p := \max_i \text{diam}(S_i)$ ($\text{diam}(S_i)$ denotes the diameter of $S_i$), or $b^2 := d^2 - (\text{diam}_q / 2)^2 > 0$ where $d := \min_{i,j}\{d(\vec{v}_i, Q), d(\vec{v}'_j, P)\} =$ $d(\vec{v}'_{j_0}, P)$ and $\text{diam}_q := \max_j \text{diam}(S'_j)$ .

## 6.  Interval Data Input

Now we will assume that the vertices of our polyhedra are elements of the three-dimensional interval vector space $IR^3$. We define the set $\mathcal{P}$ of all polyhedra with vertices belonging to the cubical boxes $C(\vec{v}_i)$ which are enclosed in spheres $B(\vec{v}_i, r)$, $i = 1, ..., n$. Suppose that the radius $r$ can be chosen so that the spheres do not overlap. Evidently, the case $r = 0$ corresponds to a classical polyhedron with vertices $\vec{v}_i$, $i = 1, ..., n$. We construct a polyhedron which contains all elements of set $\mathcal{P}$. For this purpose we use a common method in computer graphics introducing new edges so that all facets of the polyhedra are triangles.

Given three spheres with centers $\vec{v}_1$, $\vec{v}_2$, $\vec{v}_3$ and radii $r$ there is a tangential plane $\vec{n} \cdot \vec{x} - \vec{n} \cdot \vec{z} = 0$ with $\vec{z} = \vec{n}r + \vec{v}_1$ and

$$\vec{n} = \frac{(\vec{v}_2 - \vec{v}_1) \times (\vec{v}_3 - \vec{v}_1)}{\|(\vec{v}_2 - \vec{v}_1) \times (\vec{v}_3 - \vec{v}_1)\|} \, .$$

The vector $\vec{n}$ is oriented to the outside of the original polyhedron. For each triangular facet with $\vec{v}_i$, $\vec{v}_j$, $\vec{v}_k$ we construct these planes and form the intersection of all adjacent half-spaces.

Figure 2 shows a polyhedron with interval-vertices enclosed in spheres, the tangential planes and a cubical box around a point. We are interested in lower bounds for the distance between the point and set $\mathcal{P}$. However, an unwanted fact arises if more than three triangles are adjacent to a vertex. Here the common points of each choice of three tangential surfaces are different in general. There is a way to cut these different vertices by introducing a tangential plane at the point $\vec{v} + r\vec{m}/\|\vec{m}\|$

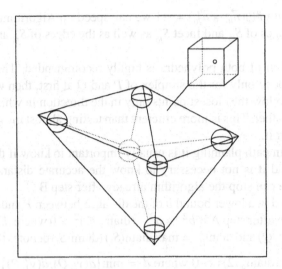

*Figure 2.* Distance cubical box to set of polyhedra.

with the normal vector $\vec{m}$ given by the arithmetic mean $\sum_{i=1}^{k} \vec{n}_i$ of the normal vectors of all adjacent facets belonging to the vertex $\vec{v}$. If we consider a curve joining all normal vectors $\vec{n}_i$ starting from $\vec{v}$, then $\vec{m}$ stays inside this curve. However, this leads to $k$ extra vertices replacing the original vertex $\vec{v}$. Thus the enclosing polyhedron is built by hexagons instead of triangles and $n$ convex polygons with $\deg(\vec{v})$ corners corresponding to the original vertices.

Now we can calculate a verified lower bound for the distance between a point and a polyhedron with interval data using the distance between a cube and the enclosing polyhedron as derived above.

In a similar way we can derive an upper bound. For this purpose we must construct an inner polyhedron using the inverse normal vectors. Here the cutting plane is not tangential to the sphere and is to be moved along $\vec{m}$ so that the $k$ extra vertices replacing the original vertex $\vec{v}$ have the right orientation.

There is another possibility to calculate sharp lower and upper bounds of the distance between two polyhedra with uncertain data.

THEOREM 6.1. *Given two polyhedra $P$ and $Q$ bounded by facets and vertices $\partial P = \{S_i, i = 1, ..., m; \vec{v}_k, k = 1, ..., n\}$ and $\partial Q = \{S'_j, j = 1, ..., m'; \vec{v}'_l, l = 1, ..., n'\}$. Assume that the vertices $\vec{v}_k$ and $\vec{v}'_l$ lie in the non-overlapping spheres $B(\vec{x}_k, r)$, $k = 1, ..., n$, and $B(\vec{x}'_l, r)$, $l = 1, ..., n'$, which centers $\vec{x}_k$ and $\vec{x}'_l$ form the vertices of two convex polyhedrons $P_1$ and $Q_1$. Moreover, let $d(\vec{t}, \vec{u})$ denote the distance and $\vec{t}, \vec{u}$ the end-points of the closest vector between the polyhedra $P$ and $Q$.*

*If the distance $d(\vec{x}, \vec{y})$ and starting and landing points $\vec{x}$, $\vec{y}$ between the polyhedra $P_1$ and $Q_1$ are calculated then the following inequalities hold:*

$$d(\vec{x}, \vec{y}) - 2r \leq d(\vec{t}, \vec{u}) \leq d(\vec{x}, \vec{y}) + 2r.$$

*Proof.* Construct two spheres $B(\vec{x}, r)$ and $B(\vec{y}, r)$ and two parallel tangential planes to these spheres with a normal vector in the direction of $\vec{x} - \vec{y}$. Then the polyhedra $P$ and $Q$ do not meet the space between the two planes. It follows that the lower bound is valid. The proof for deriving the upper bound is analogous.   □

For computing an inclusion of the distance of $P_1$ and $Q_1$ a procedure based on algorithms suggested in this paper can be used.

Theorem 6.1 shows that only local properties of the involved polyhedra with interval-vertices have an influence on the derived bounds. If $r$ is sufficiently small then only spheres and facets adjacent to the end-points of the closest vector between the polyhedra $P_1$ and $Q_1$ have to be considered.

## Acknowledgements

We thank the referees for their valuable comments.

## References

1. Bobrow, J. E. : A Direct Minimization Approach for Obtaining the Distance Between Convex Polyhedra, *The International Journal of Robotics Research* **8** (3) (1989), pp. 65–76.
2. Chen, Liang, Huang, Wen Qi, and Song, En Min: A Fast Algorithm for Computing the Distance between Two Convex Polyhedra, *Math. J. Chinese Univ.* **16** (4) (1994), pp. 345–359.
3. Dobkin, D. and Kirkpatrick, D.: Determining the Separation of Preprocessed Polyhedra—A Unified Approach, in: *Automata, Languages and Programming* (Coventry, 1990), *Lecture Notes in Computer Science* **443**, Springer, New York, 1990, pp. 400–413.
4. Dobkin, D., Hershberger, J., Kirkpatrick, D., and Suri, S.: Computing the Intersection–Depth of Polyhedra, *Algorithmica* **9** (6) (1993), pp. 528–533.
5. Edelsbrunner, H.: *Algorithms in Combinatorial Geometry*, Springer, Berlin, 1987.
6. Gilbert, E. G., Johnson, D. W., and Keerthi S. S.: A Fast Procedure for Computing the Distance Between Complex Objects in Three-Dimensional Space, *IEEE Journal of Robotics and Automation* **4** (2) (1988), pp. 193–203.
7. Kulisch, U. (Hrsg.): *Wissenschaftliches Rechnen mit Ergebnisverifikation—Eine Einführung*, Akademie-Verlag Berlin und Vieweg-Verlag Wiesbaden, 1989.
8. Quinlan, S.: Efficient Distance Computation Between Non-Convex Objects, in: *IEEE Intern. Conf. on Robotics and Automation*, 1994, pp. 3324–3329.
9. Sato, Y., Hirata, M., Maruyama, T., and Arita, Y.: Efficient Collision Detection Using Fast Distance-Calculation Algorithms for Convex and Non-Convex Objects, in: *Proc. IEEE Intern. Conf. on Robotics and Automation*, Minneapolis, Minnesota, 1996, pp. 771–778.
10. Werner, K.: *Verifizierte Berechnung der inversen Weierstraß-Funktion und der elliptischen Funktion von Jacobi in beliebigen Maschinenarithmetiken*, PhD-Thesis, Duisburg, 1996.

*Reliable Computing* **5**: 255–267, 1999.
© 1999 *Kluwer Academic Publishers*.

# Verified Error Bounds for Linear Systems Through the Lanczos Process

ANDREAS FROMMER and ANDRE WEINBERG
*Fachbereich Mathematik, Bergische Universität Wuppertal, D-42097 Wuppertal, Germany,
e-mail: {frommer,weinberg}@math.uni-wuppertal.de*

(Received: 5 December 1998; accepted: 20 December 1998)

**Abstract.** We use verified computations and the Lanczos process to obtain guaranteed lower and upper bounds on the 2-norm and the energy-norm error of an approximate solution to a symmetric positive definite linear system. The upper bounds require the a priori knowledge of a lower bound on the smallest eigenvalue.

## 1. Introduction

Throughout this note we consider a linear system

$$Az = b \tag{1.1}$$

with $A$ being a symmetric positive definite matrix from $\mathbb{R}^{n \times n}$, $b \in \mathbb{R}^n$. Moreover, we assume that we are already given an approximate solution $x$ of (1.1) which has been obtained through some arbitrary (direct or iterative) solver. Our goal is to get quantitative bounds on the error $x - x^*$, where $x^* = A^{-1}b$. Note that the residual $r = b - Ax$ can easily be computed from $x$ (but cancellation will necessarily occur if one uses standard floating point arithmetic). Now, $\|r\|_2$ is a special norm for $x - x^*$ since

$$\|r\|_2^2 = \langle r, r \rangle = \langle A(x - x^*), A(x - x^*) \rangle = \|x - x^*\|_{A^2}^2.$$

Here, $\|x\|_B = \langle x, Bx \rangle^{1/2}$ denotes the energy norm with respect to a (symmetric positive definite) matrix $B$.

Usually, however, the norms of interest are the 2-norm $\|x - x^*\|_2$ and the $A$-energy norm $\|x - x^*\|_A$ of the error. We have

$$\|x - x^*\|_A^2 = \langle A^{-1}r, AA^{-1}r \rangle = \langle r, A^{-1}r \rangle,$$
$$\|x - x^*\|_2^2 = \langle A^{-1}r, A^{-1}r \rangle = \langle r, A^{-2}r \rangle.$$

Let $r = \sum_{i=1}^{n} \gamma_i q_i$ be an expansion of $r$ in terms of orthonormal eigenvectors $q_i$ of $A$ with corresponding eigenvalues $0 < \lambda_1 \le \lambda_2 \le \cdots \le \lambda_n$. Then

$$\|x - x^*\|_A^2 = \langle r, A^{-1}r \rangle = \sum_{i=1}^{n} \frac{1}{\lambda_i} \gamma_i^2 \begin{cases} \le \frac{1}{\lambda_1} \sum_{i=1}^{n} \gamma_i^2 = \frac{1}{\lambda_1} \|r\|_2^2, \\ \ge \frac{1}{\lambda_n} \sum_{i=1}^{n} \gamma_i^2 = \frac{1}{\lambda_n} \|r\|_2^2; \end{cases} \tag{1.2}$$

$$\|x - x^*\|_2^2 = \langle r, A^{-2}r \rangle = \sum_{i=1}^{n} \frac{1}{\lambda_i^2} \gamma_i^2 \begin{cases} \le \frac{1}{\lambda_1^2} \sum_{i=1}^{n} \gamma_i^2 = \frac{1}{\lambda_1^2} \|r\|_2^2, \\ \ge \frac{1}{\lambda_n^2} \sum_{i=1}^{n} \gamma_i^2 = \frac{1}{\lambda_n^2} \|r\|_2^2. \end{cases} \tag{1.3}$$

These estimates are worst case bounds which do not make any use of the extent to which each of the eigenvectors $q_i$ is actually present in $r$, i.e. of the size of the $\gamma_i$.

## 2. Quadrature Rules and the Lanczos Process

The coefficients $\gamma_i$ define a positive, discrete measure $\gamma$ on $\mathbb{R}$ via

$$\gamma(\lambda) = \begin{cases} 0 & \text{if } \lambda < \lambda_1, \\ \sum_{j=1}^{i} \gamma_j^2 & \text{if } \lambda_i \le \lambda < \lambda_{i+1}, \\ \sum_{j=1}^{n} \gamma_j^2 & \text{if } \lambda_n \le \lambda. \end{cases}$$

Given an analytic function $f$ on an open interval containing $[a, b] \supseteq [\lambda_1, \lambda_n]$ we can therefore write

$$\langle r, f(A)r \rangle = \sum_{i=1}^{n} f(\lambda_i)\gamma_i^2 = \int_a^b f(\lambda) \, d\gamma.$$

Because of (1.2) and (1.3), our primary interest is $f(\lambda) = \lambda^{-1}$ and $f(\lambda) = \lambda^{-2}$.

As was shown in [6], [7] we can now use the Gauss quadrature rule and the Gauss-Radau quadrature rule to compute approximations $I_G$ and $I_{GR}$, resp., to the above integral. Using $i$ as a superscript to denote the number of (free) nodes in the respective rules on gets

$$\int_a^b f(\lambda) \, d\gamma = I_G^{(i)}[f] + R_G^{(i)}[f] = I_{GR}^{(i)}[f] + R_{GR}^{(i)}[f],$$

$\beta_0 = \|r\|_2; \quad r^{(0)} = r / \beta_0; \quad r^{(-1)} = 0$

for $j = 1, 2, \ldots$

$\quad \alpha_j = \langle r^{(j-1)}, A r^{(j-1)} \rangle$

$\quad \tilde{r}^{(j)} = A r^{(j-1)} - \alpha_j r^{(j-1)} - \beta_{j-1} r^{(j-2)}$

$\quad \beta_j = \|\tilde{r}^{(j)}\|_2$

$\quad r^{(j)} = \tilde{r}^{(j)} / \beta_j$

*Figure 1.*   The Lanczos process.

where the remainder terms can be represented as (see [6])

$$R_G^{(i)}[f] = \frac{f^{(2i)}(\eta)}{(2i)!} \cdot \int_a^b \prod_{j=1}^{i} (\lambda - \xi_j^{(i)})^2 \, d\gamma, \qquad a < \eta < b,$$

$$R_{GR}^{(i)}[f] = \frac{f^{(2i+1)}(\hat{\eta})}{(2i+1)!} \int_a^b (\lambda - a) \prod_{j=1}^{i} (\lambda - \hat{\xi}_j^{(i)})^2 \, d\gamma, \qquad a < \hat{\eta} < b,$$

$\xi_j^{(i)}$ and $\hat{\xi}_j^{(i)}$ being the nodes of the quadrature rule. Note that for $f(\lambda) = \lambda^{-1}$ and $f(\lambda) = \lambda^{-2}$ we have $R_G[f] \geq 0$ and $R_{GR}[f] \leq 0$. In other words, for these functions $f$ the quantities $I_G^{(i)}[f]$ represent lower bounds, the quantities $I_{GR}^{(i)}[f]$ represent upper bounds on $\int_a^b f(\lambda) \, d\gamma$.

As a final observation we note that $I_G^{(i)}[f]$ and $I_{GR}^{(i)}[f]$ can actually be computed via the Lanczos process. More precisely, the following has been shown in [6]: Denote $J^{(i)}$ the tridiagonal matrix

$$J^{(i)} = \begin{pmatrix} \alpha_1 & \beta_1 & & & \\ \beta_1 & \alpha_2 & \beta_2 & & \\ & \ddots & \ddots & \ddots & \\ & & \beta_{i-2} & \alpha_{i-1} & \beta_{i-1} \\ & & & \beta_{i-1} & \alpha_i \end{pmatrix},$$

where the entries $\alpha_j$, $\beta_j$ are given by the Lanczos process as given in Figure 1. Then

$$I_G^{(i)}[f] = \left( f(J^{(i)}) \right)_{11} \cdot \beta_0^2,$$

$(f(J^{(i)}))_{11}$ being the (1, 1) entry of the $i \times i$ matrix $f(J^{(i)})$. Moreover, define

$$J^{(i)} = \begin{pmatrix} \alpha_1 & \beta_1 & & & \\ \beta_1 & \alpha_2 & \beta_2 & & \\ & \ddots & \ddots & \ddots & \\ & & \beta_{i-2} & \alpha_{i-1} & \beta_{i-1} \\ & & & \beta_{i-1} & a + \delta_i \end{pmatrix},$$

where the $\delta_i$ is the last component of the solution $\delta^{(i)}$ of the linear system

$$(J^{(i)} - aI)\delta^{(i)} = (0, ..., 0, \beta_i^2)^T.$$

Then

$$I_{GR}^{(i)}[f] = (f(J^{(i)}))_{11} \cdot \beta_0^2.$$

Summarizing for $f(\lambda) = \lambda^{-1}$ and $f(\lambda) = \lambda^{-2}$ we get the following theorem.

THEOREM 2.1. *For $i = 1, 2, ...$ we have*

$$((J^{(i)})^{-1})_{11} \cdot \|r\|_2^2 \leq \|x - x^*\|_A^2 \leq ((\hat{J}^{(i)})^{-1})_{11} \cdot \|r\|_2^2,$$

$$((J^{(i)})^{-2})_{11} \cdot \|r\|_2^2 \leq \|x - x^*\|_2^2 \leq ((\hat{J}^{(i)})^{-2})_{11} \cdot \|r\|_2^2.$$

The following proposition shows that the bounds from Theorem 2.1 are improvements upon (1.2) and (1.3).

PROPOSITION 2.1. *For $i = 1, 2, ...$ one has*

$$((J^{(i)})^{-1})_{11} \geq \frac{1}{\lambda_n}, \qquad ((\hat{J}^{(i)})^{-1})_{11} \leq \frac{1}{a}, \qquad (2.1)$$

$$((J^{(i)})^{-2})_{11} \geq \frac{1}{\lambda_n^2}, \qquad ((\hat{J}^{(i)})^{-2})_{11} \leq \frac{1}{a^2}. \qquad (2.2)$$

*Proof.* Each matrix $J^{(i)}$ is a representation of $A$ projected onto the subspace spanned by $r, Ar, ..., A^{i-1}r$, see [14]. Therefore, the eigenvalues of each $J^{(i)}$ are all in the interval $[\lambda_1, \lambda_n]$. Now, any diagonal entry of the symmetric positive definite matrix $(J^{(i)})^{-1}$ lies between its smallest and largest eigenvalue which shows $((J^{(i)})^{-1})_{11} \geq \frac{1}{\lambda_n}$. Moreover, as is shown in [6], $a$ is an eigenvalue of each $\hat{J}^{(i)}$, and all other eigenvalues are larger than $a$. Therefore, the same argument as before shows $((\hat{J}^{(i)})^{-1})_{11} \leq \frac{1}{a}$. This proves (2.1). For (2.2) one proceeds similarly. $\square$

## 3. Verified Computation

The results of the preceding section assumed that all computations are done in *exact* arithmetic. In practical computations this is hardly ever the case because of the use of floating point arithmetic. So it could happen that the computed values for

$$((J^{(i)})^{-1})_{11}, ((\hat{J}^{(i)})^{-1})_{11}, ((J^{(i)})^{-2})_{11}, ((\hat{J}^{(i)})^{-2})_{11} \qquad (3.1)$$

will not yield correct lower and upper bounds in (2.1) and (2.2).

In this section we describe and discuss the use of (machine) interval arithmetic as a means of getting compact intervals known to contain the correct values for

$$[r] = b - Ax, \quad [\beta]_0 = \left( \sum_{i=1}^{n} [r]_i^2 \right)^{1/2}$$

$$[r]^{(0)} = [r] / [\beta]_0, \quad [r]^{(-1)} = 0$$

for $j = 1, 2, \ldots$

$$[\alpha]_j = \sum_{i=1}^{n} [r]_i^{(j-1)} \left( \sum_{k=1}^{n} A_{ik} [r]_k^{(j-1)} \right)$$

$$[\tilde{r}]^{(j)} = (A - [\alpha]_j I)[r]^{(j-1)} - [\beta]_j [r]^{(j-2)}$$

$$[\beta]_j = \left( \sum_{i=1}^{n} ([\tilde{r}]_i^{(j)})^2 \right)^{1/2}$$

$$[r]^{(j)} = [\tilde{r}]^{(j)} / [\beta_j]$$

*Figure 2.*  The interval Lanczos process.

the matrix entries from (3.1). Replacing these quantities in (2.1) and (2.2) by the adequate endpoints of the corresponding intervals will then produce guaranteed bounds for the error norms.

We assume that the reader is familiar with the elementary rules of interval arithmetic and its correct machine implementation using outward rounding; see [1], e.g. We denote interval quantities with square brackets.

Given an approximate solution $x$ we have to start by computing an enclosure for the residual $[r] = b - Ax$, which can be done using machine interval arithmetic on the point quantities $x, A$ and $b$. If a precise scalar product in the sense of Kulisch [10] is available, each component of $[r]$ can be computed to maximum accuracy. We then start a straightforward interval version of the Lanczos process as given in Figure 2.

Note that this naive interval Lanczos algorithm will suffer from increasing interval widths for mainly two reasons: Firstly, in the expression for $[\alpha_j]$, each component $[r]_i^{(j-1)}$ occurs several times, so that the interval arithmetic evaluation will overestimate the range $\{\langle r^{(j-1)}, Ar^{(j-1)} \rangle, \ r^{(j-1)} \in [r]^{(j-1)}\}$. Secondly, the computation of $[\tilde{r}]^{(j)}$ is affected by the wrapping effect. Practically this means that we will have to stop the interval Lanczos process when the (relative) diameters of the intervals become too large. Since we want to apply only a few steps of the Lanczos process anyway (see below), this is not as a severe drawback as it might seem at first sight. As an illustration, Figure 3 shows the relative width $\text{diam}[\alpha]_j / |[\alpha]_j|$ of the $[\alpha]_j$ in a typical computation (matrix s1rmq4m1, see Section 4).

As in the previous section let us collect the coefficients of the interval Lanczos process into a tridiagonal interval matrix $[J]^{(i)}$. Clearly, $J^{(i)} \in [J]^{(i)}$. Therefore, we can use interval Gaussian elimination to compute an interval containing $((J^{(i)})^{-1})_{11}$ and $((J^{(i)})^{-2})_{11}$. To be precise, denote $\text{IGA}([B], [c])$ the interval vector resulting from interval Gaussian elimination (without pivoting) for the matrix $[B]$ and the

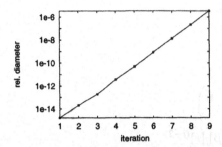

*Figure 3.*   Relative diameters of $\alpha_j$ for example s1rmq4m1 (Section 4).

right hand side $[c]$, and denote $e^1$ the first unit vector. Setting $[y]^{(i)} = \text{IGA}([J]^{(i)}, e^1)$ we have

$$((J^{(i)})^{-1})_{11} \in [y]_1^{(i)} =: [l_A]^{(i)},$$

$$((J^{(i)})^{-2})_{11} \in \sum_{j=1}^{n}([y]_j^{(i)})^2 =: [l_2]^{(i)}.$$

Similarly, let $e_i$ denote the $i$-th unit vector in $\mathbb{R}^i$, $[\delta]^{(i)} = \text{IGA}([J]^{(i)} - aI, [\beta_i]^2 e^i)$ and $[\hat{J}]^{(i)}$ the matrix obtained from $[J^{(i)}]$ by replacing its $(i, i)$ entry with $a + [\delta]_i^{(i)}$. Setting $[z]^{(i)} = \text{IGA}([\hat{J}]^{(i)}, e^1)$ we then have

$$((\hat{J}^{(i)})^{-1})_{11} \in [z]_1^{(i)} =: [u_A]^{(i)},$$

$$((\hat{J}^{(i)})^{-2})_{11} \in \sum_{j=1}^{n}([z]_j^{(i)})^2 =: [u_2]^{(i)}.$$

Very interestingly, interval Gaussian elimination does not suffer from overestimation for the interval linear systems considered here. This will be proved in the appendix.

Summarizing, we may state a computationally verifiable version of Theorem 2.1 as follows:

$$\text{lb}([l_A]^{(i)}[\beta]_0^2) \le \|x - x^*\|_A^2 \le \text{ub}([u_A]^{(i)}[\beta]_0^2), \tag{3.2}$$

$$\text{lb}([l_2]^{(i)}[\beta]_0^2) \le \|x - x^*\|_2^2 \le \text{ub}([u_2]^{(i)}[\beta]_0^2). \tag{3.3}$$

Here, $\text{lb}[a]$ and $\text{ub}[a]$ denote the smallest and largest element of the interval $[a]$, resp.

We conclude this section with a general discussion of the proposed method. Basically, one should view the estimates (3.2) and (3.3) as computationally cheap improvements upon the standard estimates

$$\frac{1}{b}\|r\|_A^2 \le \|x - x^*\|_A^2 \le \frac{1}{a}\|r\|_A^2, \tag{3.4}$$

$$\left(\frac{1}{b}\right)^2 \|r\|_2^2 \le \|x - x^*\|_2^2 \le \left(\frac{1}{a}\right)^2 \|r\|_2^2. \tag{3.5}$$

which arise from (1.2), (1.3) if one knows $a$ and $b$ such that $0 < a \le \lambda_1 \le \lambda_n \le b$. Strikingly, the improved *lower* bounds in (3.2) and (3.3) do not require any knowledge of $b$. The *upper* bounds, however, in both the standard estimates (3.4), (3.5) and the improved estimates (3.2), (3.3) need the bound $a$ with $a \le \lambda_1$. In special situations (see Section 4) such a bound might be known *a priori*, but in general, it is a non-trivial task to obtain $a$. To our knowledge, the only method delivering guaranteed results is due to Rump [13]. His method relies on a (floating point) Cholesky factorization of $A - aI$ for some candidate $a$ yielding the triangular matrix $L$ and the careful evaluation of the residual $\|A - aI - LL^T\|_\infty$ using directed rounding or interval arithmetic.

In order to keep the computational effort low, one should not try to perform too many Lanczos steps. The arithmetic work for one step is dominated by the (point) matrix (interval) vector product $A[r]^{(i)}$ which should be far less than the work that was invested beforehand to get the approximate solution $x$. As a rule, we suggest not to go farther than 10 Lanczos steps. Another reason for this rule is the fact that the interval matrix $[\hat{J}^{(i)}]$ tends to contain singular elements (due to the increasing width of its interval entries) as $i$ increases, so that interval Gaussian elimination will break down.

## 4. Numerical Results

Our numerical experiments used the PASCAL–XSC language to perform (machine) interval arithmetic, see [9]. We used the available precise scalar products for computing the residual $[r] = b - Ax$ and for $[\alpha]_j$ and $[\beta]_j$ in the interval Lanczos process which we ran up to $j = 9$. The approximate solutions $x$ were computed in Matlab via a (sparse) Cholesky factorization of the matrix $A$.

Our first set of experiments was for all 40 sparse symmetric positive definite matrices available at the Matrix Market [11]. The right hand sides for these systems were generated randomly. Rump's method from [13] was used to compute a guaranteed lower bound $a$ for $\lambda_1$.

Figure 4 shows the computed bounds (3.2) for the $A$-norm and (3.3) for the 2-norm as a function of the number of Lanczos steps. The straight line on top represents the standard upper bound from (3.4), (3.5). The figure plots the results for two different matrices: gr_30_30, a 9 point discrete Laplacian on a 30 × 30 mesh and s1rmq4m1, a sparse matrix of dimension 5489 arising from the finite element modelling of cylindrical shells. The condition number for the first example is approximately 380, for the second example it is $1.8 \cdot 10^6$. We see that in the first case we get quite a good enclosure for the $A$-norm and the 2-norm of the error. In the second example, the computed lower and upper bounds are not as narrow as in the first example.

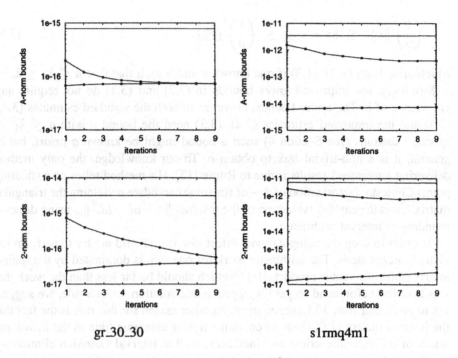

gr_30_30                                                    s1rmq4m1

*Figure 4.*   Verified lower and upper bounds for $\|x - x^*\|_A$ (top) and $\|x - x^*\|_2$ (bottom) for two matrices from Matrix Market.

The results given in the left part of Figure 5 allow to appreciate the improvement of the upper bound $\sqrt{\mathrm{ub}[u_A]^{(i_{opt})}} \cdot \|r\|_2$ for $\|x - x^*\|_A$ over the standard bound $\sqrt{1/a} \cdot \|r\|_2$. The index $i_{opt}$ denotes that index from $i = 1, ..., 9$ for which $\mathrm{ub}[u_A]^{(i)}$ is smallest. We plot $\sqrt{a/\mathrm{ub}([u_A]^{(i_{opt})})}$ for all 40 matrices, ordered by increasing dimension $n$. The improvement is mostly in the range of 2 to 6, but it sometimes achieves a factor of up to 16. There seems to be no correlation between the problem size and the improvement factor. For the 2-norm the situation is very similar, so we do not reproduce these results here.

The right part of Figure 5 gives information on the "narrowness" of the computed enclosures for $\|x - x^*\|_A$. Here, we plot the square root of the ratio of the best upper bound from (3.2) to the best lower bound from (3.3) as a function of the (approximate) condition number $\kappa$ of the matrices. $\kappa$ was obtained as $b'/a$ where $b'$ is an approximation for $\lambda_n$ obtained by a few steps of the power iteration. The dotted line is $\sqrt{\kappa}$, i.e. it represents the relation of the standard upper bound to the standard lower bound from (3.4). We see that the ratio for the improved estimates is almost always roughly one order of magnitude better than that for the standard estimate and that it increases with the condition number.

As a last example we consider Tikhonov-Phillips regularization for ill-posed problems. In this case the matrix $A$ is of the form $A = \mu I + B^T B$ with $\mu > 0$ an

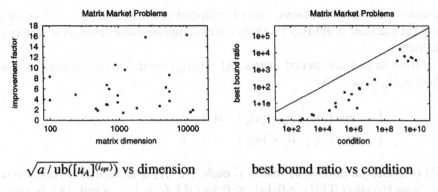

$\sqrt{a \, / \, \mathrm{ub}([u_A]^{(i_{opt})})}$ vs dimension    best bound ratio vs condition

*Figure 5.* Summary for 40 matrices from Matrix Market, A-norms only.

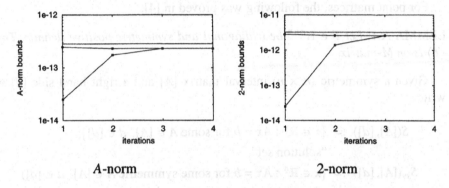

A-norm    2-norm

*Figure 6.* Lower and upper bounds for Tikhonov-Phillips regularization, problem foxgood.

adequate regularization parameter so that we can take $a = \mu$ *a priori*. In practical situations, typical values for $\mu$ are such that the condition number of $A$ is relatively small. This represents a favorable setting for our methods. As a specific example we took problem foxgood from Hansen's toolbox for ill-posed problems [8] with $\mu$ chosen to be an optimal regularization parameter for a relative noise level of 1%. The approximate solution $x$ together with the optimal regularization parameter $\mu = 2^{-7}$ were obtained through the shifted cg variant described in [5]. The matrix dimension was 1024. Figure 6 shows that in this case we get very close lower and upper bounds for the $A$-norm and the 2-norm of the error with just 4 iterations.

## Appendix. IGA for Tridiagonal Matrices

The purpose of this appendix is to show that interval Gaussian elimination produces best possible results for those tridiagonal interval systems which arose in Section 3.

We need additional terminology: We write $[a] \geq 0 \ (\leq 0)$ for some interval $[a]$ if each element is non-negative (non-positive). For interval vectors $[x] \geq 0, [x] \leq 0$

is understood componentwise. The left endpoint of an interval $[a]$ is denoted $lb[a]$, its right endpoint is $ub[a]$. This carries over componentwise to interval vectors and matrices.

Given an $(n \times n)$ interval matrix $[A] = ([a]_{ij})$ we define its comparison matrix $\langle [A] \rangle = (\alpha_{ij}) \in \mathbb{R}^{n \times n}$ by

$$
\alpha_{ij} = \begin{cases} \min\{|a_{ij}| : a_{ij} \in [a]_{ij}\} & \text{if } i = j, \\ -\max\{|a_{ij}| : a_{ij} \in [a]_{ij}\} & \text{if } i \neq j. \end{cases}
$$

$[A]$ is termed an interval $M$-matrix if each $A \in [A]$ is an $M$-matrix. Equivalently, $[A]$ is an $M$-matrix if $[a]_{ii} \geq 0$, $[a]_{ij} \leq 0$ for $i \neq j$, $i, j = 1, ..., n$ and $\langle [A] \rangle$ is a (point) $M$-matrix, see [12].

For point matrices, the following was proved in [4].

LEMMA A.1. *Let $T \in \mathbb{R}^{n \times n}$ be tridiagonal and symmetric positive definite. Then $\langle T \rangle$ is an $M$-matrix.*

Given a symmetric $(n \times n)$ interval matrix $[A]$ and a right hand side $[d]$ we write

$$
S([A], [d]) := \{x \in \mathbb{R}^n : Ax = b \text{ for some } A \in [A], \, d \in [d]\}
$$
$$
\text{"solution set"},
$$
$$
S_{sy}([A], [d]) := \{x \in \mathbb{R}^n : Ax = b \text{ for some symmetric } A \in [A], \, d \in [d]).
$$
$$
\text{"symmetric solution set"}.
$$

Clearly, $S_{sy} \subseteq S$. Denote $I(S)$ the interval hull of $S$, i.e. the intersection of all interval vectors containing $S$. Due to the inclusion isotonicity of interval arithmetic we have

$$
I(S_{sy}([A], [d])) \subseteq I(S([A], [d])) \subseteq IGA([A], [d]),
$$

and both inclusions "$\subseteq$" are proper in general. For interval $M$-matrices, however the following holds.

THEOREM A.1. *Assume that $[A]$ is an interval $M$-matrix and that either $[d] \leq 0$ or $0 \in [d]$ or $[d] \geq 0$. Then interval Gaussian elimination is feasible without pivoting. Moreover*

a) $IGA([A], [d]) = I(S([A], [d]))$.

b) *If, in addition, $[A]$ is symmetric, then $I(S_{sy}[A], [d])) = I(S([A], [d])) = IGA([A], [d])$.*

*Proof.* Part a) is from [3], part b) was observed in [2].                                    □

for $i = 2$ to $n$        {elimination}
$$[c]_i = [a]_i - [b]_{i-1} \cdot [b]_{i-1} / [c]_{i-1}$$
$$[e]_i = [d]_i - [b]_{i-1} \cdot [e]_{i-1} / [c]_{i-1}$$
$$[x]_n = [e]_n / [c]_n$$
for $i = n - 1$ to $1$      {back substitution}
$$[x]_i = ([e]_i - [b]_i \cdot [x]_{i+1}) / [c]_i$$

*Figure A1.*   Computation of $[x] = IGA([T], [d])$.

As a final preparation, let us adopt the notation $[T] = \text{tridiag}([b]_{i-1}, [a]_i, [b]_i)$ for the symmetric tridiagonal matrix

$$[T] = \begin{pmatrix} [a]_1 & [b]_1 & & & \\ [b]_1 & [a]_2 & [b]_2 & & \\ & \ddots & \ddots & \ddots & \\ & & [b]_{n-2} & [a]_{n-1} & [b]_{n-1} \\ & & & [b]_{n-1} & [a]_n \end{pmatrix} \qquad (A.1)$$

The interval Gaussian elimination process for $[T]$ and a right hand side $[d]$ is given in Figure A1 where $[x] = IGA([T], [d])$. Our central result is the following

THEOREM A.2. *Assume that $[T]$ from (A.1) satisfies $\text{lb}([a]_i) \geq 0$, $\text{lb}([b]_i) \geq 0$ for all $i$.*

a) *If every symmetric matrix $T \in [T]$ is positive definite, then interval Gaussian elimination is feasible without pivoting for any right hand side $[d]$. Moreover, if $0 \in [d]$ or $[d]_i \cdot [d]_{i+1} \leq 0$ for $i = 1, ..., n - 1$, then*

$$IGA([T], [d]) = I(S_{sy}([T], [d])).$$

b) *If $[T]$ contains a symmetric positive definite matrix $T_1$ and an indefinite symmetric matrix $T_2$ we have $0 \in [c]_i$ ($[c]_i$ from Figure A1) for at least one $i$.*

*Proof.* a) Let $D$ denote the diagonal "signature" matrix

$$D = \text{diag}(1, -1, 1, -1, ...) \in \mathbb{R}^{n \times n}$$

and set

$$[T]' = D \cdot [T] \cdot D = \text{tridiag}(-[b]_{i-1}, [a]_i, -[b]_i) =: \text{tridiag}([b]'_{i-1}, [a]'_i, [b]'_i).$$

The tridiagonal matrix $[T]'$ has the sign pattern of an $M$-matrix and $\langle [T]' \rangle = \langle [T] \rangle = \langle T \rangle$ where $T = \text{tridiag}(\text{ub}[b]_{i-1}, \text{lb}[a]_i, \text{ub}[b]_i) \in [T]$. Since $T$ is symmetric positive definite, $\langle T \rangle$ is an $M$-matrix by Lemma A.1. Consequently, $[T]'$ is an interval $M$-matrix. Given a right hand side $[d]$, denote $[d]' = D \cdot [d]$ and define $[c]'_i, [e]'_i$ and

$[x]_i'$ in analogy to Figure A1. Note that due to Theorem A.1 all quantities $[c]_i'$, $[e]_i'$, $[x]_i'$, $i = 1, \ldots, n$ exist. A trivial induction shows that for $i = 1, \ldots, n$

$$[c]_i = [c]_i',$$
$$[e]_i = (-1)^{i-1}[e]_i',$$
$$[x]_i = (-1)^{i-1}[x]_i'. \tag{A.2}$$

In particular, $[c]_i$, $[e]_i$, and $[x]_i$ all exist, i.e. interval Gaussian elimination is feasible for $[T]$ and any right hand side $[d]$. Moreover, by the assumptions on $[d]$ we have $0 \in [d]'$ or $0 \le [d]'$ or $0 \ge [d]'$, so that by Theorem A.1 b)

$$[x]' = \text{IGA}([T]', [d]') = I(S_{sy}([T]', [d]')). \tag{A.3}$$

But $S_{sy}([T]', [d]') = D \cdot S_{sy}([T], [d])$ so that (A.3) yields $D \cdot [x]' = I(S_{sy}([T]', [d]'))$. Since $D \cdot [x]' = [x] = \text{IGA}([T], [d])$ by (A.2) this completes the proof for a).

To prove b) note that there exists $\theta \in [0, 1]$ such that $\theta T_1 + (1 - \theta)T_2 \in [T]$ is singular. So (point) Gaussian elimination on that matrix will produce some zero diagonal entry $c_i$. By the inclusion property of interval arithmetic this implies $0 \in [c]_i$ for some $i$.                                                                              □

A careful inspection of the proof shows that the above theorem remains valid if one assumes only $[a]_i \ge 0$ and $0 \notin [b]_i$ for all $i$. The signature matrix $D$ has then to be chosen according to the signs of the $[b]_i$ and one needs $0 \in D[d]$ or $D[d] \ge 0$ or $D[d] \le 0$.

As a final remark let us just give an example showing that $\text{IGA}([T], [d])$ may be larger than the solution set for general symmetric tridiagonal matrices $[T]$. Take

$$[T] = \begin{pmatrix} 4 & [-2, 0] \\ [-2, 0] & -4 \end{pmatrix}, \quad [d] = \begin{pmatrix} 1 \\ 0 \end{pmatrix}.$$

Then

$$\text{IGA}([T], [d]) = \begin{pmatrix} [-3, 4] \\ [-2, 0] \end{pmatrix},$$

$$S([T], [d]) = \begin{pmatrix} [-\frac{16}{5}, 4] \\ [-2, 0] \end{pmatrix}, \quad S_{sy}([T], [d]) = \begin{pmatrix} [-\frac{16}{5}, 4] \\ [-\frac{8}{5}, 0] \end{pmatrix}.$$

## References

1. Alefeld, G. and Herzberger, J.: *Introduction to Interval Computations*, Academic Press, 1983.
2. Alefeld, G. and Mayer, G.: The Cholesky Method for Interval Data, *Linear Algebra Appl.* **194** (1993), pp. 161–182.
3. Barth, W. and Nuding, E.: Optimale Lösung von Intervallgleichungssystemen, *Computing* **12** (1974), pp. 117–125.
4. Frommer, A.: *Lösung linearer Gleichungssysteme auf Parallelrechnern*, Vieweg-Verlag, 1990.

5. Frommer, A. and Maass, P.: Fast CG-Based Methods for Thikonov-Phillips Regularization, *SIAM J. Sc. Comp.*, to appear, also available at http://www.math.uni-wuppertal.de/SciComp/Preprints.html as preprint BUGHW-SC 96/10.
6. Golub, G. and Meurant, G.: Matrices, Moments and Quadrature, in: *Numerical Analysis 1993 (Dundee 1993) (Pitman Res. Notes Math. Ser.* **302**), Longman Sci. Tech., Harlow, 1994.
7. Golub, G. and Meurant, G.: Matrices, Moments and Quadrature II. How to Compute the Error in Iterative Methods, *BIT* **37** (1997), pp. 687–705.
8. Hansen, P.: *Regularization Tool, a MATLAB Package for the Analysis and Solution of Discrete Ill-Posed Problems; Version 2.0 for MATLAB 4.0*, Tech.Rep. UNIC, 92-03, 1993.
9. Klatte, R., Kulisch, U., Neaga, M., Ratz, D., and Ullrich, Ch.: *PASCAL–XSC Sprachbeschreibung mit Beispielen*, Springer-Verlag, 1991.
10. Kulisch, U. and Miranker, W.: The Arithmetic of the Digital Computer, *SIAM Rev.* **28** (1986).
11. Matrix Market, http://math.nist.gov/MatrixMarket/.
12. Neumaier, A.: *Interval Methods for Linear Systems of Equations*, Cambridge University Press, 1990.
13. Rump, S.: Verification Methods for Dense and Sparse Systems of Equations, in: Herzberger, J. (ed.), *Topics in Validated Computations*, North-Holland, 1993.
14. Saad, Y.: *Iterative Methods for Sparse Linear Systems*, PWS, Boston, 1996.

*Reliable Computing* **5**: 269–278, 1999.
© 1999 *Kluwer Academic Publishers.*                                              269

# A Representation of the Interval Hull of a Tolerance Polyhedron Describing Inclusions of Function Values and Slopes

GERHARD HEINDL
*Fachbereich Mathematik, Universität Wuppertal, Gaußstraße 20, 42097 Wuppertal, Germany,*
*e-mail: heindl@math.uni-wuppertal.de*

(Received: 30 October 1998; accepted: 27 January 1999)

**Abstract.** Given a nonempty set of functions

$$F = \{f : [a, b] \to \mathbb{R} :$$

$$f(x_i) \in w_i, \quad i = 0, \ldots, n, \quad \text{and}$$

$$f(x) - f(y) \in d_i(x - y) \ \forall x, y \in [x_{i-1}, x_i], \quad i = 1, \ldots, n\},$$

where $a = x_0 < \cdots < x_n = b$ are known nodes and $w_i$, $i = 0, \ldots, n$, $d_i$, $i = 1, \ldots, n$, known compact intervals, the main aim of the present paper is to show that the functions

$$\underline{f} : x \mapsto \min\{f(x) : f \in F\}, \ x \in [a, b], \quad \text{and}$$

$$\overline{f} : x \mapsto \max\{f(x) : f \in F\}, \ x \in [a, b],$$

exist, are in $F$, and are easily computable. This is achieved essentially by giving simple formulas for computing two vectors $\tilde{l}, \tilde{u} \in \mathbb{R}^{n+1}$ with the properties

- $\tilde{l} \le \tilde{u}$ implies

$$\tilde{l}, \tilde{u} \in T := \{\xi = (\xi_0, \ldots, \xi_n)^T \in \mathbb{R}^{n+1} :$$

$$\xi_i \in w_i, \ i = 0, \ldots, n, \quad \text{and}$$

$$\xi_i - \xi_{i-1} \in d_i(x_i - x_{i-1}), \ i = 1, \ldots, n \}$$

and that $[\tilde{l}, \tilde{u}]$ is the interval hull of (the tolerance polyhedron) $T$;

- $\tilde{l} \le \tilde{u}$ iff $T \ne \emptyset$ iff $F \ne \emptyset$.

$\underline{f}, \overline{f}$ can serve for solving the following problem:

Assume that $\mu$ is a monotonically increasing functional on the set of Lipschitz-continuous functions $f : [a, b] \to \mathbb{R}$ (e.g. $\mu(f) = \int_a^b f(x)\, dx$ or $\mu(f) = \min f([a, b])$ or $\mu(f) = \max f([a, b])$), and that the available information about a function $g : [a, b] \to \mathbb{R}$ is "$g \in F$," then the problem is to find the best possible interval inclusion of $\mu(g)$. Obviously, this inclusion is given by the interval $[\mu(\underline{f}), \mu(\overline{f})]$. Complete formulas for computing this interval are given for the case $\mu(f) = \int_a^b f(x)\, dx$.

## Introduction

The investigations of the present paper have their origin in the concrete problem how to compute an inclusion of the integral $\int_0^1 g(x)\, dx$ of the function

$$g : x \mapsto \left(0.5(|\max\{\sin(10x), \cos(10x)\}| + 1)\right)^{\frac{1}{2}}, \quad x \in [0, 1].$$

Since $g$ is only piecewise differentiable, and since it is very costly to derive inclusions of the points at which $g$ is not differentiable, traditional methods based on inclusions of derivatives cannot be applied. A crude method is to program an interval extension of $g$ and to use it for computing step functions $\underline{g}$, $\overline{g}$ such that $\underline{g} \le g \le \overline{g}$. Then for any computed $\underline{i} \le \int_0^1 \underline{g}(x)\,dx$ and $\overline{i} \ge \int_0^1 \overline{g}(x)\,dx$ obviously $\int_0^1 g(x)\,dx \in [\underline{i}, \overline{i}]$.

But one can do better. $g$ is a function for which inclusions of slopes can be computed automatically. Using inclusions of the values of $g$ at the nodes of a partition of $[0, 1]$ together with inclusions of the slopes of $g$ in the intervals between the nodes results in a considerably improved method. This was demonstrated in [2]. In the present paper it will be shown that it is possible to use the information given by inclusions of function values and slopes in an optimal way.

## 1. A Tolerance Polyhedron Describing Inclusions of Function Values and Slopes

Let us assume that the available information about a function $g : [a, b] \to \mathbb{R}$ is "$g \in F$," where

$$F : = \{f : [a, b] \to \mathbb{R} :$$
$$f(x_i) \in w_i, \quad i = 0, \dots, n, \qquad \text{and}$$
$$f(x) - f(y) \in d_i(x - y) \ \forall x, y \in [x_{i-1}, x_i], \quad i = 1, \dots, n\},$$

with known data $a = x_0 < \cdots < x_n = b$, $w_i = [\underline{w}_i, \overline{w}_i]$, $i = 0, \dots, n$, and $d_i = [\underline{d}_i, \overline{d}_i]$, $i = 1, \dots, n$.

Then, introducing the intervals $\delta_i = [\underline{\delta}_i, \overline{\delta}_i] := d_i(x_i - x_{i-1})$, $i = 1, \dots, n$, and the tolerance polyhedron

$$T : = \{\xi = (\xi_0, \dots, \xi_n)^{\mathrm{T}} \in \mathbb{R}^{n+1} :$$
$$\xi_i \in w_i, \quad i = 0, \dots, n, \qquad \text{and}$$
$$\xi_i - \xi_{i-1} \in \delta_i, \quad i = 1, \dots, n\},$$

it is evident that $(f(x_0), \dots, f(x_n))^{\mathrm{T}} \in T$ for all $f \in F$.

On the other hand, if $(\xi_0, \dots, \xi_n)^{\mathrm{T}} \in T$, then the unique polygonal function $f : [a, b] \to \mathbb{R}$ satisfying

$$f(x) = \xi_{i-1} + \frac{\xi_i - \xi_{i-1}}{x_i - x_{i-1}}(x - x_{i-1}) \ \forall x \in [x_{i-1}, x_i], \quad i = 1, \dots, n,$$

has the properties

$$f(x_i) = \xi_i \in w_i, \ i = 0, \dots, n, \qquad \text{and}$$
$$f(x) - f(y) = \frac{\xi_i - \xi_{i-1}}{x_i - x_{i-1}}(x - y) \in d_i(x - y) \ \forall x, y \in [x_{i-1}, x_i], \quad i = 1, \dots, n.$$

Hence we have shown

- $f \in F$ implies $(f(x_0), ..., f(x_n))^T \in T$,
- $\xi \in T$ implies the existence of an $f \in F$ such that $f(x_i) = \xi_i$, $i = 0, ..., n$.

Consequently, $T \neq \emptyset$ iff $F \neq \emptyset$ (where the latter must be satisfied in case our information about $g$ is not wrong).

## 2. A Property of the Interval Hull of $T$

Assuming $T \neq \emptyset$, then $T$ is a nonempty compact subset of $\mathbb{R}^{n+1}$. Hence the interval hull $[l, u] = [(l_0, ..., l_n)^T, (u_0, ..., u_n)^T]$ of $T$ exists, where

$$l_i = \min\{\xi_i : \xi \in T\} \leq u_i = \max\{\xi_i : \xi \in T\}, \quad i = 0, ..., n.$$

A somewhat surprising property of $[l, u]$ is that $l$ and $u$ belong to $T$, i.e. we have

THEOREM 2.1. $l, u \in T$.

*Proof.* It is obvious that

$$l_i, u_i \in w_i, \quad i = 0, ..., n.$$

Hence it remains to prove

$$l_i - l_{i-1}, u_i - u_{i-1} \in \delta_i, \quad i = 1, ..., n.$$

We will show $u_i - u_{i-1} \in \delta_i$, $i = 1, ..., n$. The proof of $l_i - l_{i-1} \in \delta_i$, $i = 1, ..., n$, is completely analogous.

Let $i \in \{1, ..., n\}$ be fixed. There are $\xi$, $\eta \in T$ such that $\xi_{i-1} = u_{i-1}$, $\eta_i = u_i$.

$$\xi \in T \text{ implies } \underline{\delta}_i \leq \xi_i - \xi_{i-1} = \xi_i - u_{i-1} \leq u_i - u_{i-1},$$
$$\eta \in T \text{ implies } \overline{\delta}_i \geq \eta_i - \eta_{i-1} = u_i - \eta_{i-1} \geq u_i - u_{i-1}.$$

Hence $u_i - u_{i-1} \in \delta_i$. □

The main consequence of Theorem 2.1 and the relations between $T$ and $F$ is that in case $T \neq \emptyset$ (equivalent to $F \neq \emptyset$), $[l_i, u_i]$ is the best possible inclusion of $g(x_i)$ which can be derived from the information "$g \in F$." $[l_i, u_i] = \{\xi_i : \xi \in T\} = \{f(x_i) : f \in F\}$, $i = 0, ..., n$. The bounds $l_i$ and $u_i$ can be computed in principle by Linear Programming methods. However, there is a much easier possibility as we will show now.

## 3. Computing the Interval Hull of $T$

The main result of the present paper is

THEOREM 3.1. *Defining* $\tilde{l} = (\tilde{l}_0, ..., \tilde{l}_n)^T \in \mathbb{R}^{n+1}$ *and* $\tilde{u} = (\tilde{u}_0, ..., \tilde{u}_n)^T \in \mathbb{R}^{n+1}$ *by*

$$\tilde{l}_j := \max\left\{ \max_{i=0}^{j-1}\left(\underline{w}_i + \sum_{v=i+1}^{j}\underline{\delta}_v\right), \underline{w}_j, \max_{k=j+1}^{n}\left(\underline{w}_k - \sum_{v=j+1}^{k}\overline{\delta}_v\right)\right\} \quad and$$

$$\tilde{u}_j := \min\left\{ \min_{i=0}^{j-1}\left(\overline{w}_i + \sum_{v=i+1}^{j}\overline{\delta}_v\right), \overline{w}_j, \min_{k=j+1}^{n}\left(\overline{w}_k - \sum_{v=j+1}^{k}\underline{\delta}_v\right)\right\}, \quad j = 0, ..., n,$$

*(assuming* $\max_{i=s}^{k}(...) = -\infty$ *and* $\min_{i=s}^{k}(...) = \infty$ *if* $k < s$), *then*

1. $T \neq \emptyset$ *iff* $\tilde{l} \leq \tilde{u}$.
2. *If* $\tilde{l} \leq \tilde{u}$ *then* $\tilde{l} = l$ *and* $\tilde{u} = u$ *(implying* $\tilde{l}, \tilde{u} \in T$ *(by Theorem 2.1) and that* $[\tilde{l}, \tilde{u}]$ *is the interval hull of* $T$).

*Proof.* In a first step it will be shown that $\tilde{l} \leq \xi \leq \tilde{u}$ for all $\xi \in T$. Hence let us assume that $\xi = (\xi_0, ..., \xi_n)^T \in T$, i.e.

$$\underline{w}_i \leq \xi_i \leq \overline{w}_i, \qquad i = 0, ..., n, \qquad and$$
$$\underline{\delta}_i \leq \xi_i - \xi_{i-1} \leq \overline{\delta}_i, \qquad i = 1, ..., n.$$

Since $\xi_j = \xi_i + \sum_{v=i+1}^{j}(\xi_v - \xi_{v-1})$ if $i < j$, we have

$$\underline{w}_i + \sum_{v=i+1}^{j}\underline{\delta}_v \leq \xi_j \leq \overline{w}_i + \sum_{v=i+1}^{j}\overline{\delta}_v \qquad if \ i < j.$$

Since $\xi_j = \xi_k - \sum_{v=j+1}^{k}(\xi_v - \xi_{v-1})$ if $k > j$, we have

$$\underline{w}_k - \sum_{v=j+1}^{k}\overline{\delta}_v \leq \xi_j \leq \overline{w}_k - \sum_{v=j+1}^{k}\underline{\delta}_v \qquad if \ k > j.$$

Consequently,

$$\max_{i=0}^{j-1}\left(\underline{w}_i + \sum_{v=i+1}^{j}\underline{\delta}_v\right) \leq \xi_j \leq \min_{i=0}^{j-1}\left(\overline{w}_i + \sum_{v=i+1}^{j}\overline{\delta}_v\right), \qquad j = 1, ..., n,$$

$$\max_{k=j+1}^{n}\left(\underline{w}_k - \sum_{v=j+1}^{k}\overline{\delta}_v\right) \leq \xi_j \leq \min_{k=j+1}^{n}\left(\overline{w}_k - \sum_{v=j+1}^{k}\underline{\delta}_v\right), \qquad j = 0, ..., n-1.$$

Together with $\underline{w}_j \leq \xi_j \leq \overline{w}_j, j = 0, ..., n$, this implies $\tilde{l}_j \leq \xi_j \leq \tilde{u}_j, j = 0, ..., n$, i.e. $\tilde{l} \leq \xi \leq \tilde{u}$. An important consequence of this result is that $T \neq \emptyset$ implies $\tilde{l} \leq l \leq u \leq \tilde{u}$.

In the second step it will be shown that $\tilde{l} \leq \tilde{u}$ implies $\tilde{l}, \tilde{u} \in T$. Hence let us assume $\tilde{l}_i \leq \tilde{u}_i$, $i = 0, ..., n$. Then

$$\underline{w}_i \leq \tilde{l}_i \leq \tilde{u}_i \leq \overline{w}_i, \quad i = 0, ..., n,$$

by the definition of $\tilde{l}$ and $\tilde{u}$ and the assumption $\tilde{l} \leq \tilde{u}$.

It remains to prove

$$\underline{\delta}_j \leq \tilde{l}_j - \tilde{l}_{j-1} \leq \overline{\delta}_j \quad \text{and}$$
$$\underline{\delta}_j \leq \tilde{u}_j - \tilde{u}_{j-1} \leq \overline{\delta}_j, \quad j = 1, ..., n.$$

We will show $\underline{\delta}_j \leq \tilde{u}_j - \tilde{u}_{j-1} \leq \overline{\delta}_j$, $j = 1, ..., n$. The proof of $\underline{\delta}_j \leq \tilde{l}_j - \tilde{l}_{j-1} \leq \overline{\delta}_j$, $j = 1, ..., n$, is completely analogous, or can be carried out by applying to $-T$ what we will prove for $T$.

Let $j \in \{1, ..., n\}$ be fixed.

Proof of $\tilde{u}_j - \tilde{u}_{j-1} \leq \overline{\delta}_j$:

$$\tilde{u}_{j-1} + \overline{\delta}_j$$
$$= \min\left\{ \min_{i=0}^{j-2}\left(\overline{w}_i + \sum_{v=i+1}^{j-1} \overline{\delta}_v\right), \overline{w}_{j-1}, \overline{w}_j - \underline{\delta}_j, \min_{k=j+1}^{n}\left(\overline{w}_k - \sum_{v=j}^{k} \underline{\delta}_v\right)\right\} + \overline{\delta}_j$$
$$= \min\left\{ \min_{i=0}^{j-2}\left(\overline{w}_i + \sum_{v=i+1}^{j} \overline{\delta}_v\right), \overline{w}_{j-1} + \overline{\delta}_j, \overline{w}_j + (\overline{\delta}_j - \underline{\delta}_j), \min_{k=j+1}^{n}\left(\overline{w}_k - \sum_{v=j}^{k} \underline{\delta}_v\right) + \overline{\delta}_j\right\}$$
$$= \min\left\{ \min_{i=0}^{j-1}\left(\overline{w}_i + \sum_{v=i+1}^{j} \overline{\delta}_v\right), \overline{w}_j + (\overline{\delta}_j - \underline{\delta}_j), \min_{k=j+1}^{n}\left(\overline{w}_k - \sum_{v=j+1}^{k} \underline{\delta}_v\right) + (\overline{\delta}_j - \underline{\delta}_j)\right\}$$
$$\geq \tilde{u}_j, \quad \text{since } \overline{\delta}_j \geq \underline{\delta}_j.$$

Proof of $\underline{\delta}_j \leq \tilde{u}_j - \tilde{u}_{j-1}$:

$$\tilde{u}_j - \underline{\delta}_j$$
$$= \min\left\{ \min_{i=0}^{j-2}\left(\overline{w}_i + \sum_{v=i+1}^{j} \overline{\delta}_v\right), \overline{w}_{j-1} + \overline{\delta}_j, \overline{w}_j, \min_{k=j+1}^{n}\left(\overline{w}_k - \sum_{v=j+1}^{k} \underline{\delta}_v\right)\right\} - \underline{\delta}_j$$
$$= \min\left\{ \min_{i=0}^{j-2}\left(\overline{w}_i + \sum_{v=i+1}^{j} \overline{\delta}_v\right) - \underline{\delta}_j, \overline{w}_{j-1} + (\overline{\delta}_j - \underline{\delta}_j), \overline{w}_j - \underline{\delta}_j, \min_{k=j+1}^{n}\left(\overline{w}_k - \sum_{v=j}^{k} \underline{\delta}_v\right)\right\}$$
$$= \min\left\{ \min_{i=0}^{j-2}\left(\overline{w}_i + \sum_{v=i+1}^{j-1} \overline{\delta}_v\right) + (\overline{\delta}_j - \underline{\delta}_j), \overline{w}_{j-1} + (\overline{\delta}_j - \underline{\delta}_j), \min_{k=j}^{n}\left(\overline{w}_k - \sum_{v=j}^{k} \underline{\delta}_v\right)\right\}$$
$$\geq \tilde{u}_{j-1}, \quad \text{since } \overline{\delta}_j \geq \underline{\delta}_j. \qquad \square$$

*Remark.* A very simple case arises if $\underline{d}_i \leq \dfrac{w_i - w_{i-1}}{x_i - x_{i-1}} \leq \overline{d}_i$ and $\underline{d}_i \leq \dfrac{\overline{w}_i - \overline{w}_{i-1}}{x_i - x_{i-1}}$ $\leq \overline{d}_i$, $i = 1, ..., n$, i.e. $(\underline{w}_0, ..., \underline{w}_n)^{\mathrm{T}}$, $(\overline{w}_0, ..., \overline{w}_n)^{\mathrm{T}} \in T$. Then obviously $l = \tilde{l} = (\underline{w}_0, ..., \underline{w}_n)^{\mathrm{T}} \leq (\overline{w}_0, ..., \overline{w}_k)^{\mathrm{T}} = \tilde{u} = u$.

## 4. The Extremal Functions $\underline{f}, \overline{f}$

Using the notations intoduced in Theorem 3.1, we define the functions $\underline{f}_i, \overline{f}_i :$
$[x_{i-1}, x_i] \to \mathbb{R}$, $i = 1, \dots, n$, by

$$\underline{f}_i(x) := \max\{\tilde{l}_{i-1} + \underline{d}_i(x - x_{i-1}), \; \tilde{l}_i + \overline{d}_i(x - x_i)\},$$

$$\overline{f}_i(x) := \min\{\tilde{u}_{i-1} + \overline{d}_i(x - x_{i-1}), \; \tilde{u}_i + \underline{d}_i(x - x_i)\} \quad \forall x \in [x_{i-1}, x_i].$$

It is easily seen that

$$\underline{f}_i(x) - \underline{f}_i(y), \quad \overline{f}_i(x) - \overline{f}_i(y) \in d_i(x - y) \quad \forall x, y \in [x_{i-1}, x_i]. \tag{$*$}$$

Let us assume now $\tilde{l} \leq \tilde{u}$ (i.e. $F \neq \emptyset$). Then $\tilde{l} = l \in T$ and $\tilde{u} = u \in T$ (by
Theorem 3.1 and Theorem 2.1) and, as a consequence of $\tilde{l}, \tilde{u} \in T$:
If $\underline{d}_i = \overline{d}_i$ (an exceptional case) then

$$\underline{f}_i(x) = \tilde{l}_{i-1} + \underline{d}_i(x - x_{i-1}) = \tilde{l}_i + \overline{d}_i(x - x_i),$$

$$\overline{f}_i(x) = \tilde{u}_{i-1} + \overline{d}_i(x - x_{i-1}) = \tilde{u}_i + \underline{d}_i(x - x_i) \quad \forall x \in [x_{i-1}, x_i].$$

If $\underline{d}_i < \overline{d}_i$ then

$$\underline{f}_i(x) = \begin{cases} \tilde{l}_{i-1} + \underline{d}_i(x - x_{i-1}) & \text{if } x \in [x_{i-1}, \underline{x}_i], \\ \tilde{l}_i + \overline{d}_i(x - x_i) & \text{if } x \in [\underline{x}_i, x_i], \end{cases}$$

where

$$\underline{x}_i := \frac{\dfrac{\tilde{l}_i - \tilde{l}_{i-1}}{x_i - x_{i-1}} - \underline{d}_i}{\overline{d}_i - \underline{d}_i} x_{i-1} + \frac{\overline{d}_i - \dfrac{\tilde{l}_i - \tilde{l}_{i-1}}{x_i - x_{i-1}}}{\overline{d}_i - \underline{d}_i} x_i \quad (\in [x_{i-1}, x_i]).$$

Especially,

$$\underline{f}_i(x_{i-1}) = \tilde{l}_{i-1}, \quad \underline{f}_i(x_i) = \tilde{l}_i. \tag{$\underline{*}$}$$

$$\overline{f}_i(x) = \begin{cases} \tilde{u}_{i-1} + \overline{d}_i(x - x_{i-1}) & \text{if } x \in [x_{i-1}, \overline{x}_i], \\ \tilde{u}_i + \underline{d}_i(x - x_i) & \text{if } x \in [\overline{x}_i, x_i], \end{cases}$$

where

$$\overline{x}_i := \frac{\overline{d}_i - \dfrac{\tilde{u}_i - \tilde{u}_{i-1}}{x_i - x_{i-1}}}{\overline{d}_i - \underline{d}_i} x_{i-1} + \frac{\dfrac{\tilde{u}_i - \tilde{u}_{i-1}}{x_i - x_{i-1}} - \underline{d}_i}{\overline{d}_i - \underline{d}_i} x_i \quad (\in [x_{i-1}, x_i]).$$

Especially,

$$\overline{f}_i(x_{i-1}) = \tilde{u}_{i-1}, \quad \overline{f}_i(x_i) = \tilde{u}_i. \tag{$\overline{*}$}$$

$(\underline{*})$ and $(\overline{*})$ show that there are unique functions $\underline{f}, \overline{f} : [a, b] \to \mathbb{R}$ such that

$$\underline{f} \mid [x_{i-1}, x_i] = \underline{f}_i, \quad \overline{f} \mid [x_{i-1}, x_i] = \overline{f}_i, \quad i = 1, \dots, n.$$

From $\underline{f}(x_i) = \tilde{l}_i \in w_i$, $\overline{f}(x_i) = \tilde{u}_i \in w_i$, $i = 0, \ldots, n$, and $(*)$ we can conclude $\underline{f}, \overline{f} \in F$.

Let us consider now any $f \in F$. Then (by the definition of $F$),

$$f(x) - f(y) \in d_i(x - y) \quad \forall x, y \in [x_{i-1}, x_i].$$

Hence

$$f(x) \in f(x_{i-1}) + d_i(x - x_{i-1}) \quad \text{and}$$
$$f(x) \in f(x_i) + d_i(x - x_i) \quad \forall x \in [x_{i-1}, x_i].$$

But this implies

$$f(x) \geq f(x_{i-1}) + \underline{d}_i(x - x_{i-1}) \geq \tilde{l}_{i-1} + \underline{d}_i(x - x_{i-1}),$$
$$f(x) \geq f(x_i) + \overline{d}_i(x - x_i) \geq \tilde{l}_i + \overline{d}_i(x - x_i),$$
$$f(x) \leq f(x_{i-1}) + \overline{d}_i(x - x_{i-1}) \leq \tilde{u}_{i-1} + \overline{d}_i(x - x_{i-1}),$$
$$f(x) \leq f(x_i) + \underline{d}_i(x - x_i) \leq \tilde{u}_i + \underline{d}_i(x - x_i) \quad \forall x \in [x_{i-1}, x_i].$$

Consequently,

$$f(x) \geq \underline{f}_i(x) = \underline{f}(x) \quad \text{and}$$
$$f(x) \leq \overline{f}_i(x) = \overline{f}(x) \quad \forall x \in [x_{i-1}, x_i].$$

Thus we have

**THEOREM 4.1.** *If $\tilde{l} \leq \tilde{u}$ then $\underline{f}, \overline{f} \in F$ and $\underline{f}(x) = \min\limits_{f \in F} f(x)$, $\overline{f}(x) = \max\limits_{f \in F} f(x)$ $\forall x \in [a, b].$*

In addition, the definition of the $\tilde{l}_i, \tilde{u}_i$ and the given representations of the $\underline{f}_i, \overline{f}_i$ show up an easy way for computing $\underline{f}$ and $\overline{f}$.

*Remark.* Using the concept of convex lower and concave upper bound functions introduced by C. Jansson in [3], we can state also: If the available information about $g : [a, b] \to \mathbb{R}$ is "$g \in F$," then $\underline{f}_i$ is the greatest verifiable convex lower bound function, $\overline{f}_i$ the least verifiable concave upper bound function of $g|[x_{i-1}, x_i]$, $i = 1, \ldots, n$.

## 5. A Simple Illustrating Example

In order to demonstrate the main steps for computing $\underline{f}$ and $\overline{f}$, let us consider the following simple example:

$$n := 2, \qquad a := 0 = x_0 < x_1 := 9 < x_2 = b := 22,$$
$$w_0 := [4, 9], \qquad w_1 := [5, 8], \qquad w_2 := [2, 4],$$
$$d_1 := \left[0, \frac{5}{9}\right], \qquad d_2 := \left[-\frac{2}{13}, \frac{2}{13}\right].$$

Then

$$\delta_1 = d_1 \cdot 9 = [0,5], \qquad \delta_2 = d_2 \cdot 13 = [-2,2],$$
$$\tilde{l}_0 = \max\{4, \ \max\{5-5, \ 2-(5+2)\}\} = 4,$$
$$\tilde{l}_1 = \max\{4+0, \ 5, \ 2-2\} = 5,$$
$$\tilde{l}_2 = \max\{\max\{4+0-2, \ 5-2\},2\} = 3,$$
$$\tilde{u}_0 = \min\{9, \ \min\{8-0, \ 4-(0-2)\}\} = 6,$$
$$\tilde{u}_1 = \min\{9+5, \ 8, \ 4-(-2)\} = 6,$$
$$\tilde{u}_2 = \min\{\min\{9+(5+2), \ 8+2\}, \ 4\} = 4.$$

Hence $\tilde{l} = (4,5,3)^{\mathrm{T}} < \tilde{u} = (6,6,4)^{\mathrm{T}}$, implying $F \neq \emptyset$, $\tilde{l} = l(\in T)$, $\tilde{u} = u(\in T)$.

$$\underline{x}_1 = \frac{\frac{5-4}{9-0}-0}{\frac{5}{9}-0} \cdot 0 + \frac{\frac{5}{9}-\frac{5-4}{9-0}}{\frac{5}{9}-0} \cdot 9 = \frac{36}{5}, \quad \text{implying}$$

$$\underline{f}(x) = \underline{f}_1(x) = \begin{cases} 4+0(x-0) = 4 & \text{if } x \in \left[0, \frac{36}{5}\right], \\[2mm] 5+\frac{5}{9}(x-9) & \text{if } x \in \left[\frac{36}{5}, 9\right]; \end{cases}$$

$$\underline{x}_2 = \frac{\frac{3-5}{22-9}-\left(-\frac{2}{13}\right)}{\frac{2}{13}-\left(-\frac{2}{13}\right)} \cdot 9 + \frac{\frac{2}{13}-\frac{3-5}{22-9}}{\frac{2}{13}-\left(-\frac{2}{13}\right)} \cdot 22 = 22, \quad \text{implying}$$

$$\underline{f}(x) = \underline{f}_2(x) = 5+\left(-\frac{2}{13}\right)(x-9) \quad \text{if } x \in [9,22];$$

$$\overline{x}_1 = \frac{\frac{5}{9}-\frac{6-6}{9-0}}{\frac{5}{9}-0} \cdot 0 + \frac{\frac{6-6}{9-0}-0}{\frac{5}{9}-0} \cdot 9 = 0, \quad \text{implying}$$

$$\overline{f}(x) = \overline{f}_1(x) = 6+0(x-9) = 6 \quad \text{if } x \in [0,9];$$

$$\overline{x}_2 = \frac{\frac{2}{13}-\frac{4-6}{22-9}}{\frac{2}{13}-\left(-\frac{2}{13}\right)} \cdot 9 + \frac{\frac{4-6}{22-9}-\left(-\frac{2}{13}\right)}{\frac{2}{13}-\left(-\frac{2}{13}\right)} \cdot 22 = 9, \quad \text{implying}$$

$$\overline{f}(x) = \overline{f}_2(x) = 4+\left(-\frac{2}{13}\right)(x-22) \quad \text{if } x \in [9,22].$$

## 6. Applications

The functions $\underline{f}$ and $\overline{f}$ can serve mainly for the solution of problems of the following type:

Let $\mu$ be a monotonic functional on the set of Lipschitz-continuous functions $f : [a, b] \to \mathbb{R}$ (e.g. $\mu(f) = f(x)$ for a given $x \in [a, b]$, or $\mu(f) = \min f([a, b])$, or $\mu(f) = \max f([a, b])$, or $\mu(f) = \int_\alpha^\beta f(x)\, dx$, $\alpha, \beta \in [a, b]$). Then, assuming that the available information about a function $g : [a, b] \to \mathbb{R}$ is "$g \in F$," the problem is to derive the interval hull of the set $\{\mu(f) : f \in F\}$ of all possible values of $\mu(g)$.

Clearly, this hull is $[\mu(\underline{f}), \mu(\overline{f})]$ if $\mu$ is monotonically increasing, $[\mu(\overline{f}), \mu(\underline{f})]$ if $\mu$ is monotonically decreasing.

For the example considered in the last section, we have e.g.

$$g(8) \in [\underline{f}(8), \overline{f}(8)] = \left[\frac{40}{9}, 6\right],$$

$$\min(g) \in [\min \underline{f}, \min \overline{f}] = [3, 4],$$

$$\max(g) \in [\max \underline{f}, \max \overline{f}] = [5, 6], \qquad \text{hence}$$

$$[4, 5] \subset g([0, 22]) \subset [3, 6],$$

$$\int_8^0 g(x)\, dx \in \left[\int_8^0 \overline{f}(x)\, dx, \int_8^0 \underline{f}(x)\, dx\right] = \left[-48, -\frac{1448}{45}\right].$$

The computation (inclusion) of the bounds of $[\underline{f}(x), \overline{f}(x)]$, $[\min \underline{f}, \min \overline{f}]$, and $[\max \underline{f}, \max \overline{f}]$ can be carried out easily also in the general case. For the functional $\mu : f \mapsto \int_a^b f(x)\, dx$ the interval $[\mu(\underline{f}), \mu(\overline{f})]$ can be derived as follows:

Assuming $\tilde{l} \leq \tilde{u}$ (i.e. $F \neq \emptyset$) we have

$$\mu(\underline{f}) = \sum_{i=1}^n \int_{x_{i-1}}^{x_i} \underline{f}_i(x)\, dx, \quad \mu(\overline{f}) = \sum_{i=1}^n \int_{x_{i-1}}^{x_i} \overline{f}_i(x)\, dx,$$

and an elementary calculation shows that with the slopes

$$\underline{s}_i := \frac{\underline{l}_i - \underline{l}_{i-1}}{x_i - x_{i-1}}, \quad \overline{s}_i := \frac{\tilde{u}_i - \tilde{u}_{i-1}}{x_i - x_{i-1}}, \quad i = 1, \dots, n,$$

we get

$$\int_{x_{i-1}}^{x_i} \underline{f}_i(x)\, dx = \frac{\underline{l}_{i-1} + \underline{l}_i}{2}(x_i - x_{i-1})$$

$$-\frac{(x_i - x_{i-1})^2}{8} \cdot \begin{cases} 4(\underline{s}_i - \underline{d}_i) & \text{if } \underline{s}_i \leq \underline{d}_i, \\[2mm] (\overline{d}_i - \underline{d}_i)\left(1 - \left(\dfrac{\underline{s}_i - \dfrac{\underline{d}_i + \overline{d}_i}{2}}{\dfrac{\overline{d}_i - \underline{d}_i}{2}}\right)^2\right) & \text{if } \underline{d}_i < \underline{s}_i < \overline{d}_i, \\[2mm] 4(\overline{d}_i - \underline{s}_i) & \text{if } \overline{d}_i \leq \underline{s}_i \end{cases}$$

$$\geq \frac{\bar{l}_{i-1} + \bar{l}_i}{2}(x_i - x_{i-1}) - \frac{1}{8}(\bar{d}_i - \underline{d}_i)(x_i - x_{i-1})^2,$$

$$\int_{x_{i-1}}^{x_i} \bar{f}_i(x)\,dx = \frac{\tilde{u}_{i-1} + \tilde{u}_i}{2}(x_i - x_{i-1})$$

$$+ \frac{(x_i - x_{i-1})^2}{8} \cdot \begin{cases} 4(\bar{s}_i - \underline{d}_i) & \text{if } \bar{s}_i \leq \underline{d}_i, \\[2mm] (\bar{d}_i - \underline{d}_i)\left(1 - \left(\dfrac{\bar{s}_i - \frac{\underline{d}_i + \bar{d}_i}{2}}{\frac{\bar{d}_i - \underline{d}_i}{2}}\right)^2\right) & \text{if } \underline{d}_i < \bar{s}_i < \bar{d}_i, \\[4mm] 4(\bar{d}_i - \bar{s}_i) & \text{if } \bar{d}_i \leq \bar{s}_i \end{cases}$$

$$\leq \frac{\tilde{u}_{i-1} + \tilde{u}_i}{2}(x_i - x_{i-1}) + \frac{1}{8}(\bar{d}_i - \underline{d}_i)(x_i - x_{i-1})^2, \qquad \forall i = 1, \ldots, n.$$

These formulas were used to solve the following problem:

Given $g : x \mapsto (0.5(|\max\{\sin(10x), \cos(10x)\}| + 1))^{\frac{1}{2}}$, $x \in [0, 1]$, compute the best possible interval inclusion $[\underline{i}, \bar{i}]$ of $\int_0^1 g(x)\,dx$ which can be derived from the inclusions

$$g\left(\frac{i}{200}\right) \in w_i, \qquad i = 0, \ldots, 200, \quad \text{and}$$

$$g(x) - g(y) \in d_i(x - y) \qquad \forall x, y \in \left[\frac{i-1}{200}, \frac{i}{200}\right], \quad i = 1, \ldots, 200,$$

where the intervals $w_i$ and $d_i$ are obtained by applying the PASCAL–XSC module vsl_ari (based on [1], [4], and available under the URL http://www.math.uni-wuppertal.de/~heindl) to $g$. It turned out that $\bar{l} = (\underline{w}_0, \ldots, \underline{w}_{200})^T$, $\tilde{u} = (\bar{w}_0, \ldots, \bar{w}_{200})^T$, and $[0.90597, 0.90627] \subset [\underline{i}, \bar{i}] \subset [0.90596, 0.90628]$. If we use only the given lower (upper) bounds of the integrals $\int_{x_{i-1}}^{x_i} \underline{f}_i(x)\,dx$ ($\int_{x_{i-1}}^{x_i} \bar{f}_i(x)\,dx$), as suggested in [2], it is not possible to derive an inner approximation of $[\underline{i}, \bar{i}]$. The resulting inclusion is $[\underline{i}, \bar{i}] \subset [0.90594, 0.90630]$.

## Acknowledgements

I am much obliged to my assistant A. Rogat for writing the final TEX version of the paper. Many thanks also to the unknown referees for their valuable suggestions. One of them helped to shorten the proof of Theorem 3.1 considerably.

## References

1. Hammer, R., Hocks, M., Kulisch, U., and Ratz, D.: *Numerical Toolbox for Verified Computing I–Basic Numerical Problems*, Springer, Berlin, 1993.
2. Heindl, G.: Zur Einschließung der Werte von Peanofunktionalen, *Z. angew. Math. Mech.* **75** (II) (1995), pp. 637-638.
3. Jansson, C.: *Construction of Convex Lower and Concave Upper Bound Functions*, Bericht 98.1 des Forschungsschwerpunktes Informations- und Kommunikationstechnik der Technischen Universität Hamburg-Harburg, März, 1998.
4. Klatte, R., Kulisch, U., Neaga, M., Ratz, D., and Ullrich, C. P.: *PASCAL–XSC Language Reference with Examples*, Springer, Berlin, 1992.

*Reliable Computing* **5:** 279–288, 1999.
© 1999 *Kluwer Academic Publishers.*

# A Few Results on Table-Based Methods

JEAN-MICHEL MULLER
*CNRS, Laboratoire LIP, Projet ARENAIRE, Ecole Normale Supérieure de Lyon, France,*
*e-mail: Jean-Michel.Muller@ens-lyon.fr*

(Received: 28 September 1998; accepted: 21 December 1998)

**Abstract.** Table-based methods are frequently used to implement functions. We examine some methods introduced in the literature, and we introduce a generalization of the bipartite table method, named the **multipartite table method**.

## 1. Introduction

Throughout the paper, $f$ is the function to be evaluated. We assume $n$-bit, fixed-point arguments, between $1/2$ and $1$ (that is, they are mantissas of floating-point numbers).

Table-based methods have frequently been suggested and used to implement some arithmetic (reciprocal, square root) and transcendental functions. One can distinguish three different classes of methods:

- **compute-bound methods:** these methods use table-lookup in a small table to find parameters used afterward for a polynomial or rational evaluation. The main part of the evaluation of $f$ consists in arithmetic computations;

- **table-bound methods:** The main part of the evaluation of $f$ consists in looking up in a generally rather large table. The computational part of the function evaluation is rather small (e.g., a few additions);

- **in-between methods:** these methods use the combination of table lookup in a medium-size table and a significant yet reduced amount of computation (e.g. one or two multiplications, or several "small multiplications" that use rectangular—fast and/or small—multipliers).

Many methods currently used on general-purpose systems belong to the first class (e.g. Tang's methods [8]–[11]. The third class of methods has been widely studied since 1981 [2]. The use of small (e.g., rectangular) multipliers to fasten the computational part of the evaluation has been suggested by several authors (see for instance Wong and Goto's algorithms for double precision calculations [13], or Ercegovac et al.'s methods [1]).

In this paper, we examine some table-bound methods. Of course, the straightforward method, consisting in building a table with $n$ address bits, cannot be used unless $n$ is very small. The first useful table-bound methods have been introduced in the last decade: they have become implementable thanks to progress in VLSI technology. Wong and Goto [12] have suggested the following method. We split the binary representation of the input-number $x$ into four $k$-bit numbers, where $k = n/4$. That is, we write*:

$$= x_1 + x_2 2^{-k} + x_3 2^{-2k} + x_4 2^{-3k}$$

where $0 \le x_i \le 1 - 2^{-k}$ is a multiple of $2^{-k}$.
Then $f(x)$ is approximated by:

$$f(x_1 + x_2 2^{-k}) + \frac{1}{2} 2^{-k} \{ f(x_1 + x_2 2^{-k} + x_3 2^{-k}) - f(x_1 + x_2 2^{-k} - x_3 2^{-k}) \}$$

$$+ \frac{1}{2} 2^{-2k} \{ f(x_1 + x_2 2^{-k} + x_4 2^{-k}) - f(x_1 + x_2 2^{-k} - x_4 2^{-k}) \}$$

$$+ 2^{-4k} \left\{ \frac{x_3^2}{2} f^{(2)}(x_1) - \frac{x_3^3}{6} f^{(3)}(x_1) \right\}.$$

The approximation error due to the use of this approximation is about $2^{-5k}$.

The **bipartite table method** was first suggested by Das Sarma and Matula [4] for quickly computing reciprocals. A slight improvement, the **symmetric bipartite table method** was introduced by Schulte and Stine [6]. Due to the importance of the bipartite table method (BTM), we will present it in detail in the next section. Compared to Wong and Goto's method, it requires larger tables. And yet, the amount of computation required by the BTM is reduced to one addition.

The problem of evaluating a function given by a converging series can be reduced to the evaluation of a partial product array (PPA). Schwarz [7] suggested to use multiplier structures to sum up PPAs. Hassler and Takagi [3] use PPAs to evaluate functions by table look-up and addition.

## 2. Order-1 Methods

The methods described in this section use an order-1 Taylor approximation of $f$. This leads to very simple computations (mere additions), but the size of the required tables may be quite large.

### 2.1. THE BIPARTITE TABLE METHOD

This method was first suggested by DasSarma and Matula [4] for computing reciprocals. We split the binary representation of the input number $x$ into 3 $k$-bit numbers, where $k = n / 3$. That is, we write:

$$= x_1 + x_2 2^{-k} + x_3 2^{-2k}$$

---

* To make the paper easier to read and more consistent, we do not use Wong and Goto's notations here. We use the same notations as in the sequel of this paper.

where $0 \leq x_i \leq 1 - 2^{-k}$ is a multiple of $2^{-k}$.

| $x =$ | $x_1$ | $x_2$ | $x_3$ |
|---|---|---|---|

We then write the order-1 Taylor expansion of $f$ at $x_1 + x_2 2^{-k}$. This gives:

$$f(x) = f(x_1 + x_2 2^{-k}) + x_3 2^{-2k} f'(x_1 + x_2 2^{-k}) + \varepsilon_1 \qquad (2.1)$$

with $\varepsilon_1 = \frac{1}{2} x_3^2 2^{-4k} f''(\xi_1)$, where $\xi_1 \in [x_1 + x_2 2^{-k}, x]$. Now, we approximate the value $f'(x_1 + x_2 2^{-k})$ by its order-0 Taylor expansion at $x_1$ (that is, by $f'(x1)$). This gives:

$$f(x) = f(x_1 + x_2 2^{-k}) + x_3 2^{-2k} f'(x_1) + \varepsilon_1 + \varepsilon_2 \qquad (2.2)$$

with $\varepsilon_2 = x_2 x_3 2^{-3k} f''(\xi_2)$, where $\xi_2 \in [x_1, x_1 + x_2 2^{-k}]$. This gives the **bipartite formula**:

$$f(x) = \alpha(x_1, x_2) + \beta(x_1, x_3) + \varepsilon \qquad (2.3)$$

where

$$\begin{cases} \alpha(x_1, x_2) = f(x_1 + x_2 2^{-k}), \\ \beta(x_1, x_3) = x_3 2^{-2k} f'(x_1), \\ \varepsilon \leq \left(\frac{1}{2} 2^{-4k} + 2^{-3k}\right) \max f'' \approx 2^{-3k} \max f''. \end{cases}$$

Hence, $f(x)$ can be approximated, with approximately $n$ bits of accuracy (by this, we mean "with error $\approx 2^{-n}$") by the sum of two terms ($\alpha$ and $\beta$) that can be looked-up in $2n/3$-address bit tables, as illustrated by Figure 1. Moreover, whereas the first table (function $\alpha$) must contain $n$-bit words, we can take into account the fact that approximately* $2k$ most significant bits of $\beta(x_1, x_3)$ are zero: there is no need to store them.

Assume we wish to compute the sine function, with 24-bit input numbers between $1/2$ and 1, and error less than $2^{-24}$. Functions $f'$ and $f''$ are always less than 1. A straightforward use of what we have presented leads to choose $k = 9$, and to perform the summation (2.3) with 25-bit words. Hence the error $\varepsilon$ of (2.3) is approximately $2^{-27}$, and the error due to the truncation to 25 bits of $\alpha$ and $\beta$** is less than $2^{-25}$. The first table has size $25 \times 2^{17}$ (the first bit of $x$ is a one: there is no need to use it as an address bit). The second table has size $7 \times 2^{15}$. All this requires 428 Kbytes of memory. It is worth noticing that one can get smaller tables by using symmetries and splitting $n$ into sub-words of slightly different sizes. All this has been suggested and implemented by Schulte and Stine [5], [6].

---

   * This depends on the values of $f'$.

  ** Assuming that the values stored in the tables are rounded to the nearest.

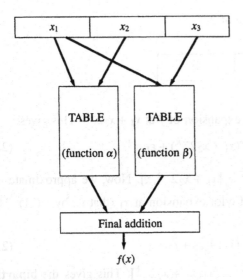

*Figure 1.*   The bipartite table method.

The BTM still leads to large tables in single precision, and thus it is far from being implementable in double precision. And yet, this leads to another idea: we should try to generalize the bipartite method, by splitting the input word into more than three parts. Let us first try a splitting into five parts, we will after that generalize to an arbitrary odd number of parts. It is worth noticing that Schulte and Stine also have proposed [5] a method based on a splitting into more than 3 parts. Their method is called the *Symmetric table addition method* (STAM). We will compare our method and the STAM method in Section 2.4.

## 2.2.  THE TRIPARTITE TABLE METHOD

Now, we split the input $n$-bit fixed-point number $x$ into five $k$-bit parts $x_1, x_2, ..., x_5$. That is, we write:

$$= x_1 + x_2 2^{-k} + x_3 2^{-2k} + x_4 2^{-3k} + x_5 2^{-4k}$$

where $0 \leq x_i \leq 1 - 2^{-k}$ is a multiple of $2^{-k}$.

| $x =$ | $x_1$ | $x_2$ | $x_3$ | $x_4$ | $x_5$ |
|---|---|---|---|---|---|

We use the order-1 Taylor expansion of $f$ at $= x_1 + x_2 2^{-k} + x_3 2^{-2k}$:

$$\begin{aligned}
f(x) &= f(x_1 + x_2 2^{-k} + x_3 2^{-2k}) \\
&\quad + (x_4 2^{-3k} + x_5 2^{-4k}) f'(x_1 + x_2 2^{-k} + x_3 2^{-2k}) \\
&\quad + \varepsilon_3 + \varepsilon_1
\end{aligned} \qquad (2.4)$$

with $\varepsilon_3 = \frac{1}{2}(x_4 2^{-3k} + x_5 2^{-4k})^2 f''(\xi_3)$, with $\xi_3 \in [x_1 + x_2 2^{-k} + x_3 2^{-2k}, x]$, which

gives $\varepsilon_3 \leq \frac{1}{2} 2^{-6k} \max f''$.

In (2.4), we expand the term $(x_4 2^{-3k} + x_5 2^{-4k}) f'(x_1 + x_2 2^{-k} + x_3 2^{-2k})$ as follows:

- $x_4 2^{-3k} f'(x_1 + x_2 2^{-k} + x_3 2^{-2k})$ is replaced by $x_4 2^{-3k} f'(x_1 + x_2 2^{-k})$. The error committed is $\varepsilon_4 = x_3 x_4 2^{-5k} f''(\xi_4)$, where $\xi_4 \in [x_1 + x_2 2^{-k}, x_1 + x_2 2^{-k} + x_3 2^{-2k}]$. We easily get $\varepsilon_4 \leq 2^{-5k} \max f''$.

- $x_5 2^{-4k} f'(x_1 + x_2 2^{-k} + x_3 2^{-2k})$ is replaced by $x_5 2^{-4k} f'(x_1)$. The error committed is $\varepsilon_5 = (x_2 2^{-k} + x_3 2^{-2k}) x_5 2^{-4k} f''(\xi_5)$, where $\xi_5 \in [x_1, x_1 + x_2 2^{-k} + x_3 2^{-2k}]$. We get $\varepsilon_5 \leq 2^{-5k} \max f''$.

This gives the **tripartite formula**:

$$f(x) = \gamma(x_1, x_2, x_3) + \delta(x_1, x_2, x_4) + \theta(x_1, x_5) + \varepsilon, \tag{2.5}$$

where

$$\begin{aligned}
\gamma(x_1, x_2, x_3) &= f(x_1 + x_2 2^{-k} + x_3 2^{-2k}), \\
\delta(x_1, x_2, x_4) &= x_4 2^{-3k} f'(x_1 + x_2 2^{-k}), \\
\theta(x_1, x_5) &= x_5 2^{-4k} f'(x_1), \\
\varepsilon &\leq \left(\frac{1}{2} 2^{-6k} + 2 \times 2^{-5k}\right) \max f'' \approx 2^{-5k+1} \max f''.
\end{aligned}$$

Hence, $f(x)$ can be obtained by adding three terms, each of them being looked-up in a table with (at most) $3n/5$ address bits. This is illustrated by Figure 2.

## 2.3. GENERALIZATION: THE MULTIPARTITE TABLE METHOD

The previous approach is straightforwardly generalized. We now assume that the $n$-bit input number $x$ is split into $2p + 1$ $k$-bit values $x_1, x_2, \ldots, x_{2p+1}$. That is,

$$x = \sum_{i=1}^{2p+1} x_i 2^{(i-1)k},$$

where the $x_i$'s are multiples of $2^{-k}$ and satisfy $0 \leq x_i < 1$. As in the previous sections, we use the order-1 Taylor expansion:

$$\begin{aligned}
f(x) = {}& f(x_1 + x_2 2^{-k} + \cdots + x_{p+1} 2^{-pk}) \\
&+ (x_{p+2} 2^{(-p-1)k} + \cdots + x_{2p+1} 2^{-2pk}) f'(x_1 + x_2 2^{-k} + \cdots + x_{p+1} 2^{-pk}) \\
&+ \varepsilon_{p+1}
\end{aligned}$$

with $\varepsilon_{p+1} \leq \frac{1}{2} 2^{-2(p+1)k} \max f''$. We expand the term

$$(x_{p+2} 2^{(-p-1)k} + \cdots + x_{2p+1} 2^{-2pk}) f'(x_1 + x_2 2^{-k} + \cdots + x_{p+1} 2^{-pk}),$$

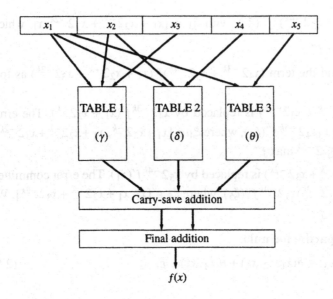

*Figure 2.*   The tripartite table method.

and perform Taylor approximations to $f'(x_1 + x_2 2^{-k} + \cdots + x_{p+1} 2^{-pk})$. We then get:

$$
\begin{aligned}
f(x) = {} & f(x_1 + x_2 2^{-k} + \cdots + x_{p+1} 2^{-pk}) + \varepsilon_{p+1} \\
& + x_{p+2} 2^{(-p-1)k} f'(x_1 + x_2 2^{-k} + \cdots + x_p 2^{(-p+1)k}) + \varepsilon_{p+2} \\
& + x_{p+3} 2^{(-p-2)k} f'(x_1 + x_2 2^{-k} + \cdots + x_{p-1} 2^{(-p+2)k}) + \varepsilon_{p+3} \\
& + x_{p+4} 2^{(-p-3)k} f'(x_1 + x_2 2^{-k} + \cdots + x_{p-2} 2^{(-p+3)k}) + \varepsilon_{p+4} \\
& \cdots \\
& + x_{2p+1} 2^{-2pk} f'(x_1) + \varepsilon_{2p+1}
\end{aligned}
$$

where $\varepsilon_{p+2}, \varepsilon_{p+3}, \ldots, \varepsilon_{2p+1}$ are less than $2^{(-2p-1)k} \max f''$.

This gives the **multipartite (or $(p+1)$-partite) formula:**

$$
\begin{aligned}
f(x) = {} & \alpha_1(x_1, x_2, \ldots, x_{p+1}) \\
& + \alpha_2(x_1, x_2, \ldots, x_{p-1}, x_{p+3}) \\
& + \alpha_3(x_1, x_2, \ldots, x_{p-2}, x_{p+4}) \\
& + \alpha_4(x_1, x_2, \ldots, x_{p-3}, x_{p+5}) \\
& \cdots \\
& + \alpha_{p+1}(x_1, x_{2p+1}) \\
& + \varepsilon
\end{aligned}
\tag{2.6}
$$

where

$$
\begin{cases}
\alpha_1(x_1, ..., x_{p+1}) = f(x_1 + x_2 2^{-k} + \cdots + x_{p+1} 2^{-pk}), \\
\alpha_i(x_1, ..., x_{p-i+2}, x_{p+i}) = x_{p+i} 2^{-p-i+1} f'(x_1 + x_2 2^{-k} + \cdots + x_{p-i+2} 2^{(-p+i-1)k}), \\
\quad \varepsilon \leq \left( \tfrac{1}{2} 2^{(-2p-2)k} + p 2^{(-2p-1)k} \right) \max f'' \\
\quad\quad \approx p 2^{(-2p-1)k} \max f''.
\end{cases}
$$

Too large values of $p$ are unrealistic: performing many additions to avoid a few multiplications is not reasonable.

Let us try to calculate the amount of memory that would be required for implementing the double-precision sine function ($n^{\cdot} = 53$), and $2p + 1 = 7$. We would choose $k = 8$, and perform the final summation with 55-bit accuracy. Hence the total amount of memory would be $(55 + 23 + 15) \times 2^{31} + 7 \times 2^5$ bits, that is, around 20 Gbytes, which is far from being feasible: the order-1 methods do not seem to be applicable beyond single-precision.

## 2.4. COMPARISON WITH THE STAM METHOD

The Symmetric table addition method (STAM) was suggested by Schulte and Stine [5]. They split the binary representation of the input number $x$ into $m$ subwords (whose sizes may be different), $x_1, x_2, ..., x_m$ and they use the order-1 Taylor expansion of $f$ at the midpoint of the interval of numbers whose binary representation starts with $x_1$ and $x_2$. They expand the order-1 part of the Taylor approximation so that each term is a function of $x_1$ and a sub-word $x_i$ only. Hence they approximate the result by the sum of $m - 1$ values, each of them being looked-up in a table that receives two sub-words as an input address.

At a first glance, if we examine the multipartite method and the STAM assuming in both cases a splitting into $m$ sub-words, the STAM method requires smaller tables: the STAM method use 2-sub-word tables, and the multipartite method uses $p+1$-sub word tables.

And yet, assuming that the first two words have size $n_0$ and $n_1$, the order-1 Taylor approximation used by the STAM method has an error whose order of magnitude is

$$
\frac{1}{2} 2^{-2(n_0 + n_1 + 1)} f''(x)
$$

therefore, to reach an error equal to approximately $2^{-n}$, we need

$$
n_0 + n_1 \geq \frac{n - 3}{2}.
$$

Hence, the first two sub-words must represent approximately half the global input-word. As a consequence, the tables will have around $n/2$ address bits.

Table 1 gives table sizes for various order-1 methods, assuming that we implement a 24-bit sine function. This table shows that both methods lead to a similar

Table 1. Table sizes for various order-1 methods, assuming that we implement a 24-bit sine function.

| method | memory size (Kbytes) | error bound |
|---|---|---|
| bipartite (straightforward) | 428 | $0.625 \times 2^{-24}$ |
| bipartite (Schulte and Stine) | 244 | |
| tripartite (ours) | 74 | $1.38 \times 2^{-24}$ |
| splitting into 4 terms (Schulte and Stine) | 92 | |
| splitting into 5 terms (Schulte and Stine) | 74.5 | |

amount of memory. This is not surprising, since both methods are very similar. Now, let us try to get smaller tables by using approximations of order larger than 1.

## 3. Higher-Order Methods

In the previous section, we have used order-1 Taylor expansions only. We also have seen that these methods are not applicable for precisions significantly larger than single precision. Now, let us give an example of the use of an order-2 expansion. As in Section 2.2, we split the input $n$-bit fixed-point number $x$ into five $k$-bit parts $x_1, x_2, \ldots, x_5$. That is, we write:

$$= x_1 + x_2 2^{-k} + x_3 2^{-2k} + x_4 2^{-3k} + x_5 2^{-4k}$$

where $0 \leq x_i \leq 1 - 2^{-k}$ is a multiple of $2^{-k}$.

$$\alpha_1(x_1, x_2) = f(x_1 + x_2 2^{-k}) - 3f(x_1) - \frac{1}{2}(2^{-3k} + 2^{-4k})x_2^2 f''(x_1),$$

$$\alpha_2(x_1, x_3) = f(x_1 + x_3 2^{-2k}) - \frac{1}{2}2^{-3k}x_3^2 f''(x_1) - \frac{1}{6}2^{-4k}x_3^3 f'''(x_1),$$

$$\alpha_3(x_1, x_4) = f(x_1 + x_4 2^{-3k}) - \frac{1}{2}2^{-4k}x_4^2 f''(x_1),$$

$$\alpha_4(x_1, x_5) = f(x_1 + x_5 2^{-4k}),$$

$$u = x_2 + x_3,$$

$$v = x_2 - x_3,$$                                                                        (3.1)

$$w = x_2 + x_4,$$

$$z = x_2 - x_4,$$

$$\beta_1(u, x_1) = \frac{1}{12}2^{-4k}u^3 f'''(x_1) + \frac{1}{2}2^{-3k}u^2 f''(x_1),$$

$$\beta_2(v, x_1) = -\frac{1}{12}2^{-4k}v^3 f'''(x_1) - \frac{1}{2}2^{-3k}v^2 f''(x_1),$$

$$\beta_3(w, x_1) = \frac{1}{4}w^2 f''(x_1),$$

$$\beta_4(z, x_1) = -\frac{1}{4}z^2 f''(x_1).$$

Then

$$f(x) \approx \alpha_1(x_1, x_2) + \alpha_2(x_1, x_3) + \alpha_3(x_1, x_4) + \alpha_4(x_1, x_5)$$
$$+ \beta_1(u, x_1) + \beta_2(v, x_1) + \beta_3(w, x_1) + \beta_4(z, x_1)$$

with an error of the order of $2^{-5k}$. Hence, with this method, we can use tables with $2n/5$ address bits. Four additions are used to generate $u$, $v$, $w$ and $z$, and after the table-lookups, 8 terms are added with a carry-save addition tree. This method would require around 20 Kbytes of table for single-precision. It would require around 100 Mbytes for double precision, which is still a lot.

## Conclusion

Various table-based methods have been suggested during the last decade. When single-precision implementation is at stake, table-bound methods seem to be a good candidate for implementing fast functions. Unless there is a technology break-through, these methods are not suitable for double precision.

## References

1. Ercegovac, M. D., Lang, T., Muller, J. M., and Tisserand, A.: *Reciprocation, Square Root, Inverse Square Root, and Some Elementary Functions Using Small Multipliers*, Technical Report RR97-47, LIP, École Normale Supérieure de Lyon, November 1997, available at `ftp://ftp.lip.ens-lyon.fr/pub/Rapports/RR/RR97/RR97-47.ps.Z`.
2. Farmwald, P. M.: High Bandwidth Evaluation of Elementary Functions, in: Trivedi, K. S. and Atkins, D. E. (eds), *Proceedings of the 5th IEEE Symposium on Computer Arithmetic*, IEEE Computer Society Press, Los Alamitos, CA, 1981.
3. Hassler, H. and Takagi, N.: Function Evaluation by Table Look-Up and Addition, in: Knowles, S. and McAllister, W. (eds), *Proceedings of the 12th IEEE Symposium on Computer Arithmetic*, Bath, UK, July 1995. IEEE Computer Society Press, Los Alamitos, CA.
4. Das Sarma, D. and Matula. D. W.: Faithful Bipartite Rom Reciprocal Tables, in: Knowles, S. and McAllister, W. H. (eds), *Proceedings of the 12th IEEE Symposium on Computer Arithmetic*, Bath, UK, 1995, IEEE Computer Society Press, Los Alamitos, CA, pp. 17–28.
5. Schulte, M. and Stine, J.: Accurate Function Approximation by Symmetric Table Lookup and Addition, in: *Proceedings of ASAP'97*, IEEE Computer Society Press, Los Alamitos, CA, 1997.
6. Schulte, M. and Stine, J.: Symmetric Bipartite Tables for Accurate Function Approximation, in: Lang, T., Muller, J. M., and Takagi, N. (eds), *Proceedings of the 13th IEEE Symposium on Computer Arithmetic*, IEEE Computer Society Press, Los Alamitos, CA, 1997.
7. Schwarz. E.: *High-Radix Algorithms for High-Order Arithmetic Operations*, PhD thesis, Dept. of Electrical Engineering, Stanford University, 1992.
8. Tang, P. T. P.: Table Lookup Algorithms for Elementary Functions and Their Error Analysis, in: Kornerup, P. and Matula, D. W. (eds), *Proceedings of the 10th IEEE Symposium on Computer Arithmetic*, Grenoble, France, June 1991, IEEE Computer Society Press, Los Alamitos, CA, pp. 232–236.
9. Tang, P. T. P.: Table-Driven Implementation of the Expm1 Function in IEEE Floating-Point Arithmetic, *ACM Transactions on Mathematical Software* **18** (2) (1992), pp. 211–222.
10. Tang, P. T. P.: Table-Driven Implementation of the Exponential Function in IEEE Floating-Point Arithmetic, *ACM Transactions on Mathematical Software* **15** (2) (1989), pp. 144–157.

11. Tang, P. T. P.: Table-Driven Implementation of the Logarithm Function in IEEE Floating-Point Arithmetic, *ACM Transactions on Mathematical Software* **16** (4) (1990), pp. 378–400.
12. Wong, W. F. and Goto, E.: Fast Evaluation of the Elementary Functions in Single Precision, *IEEE Transactions on Computers* **44** (3) (1995), pp. 453–457.
13. Wong, W. F. and Goto, E.: Fast Hardware-Based Algorithms for Elementary Function Computations Using Rectangular Multipliers, *IEEE Transactions on Computers* **43** (3) (1994), pp. 278–294.

*Reliable Computing* 5: 289–310, 1999.
© 1999 *Kluwer Academic Publishers.*

# An Interval Hermite-Obreschkoff Method for Computing Rigorous Bounds on the Solution of an Initial Value Problem for an Ordinary Differential Equation*

NEDIALKO S. NEDIALKOV and KENNETH R. JACKSON
*Department of Computer Science, University of Toronto, Toronto, Ontario, Canada, M5S 3G4,*
*e-mail:* {ned,krj}@cs.toronto.edu

(Received: 2 November 1998; accepted: 5 February 1999)

**Abstract.** To date, the only effective approach for computing guaranteed bounds on the solution of an initial value problem (IVP) for an ordinary differential equation (ODE) has been interval methods based on Taylor series. This paper derives a new approach, an interval Hermite-Obreschkoff (IHO) method, for computing such enclosures. Compared to interval Taylor series (ITS) methods, for the same stepsize and order, our IHO scheme has a smaller truncation error, better stability, and requires fewer Taylor coefficients and high-order Jacobians.

The stability properties of the ITS and IHO methods are investigated. We show as an important by-product of this analysis that the stability of an interval method is determined not only by the stability function of the underlying formula, as in a standard method for an IVP for an ODE, but also by the associated formula for the truncation error.

## 1. Introduction

We consider the set of autonomous initial value problems (IVPs)

$$y'(t) = f(y) \tag{1.1}$$
$$y(t_0) = [y_0], \tag{1.2}$$

where $t \in [t_0, T]$ for some $T > t_0$. Here $t_0, T \in \mathbb{R}$, $f \in C^{k-1}(\mathcal{D})$, $\mathcal{D} \subseteq \mathbb{R}^n$ is open, $f : \mathcal{D} \to \mathbb{R}^n$, and $[y_0] \subseteq \mathcal{D}$. The condition (1.2) permits the initial value $y(t_0)$ to be in an interval, rather than specifying a particular value. We assume that the representation of $f$ contains only a finite number of constants, variables, elementary operations, and standard functions. Since we assume $f \in C^{k-1}(\mathcal{D})$, we exclude functions that contain, for example, branches, abs, or min. For expositional convenience, we consider only autonomous systems. This is not a restriction of consequence since a nonautonomous system of ordinary differential equations (ODEs) can be converted

* This work was supported in part by the Natural Sciences and Engineering Research Council of Canada, the Information Technology Research Centre of Ontario, and Communications and Information Technology Ontario.

into an autonomous system. Moreover, the methods discussed here can be extended easily to nonautonomous systems.

We consider a grid $t_0 < t_1 < \cdots < t_m = T$, which is not necessarily equally spaced, and denote the stepsize from $t_j$ to $t_{j+1}$ by $h_j$, $(h_j = t_{j+1} - t_j)$. The step from $t_j$ to $t_{j+1}$ is referred to as the $(j + 1)$-st step. We denote the solution of (1.1) with an initial condition $y_j$ at $t_j$ by $y(t; t_j, y_j)$. For an interval vector $[y_j]$, we denote by $y(t; t_j, [y_j])$ the set of solutions

$$\{y(t; t_j, y_j) \mid y_j \in [y_j]\}.$$

Our goal is to compute interval vectors, $[y_j]$, $j = 1, ..., m$, that are guaranteed to contain the solution of (1.1)–(1.2) at $t_1, t_2, ..., t_m$. That is,

$$y(t_j; t_0, [y_0]) \subseteq [y_j], \quad \text{for } j = 1, ..., m.$$

Usually, validated methods for IVPs for ODEs are one-step methods, where the $(j + 1)$-st step consists of two phases, which we refer to as Algorithm I and Algorithm II [25]:

Algorithm I: validate existence and uniqueness of the solution in $[t_j, t_{j+1}]$ and compute an enclosure $[\tilde{y}_j]$ such that

$$y(t; t_j, [y_j]) \subseteq [\tilde{y}_j], \quad \text{for all } t \in [t_j, t_{j+1}];$$

Algorithm II: compute a tighter enclosure $[y_{j+1}] \subseteq [\tilde{y}_j]$, for which

$$y(t_{j+1}; t_0, [y_0]) \subseteq [y_{j+1}].$$

Significant developments in the area of validated solutions of IVPs for ODEs are the interval methods of Moore [22]–[24], Krückeberg [18], Eijgenraam [9], and Lohner [1], [20], [21]. All these methods are based on Taylor series. One reason for the popularity of the Taylor series approach is the simple form of the error term. In addition, the Taylor series coefficients can be readily generated by automatic differentiation [23, pp. 107–130], the order of the method can be changed easily by adding or deleting Taylor series terms, and the stepsize can be changed without recomputing Taylor series coefficients.

In this paper, we develop an interval Hermite-Obreschkoff (IHO) method for computing tight enclosures of the solution. Validated methods based on the Hermite-Obreschkoff formula [15], [27], [28] have not been derived or considered before. Although Taylor series methods can be viewed as a special case of the more general Hermite-Obreschkoff methods, the method we propose is conceptually different from the interval Taylor series (ITS) methods. We show that compared with ITS methods, for the same stepsize and order, our IHO scheme has a smaller truncation error, better stability, and may be less expensive for problems for which the computation of $f(y)$ involves many terms.

This paper also shows that the stability of the ITS and IHO methods depends not only on the stability function of the underlying formula, as in standard numerical methods for IVPs for ODEs, but also on the associated formula for the truncation error. In standard numerical methods, Hermite-Obreschkoff methods are known to be suitable for stiff systems [11], [12], [34], [35], but in the interval case, they still have a restriction on the stepsize. To develop an interval method for stiff problems, we need not only a formula with a good stability function, but also a better associated formula for the truncation error.

An outline of this paper follows. Section 2 explains briefly how ITS methods for computing tight enclosures of the solution work and reviews Lohner's method. Section 3 derives the IHO method. First, in Section 3.1, we show how the point Hermite-Obreschkoff method can be obtained. Then in Section 3.2, we derive the interval version and show how to represent the enclosure in a manner that reduces the wrapping effect in propagating the error. Section 4 compares the IHO and ITS methods in the constant coefficient case and gives estimates for the amount of work per step required by these methods in the general case. This section also shows that the IHO method is more stable than an ITS method. Section 5 contains numerical results comparing the two methods.

NOTATION

An interval or interval vector "$a$" is denoted by $[a]$; an interval matrix "$A$" is denoted by $[A]$. We use small letters for intervals and interval vectors and capital letters for interval matrices. For an interval $[a] = [\underline{a}, \bar{a}]$, we define

width: $w([a]) = \bar{a} - \underline{a}$;

midpoint: $m([a]) = (\bar{a} + \underline{a})/2$; and

magnitude: $|[a]| = \max\{|\bar{a}|, |\underline{a}|\}$.

We define width, midpoint, and magnitude componentwise for interval vectors and matrices.

We introduce the sequence of functions

$$f^{[0]}(y) = y, \tag{1.3}$$

$$f^{[i]}(y) = \frac{1}{i}\left(\frac{\partial f^{[i-1]}}{\partial y} f\right)(y), \quad \text{for } i \geq 1. \tag{1.4}$$

For the autonomous IVP $y'(t) = f(y)$, $y(t_j) = y_j$, $f^{[i]}(y_j) = y^{(i)}(t_j)/i!$. That is, $f^{[i]}(y_j)$ denotes the $i$-th Taylor coefficient of $y(t; t_j, y_j)$.

## 2. Taylor Series Methods

Consider the Taylor series expansion

$$y_{j+1} = y_j + \sum_{i=1}^{k-1} h_j^i f^{[i]}(y_j) + h_j^k f^{[k]}(y; t_j, t_{j+1}), \tag{2.1}$$

where $y_j \in [y_j]$ and $f^{[k]}(y; t_j, t_{j+1})$ denotes $f^{[k]}$ with its $l$-th component evaluated at $y(\xi_{jl})$, for some $\xi_{jl} \in [t_j, t_{j+1}]$. If (2.1) is evaluated in interval arithmetic with $y_j$ replaced by $[y_j]$ and $f^{[k]}(y; t_j, t_{j+1})$ replaced by $f^{[k]}([\tilde{y}_j])$, we obtain

$$[y_{j+1}] = [y_j] + \sum_{i=1}^{k-1} h_j^i f^{[i]}([y_j]) + h_j^k f^{[k]}([\tilde{y}_j]). \tag{2.2}$$

With (2.2), we can compute enclosures of the solution, but the width of $[y_j]$ always increases with $j$, even if the true solution contracts.

By applying the mean-value theorem to $f^{[i]}$ at some $\hat{y}_j \in [y_j]$, we have

$$f^{[i]}(y_j) = f^{[i]}(\hat{y}_j) + J(f^{[i]}; y_j, \hat{y}_j)(y_j - \hat{y}_j), \tag{2.3}$$

where $J(f^{[i]}; y_j, \hat{y}_j)$ is the Jacobian of $f^{[i]}$ with its $l$-th row evaluated at $y_j + \theta_{il}(\hat{y}_j - y_j)$ for some $\theta_{il} \in [0, 1]$, $(l = 1, \ldots, n)$. Then from (2.1) and (2.3),

$$y_{j+1} = \hat{y}_j + \sum_{i=1}^{k-1} h_j^i f^{[i]}(\hat{y}_j) + h_j^k f^{[k]}(y; t_j, t_{j+1})$$

$$+ \left\{ I + \sum_{i=1}^{k-1} h_j^i J(f^{[i]}; y_j, \hat{y}_j) \right\}(y_j - \hat{y}_j). \tag{2.4}$$

This formula is the basis of the interval Taylor series methods of Moore [22]–[24], Eijgenraam [9], Lohner [1], [20], [21], and Rihm [30] (see also [25]). If we evaluate (2.4) in interval arithmetic, we can often obtain enclosures with smaller widths than in (2.2). A serious problem that may occur in this evaluation is the wrapping effect [23]. Here, we briefly describe Lohner's method for reducing it.

## 2.1. LOHNER'S METHOD

Let

$$z_{j+1} = h_j^k f^{[k]}(y; t_j, t_{j+1}) \in h_j^k f^{[k]}([\tilde{y}_j]) \equiv [z_{j+1}], \tag{2.5}$$

$$s_{j+1} = m([z_{j+1}]), \tag{2.6}$$

$$\hat{y}_{j+1} = \hat{y}_j + \sum_{i=1}^{k-1} h_j^i f^{[i]}(\hat{y}_j) + s_{j+1}, \quad \text{and} \tag{2.7}$$

$$S_j = I + \sum_{i=1}^{k-1} h_j^i J(f^{[i]}; y_j, \hat{y}_j) \in I + \sum_{i=1}^{k-1} h_j^i J(f^{[i]}; [y_j]) \equiv [S_j]. \tag{2.8}$$

Also let

$$A_0 = I, \quad \hat{y}_0 = m([y_0]), \quad \text{and} \quad r_0 = y_0 - \hat{y}_0 \in [r_0] = [y_0] - \hat{y}_0, \tag{2.9}$$

where $I$ is the identity matrix. From (2.4) and (2.5)–(2.9), we compute

$$[y_{j+1}] = \hat{y}_j + \sum_{i=1}^{k-1} h_j^i f^{[i]}(\hat{y}_j) + ([S_j]A_j)[r_j] + [z_{j+1}] \tag{2.10}$$

and propagate for the next step the error vector

$$[r_{j+1}] = \left(A_{j+1}^{-1}([S_j]A_j)\right)[r_j] + A_{j+1}^{-1}([z_{j+1}] - s_{j+1}). \tag{2.11}$$

Here $A_j \in \mathbb{R}^{n \times n}$, $(j \geq 0)$ is a nonsingular matrix. Usually, a good choice for $A_{j+1}$ is the Q-factor from the QR-factorization of $\widehat{A}_{j+1} = m([S_j]A_j)$. A detailed explanation of the reasons for this and other choices for $A_{j+1}$ can be found in [21] and [25].

## 3. Derivation of the Interval Hermite-Obreschkoff Method

### 3.1. THE POINT METHOD

Let

$$P_{p,q}(s) = \frac{s^q(s-1)^p}{(p+q)!}, \tag{3.1}$$

$$c_i^{q,p} = \frac{q!}{(p+q)!} \frac{(q+p-i)!}{(q-i)!}, \quad \text{and} \tag{3.2}$$

$$g_i(s) = \frac{g^{(i)}(s)}{i!}, \tag{3.3}$$

where $p \geq 0$, $q \geq 0$, $0 \leq i \leq q$, and $g(t)$ is $(p+q+1)$ times differentiable. If we integrate $\int_0^1 P_{p,q}(s)g^{(p+q+1)}(s)\,ds$ repeatedly by parts, we find

$$(-1)^{(p+q)} \int_0^1 P_{p,q}(s)g^{(p+q+1)}(s)\,ds = \sum_{i=0}^{q}(-1)^i c_i^{q,p} g_i(1) - \sum_{i=0}^{p} c_i^{p,q} g_i(0). \tag{3.4}$$

This derivation is sometimes attributed to Darboux [7] and Hermite [15].

If $y(t)$ is the solution to

$$y' = f(y), \quad y(t_j) = y_j,$$

and we set $g(s) = y(t_j + sh_j)$, then

$$g^{(p+q+1)}(s) = h_j^{p+q+1} y^{(p+q+1)}(t_j + sh_j), \tag{3.5}$$

$$g_i(0) = \frac{g^{(i)}(0)}{i!} = h_j^i \frac{y^{(i)}(t_j)}{i!} = h_j^i f^{[i]}(y_j), \quad \text{and} \tag{3.6}$$

$$g_i(1) = \frac{g^{(i)}(1)}{i!} = h_j^i \frac{y^{(i)}(t_j + h_j)}{i!} = h_j^i f^{[i]}(y_{j+1}), \tag{3.7}$$

where $y_{j+1} = y(t_j + h_j)$, and the functions $f^{[i]}$ are defined in (1.3)–(1.4). From (3.4) and (3.5)–(3.7),

$$\sum_{i=0}^{q}(-1)^i c_i^{q,p} h_j^i f^{[i]}(y_{j+1}) = \sum_{i=0}^{p} c_i^{p,q} h_j^i f^{[i]}(y_j)$$

$$+(-1)^q \frac{q!p!}{(p+q)!} h_j^{p+q+1} \frac{y^{(p+q+1)}(t; t_j, t_{j+1})}{(p+q+1)!}, \qquad (3.8)$$

where the $l$-th component of $y^{(p+q+1)}(t; t_j, t_{j+1})$ is evaluated at some $\xi_{jl} \in [t_j, t_{j+1}]$.

For a given $y_j$, if we solve the nonlinear (in general) system of equations

$$\sum_{i=0}^{q}(-1)^i c_i^{q,p} h_j^i f^{[i]}(y_{j+1}) = \sum_{i=0}^{p} c_i^{p,q} h_j^i f^{[i]}(y_j) \qquad (3.9)$$

for $y_{j+1}$, we obtain an approximation of local order $O(h_j^{p+q+1})$ to the solution $y(t; t_j, y_j)$ at $t_{j+1}$. The system (3.9) defines the point $(q, p)$ Hermite-Obreschkoff method [11], [12], [14, p. 277], [34], [35].

If $p > 0$ and $q = 0$, then we obtain from (3.8) an explicit Taylor series method; if $p = 0$ and $q > 0$, then (3.8) becomes an implicit Taylor series formula. Hence, we can consider the Hermite-Obreschkoff methods that we obtain from (3.8) as a generalization of Taylor series methods.

## 3.2. THE INTERVAL METHOD

Suppose that we have computed $[y_j]$, $\hat{y}_j$, $A_j$, and $[r_j]$ at $t_j$ such that

$y(t_j; t_0, [y_0]) \subseteq [y_j]$ and
$y(t_j; t_0, [y_0]) \subseteq \{\hat{y}_j + A_j r_j \mid r_j \in [r_j]\} \equiv \mathcal{U}_j,$

where $\hat{y}_j = m([y_j])$, $A_j \in \mathbb{R}^{n \times n}$ is nonsingular, and $[r_j]$ is an interval vector. The interval vectors $[y_j]$ and $\hat{y}_j + A_j[r_j]$ are not necessarily the same. We use the representation $\mathcal{U}_j$ to reduce the wrapping effect in propagating the error and the representation $[y_j]$ in generating the Jacobians of the Taylor coefficients $f^{[i]}$. Suppose also that we have verified existence and uniqueness of the solution on $[t_j, t_{j+1}]$ and have computed an a priori enclosure $[\tilde{y}_j]$ on $[t_j, t_{j+1}]$.

The method we propose consists of two phases, which can be considered as a predictor and a corrector. The predictor computes an enclosure $[y_{j+1}^{(0)}]$ of the solution at $t_{j+1}$, and using this enclosure, the corrector computes a tighter enclosure $[y_{j+1}] \subseteq [y_{j+1}^{(0)}]$ at $t_{j+1}$. In the predictor, we compute the enclosure $[y_{j+1}^{(0)}]$ with an ITS method of order $(q + 1)$; for example, we can use Lohner's method from Section 2. Here, we show how to construct a corrector based on (3.8).

Our goal is to compute a tighter enclosure $[y_{j+1}]$ of the solution than $[y_{j+1}^{(0)}]$ and a representation of the enclosure set in the form

$$\mathcal{U}_{j+1} = \{\hat{y}_{j+1} + A_{j+1} r_{j+1} \mid r_{j+1} \in [r_{j+1}]\}.$$

That is, we have to compute $[y_{j+1}]$, $\hat{y}_{j+1}$, $A_{j+1}$, and $[r_{j+1}]$ for the next step.

Let

$$y_j = y(t_j; t_0, y_0) \quad \text{and} \quad y_{j+1} = y(t_{j+1}; t_0, y_0)$$

for some $y_0 \in [y_0]$, and $\hat{y}_{j+1}^{(0)} = m([y_{j+1}^{(0)}])$. Since

$$y_{j+1}, \hat{y}_{j+1}^{(0)} \in [y_{j+1}^{(0)}] \quad \text{and} \quad y_j, \hat{y}_j \in [y_j],$$

we can apply the mean-value theorem to the two sums in (3.8) to obtain

$$\left( \sum_{i=0}^{q} (-1)^i c_i^{q,p} h_j^i J(f^{[i]}; y_{j+1}, \hat{y}_{j+1}^{(0)}) \right) (y_{j+1} - \hat{y}_{j+1}^{(0)})$$

$$= \sum_{i=0}^{p} c_i^{p,q} h_j^i f^{[i]}(\hat{y}_j) - \sum_{i=0}^{q} (-1)^i c_i^{q,p} h_j^i f^{[i]}(\hat{y}_{j+1}^{(0)})$$

$$+ \left( \sum_{i=0}^{p} c_i^{p,q} h_j^i J(f^{[i]}; y_j, \hat{y}_j) \right) (y_j - \hat{y}_j) + \varepsilon_{j+1}, \qquad (3.10)$$

where $J(f^{[i]}; y_{j+1}, \hat{y}_{j+1}^{(0)})$ is the Jacobian of $f^{[i]}$ with its $l$-th row evaluated at $y_{j+1} + \theta_{il}(\hat{y}_{j+1}^{(0)} - y_{j+1})$ for some $\theta_{il} \in [0, 1]$, $J(f^{[i]}; y_j, \hat{y}_j)$ is the Jacobian of $f^{[i]}$ with its $l$-th row evaluated at $y_j + \eta_{il}(\hat{y}_j - y_j)$ for some $\eta_{il} \in [0, 1]$, $(l = 1, ..., n)$, and

$$\varepsilon_{j+1} = (-1)^q \frac{q!p!}{(p+q)!} h_j^{p+q+1} \frac{y^{(p+q+1)}(t; t_j, t_{j+1})}{(p+q+1)!}. \qquad (3.11)$$

Using (3.10), we show how to compute a better enclosure than $[y_{j+1}^{(0)}]$ at $t_{j+1}$. To this end, we denote by $J(f^{[i]}; [y_{j+1}^{(0)}])$ the Jacobian of $f^{[i]}$ evaluated over $[y_{j+1}^{(0)}]$ and by $J(f^{[i]}; [y_j])$ the Jacobian of $f^{[i]}$ evaluated over $[y_j]$. Let

$$S_{j+1,-} = \sum_{i=0}^{q} (-1)^i c_i^{q,p} h_j^i J(f^{[i]}; y_{j+1}, \hat{y}_{j+1}^{(0)})$$

$$\in \sum_{i=0}^{q} (-1)^i c_i^{q,p} h_j^i J(f^{[i]}; [y_{j+1}^{(0)}]) \equiv [S_{j+1,-}], \qquad (3.12)$$

$$\hat{S}_{j+1,-} = m([S_{j+1,-}]), \qquad (3.13)$$

$$S_{j,+} = \sum_{i=0}^{p} c_i^{p,q} h_j^i J(f^{[i]}; y_j, \hat{y}_j)$$

$$\in \sum_{i=0}^{p} c_i^{p,q} h_j^i J(f^{[i]}; [y_j]) \equiv [S_{j,+}], \qquad (3.14)$$

$$[B_j] = (\hat{S}_{j+1,-}^{-1}[S_{j,+}])A_j, \qquad (3.15)$$

$$[C_j] = I - \widehat{S}_{j+1,\,-}^{-1}[S_{j+1,\,-}], \tag{3.16}$$

$$[v_j] = [y_{j+1}^{(0)}] - \hat{y}_{j+1}^{(0)}, \tag{3.17}$$

$$\varepsilon_{j+1} \in (-1)^q \frac{q!p!}{(p+q)!} h_j^{p+q+1} f^{[p+q+1]}([\tilde{y}_j]) \equiv [\varepsilon_{j+1}], \tag{3.18}$$

$$g_{j+1} = \sum_{i=0}^{p} c_i^{p,\,q} h_j^i f^{[i]}(\hat{y}_j) - \sum_{i=0}^{q}(-1)^i c_i^{q,\,p} h_j^i f^{[i]}(\hat{y}_{j+1}^{(0)}), \quad \text{and} \tag{3.19}$$

$$\delta_{j+1} = g_{j+1} + \varepsilon_{j+1} \in g_{j+1} + [\varepsilon_{j+1}] \equiv [\delta_{j+1}]. \tag{3.20}$$

We can derive from (3.10)–(3.20) that

$$\begin{aligned}
y_{j+1} - \hat{y}_{j+1}^{(0)} \in\ & ((\widehat{S}_{j+1,\,-}^{-1}[S_{j,\,+}]A_j)[r_j] + \widehat{S}_{j+1,\,-}^{-1}[\delta_{j+1}] \\
& + (I - \widehat{S}_{j+1,\,-}^{-1}[S_{j+1,\,-}])([y_{j+1}^{(0)}] - \hat{y}_{j+1}^{(0)}) \\
=\ & [B_j][r_j] + [C_j][v_j] + \widehat{S}_{j+1,\,-}^{-1}[\delta_{j+1}].
\end{aligned}$$

Note that for small $h_j$, we can compute the inverse of $\widehat{S}_{j+1,\,-}$. Since

$$y_{j+1} = y(t_{j+1}; t_0, y_0) \in \hat{y}_{j+1}^{(0)} + [B_j][r_j] + [C_j][v_j] + \widehat{S}_{j+1,\,-}^{-1}[\delta_{j+1}],$$

for an arbitrary $y_0 \in [y_0]$, we compute an interval vector that is a tight enclosure of the solution at $t_{j+1}$ by

$$[y_{j+1}] = (\hat{y}_{j+1}^{(0)} + [B_j][r_j] + [C_j][v_j] + \widehat{S}_{j+1,\,-}^{-1}[\delta_{j+1}]) \cap [y_{j+1}^{(0)}], \tag{3.21}$$

where "$\cap$" denotes intersection of interval vectors. For the next step, we propagate $\hat{y}_{j+1} = m([y_{j+1}])$, $A_{j+1}$, which is the Q-factor from the QR-factorization of $m([B_j])$, and

$$\begin{aligned}
[r_{j+1}] =\ & (A_{j+1}^{-1}[B_j])[r_j] + (A_{j+1}^{-1}[C_j])[v_j] \\
& + (A_{j+1}^{-1}\widehat{S}_{j+1,\,-}^{-1})[\delta_{j+1}] + A_{j+1}^{-1}(\hat{y}_{j+1}^{(0)} - \hat{y}_{j+1}).
\end{aligned} \tag{3.22}$$

## REMARKS

1. Once we obtain $[y_{j+1}]$, we can set $[y_{j+1}^{(0)}] = [y_{j+1}]$ and compute another enclosure, hopefully better than $[y_{j+1}]$, by repeating the same procedure. Thus, we can improve this enclosure iteratively. The experiments that we have performed show, however, that this iteration does not improve the results significantly, but increases the cost.

2. We could use the a priori enclosure $[\tilde{y}_j]$ computed in Algorithm I, instead of computing $[y_{j+1}^{(0)}]$. We briefly explain the reasons for computing $[y_{j+1}^{(0)}]$.

   a) The a priori enclosure $[\tilde{y}_j]$ may be too wide and the corrector phase may not produce a tight enough enclosure in one iteration. As a result, the corrector, which is the expensive part, may need more than one iteration to obtain a tight enough enclosure.

b) Predicting a reasonably tight enclosure $[y_{j+1}^{(0)}]$ is not expensive: we need to generate the coefficients $f^{[i]}(\hat{y}_j)$ and $J(f^{[i]}; [y_j])$ for $i = 1, ..., q$. We need them in the corrector, but for $i = 1, ..., p$. Usually, a good choice for $q$ is $q \in \{p, p+1, p+2\}$ (see Section 4.1). Therefore, we do not create extra work when generating these terms in the predictor.

3. We could use the inverse of the interval matrix $[S_{j+1}, -]$ instead of $\hat{S}_{j+1, -}^{-1}$. However, it is easier to compute the enclosure of the inverse of a point matrix than of an interval matrix. In fact, computing an enclosure of an interval matrix is NP hard in general [31].

4. If we intersect the computed enclosure as in (3.21), it is important to choose $\hat{y}_{j+1} \in [y_{j+1}]$. If we set $\hat{y}_{j+1} = \hat{y}_{j+1}^{(0)}$, it might happen that $\hat{y}_{j+1} = \hat{y}_{j+1}^{(0)} \notin [y_{j+1}]$, because $\hat{y}_{j+1}^{(0)}$ is the midpoint of $[y_{j+1}^{(0)}]$, which is generally a wider enclosure than $[y_{j+1}]$.

## 4. Interval Hermite-Obreschkoff Versus Interval Taylor Series Method

For the remainder of the paper, we compare an ITS method with order $k$ truncation error and our IHO scheme with the same order of the truncation error; that is, with $p$ and $q$ such that $k = p + q + 1$. In Sections 4.1 and 4.2, we consider these methods with constant stepsize $h$.

### 4.1. THE ONE-DIMENSIONAL CONSTANT COEFFICIENT CASE. INSTABILITY RESULTS

Consider the problem

$$y' = \lambda y, \quad y(0) \in [y_0], \tag{4.1}$$

where $\lambda \in \mathbb{R}$ and $\lambda < 0$. Since we have not defined complex interval arithmetic, we do not consider problems with $\lambda$ complex.

DEFINITION 4.1. We say that an interval method with a constant stepsize is *asymptotically unstable*, if, when applied to the problem (4.1),

$$w([y_j]) \rightarrow \infty, \quad \text{as} \quad j \rightarrow \infty.$$

In this subsection, we look at the stability of the ITS and IHO methods. Note that the wrapping effect does not occur in one-dimensional problems.

### 4.1.1. *The Interval Taylor Series Method*

Suppose that at $t_j > 0$, we have computed a tight enclosure $[y_j^{ITS}]$ of the solution at $t_j$ with an ITS method, and $[\tilde{y}_j^{ITS}]$ is an a priori enclosure of the solution on $[t_j, t_{j+1}]$, for all $y_j \in [y_j^{ITS}]$, where $[y_0^{ITS}] = [y_0]$. Let

$$T_r(z) = \sum_{i=0}^{r} \frac{z^i}{i!}. \tag{4.2}$$

Using (4.2), an interval Taylor series method for computing tight enclosures of the solution to (4.1) can be written as

$$[y_{j+1}^{ITS}] = T_{k-1}(\lambda h)[y_j^{ITS}] + \frac{(\lambda h)^k}{k!}[\bar{y}_j^{ITS}]; \tag{4.3}$$

cf. (2.4). Since $w([\bar{y}_j^{ITS}]) \geq w([y_j^{ITS}])$, we derive from (4.2)–(4.3)

$$w([y_{j+1}^{ITS}]) \geq \left( |T_{k-1}(\lambda h)| + \frac{|\lambda h|^k}{k!} \right) w([y_j^{ITS}]).$$

Therefore, the ITS method given by (4.3) is asymptotically unstable for stepsizes $h$ such that

$$|T_{k-1}(\lambda h)| + \frac{|\lambda h|^k}{k!} > 1. \tag{4.4}$$

This result implies that we have restrictions on the stepsize not only from the function $T_{k-1}(\lambda h)$, as in point methods for IVPs for ODEs, but also from the factor $|\lambda h|^k / k!$ in the remainder term.

### 4.1.2. *The Interval Hermite-Obreschkoff Method*

Let $y_j \in [y_j^{IHO}]$, where we assume that $[y_j^{IHO}]$ is computed with an IHO method and $[y_0^{IHO}] = [y_0]$. From (3.8), the true solution $y_{j+1}$ corresponding to the point $y_j$ satisfies

$$\left( \sum_{i=0}^{q} (-1)^i c_i^{q,p} \frac{(\lambda h)^i}{i!} \right) y_{j+1} = \left( \sum_{i=0}^{p} c_i^{p,q} \frac{(\lambda h)^i}{i!} \right) y_j$$

$$+ (-1)^q \frac{q!p!}{(p+q)!} \frac{(\lambda h)^{p+q+1}}{(p+q+1)!} y(\xi), \tag{4.5}$$

where $\xi \in [t_j, t_{j+1}]$. Let

$$R_{p,q}(z) \equiv \frac{\sum_{i=0}^{p} c_i^{p,q} \frac{z^i}{i!}}{\sum_{i=0}^{q} c_i^{q,p} \frac{(-z)^i}{i!}} \quad \text{and} \quad Q_{p,q}(z) = \sum_{i=0}^{q} c_i^{q,p} \frac{(-z)^i}{i!}, \tag{4.6}$$

where $c_i^{q,p}$ ($c_i^{p,q}$) are defined in (3.2). Also let $[\bar{y}_j^{IHO}]$ be an a priori enclosure of the solution on $[t_j, t_{j+1}]$ for any $y_j \in [y_j^{IHO}]$. From (4.5)–(4.6), we compute an enclosure $[y_{j+1}^{IHO}]$ by

$$[y_{j+1}^{IHO}] = R_{p,q}(\lambda h)[y_j^{IHO}] + (-1)^q \frac{\gamma_{p,q}}{Q_{p,q}(\lambda h)} \frac{|\lambda h|^k}{k!} [\bar{y}_j^{IHO}], \tag{4.7}$$

where $\gamma_{p,q} = q!p! / (p+q)!$. From (4.7),

$$w([y_{j+1}^{\text{IHO}}]) \geq \left(|R_{p,q}(\lambda h)| + \frac{\gamma_{p,q}}{|Q_{p,q}(\lambda h)|} \frac{|\lambda h|^k}{k!}\right) w([y_j^{\text{IHO}}]).$$

Therefore, the IHO method is asymptotically unstable for $h$ such that

$$|R_{p,q}(\lambda h)| + \frac{\gamma_{p,q}}{|Q_{p,q}(\lambda h)|} \frac{|\lambda h|^k}{k!} > 1. \tag{4.8}$$

In (4.4) and (4.8),

$$T_{k-1}(z) = e^z + O(z^k) \quad \text{and} \quad R_{p,q}(z) = e^z + O(z^{p+q+1}) = e^z + O(z^k)$$

are approximations to $e^z$ of the same order. However, $R_{p,q}(z)$ is the Padé rational approximation to $e^z$ (see for example [29]). If $z$ is complex with $\text{Re}(z) < 0$, the following results are known:

- if $p = q$, then $|R_{p,q}(z)| < 1$ [5]; and
- if $q = p + 1$ or $q = p + 2$, then $|R_{p,q}(z)| < 1$, and $|R_{p,q}(z)| \to 0$ as $|z| \to \infty$ [8]

(see also [19, pp. 236–237]).

Consider (4.4) and (4.8). For the ITS method, $|T_{k-1}(\lambda h)| < 1$ when $\lambda h$ is in the stability region of $T_{k-1}(z)$. However, for the IHO method, $|R_{p,q}(\lambda h)| < 1$ for any $h > 0$ when $q \geq p$ [33], and $|R_{p,q}(\lambda h)| \to 0$ as $\lambda h \to -\infty$ when $q > p$. Roughly speaking, the stepsize in the ITS method is restricted by both

$$|T_{k-1}(\lambda h)| \quad \text{and} \quad \frac{|\lambda h|^k}{k!},$$

while in the IHO method, the stepsize is limited mainly by

$$\frac{\gamma_{p,q}}{|Q_{p,q}(\lambda h)|} \frac{|\lambda h|^k}{k!}.$$

In the latter case, $\gamma_{p,q} / Q_{p,q}(\lambda h)$ is usually much smaller than one; thus, the stepsize limit for the IHO method is usually much larger than for the ITS method.

An important point to note here is that an interval version of a standard numerical method that is suitable for stiff problems may still have a restriction on the stepsize. To obtain an interval method without a stepsize restriction, we must find a stable formula not only for the propagated error, but also for the associated truncation error.

In Section 5.1, we show numerical results comparing the ITS and IHO methods on problem (4.1) for $\lambda = -10$.

### 4.2. THE $n$-DIMENSIONAL CONSTANT COEFFICIENT CASE

Consider the IVP

$$y' = By, \quad y_0 \in [y_0], \tag{4.9}$$

where $B \in \mathbb{R}^{n \times n}$ and $n > 1$.

We compare one step of an ITS method, which uses Lohner's technique for reducing the wrapping effect, and one step of the IHO method, which uses a similar technique for reducing the wrapping effect. We assume that in addition to an enclosure $[y_j]$ of the solution at $t_j$, we also have a representation of the enclosure in the form

$$\{\hat{y}_j + A_j r_j \mid r_j \in [r_j]\},$$

where $\hat{y}_j \in [y_j]$, $A_j \in \mathbb{R}^{n \times n}$ is nonsingular, and $[r_j]$ is an interval vector. We also assume that we have an a priori enclosure of the solution $[\tilde{y}_j]$ on $[t_j, t_{j+1}]$, where $h = t_{j+1} - t_j$.

### 4.2.1. The Interval Taylor Series Method

Using (4.2), we can write an ITS method, with Lohner's coordinate transformation, as

$$[y_{j+1}^{ITS}] = T_{k-1}(hB)\hat{y}_j + (T_{k-1}(hB)A_j)[r_j] + [z_{j+1}],$$

where

$$[z_{j+1}] = \frac{h^k}{k!}B^k[\tilde{y}_j]. \tag{4.10}$$

The width of $[y_{j+1}^{ITS}]$ is

$$w([y_{j+1}^{ITS}]) = |T_{k-1}(hB)A_j|w([r_j]) + w([z_{j+1}]). \tag{4.11}$$

### 4.2.2. The Interval Hermite-Obreschkoff method

Let

$$R_{p,q}(hB) = \left( \sum_{i=0}^{q} c_i^{q,p} \frac{(-hB)^i}{i!} \right)^{-1} \left( \sum_{i=0}^{p} c_i^{p,q} \frac{(hB)^i}{i!} \right) \quad \text{and} \tag{4.12}$$

$$Q_{p,q}(hB) = \sum_{i=0}^{q} c_i^{q,p} \frac{(-hB)^i}{i!}. \tag{4.13}$$

Using (4.10) and (4.12)–(4.13), the IHO method can be expressed by

$$[y_{j+1}^{IHO}] = R_{p,q}(hB)\hat{y}_j + (R_{p,q}(hB)A_j)[r_j] + (-1)^q \gamma_{p,q}(Q_{p,q}(hB))^{-1}[z_{j+1}].$$

(Note that for $h$ small, we can compute the inverse of the matrix $Q_{p,q}(hB)$.) The width of $[y_{j+1}^{IHO}]$ is given by

$$w([y_{j+1}^{IHO}]) = |R_{p,q}(hB)A_j|w([r_j]) + \gamma_{p,q}|(Q_{p,q}(hB))^{-1}|w([z_{j+1}]). \tag{4.14}$$

Comparing (4.14) and (4.11), we see that in the IHO method we multiply the width of the error term, $w([z_{j+1}])$, from the ITS method by $\gamma_{p,q}|(Q_{p,q}(hB))^{-1}|$. If, for example, $p = q = 8$, then $\gamma_{8,8} \approx 7.8 \times 10^{-5}$. Consider $(Q_{p,q}(hB))^{-1}$ and suppose that $q > 0$. For small $h$,

$$(Q_{p,q}(hB))^{-1} \approx (I - c_1^{q,p} hB)^{-1} \approx I + c_1^{q,p} hB.$$

This implies that for small $h$, multiplying by the matrix $|(Q_{p,q}(hB))^{-1}|$ does not significantly increase $w([z_{j+1}])$. Moreover, it often happens that

$$\|(Q_{p,q}(hB))^{-1}\| < 1.$$

Hence, multiplying by this matrix may reduce $w([z_{j+1}])$ still further.

In Lohner's method, we propagate $(T_{k-1}(hB)A_j)[r_j]$, where $T_{k-1}(hB)$ is an approximation of the matrix exponential of order $O(h^k)$. In the IHO method, we propagate $(R_{p,q}(hB)A_j)[r_j]$, where $R_{p,q}(hB)$ is a rational approximation to the matrix exponential of order $O(h^{p+q+1}) = O(h^k)$. If the width of $[r_j]$ is small and $hB$ is small, then we do not have a significant difference in propagating the error in both methods. For example, if $w([r_j]) = O(h^r)$, then this difference would be $O(h^{k+r})$.

Furthermore, $R_{p,q}(hB)$ has better stability properties than $T_{k-1}(hB)$. We briefly discuss them here. For simplicity, assume that the matrix $B$ in (4.9) is diagonalizable and can be represented in the form

$$B = X^{-1}DX,$$

where $D = \text{diag}(\lambda_1, \lambda_2, ..., \lambda_n)$ is a diagonal matrix and $\{\lambda_1, \lambda_2, ..., \lambda_n\}$ are the eigenvalues of $B$. Suppose that $\text{Re}(\lambda_i) < 0$ for $i = 1, ..., n$. The matrices $T_{k-1}(hD)$ and $R_{p,q}(hD)$ are diagonal with diagonal elements $T_{k-1}(h\lambda_i)$ and $R_{p,q}(h\lambda_i)$, respectively. As $h$ increases, $\|T_{k-1}(hD)\|$ will eventually become greater than one, and then the ITS method is asymptotically unstable. However, for any $h > 0$, $\|R_{p,q}(hD)\| < 1$ for $q = p, p + 1$, or $p + 2$, and $\|R_{p,q}(hD)\| \to 0$ as $h \to \infty$ for $q = p + 1$ or $q = p + 2$ (see Section 4.1).

We can also show for the ITS method that

$$w([y_{j+1}^{ITS}]) \geq \frac{h^k}{k!}|B^k|w([y_j^{ITS}]),$$

and for the IHO method that

$$w([y_{j+1}^{IHO}]) \geq \left(\gamma_{p,q}|(Q_{p,q}(hB))^{-1}|\right)\left(\frac{h^k}{k!}|B^k|w([y_j^{ITS}])\right).$$

In the latter case, the stability restriction on the stepsize usually occurs at values significantly larger than in the former case.

To summarize, we have shown that on constant coefficient problems, our IHO scheme has a much smaller local error and better stability than an ITS method with the same order and stepsize.

In the nonlinear case, we must enclose the solution of a nonlinear system (see Section 3.2). Computing such an enclosure normally introduces additional errors. It can be shown that if the predicted enclosure is computed with an ITS method of order $(q + 1)$, then these errors are sufficiently small [26]. As in the constant coefficient case, the error propagation in the IHO method is similar to that of Lohner's method, but the local error term in the former is much smaller than in the latter [26]. Furthermore, the IHO method requires roughly half the work for generating high-order Jacobians, the computation of which is usually the most expensive part in the IHO and ITS methods (see Section 4.3).

In Section 5.2, we show numerical results comparing the two methods on two nonlinear problems.

### 4.3. WORK PER STEP

We briefly discuss the most expensive parts of the ITS and IHO methods: generating high-order Jacobians, matrix-matrix multiplications, and enclosing the inverse of a point matrix. We measure the work by the number of floating-point operations. However, the time spent on memory operations may not be insignificant for the following reasons.

- The packages for automatic differentiation are often implemented through operator overloading [3], [4], [13], which may involve many memory allocations and deallocations.

- In generating Taylor coefficients, there may be a significant overhead caused by reading and storing the Taylor coefficients, $f^{[i]}$, and their Jacobians [11].

#### 4.3.1. *Generating High-Order Jacobians*

To obtain an approximate bound for the number of floating point operations to generate $k - 1$ Jacobians, $\partial f^{[i]} / \partial y$ for $i = 1, ..., k - 1$, we assume that they are computed by differentiating the code list of the corresponding $f^{[i]}$ and using information from the previously computed $\partial f^{[l]} / \partial y$, for $l = 1, ..., i - 1$. The FADBAD/TADIFF [3], [4] and IADOL-C [16] packages compute $\partial f^{[i]} / \partial y$ by differentiating the code list of $f^{[i]}$ (IADOL-C is an interval version of ADOL-C [13]). We also assume that the cost of evaluating $\partial f^{[i]} / \partial y$ is roughly $n$ times the cost of evaluating $f^{[i]}$ [11].

For simplicity, suppose that $f$ contains only arithmetic operations. If $N$ is the number of these operations, and $c_f$ is the ratio of multiplications and divisions in $N$, then to generate $k - 1$ coefficients $f^{[i]}$, $(i = 1, ..., k - 1)$, we need $c_f N k^2 + O(Nk)$ operations [23, pp. 111–112]; to generate $k - 1$ Jacobians in an ITS method, we use

$$c_f n N k^2 + O(nNk) \tag{4.15}$$

arithmetic operations [26]. Let $p = q$, $(k = p+q+1)$. In the IHO method we generate $p = (k - 1) / 2$ terms for the forward solution and $q = p = (k - 1) / 2$ terms for the backward one—the corresponding work is

$$c_f n N k^2 / 2 + O(nNk). \tag{4.16}$$

### 4.3.2. Matrix Inverses and Matrix-Matrix Multiplications

In Lohner's method and in the IHO method with the QR-factorization technique, we compute an enclosure of the inverse of a point matrix, which is a floating-point approximation to an orthogonal matrix. However, in the IHO method, we also enclose the inverse of a point matrix (see Section 3.2). In general, enclosing the inverse of an arbitrary point matrix is more expensive than enclosing the inverse of a floating-point approximation to an orthogonal matrix. However, we can still enclose the inverse of an arbitrary point matrix in $O(n^3)$ operations [2].

Lohner's method has 2 matrix-matrix multiplications, while the IHO method has 6 matrix-matrix multiplications.

To summarize, in the IHO method, we reduce the work for generating Jacobians, but increase the number of matrix operations. Suppose that $N \approx n^2$. This number can be easily achieved if, for example, each component of $f$ contains approximately $n$ operations. Then, (4.15) and (4.16) become

$$c_f n^3 k^2 + O(n^3 k) \quad \text{and} \quad c_f n^3 k^2 / 2 + O(n^3 k).$$

Therefore, we should expect the IHO method to outperform ITS methods in terms of the amount of work per step when the right side of the problem contains many terms. If the right side contains a few terms only, an ITS method may be less expensive for low orders, but we expect that the IHO method will perform better for higher orders. Note also that we expect the IHO method to allow larger stepsizes for the same order of the truncation error, thus saving computation time during the whole integration. In addition, the IHO method (with $p = q$) needs half the memory for storing the point Taylor coefficients and the high-order Jacobians.

In Section 5.2.1, we study empirically the amount of work per step on Van der Pol's equation.

## 5. Numerical Results

The numerical results in this section are produced with the VNODE (Validated Numerical ODE) package [26], which we are developing at the University of Toronto. VNODE is a set of C++ classes that implement existing and new techniques (including the IHO method) used in validated ODE solving. The underlying interval-arithmetic and automatic differentiation packages are PROFIL/BIAS [17] and FADBAD/TADIFF [3], [4], respectively. We compiled VNODE with the GNU C++ compiler version 2.7.2 on a Sun Ultra 2/2170 workstation with an 168 MHz UltraSPARC CPU.

Table 1. ITS(17) and IHO(8, 8) on $y' = -10y$, $y(0) = 1$, $t \in [0, 10]$.

| $h$ | Error | | Reductions | | Time | |
|------|-------|-----|---------|-------------------------|--------|--------|
|      | ITS     | IHO     | IHO/ITS   | $\gamma_{8,8} / Q_{8,8}(-10h)$ | ITS      | IHO      |
| 0.2  | 4.4e−51 | 1.3e−55 | 3.0e−05   | 3.0e−05                  | 3.5e−02  | 2.1e−02  |
| 0.3  | 8.5e−48 | 1.6e−52 | 1.9e−05   | 1.9e−05                  | 2.3e−02  | 1.4e−02  |
| 0.4  | 2.5e−45 | 2.9e−50 | 1.1e−05   | 1.2e−05                  | 1.8e−02  | 1.1e−02  |
| 0.5  | 1.2e−40 | 1.8e−48 | 1.5e−08   | 7.8e−06                  | 1.3e−02  | 8.8e−03  |
| 0.6  | 8.5e−20 | 5.9e−47 | 6.9e−28   | 5.2e−06                  | 1.1e−02  | 7.0e−03  |
| 0.7  | 4.0e−01 | 1.3e−45 | 3.2e−45   | 3.5e−06                  | 1.0e−02  | 6.3e−03  |
| 0.8  | 2.4e+10 | 2.0e−44 | 8.4e−55   | 2.4e−06                  | 8.9e−03  | 5.4e−03  |

For each of the examples, we measured the (global) error at the end of the interval of integration by taking the maximum norm of the width of the enclosure. Since we are interested mainly in the performance of Algorithm II, we report only the CPU time in seconds spent in this algorithm. For convenience, we denote an ITS method with $k$ terms by ITS($k$) and an IHO method with parameters $p$ and $q$ by IHO($q, p$). In this section $p = q$, ($k = p + q + 1$).

## 5.1. Constant Coefficient Problem

### 5.1.1. Constant Stepsizes

We integrated

$$y' = -10y, \quad y(0) = 1 \tag{5.1}$$

on $[0, 10]$ with constant stepsizes $h = 0.2, 0.3, ..., 0.8$. To avoid possible stepsize reductions in Algorithm I [25], we computed a priori enclosures on each step by $[\bar{y}_j] = [e^{-10h}y_j, \bar{y}_j]$. In Algorithm II, we used the ITS(17) and IHO(8, 8) methods.

For stepsizes $0.2, 0.3, ..., 0.8$, Table 1 shows the errors at $T = 10$, the ratio of the error of the IHO method to the error of the ITS method, $\gamma_{8,8} / Q_{8,8}(-10h)$, and the CPU time spent in Algorithm II. We compute the base-10 logarithm of the data and plot in Figure 1 the error versus the stepsize and the time versus the error.

Consider Table 1 and Figure 1(a). For "small" stepsizes, $h = 0.2, 0.3, 0.4$, the error in the IHO method is approximately $\gamma_{8,8} / Q_{8,8}(-10h) \approx 10^{-5}$ times the error in the ITS method. As $h$ increases beyond 0.4, the ITS method produces enclosures with rapidly increasing widths, while the IHO method computes good enclosures for those stepsizes.

### 5.1.2. Variable Stepsizes

We integrated (5.1) for $t \in [0, 100]$ with a simple stepsize selection scheme [26] based on controlling the width of the enclosure of the truncation error per unit step.

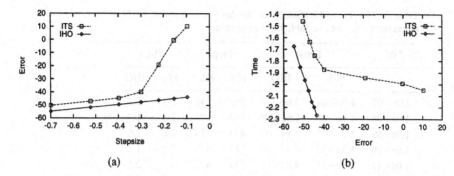

*Figure 1.* ITS(17) and IHO(8, 8) on $y' = -10y$, $y(0) = 1$, $t \in [0, 10]$.

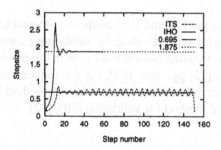

*Figure 2.* ITS(17) and IHO(8, 8) on $y' = -10y$, $y(0) = 1$, $t \in [0, 100]$, variable stepsize control with $Tol = 10^{-10}$.

We used an absolute tolerance of $10^{-10}$. In Figure 2, we plot the stepsizes against the step number for the two methods. The reduction in the stepsize on the last step for the ITS method is because our program decreases the stepsize to hit the endpoint exactly.

The ITS method is asymptotically unstable for stepsizes $h$ such that

$$|T_{16}(-10h)| + \frac{(10h)^{17}}{17!} > 1$$

(see Section 4.1). For $h = 0.695$,

$$|T_{16}(-10h)| + \frac{(10h)^{17}}{17!} \approx 1.008.$$

For the IHO method, the stepsize oscillates around 1.875, which is about 2.7 times bigger than 0.695. For $h = 1.875$,

$$|R_{8,8}(-10h)| + \frac{\gamma_{8,8}}{|Q_{8,8}(-10h)|} \frac{(10h)^{17}}{17!} \approx 0.996.$$

Although the IHO method permits larger stepsizes, they are still limited by its local error term. This observation confirms the conclusions in Section 4.1.

Table 2. ITS(11) and IHO(5, 5) on Van der Pol's equation, Taylor series for validation, variable stepsize control.

| Tol | Error | | Steps | | Time | |
|---|---|---|---|---|---|---|
| | ITS | IHO | ITS | IHO | ITS | IHO |
| 1.0e−07 | 4.0e−08 | 2.8e−07 | 352 | 231 | 1.5 | 1.1 |
| 1.0e−08 | 1.3e−08 | 1.6e−08 | 384 | 274 | 1.6 | 1.3 |
| 1.0e−09 | 1.2e−09 | 1.1e−09 | 473 | 324 | 2.0 | 1.6 |
| 1.0e−10 | 1.3e−10 | 4.4e−10 | 587 | 372 | 2.6 | 1.8 |
| 1.0e−11 | 1.3e−11 | 4.8e−11 | 731 | 428 | 3.1 | 2.1 |
| 1.0e−12 | 1.8e−12 | 3.5e−12 | 912 | 531 | 3.9 | 2.5 |

In the next two examples, we compare the ITS and IHO methods with a variable stepsize control and our version of a Taylor series method for validating existence and uniqueness of the solution [26]. The stepsize control used is as in the previous subsection. The Taylor series approach [23, pp. 100–103], [6] for implementing Algorithm I often allows larger stepsizes than the constant enclosure method [9], which has been the most commonly used method for validation [20], [32].

## 5.2. NONLINEAR CASE

### 5.2.1. *Van der Pol's Equation*

We integrated Van der Pol's equation, written as a system,

$$
\begin{aligned}
y_1' &= y_2 \\
y_2' &= \mu(1 - y_1^2)y_2 - y_1,
\end{aligned}
\tag{5.2}
$$

with

$$
y(0) = (2, 0)^T,
\tag{5.3}
$$

for $t \in [0, 20]$, where $\mu = 5$. We used the ITS(11) and IHO(5, 5) methods and tolerances $10^{-7}, 10^{-8}, \ldots, 10^{-12}$.

From Table 2 and Figure 3, we see that, for approximately the same error, VNODE using the IHO method took fewer steps than the solver using the ITS method, thus saving computation time. Note that we control the local error per unit step and report the global error in Table 2. Thus the global error can be larger than the tolerance.

In Figure 3, we plot the logarithms of the error, time, and tolerance. In Figure 3(d), the stepsize corresponding to the IHO method is not as smooth as the one corresponding to the ITS method. In the regions where the stepsize is not smooth, the Taylor series method for validation could not verify existence and uniqueness with the supplied stepsizes, but verified with reduced stepsizes.

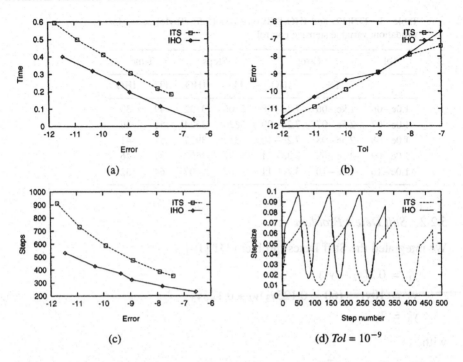

*Figure 3.* ITS(11) and IHO(5,5) on Van der Pol's equation, Taylor series for validation, variable stepsize control with $Tol = 10^{-7}, 10^{-8}, ..., 10^{-12}$.

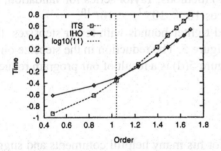

*Figure 4.* ITS and IHO with orders 3, 7, 11, 17, 25, 31, 37, 43, and 49 on Van der Pol's equation.

We also integrated (5.2)–(5.3) on $[0, 0.1]$ with an input stepsize of 0.01 to Algorithm I. We used orders $k = 3, 7, 11, 17, 25, 31, 37, 43$, and 49 for the ITS method and $p = q = (k-1)/2$ for the IHO method. Algorithm I did not reduce the input stepsize. As a result, the solver could take the same number of steps with the ITS and IHO methods. In Figure 4, we plot the logarithm of the time against the logarithm of the order for these two methods. Although on this problem, the IHO method is more expensive per step for "low" orders, including $k = 11$, we still have savings in time due to the fewer stepsizes taken.

Table 3. ITS(17) and IHO(8, 8) on (5.4)–(5.5), Taylor series for validation, variable stepsize control.

| Tol | Error | | Steps | | Time | |
|---|---|---|---|---|---|---|
| | ITS | IHO | ITS | IHO | ITS | IHO |
| 1.0e−06 | 1.9e−06 | 7.2e−08 | 5506 | 3122 | 49 | 33 |
| 1.0e−07 | 2.5e−07 | 7.2e−09 | 5829 | 3524 | 51 | 36 |
| 1.0e−08 | 2.5e−08 | 7.2e−10 | 5811 | 3977 | 51 | 41 |
| 1.0e−09 | 2.5e−09 | 7.4e−11 | 6492 | 4502 | 57 | 46 |
| 1.0e−10 | 2.6e−10 | 3.4e−11 | 7367 | 5107 | 64 | 52 |

### 5.2.2. Stiff Detest Problem

We integrated the Stiff Detest problem D1 [10],

$$
\begin{aligned}
y_1' &= 0.2(y_2 - y_1) \\
y_2' &= 10y_1 - (60 - 0.125y_3)y_2 + 0.125y_3 \\
y_3' &= 1,
\end{aligned}
\tag{5.4}
$$

with

$$y(0) = (0, 0, 0)^T, \quad \text{for } t \in [0, 400].
\tag{5.5}$$

Here, we used the ITS(17) and IHO(8, 8) methods, Taylor series for validation, and a variable stepsize control with tolerances $10^{-6}, 10^{-7}, ..., 10^{-10}$.

With the IHO method, we computed tighter bounds with fewer stepsizes, than with the ITS method; see Table 3 and Figure 5. The reduction in the stepsize on the last step for the IHO method seen in Figure 5(d) is a result of our program reducing the stepsize to hit the endpoint exactly.

### Acknowledgement

The authors thank Dr. George Corliss for his many helpful comments and suggestions on an earlier draft of this paper and for his advice and support throughout the first author's Ph.D. research.

### References

1. Adams, E., Cordes, D., and Lohner, R.: Enclosure of Solutions of Ordinary Initial Value Problems and Applications, in: Adams, E., Ansorge, R., Großmann, Ch., and Roos, H. G. (eds), *Discretization in Differential Equations and Enclosures*, Akademie-Verlag, Berlin, 1987, pp. 9–28.
2. Alefeld, G. and Herzberger, J.: *Introduction to Interval Computations*, Academic Press, New York, 1983.
3. Bendsten, C. and Stauning, O.: *FADBAD, a Flexible C++ Package for Automatic Differentiation Using the Forward and Backward Methods*, Technical Report 1996-x5-94, Department of Mathematical Modelling, Technical University of Denmark, DK-2800, Lyngby, Denmark, 1996. FADBAD is available at http://www.imm.dtu.dk/fadbad.html.

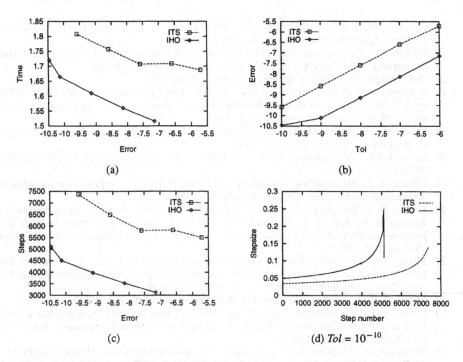

*Figure 5.* ITS(17) and IHO(8, 8) on (5.4)–(5.5), Taylor series for validation, variable stepsize control with $Tol = 10^{-6}, 10^{-7}, ..., 10^{-10}$.

4. Bendsten, C. and Stauning, O.: *TADIFF, a Flexible C++ Package for Automatic Differentiation Using Taylor Series*, Technical Report 1997-x5-94, Department of Mathematical Modelling, Technical University of Denmark, DK-2800, Lyngby, Denmark, 1997. TADIFF is available at http://www.imm.dtu.dk/fadbad.html.
5. Birkhoff, G. and Varga, R. S.: Discretization Errors for Well-Set Cauchy Problems: I, *J. Math. and Phys.* **44** (1965), pp. 1–23.
6. Corliss, G. F. and Rihm, R.: Validating an A Priori Enclosure Using High-Order Taylor Series, in: Alefeld, G. and Frommer, A. (eds), *Scientific Computing, Computer Arithmetic, and Validated Numerics*, Akademie Verlag, Berlin, 1996, pp. 228–238.
7. Darboux, G.: Sur les dèveloppements en série des fonctions d'une seule variable, *J. des Mathématique pures et appl.*, 3ème série, t. II, 1876, pp. 291–312.
8. Ehle, B. L.: On Padé Approximations to the Exponential Function and A-Stable Methods for the Numerical Solution of Initial Value Problems, *SIAM J. Math. Anal.* **4** (1973), pp. 671–680.
9. Eijgenraam, P.: *The Solution of Initial Value Problems Using Interval Arithmetic*, Mathematical Centre Tracts No. 144, Stichting Mathematisch Centrum, Amsterdam, 1981.
10. Enright, W. H., Hull, T. E., and Lindberg, B.: Comparing Numerical Methods for Stiff Systems of ODEs, *BIT* **15** (1975), pp. 10–48.
11. Griewank, A.: ODE Solving via Automatic Differentiation and Rational Prediction, in: Griffiths, D. F. and Watson, G. A. (eds), *Numerical Analysis 1995*, volume 344 of *Pitman Research Notes in Mathematics Series*, Addison-Wesley Longman Ltd, 1995.
12. Griewank, A., Corliss, G. F., Henneberger, P., Kirlinger, G., Potra, F. A., and Stetter, H. J.: High-Order Stiff ODE Solvers via Automatic Differentiation and Rational Prediction, in: *Lecture Notes in Comput. Sci.* **1196**, Springer, Berlin, 1997, pp. 114–125.
13. Griewank, A., Juedes, D., and Utke, J.: ADOL-C, a Package for the Automatic Differentiation of Algorithms Written in C/C++, *ACM Trans. Math. Software* **22** (2) (1996), pp. 131–167.

14. Haireŕ, E., Nørsett, S. P., and Wanner, G.: *Solving Ordinary Differential Equations I. Nonstiff Problems*, Springer-Verlag, 2nd revised edition, 1991.
15. Hermite, Ch.: Extrait d'une lettre de M. Ch. Hermite à M. Borchardt sur la formule d'interpolation de Lagrange, *J. de Crelle* **84** (70) (1878), Oeuvres, tome III, pp. 432–443.
16. Van Iwaarden, R.: *IADOL-C*, personal communications, 1997.
17. Knüppel, O.: PROFIL/BIAS—a Fast Interval Library, *Computing* **53** (3–4) (1994), pp. 277–287. PROFIL/BIAS is available at http://www.ti3.tu-harburg.de/Software/PROFIL/Profil.texinfo_1.html.
18. Krückeberg, F.: Ordinary Differential Equations, in: Hansen, E. (ed.), *Topics in Interval Analysis*, Clarendon Press, Oxford, 1969, pp. 91–97.
19. Lambert, J. D.: *Computational Methods in Ordinary Differential Equations*, John Wiley & Sons, 1977.
20. Lohner, R. J.: *Einschließung der Lösung gewöhnlicher Anfangs- und Randwertaufgaben und Anwendungen*, PhD thesis, Universität Karlsruhe, 1988.
21. Lohner, R. J.: Enclosing the Solutions of Ordinary Initial and Boundary Value Problems, in: Kaucher, E. W., Kulisch, U. W., and Ullrich, Ch. (eds), *Computer Arithmetic: Scientific Computation and Programming Languages*, Wiley-Teubner Series in Computer Science, Stuttgart, 1987, pp. 255–286.
22. Moore, R. E.: Automatic Local Coordinate Transformations to Reduce the Growth of Error Bounds in Interval Computation of Solutions of Ordinary Differential Equations, in: Rall, L. B. (ed.), *Error in Digital Computation, Vol. II*, Wiley, New York, 1965, pp. 103–140.
23. Moore, R. E.: *Interval Analysis*, Prentice-Hall, Englewood Cliffs, N.J., 1966.
24. Moore, R. E.: The Automatic Analysis and Control of Error in Digital Computation Based on the Use of Interval Numbers, in: Rall, L. B. (ed.), *Error in Digital Computation, Vol. I*, Wiley, New York, 1965, pp. 61–130.
25. Nedialkov, N. S., Jackson, K. R., and Corliss, G. F.: Validated Solutions of Initial Value Problems for Ordinary Differential Equations, *Applied Mathematics and Computation*, to appear. Available at http://www.cs.toronto.edu/NA/reports.html.
26. Nedialkov, N. S.: *Computing Rigorous Bounds on the Solution of an Initial Value Problem for an Ordinary Differential Equation*, PhD thesis, Department of Computer Science, University of Toronto, Toronto, Canada, M5S 3G4. Aailable at http://www.cs.toronto.edu/NA/reports.html.
27. Obreschkoff, N.: Neue Quadraturformeln, *Abh. Preuss. Akad. Wiss. Math. Nat. Kl.* **4** (1940).
28. Obreschkoff, N.: Sur le quadrature mecaniques, *Spisanie Bulgar. Akad. Nauk (Journal of the Bulgarian Academy of Sciences)* **65** (1942), pp. 191–289.
29. Ralston, A.: *A First Course in Numerical Analysis*, McGraw-Hill, New York, 2nd edition, 1978.
30. Rihm, R.: On a Class of Enclosure Methods for Initial Value Problems, *Computing* **53** (1994), pp. 369–377.
31. Rohn, J.: NP-Hardness Results for Linear Algebraic Problems with Interval Data, in: Herzberger, J. (ed.), *Topics in Validated Computations*, volume 5 of *Studies in Computational Mathematics*, North-Holland, Amsterdam, 1994, pp. 463–471.
32. Stauning, O.: *Automatic Validation of Numerical Solutions*, Technical Report IMM-PHD-1997-36, IMM, Lyngby, Denmark, 1997.
33. Varga, R. S.: On Higher Order Stable Implicit Methods for Solving Parabolic Differential Equations, *J. Math. and Phys.* **40** (1961), pp. 220–231.
34. Wanner, G.: *On the Integration of Stiff Differential Equations*, Technical report, Université de Genéve, Section de Mathematique, 1211 Genéve 24th, Suisse, 1976.
35. Wanner, G.: On the Integration of Stiff Differential Equations, in: *Proceedings of the Colloquium on Numerical Analysis*, volume 37 of *Internat. Ser. Numer. Math.*, Basel, Birkhäuser, 1977, pp. 209–226.

*Reliable Computing* **5**: 311–322, 1999.

# The Interval-Enhanced GNU Fortran Compiler

MICHAEL J. SCHULTE, VITALY ZELOV, and AHMET AKKAS
*EECS Dept., Lehigh University, Bethlehem, PA 18015, USA, e-mail: mschulte@eecs.lehigh.edu*

and

JAMES CRAIG BURLEY
*Free Software Foundation, 59 Temple Place, Suite 330, Boston, MA 02111, USA,*
*e-mail: burley@gnu.org*

(Received: 2 November 1998; accepted: 3 February 1999)

**Abstract.** Compiler support for intervals as intrinsic data types is essential for promoting the development and wide-spread use of interval software. It also plays an important role in encouraging the development of hardware support for interval arithmetic. This paper describes modifications made to the GNU Fortran Compiler to provide support for interval arithmetic. These modifications are based on a recently proposed Fortran 77 Interval Arithmetic Specification, which provides a standard for supporting interval arithmetic in Fortran. This paper also describes the design of the compiler's interval runtime libraries and the methodology used to test the compiler. The compiler and runtime libraries are designed to be portable to platforms that support the IEEE 754 floating point standard.

## 1. Introduction

Interval arithmetic provides an efficient method for performing operations on intervals of real numbers [19]. With interval arithmetic, each interval is represented by its lower and upper endpoints. More formally, an interval $X \equiv [\underline{x}, \overline{x}]$ is defined as

$$X \equiv [\underline{x}, \overline{x}] \equiv \{x \in \mathbb{R} \mid \underline{x} \leq x \leq \overline{x}\}.$$

For example, the interval $X = [1.23, 1.24]$ is a closed interval that includes all values greater than or equal to 1.23 and less than or equal to 1.24.

Although the concept of interval arithmetic is relatively straightforward, it provides a powerful mechanism for bounding the results of floating point computations. As demonstrated in [19] and [1], interval arithmetic provides a practical method for bounding errors in numerical computations including roundoff errors, approximation errors, and errors due to inexact inputs. In addition, interval arithmetic has been used to develop validated algorithms for solving problems in several areas. For example, interval arithmetic has been used to provide validated solutions in global optimization, finding roots of functions, solving systems of linear and non-linear equations, performing numerical differentiation and integration, and solving systems of ordinary and partial differential equations. These and other applications for interval arithmetic are presented in [1], [5], [9], [10], [20].

Because of its ability to provide validated solutions, several software tools have been developed to support interval arithmetic. These include interval arithmetic libraries, [12], [16], [18], language extensions with support for interval arithmetic [7], [17], [24], and interval application software [9], [13], [14]. Although these tools give programmers access to interval arithmetic, they do not conform to a specified standard.

As noted in [15], the lack of a specified standard for interval arithmetic has the following disadvantages:

- Resources are unnecessarily expended to redevelop tools for interval arithmetic.

- Diverse semantics in programming interval computations inhibit multi-person development of interval application software.

- Interval arithmetic packages written in high-level languages seldom take advantage of machine hardware, which may lead to poor runtime performance.

- Interval arithmetic packages designed on one platform may not be portable to other platforms.

Furthermore, current interval arithmetic software packages do not always guarantee containment. As noted in [22], several packages for interval arithmetic produce incorrect results due to incorrect handling of arithmetic exceptions. Some of these packages also depend on the underlying platform's built-in libraries when performing interval mathematical functions and interval input and output. Consequently, the correctness and tightness of the interval routines are dependent on the accuracy of the built-in libraries. This accuracy is often difficult to determine and can vary on different platforms [21]. Furthermore, previous software packages for interval arithmetic sometimes cause unexpected behavior. For example, with certain interval arithmetic libraries, the machine floating point rounding mode can be changed as a side-effect of performing interval arithmetic operations.

To overcome these difficulties, a standard for supporting interval arithmetic in Fortran was recently proposed [15]. This proposal specifies support for interval data types, interval arithmetic operations, interval relations, interval versions of mathematical functions, and interval I/O. The Fortran 77 Interval Arithmetic Specification builds upon [15] to provide a complete standard for supporting intervals in Fortran [4]. It adds several new operators and intrinsic functions, which are defined in [26]. It also defines a set of extended real intervals and their internal representation for IEEE 754 compliant processors. The set of extended real intervals is closed with respect to interval arithmetic operations and interval enclosures of real functions [25]. The Fortran 77 Interval Arithmetic Specification also gives algorithms for correct nonstop (i.e., without interrupts for arithmetic exceptions) handling of several interval intrinsics and operations, discusses interval expression optimization and mixed-mode evaluation, and gives several examples to illustrate the correct implementation of the specification.

To ease the burden of writing interval code and to help provide a standard for interval arithmetic, the GNU Fortran Compiler has been enhanced to support intervals as an intrinsic data type. The interval-enhanced GNU Fortran Compiler provides support for interval data types, constants, arithmetic operations, set operations, relations, special functions, mathematical functions, conversion functions, and I/O.

Support for interval arithmetic in the GNU Fortran Compiler is based on the Fortran 77 Interval Arithmetic Specification [4]. To support interval arithmetic on the GNU Fortran Compiler, the front-end of the compiler was modified and an extensive interval runtime library was developed. The interval-enhanced compiler and the routines in the interval runtime library guarantee containment, produce relatively sharp intervals, and are designed to be portable to platforms that support the IEEE 754 floating point standard. The arithmetic operations, interval special functions, and interval I/O produce minimum width intervals. Interval versions of mathematical functions can produce wider intervals.

The contents of the remainder of this paper is as follows. Section 2 gives an overview of the GNU Fortran Compiler. Section 3 describes the Fortran 77 Interval Arithmetic Specification and its implementation on the interval-enhanced GNU Fortran compiler. Section 4 discusses changes made to the compiler to provide support for interval arithmetic. Section 5 describes the interval runtime libraries that are used by the GNU Fortran compiler to support interval arithmetic. Section 6 discusses the method used for testing the compiler and the routines in the runtime library. Section 7 gives conclusions. A preliminary version of the interval-enhanced compiler is available from `http://www.eecs.lehigh.edu/~mschulte/compiler/code`.

## 2. The GNU Fortran Compiler

This section gives an overview of the GNU Fortran Compiler. The GNU Fortran Compiler, also known as g77, is a publically available Fortran compiler that is designed to run on a variety of platforms. To promote software reuse, it interfaces with other GNU compilers and tools, such as the GNU C compiler (gcc), the GNU C++ compiler (g++), and the GNU debugger (gdb).

Similar to other Fortran compilers, the g77 compiler translates Fortran programs to machine code. To perform this translation, the g77 compiler

1. Reads a program written in Fortran.
2. Checks the program for errors.
3. Translates the Fortran program to an intermediate form, which includes calls to the runtime libraries.
4. Runs the GNU code generator, which performs optimizations and converts the intermediate form to assembly language.
5. Runs the GNU assembler, which converts the assembly language to machine code.

The g77 compiler also has support for debugging, program linking, and error and warning diagnostics. Command line arguments can be used to specify various options including output control, Fortran dialect, warning messages, debugging support, optimization levels, preprocessing, directory locations, code generation, and environment variables. The GNU Fortran Compiler utilizes four main components: a modified version of the gcc compiler, the g77 interface, the libf2c runtime libraries, and the Fortran f771 compiler.

The gcc compiler is more than just a compiler for C. Based on command-line options and the names given for files on the command line, gcc determines which actions to perform, including preprocessing, compiling, assembling, and linking. In a GNU Fortran installation, gcc recognizes Fortran source files by names ending in .f or .F. For these files, it uses the f771 Fortran compiler to perform compilation.

The g77 interface is essentially just a front-end for the gcc compiler. Because of this, g77 can compile and link programs and source files written in other languages, such as C and C++. Fortran programmers will normally use g77 instead of gcc, because g77 knows how to specify the libraries that need to be linked with Fortran programs. Two libraries that are automatically linked are the Fortran library, libf2c, and the math library, lm.

The libf2c runtime library contains machine code needed to support capabilities of the Fortran language that are not directly provided by the machine code generated by the g77 compilation phase. This library includes procedures needed by Fortran programs while they are running. For example, while machine code generated by g77 performs additions, subtractions, and multiplications, it does not perform I/O or compute the trigonometric functions. Instead, Fortran statements that perform these operations are converted by the f771 compiler into function calls in the machine code. When run, the function calls in the machine code invoke functions in the libf2c runtime library.

The f771 compiler is a combination of two rather large pieces of code. One piece is called the GNU Back End (GBE), which knows how to generate fast code for a wide variety of processors. The same GBE is used by the C, C++, and Fortran compiler programs. The GBE also generates some warnings, such as those for references to undefined variables. A typical distribution of g77 contains patch files for the GBE that allow it to process Fortran code correctly and efficiently. The other piece of f771, called the Fortran Front End (FFE), is the majority of what is unique about GNU Fortran. The FFE knows how to interpret Fortran programs to determine what they are intending to do, and then communicate this knowledge to the GBE for the actual compilation of the programs. The FFE is responsible for diagnosing incorrect usage of the language in the programs it processes, and is also responsible for generating most of the warnings about questionable constructs.

## 3. Interval Arithmetic Enhancements to the GNU Fortran Compiler

The interval-enhanced version of the g77 compiler, referred to as g77-i, is based on the Fortran 77 Interval Arithmetic Specification. This section gives a summary of the interval-enhancements implemented in the g77-i compiler. Further details can be found in the Fortran 77 Interval Arithmetic Specification [4].

The g77-i compiler supports the set of extended real intervals defined in [4] and [25]. The set of extended real intervals includes the set of real intervals (i.e. intervals of the form $X \equiv [\underline{x}, \overline{x}]$, where $\underline{x}$ and $\overline{x}$ are real numbers and $\underline{x} \leq \overline{x}$), intervals with infinite endpoints, and the empty interval. The empty interval is an interval that contains no real numbers, and can be produced by inputting an empty interval, intersecting disjoint intervals, or evaluating an interval function strictly outside of its domain of definition. Results from [25] and [26] show that the set of extended real intervals is closed with respect to the interval arithmetic operations and interval enclosures of real functions.

The g77-i compiler represents intervals using their endpoints, which are either single precision or double precision IEEE 754 floating point numbers [2]. Empty intervals are represented by using a quiet not-a-number (NaN) for each interval endpoint. Double precision intervals, which are the default, are defined using either INTERVAL or INTERVAL*16. Single precision intervals are defined using INTERVAL*8. The number after the "*" corresponds to the total number of bytes used for both interval endpoints.

Interval constants may be specified using one of two formats: $[x]$ or $[\underline{x}, \overline{x}]$, where $x$, $\underline{x}$, and $\overline{x}$ are real or integer constants. An interval constant that uses the $[x]$ format is the same as the interval constant $[x, x]$. For example, the interval constants $[1.23]$ and $[1.23, 1.23]$ are equivalent. When an interval constant is converted, the internal interval contains the interval constant, regardless of the value of or number of digits in either interval endpoint. Invalid interval constants, which have a lower interval endpoint that is greater than the upper interval endpoint, result in a compile time error. When converting interval constants, the g77-i compiler always produces the minimum-width intervals that guarantee containment.

The interval arithmetic operations supported by g77-i include addition, subtraction, multiplication, and division. If either operand is the empty interval then the result is the empty interval. Unlike some interval arithmetic libraries, which can change the rounding mode as a side-effect of performing interval arithmetic operations, the interval arithmetic operations in g77-i save and then later restore the rounding mode. They also produce minimum-width intervals that guarantee containment. There are also two versions of the interval power operator: one computes the interval power of an interval base and the other computes the integer power of an interval base.

The interval set intrinsics include ISEMTPY, which returns TRUE if an interval is empty, and INF and SUP, which return the infimum and supremum of an interval, respectively. The interval set operations include .IX., which returns the intersection of two intervals, and .IH., which returns the hull of two intervals (i.e., the smallest interval that encloses both intervals). The intersection of two disjoint intervals returns the empty interval. There are also several special interval functions, which include MID, WID, MIG, MAG, ABS, MAX, MIN, NDIGITS, and INT. Definitions for each of these special functions are given in the Fortran 77 Interval Arithmetic Specification [4].

Several interval relation operations, which return logical values of TRUE or FALSE, are also supported by g77-i. These relation operations are classified as set relations, certainly relations, or possibly relations. The set relations treat intervals as sets of real numbers. The certainly relations are true if and only if the relation is true for every value in both intervals. The possibly relations are true if the relation is true for any value in both intervals.

For each real mathematical function in Fortran, a corresponding interval mathematical function is defined. Results from [25] and [26] show that the interval mathematical functions are closed with respect to the extended interval system. This is accomplished by intersecting the interval input arguments with the domain over which the function is defined. For example, SQRT([−2, 4]) returns [0, 2] and SQRT([−4, −2]) returns the empty interval. If any input argument to an interval mathematical function is the empty interval, then the result is the empty interval.

Three interval conversion functions, INTERVAL, DINTERVAL, and SINTERVAL, are defined to provide explicit conversion to interval types from integer, real, and interval types. These functions typically take either one or two integer or real arguments and produce an interval result. They can also take one interval argument and produce an interval result. The INTERVAL and DINTERVAL functions both produce double precision intervals, while the SINTERVAL function produces single precision intervals.

The g77-i compiler also supports formatted and list-directed interval I/O, as described in the Fortran 77 Interval Arithmetic Specification. With formatted interval I/O, the user specifies the total field width, $w$, the number of significant digits, $d$, and (optionally) the number of exponent digits $e$. The VE, VF, and VG edit descriptors support standard formatted interval I/O. These descriptors use Fortran's F, E, and G edit descriptors for the interval endpoints, while ensuring that the interval result is the sharpest interval that guarantees containment. The Y edit descriptor supports formatted single number interval I/O [23]. With single number interval I/O, intervals are input and output using a single decimal number. List-directed interval I/O allows intervals to be input and output without specifying the format. If the intervals are read from or written to a binary file, list-directed interval I/O avoids roundoff errors due to conversions. If intervals are read from or written to a text file, the compiler chooses suitable values for the output format, and roundoff errors due to conversions may occur.

Much of the syntax used in g77-i is similar to the ACRITH-XSC extensions to Fortran [24]. ACRITH-XSC is actually more powerful than g77-i because it provides additional features, such as dynamic arrays, operator overloading, dot product expressions, accurate vector and matrix computations, and complex intervals. The main benefits of g77-i are that it is publically available, it is portable to a wide range of platforms, it can be used in conjunction with other GNU compilers and tools, and it conforms to the Fortran 77 Interval Arithmetic Specification.

## 4. Modifying the GNU Fortran Compiler to Support Intervals

To support interval arithmetic, several modifications were made to g77. These modifications provide support for the new interval data types, interval constants, interval I/O and format statements, interval operations, and interval intrinsics.

For the new interval data types, diagnostic messages were added, internal Fortran front-end (FFE) representations of the types were created, routines for the internal GNU back-end (GBE) translations of the types were developed, and initializations of internal FFE and GBE interval type information were provided. Also, code to recognize the INTERVAL keyword as a statement or modifier and code to assign internal interval type information based on parsed keywords was added.

To support interval constants, diagnostic messages for interval constants, the ability to recognize "[" and "]" as lexemes, and code to parse such constants were added. Also, the internal representations and a new expression-parsing context for interval constants were developed. In addition, routines to build intermediate representation entries and to translate FFE interval constants to their GBE form were designed. There were also several features added to support interval constant arrays including the ability to extract an element of a constant array, create new constant arrays, copy between constant arrays, copy between a constant array and a constant scalar, and put a value in an element of a constant array.

To support interval I/O, new diagnostic messages for the interval format specifiers were added, and code to parse the format specifiers and to convert the format specifiers to their runtime form was developed. It was also necessary to provide data type information to the I/O runtime library to allow the I/O routines to distinguish between real, complex, and interval data.

To support interval arithmetic operations, new runtime routines were defined, and the code that determines whether the operands are unary or binary operators was modified to support intervals. Also, code that translates the FFE representations of such operations to their GBE counterparts was added, and the ability to define runtime routines as operating on or returning interval values was provided. These modifications allow the FFE to translate interval operations in Fortran to calls to the interval runtime library.

For interval intrinsics, the specifications of several of the intrinsics' interfaces were modified to allow intervals. Code to compile intrinsic operations to appropriate GBE representations was also added. Additionally, the ability to define an intrinsic as operating on, or returning, interval values was provided. Similar to the interval arithmetic operations, the interval intrinsics are implemented by converting the Fortran intrinsics to calls to the interval runtime library.

## 5. Interval Runtime Libraries

Much of the work to add intervals to g77 focused on developing efficient interval runtime library routines. These routines were added to the existing g77 runtime libraries. The g77 runtime directory is composed of three sub-directories: libF77, libI77, and libU77. The libF77 directory contains the non-I/0 support routines needed by the GNU Fortran Compiler, such as routines for mathematical and string functions. The libI77 directory contains support routines for Fortran I/O and file management. The libU77 directory contains support routines for UNIX system function calls. All g77 runtime library routines are written in C.

To incorporate the routines for interval arithmetic into the existing g77 runtime libraries, the non-I/O routines were added to the libF77 directory and the I/O routines were added to the libI77 directory. Support routines for changing rounding modes, using floating point numbers, and initializing intervals are accessible from both of these directories via symbolic links. Names for the interval routines in the runtime libraries are derived from the names given in the Fortran 77 Interval Arithmetic Specification [4], and function definitions are designed to be similar to those currently used in the runtime libraries.

The routines in the runtime libraries are designed to support IEEE single and double precision intervals in either big endian or little endian format. They also are designed to run on several different platforms including Sun SPARC, DEC Alpha, SGI MIPS, IBM RS/6000, Intel x86 and Motorola 68k. So far, they have been tested primarily on Sun Ultras running Solaris and Intel Pentiums running Linux. Later, they will be ported to a wider range of platforms. To provide tight intervals, the interval arithmetic routines use the underlying hardware's built-in IEEE rounding modes to produce minimum-width intervals. Changing the rounding modes is performed through platform specific function calls.

Header files and support routines were developed to assist in adding interval arithmetic to the GNU Fortran Compiler. These files include definitions for the interval data types (interval.h), definitions for various floating point values and parameters (fp.h), and support for directed rounding (round.h). Definitions and routines defined in these files are used internally by the other routines in the interval runtime libraries.

The non-I/O interval routines include those that support interval arithmetic operations, set operations, relations, and special and mathematical functions. There are two versions of each routine: one for single precision intervals and a second for double precision intervals. In the function definitions, the names of the single precision routines are prefixed with a $u$ and the names of the double precision routines are prefixed with a $v$. For example, single and double precision interval addition routines are called u_add and v_add, respectively.

As an example of the interval arithmetic routines, code for the v_add subroutine is shown in Figure 1. The interval addition routine takes as input arguments pointers to the intervals, a and b, and outputs the interval resx. Initially, the current IEEE

```
#include "round.h"
#include "interval.h"
interval v_add(interval *a, interval *b) {
    interval resx;                      /* Interval sum */
    RND_MODE_TYPE round_mode;           /* Variable for IEEE
                                            rounding mode */
    round_mode = GET_RND_SET_UP();      /* Save rounding mode
                                            and set to round up*/
    resx.inf = -(-a->inf - b->inf);     /* Compute lower endpoint
                                            of interval */
    resx.sup = a->sup + b->sup;         /* Compute upper endpoint
                                            of interval */
    SET_ROUND(round_mode);              /* Restore the IEEE
                                            rounding mode */
    return resx;                        /* Return the result */
}
```

*Figure 1.* Runtime routine for interval addition.

rounding mode is saved and the rounding mode is set to round towards plus infinity. Next, the interval endpoints of the sum are computed. At the end of the routine, the rounding mode is restored to its original value and the result is returned.

To reduce the number of rounding mode changes, sign symmetry is employed when performing interval arithmetic operations [22]. With this technique, both interval endpoints are computed with the rounding mode set to round towards positive infinity. To compute the lower interval endpoint, the additive inverse of the lower interval endpoint is computed and then negated. The upper interval endpoint is computed in the usual fashion. On most machines, in which changing the rounding mode takes several cycles and stalls the floating point pipeline, the use of sign symmetry results in an overall performance improvement.

Routines for the interval mathematical functions are based on the Fast Interval Library [11]. This library is written in ANSI-C and uses fast table look-up algorithms to evaluate the interval mathematical functions. To comply with the Fortran 77 Interval Arithmetic Specification, these routine were modified to support empty intervals, intervals with infinite endpoints, and intervals over which the function contains singularities.

The interval I/O routines are designed to interface directly with the existing I/O routines in the libI77 runtime directory. This allows them to reuse much of the code that is already available for I/O of real and complex numbers in Fortran. The routines developed for interval I/O include several interval Fortran specific routines that are designed for providing interval I/O in the VF, VE, VG, and Y formats.

In addition, there are a number of support routines for performing conversion between character strings and intervals. The support routines were developed based on routines for floating point input [8] and floating point output [3]. These routines were first modified to provide correctly rounded input and output, and then used to provide proper input and output of unformatted and formatted intervals.

## 6. Testing the Runtime Libraries and Compiler

One of the benefits of interval arithmetic algorithms is that they supply validated solutions. Therefore, it is extremely important that the underlying software used to perform the interval algorithms is reliable. To help ensure this, extensive testing of the interval compiler and runtime libraries has been performed. For each interval operation or intrinsic, interactive tests, special case tests, identity tests and random tests were performed.

Interactive tests are typically run on the routines just after they are developed. Interactive test routines allow the user to specify the input values to the interval routine. The test routine reads in the values, performs the specified function, and outputs the result. Interactive test routines are designed to catch more obvious bugs and to give the user the ability to fairly quickly test potential problem cases.

After performing interactive tests, special cases are tested to ensure that they produce correct results. This is typically done by reading input intervals from one file, writing the results to a second file, and then performing manual checks to ensure that correct results are produced. Examples of test cases include testing division by zero, multiplying very large or very small intervals together, taking the midpoint of an infinite interval, and taking the intersection of disjoint intervals. Naturally, the types of special case tests performed depend on the function being tested.

After ensuring that special cases work correctly, a large number of identity tests are run. Examples of identity tests include $X + Y = Y + X$, $X + X = 2 * X$, and $X - Y = X + -(Y)$. Several identity tests were supplied by Michael Parks of Sun Microsystems. For many of the identities, outward rounding prevents identical results from being produced. In these cases, it is often required to allow the results to differ by a unit in the last place and handle cases which lead to overflow or underflow separately. For example, $(X + Y) + Z = X + (Y + Z)$ does not always produce identical results for the left and right hand sides of the equation. Identity tests are run for both random inputs and values input from a user-specified file.

After running identity tests, a large number of random tests are performed. The random test routines are designed to test the algorithm for a very large number of pseudo-random inputs. To determine if the correct results are produced, results from these routines are compared with results from other interval software packages.

After running large numbers of random tests, the g77-i compiler was used to develop complete interval programs. Examples of interval programs that have been developed using g77-i include programs for global optimization, solving systems of linear equations, finding roots of functions, and performing vector and matrix computations. The g77-i compiler has also been used to compile interval code that is used by the GlobSol software package [6]. Replacing calls to the intlibf90 interval software package by calls to code compiled using g77-i resulted in a significant performance improvement and sharper interval results.

## 7. Conclusions

A preliminary version of the g77-i compiler has been completed and tested. This version of the compiler satisfies the requirements given in the Fortran 77 Interval Arithmetic Specification and provides an efficient tool for developing interval software in Fortran. A fast interval runtime library is used to improve the speed of interval code while guaranteeing containment and providing sharp rounding of the interval endpoints. The g77-i compiler has been extensively tested and has been used to develop several interval application programs. Similar interval-enhancements are being made to the GNU C and C++ compilers, and to versions of Sun Microsystem's Fortran compilers.

## Acknowledgements

The authors are thankful to G. William Walster and Dimitry Chiriaev for their comments and advice on the design of the interval-enhanced GNU Fortran compiler. They are also grateful to Sun Microsystems for funding this project, and to the anonymous reviewers for their valuable comments. This material is based on work supported by the National Science Foundation under Grant No. MIP-9703421.

## References

1. Alefeld, G. and Herzberger, J.: *Introduction to Interval Computations*, Academic Press, 1983.
2. *ANSI/IEEE 754-1985 Standard for Binary Floating-Point Arithmetic*, Institute of Electrical and Electronics Engineers, New York, 1985.
3. Burger, R. G. and Dybvig, R. K.: Printing Floating Point Numbers Quickly and Accurately, *Sigplan Notices* **31** (5) (1996) pp. 108–116.
4. Chiriaev, D. and Walster, G. W.: *Fortran 77 Interval Arithmetic Specification*, 1997, available at http://www.mscs.mu.edu/~globsol/Papers/spec.ps.
5. Corliss, G. F.: Industrial Applications of Interval Techniques, in: Ullrich, C. (ed.), *Computer Arithmetic and Self-Validating Numerical Methods*, Academic Press, 1990, pp. 91–113.
6. Corliss, G. F.: Rigorous Global Search: Industrial Applications, in: Csendes, T. (ed.), *Proceedings of the International Symposium on Scientific Computing, Computer Arithmetic, and Validated Numerics*, Budapest, Hungary, September, 1998.
7. Ely, J. S.: The VPI Software Package for Variable Precision Interval Arithmetic, *Interval Computations* 2 (1993), pp. 135–153.
8. Gay, D. M.: *Correctly Rounded Binary-Decimal and Decimal-Binary Conversions*, Numerical Analysis Manuscript 90-10, AT&T Bell Laboratories, 1990.
9. Hammer, R., Hocks, M., Kulisch, U., and Ratz, D.: *C++ Toolbox for Verified Computing*, Springer-Verlag, 1995.
10. Hansen, E.: *Global Optimization Using Interval Analysis*, Marcel Dekker, 1992.
11. Hofschuster, W. and Kraemer, W.: *A Fast Interval Library*, 1997, available at ftp://iamk4515.mathematik.uni-karlsruhe.de/pub/iwrmm/software.

12. Kearfott, R. B.: A FORTRAN 90 Environment for Research and Prototyping of Enclosure Algo-
    rithms for Nonlinear Equations and Global Optimization, *ACM Transactions on Mathematical
    Software* **21** (1) (1995), pp. 63–78.
13. Kearfott, R. B.: *Rigorous Global Search: Continuous Problems*, Kluwer Academic Publishers,
    Dordrecht, 1997.
14. Kearfott, R. B. and Novoa, M.: INTBIS, A Portable Interval Newton Bisection Package, *ACM
    Transactions on Mathematical Software* **16** (1990), pp. 152–157.
15. Kearfott, R. B. et al.: *A Specific Proposal for Interval Arithmetic in Fortran*, 1996, available at
    http://interval.usl.edu/F90/f96-pro.asc.
16. Kearfott, R. B. et al.: Algorithm 737: INTLIB: A Portable Fortran 77 Interval Standard Function
    Library, *ACM Transactions on Mathematical Software* **20** (1994), pp. 447–459.
17. Klatte, R. et al.: *C-XSC: A C++ Class Library for Extended Scientific Computing*, Springer-
    Verlag, 1993.
18. Knüppel, O.: PROFIL/BIAS—A Fast Interval Library, *Computing* **53** (1994), pp. 277–288.
19. Moore, R. E.: *Interval Analysis*, Prentice Hall, 1966.
20. Moore, R. E.: *Methods and Applications of Interval Analysis (SIAM Studies in Applied Mathe-
    matics)*, SIAM, 1979.
21. Priest, D.: Fast Table-Driven Algorithms for Interval Elementary Functions, *Proceedings of the
    13th Symposium on Computer Arithmetic*, 1997, pp. 168–174.
22. Priest, D.: *Handling IEEE 754 Invalid Operation Exceptions in Real Interval Arithmetic*, Manu-
    script, 1997.
23. Schulte, M. J., Zelov, V. A., Walster, G. W., and Chiriaev, D.: Single-Number Interval I/O,
    in: Csendes, T. (ed.), *Proceedings of the International Symposium on Scientific Computing,
    Computer Arithmetic, and Validated Numerics*, Budapest, Hungary, September, 1998.
24. Walter, W. V.: ACRITH-XSC: A Fortran-like Language for Verified Scientific Computing, in:
    Adams, E. and Kulisch, U. (eds), *Scientific Computing with Automatic Result Verification*, Aca-
    demic Press, 1993, pp. 45–70.
25. Walster, G. W.: *The Extended Real Interval System*, 1997, available at
    http://www.mscs.mu.edu/~globsol/Papers/extended_intervals.ps.
26. Walster, G. W. and Hansen, E. R.: *Interval Algebra, Composite Functions and
    Dependence in Compilers*, submitted to *Reliable Computing*, available at
    http://www.mscs.mu.edu/~globsol/Papers/composite.ps.

*Reliable Computing* **5**: 323–335, 1999.

# Outer Estimation of Generalized Solution Sets to Interval Linear Systems

SERGEY P. SHARY

*Institute of Computational Technologies, 6 Lavrentiev avenue, 630090 Novosibirsk, Russia,*
*e-mail: shary@ict.nsc.ru*

(Received: 2 November 1998; accepted: 11 December 1998)

**Abstract.** The work advances a numerical technique for computing enclosures of *generalized AE-solution sets* to interval linear systems of equations. We develop an approach (called *algebraic*) in which the outer estimation problem reduces to a problem of computing algebraic solutions of an auxiliary interval equation in Kaucher complete interval arithmetic.

## 1. Introduction

In our work, we will consider *generalized solution sets* for interval algebraic systems that naturally arise when interval parameters of a system express different kinds of uncertainty (ambiguity). We would like to remind that, basically, the interval data uncertainty and/or ambiguity can be understood in two ways, in accordance with the two-fold interpretation of the intervals.

In real life problems, one is hardly interested in intervals on their own, as integral and undivided objects, with no further internal structure. In most cases, we only use an interval **v** in connection with a property (let us denote it by $P$) that can be fulfilled or not for its point members. Under the circumstances, the following different situations may occur:

either the property $P(v)$ considered (that may be a point equation, inequality, etc.) holds for *all* members $v$ from the given interval **v**,

or the property $P(v)$ holds only for *some* members $v$ from the interval **v**, not necessarily all (maybe, only for one value).

In formal writing, this distinction is manifested in using the logical quantifiers— either the universal quantifier $\forall$ or the existential quantifier $\exists$:

- in the first case, we write "$(\forall v \in \mathbf{v})\ P(v)$" and shall speak of $\forall$-*type (A-type) of uncertainty,*

- in the second case, we write "$(\exists v \in \mathbf{v})\ P(v)$" and are going to speak of $\exists$-*type (E-type) of uncertainty*

(see also [13], [15], [17], [20]).

The above difference between the two uncertainty (ambiguity) types should be taken into account when strictly defining solutions and solution sets to interval equations, inequalities, etc. For instance, the most general definition of the solution set to the interval system of linear equations

$$Ax = b, \tag{1.1}$$

with an interval $m \times n$-matrix $A = (a_{ij})$ and an interval right-hand side $m$-vector $b = (b_i)$, has the form

$$\{x \in \mathbb{R}^n \mid$$
$$(Q_1 v_{\pi_1} \in \mathbf{v}_{\pi_1})(Q_2 v_{\pi_2} \in \mathbf{v}_{\pi_2}) \cdots (Q_{mn+m} v_{\pi_{mn+m}} \in \mathbf{v}_{\pi_{mn+m}}) \ (Ax = b)\}, \tag{1.2}$$

where

$$Q_1, Q_2, ..., Q_{mn+m}$$
  are the logical quantifiers $\forall$ or $\exists$,

$$(v_1, v_2, ..., v_{mn+m}) := (a_{11}, ..., a_{mn}, b_1, ..., b_m) \in \mathbb{R}^{mn+m}$$
  is the aggregated (compound) parameter vector of the system
  of equations considered,

$$(\mathbf{v}_1, \mathbf{v}_2, ..., \mathbf{v}_{mn+m}) := (a_{11}, ..., a_{mn}, b_1, ..., b_m) \in \mathbb{IR}^{mn+m}$$
  is the aggregated vector of the intervals of the possible values
  of these parameters,

$$(\pi_1, \pi_2, ..., \pi_{mn+m})$$
  is a permutation of the integers $1, 2, ..., mn + m$.

DEFINITION 1.1. The sets of the form (1.2) will be referred to as *generalized solution sets* to the interval system of equations $Ax = b$.

DEFINITION 1.2. The logical formula written out after the vertical line in the definition of the set (1.2), which determines a characteristic property of the points of this set, will be called *selecting predicate* of the corresponding solution set (1.2) to the interval system of equations.

Definition 1.1 is very general. One can easily calculate, for example, that the number of the solution sets it comprehends far much exceeds even $2^{mn+m}$. Such a great variety is, in particular, due to the fact that in logical formulas (the selecting predicates of the solution sets among them) the occurrences of the different quantifiers cannot be permuted with each other [8].

The generalized solution sets to interval equations and inequalities naturally come into being in operations research and decision making, they have interesting and significant applications. In our work, we shall not treat the solution sets of the most general form (1.2), with arbitrarily combined quantifiers at the interval parameters, but confine ourselves only to such solution sets of the interval equations for which the selecting predicate has *all the occurrences of the universal quantifier* $\forall$ *prior to the occurrences of the existential quantifier* $\exists$. To put it differently, we consider only the solution sets whose selecting predicate has AE-*form*. When interpreting in terms of systems analysis they simulate one-stage "perturbation-control" action on a system [13], [15].

DEFINITION 1.3. For the interval systems of equations, the generalized solution sets for which the selecting predicate has AE-form will be termed *AE-solution sets* (or *sets of AE-solutions*).

Such is, for example, *tolerable solution set**

$$\Sigma_{tol}(\mathbf{A}, \mathbf{b}) = \{x \in \mathbb{R}^n \mid (\forall a_{11} \in \mathbf{a}_{11})(\forall a_{12} \in \mathbf{a}_{12}) \cdots (\forall a_{mn} \in \mathbf{a}_{mn})$$
$$(\exists b_1 \in \mathbf{b}_1)(\exists b_2 \in \mathbf{b}_2) \cdots (\exists b_m \in \mathbf{b}_m)(Ax = b)\},$$

which corresponds to the case when all the entries of the matrix **A** have A-uncertainty and all the elements of the vector b have E-uncertainty. Usually, it is written in the following form

$$\Sigma_{tol}(\mathbf{A}, \mathbf{b}) = \{x \in \mathbb{R}^n \mid (\forall A \in \mathbf{A})(\exists b \in \mathbf{b})(Ax = b)\},$$
$$= \{x \in \mathbb{R}^n \mid (\forall A \in \mathbf{A})(Ax \in \mathbf{b})\}$$
$$= \{x \in \mathbb{R}^n \mid (\mathbf{A}x \subseteq \mathbf{b})\}.$$

Another example, which is also subsumed under Definition 1.3, is *united solution set* of interval systems of equation, i.e., the set of solutions to all point systems with the coefficients from given intervals. For the interval linear system (1.1), it is strictly defined as

$$\Sigma_{uni}(\mathbf{A}, \mathbf{b}) = \{x \in \mathbb{R}^n \mid (\exists a_{11} \in \mathbf{a}_{11})(\exists a_{12} \in \mathbf{a}_{12}) \cdots (\exists a_{mn} \in \mathbf{a}_{mn})$$
$$(\exists b_1 \in \mathbf{b}_1)(\exists b_2 \in \mathbf{b}_2) \cdots (\exists b_m \in \mathbf{b}_m)(Ax = b)\}$$

$$= \{x \in \mathbb{R}^n \mid (\exists A \in \mathbf{A})(\exists b \in \mathbf{b})(Ax = b)\}$$
$$= \{x \in \mathbb{R}^n \mid (\mathbf{A}x \cap \mathbf{b} \neq \emptyset)\}.$$

## 2. Quantifier Formalization

In this section we consider, for the AE-solution sets, various possible ways of describing the uncertainty types distribution with respect to the interval parameters of the system.

---

* One can find surveys of the related results in [7], [10], [19]. For dynamic systems, an analog of this solution set is the *set of viable trajectories*, while the mathematical problem statement that gives rise to it is nothing but the *viability problem*.

1. As far as the order of the quantifiers is fixed, the simplest of such ways is to directly point out which quantifier is applied to this or that element of the interval system. Namely, let us introduce an $m \times n$-matrix $\alpha = (\alpha_{ij})$ and an $m$-vector $\beta = (\beta_i)$ made up of the logical quantifiers and such that

$$\alpha_{ij} := \begin{cases} \forall, & \text{if } \mathbf{a}_{ij} \text{ has A-uncertainty,} \\ \exists, & \text{if } \mathbf{a}_{ij} \text{ has E-uncertainty,} \end{cases}$$

$$\beta_i := \begin{cases} \forall, & \text{if } \mathbf{b}_i \text{ has A-uncertainty,} \\ \exists, & \text{if } \mathbf{b}_i \text{ has E-uncertainty.} \end{cases}$$

Specifying $\alpha$ and $\beta$, along with the interval system itself, completely determines the corresponding AE-solution set.

2. Another way to represent the uncertainty types corresponding to the elements of the interval linear system (1.1) is to trace out partitions of the index sets of both the entries of the matrix $\mathbf{A}$ and components of the right-hand side $\mathbf{b}$. More precisely, let the entire set of the index pairs $(i, j)$ of the entries $a_{ij}$, that is, the set

$$\{(1, 1), (1, 2), \dots, (1, n), (2, 1), (2, 2), \dots, (2, n),$$
$$\dots, (m, 1), (m, 2), \dots, (m, n)\},$$

be divided into two nonintersecting parts $\hat{\Omega} := \{\hat{\omega}_1, \dots, \hat{\omega}_p\}$ and $\check{\Omega} := \{\check{\omega}_1, \dots, \check{\omega}_q\}$, $p + q = mn$, such that

$a_{ij}$ is of the interval A-uncertainty for $(i, j) \in \hat{\Omega}$,

$a_{ij}$ is of the interval E-uncertainty for $(i, j) \in \check{\Omega}$.

Similarly, we introduce nonintersecting sets of the integer indices $\hat{\Theta} := \{\hat{\vartheta}_1, \dots, \hat{\vartheta}_s\}$ and $\check{\Theta} := \{\check{\vartheta}_1, \dots, \check{\vartheta}_t\}$, $\hat{\Theta} \cup \check{\Theta} = \{1, 2, \dots, m\}$, such that, in the right-hand side vector,

$b_i$ is of the interval A-uncertainty for $i \in \hat{\Theta}$,

$b_i$ is of the interval E-uncertainty for $i \in \check{\Theta}$.

Also, we allow the natural possibility for some of the sets $\hat{\Omega}$, $\check{\Omega}$, $\hat{\Theta}$, $\check{\Theta}$ to be empty. It is evident that

$$\alpha_{ij} = \begin{cases} \forall, & \text{if } (i, j) \in \hat{\Omega}, \\ \exists, & \text{if } (i, j) \in \check{\Omega}, \end{cases} \qquad \beta_i = \begin{cases} \forall, & \text{if } i \in \hat{\Theta}, \\ \exists, & \text{if } i \in \check{\Theta}, \end{cases}$$

and, again, determining $\hat{\Omega}$, $\check{\Omega}$, $\hat{\Theta}$, $\check{\Theta}$ results in a complete specification of an AE-solution set to the interval linear system (1.1).

3. The third way to describe the uncertainty types distribution for an interval linear system is to fix disjoint decompositions of both the interval matrix of the system and its right-hand side. Namely, we define interval matrices $\mathbf{A}^\forall = (\mathbf{a}_{ij}^\forall)$ and $\mathbf{A}^\exists = (\mathbf{a}_{ij}^\exists)$ and interval vectors $\mathbf{b}^\forall = (\mathbf{b}_i^\forall)$ and $\mathbf{b}^\exists = (\mathbf{b}_i^\exists)$, of the same sizes

as **A** and **b**, as follows:

$$\mathbf{a}_{ij}^{\vee} := \begin{cases} \mathbf{a}_{ij}, & \text{if } \alpha_{ij} = \forall, \\ 0, & \text{otherwise,} \end{cases} \qquad \mathbf{a}_{ij}^{\exists} := \begin{cases} \mathbf{a}_{ij}, & \text{if } \alpha_{ij} = \exists, \\ 0, & \text{otherwise,} \end{cases} \quad (2.1)$$

$$\mathbf{b}_{i}^{\vee} := \begin{cases} \mathbf{b}_{i}, & \text{if } \beta_{i} = \forall, \\ 0, & \text{otherwise,} \end{cases} \qquad \mathbf{b}_{i}^{\exists} := \begin{cases} \mathbf{b}_{i}, & \text{if } \beta_{i} = \exists, \\ 0, & \text{otherwise.} \end{cases} \quad (2.2)$$

Thus

$$\mathbf{A} = \mathbf{A}^{\vee} + \mathbf{A}^{\exists}, \qquad \mathbf{a}_{ij}^{\vee} \cdot \mathbf{a}_{ij}^{\exists} = 0,$$
$$\mathbf{b} = \mathbf{b}^{\vee} + \mathbf{b}^{\exists}, \qquad \mathbf{b}_{i}^{\vee} \cdot \mathbf{b}_{i}^{\exists} = 0$$

for all $i, j$. The matrix $\mathbf{A}^{\vee}$ and vector $\mathbf{b}^{\vee}$ concentrate all the interval elements of the system that corresponds to the A-uncertainty, while the matrix $\mathbf{A}^{\exists}$ and vector $\mathbf{b}^{\exists}$ stores all the elements that correspond to the interval E-uncertainty.

It should be stressed that the three groups of the objects considered which arise in connection with an AE-solution set of an interval linear system (1.1), namely

1) the quantifier matrix $\alpha$ and vector $\beta$,
2) decompositions of the index sets of the matrix **A** and of the right-hand side vector **b** to the nonintersecting subsets $\hat{\Omega}, \check{\Omega}, \hat{\Theta}, \check{\Theta}$,
3) disjoint decompositions of the interval matrix $\mathbf{A} = \mathbf{A}^{\vee} + \mathbf{A}^{\exists}$ and of the right-hand side vector $\mathbf{b} = \mathbf{b}^{\vee} + \mathbf{b}^{\exists}$,

are in a one-to-one correspondence, so that pointing out any one item of the above triple immediately determines the other two. We will extensively use all three descriptions and change any one for another without special explanations.

Summarizing, we can give the following

DEFINITION 2.1. Let us, for an interval linear system $\mathbf{A}x = \mathbf{b}$, be given a quantifier $m \times n$-matrix $\alpha$ and an $m$-vector $\beta$ as well as the associated decompositions of the index sets of the matrix **A** and vector **b** to nonintersecting subsets $\hat{\Omega} = \{\hat{\omega}_1, ..., \hat{\omega}_p\}$ and $\check{\Omega} = \{\check{\omega}_1, ..., \check{\omega}_q\}, \hat{\Theta} = \{\hat{\vartheta}_1, ..., \hat{\vartheta}_s\}$ and $\check{\Theta} = \{\check{\vartheta}_1, ..., \check{\vartheta}_t\}, p + q = mn, s + t = m$, and disjoint decompositions $\mathbf{A} = \mathbf{A}^{\vee} + \mathbf{A}^{\exists}$ and $\mathbf{b} = \mathbf{b}^{\vee} + \mathbf{b}^{\exists}$.

*AE-solution set of the type $\alpha\beta$* to the interval linear system $\mathbf{A}x = \mathbf{b}$ is the set

$$\Sigma_{\alpha\beta}(\mathbf{A}, \mathbf{b}) := \{x \in \mathbb{R}^n \mid$$
$$(\forall a_{\hat{\omega}_1} \in \mathbf{a}_{\hat{\omega}_1}) \cdots (\forall a_{\hat{\omega}_p} \in \mathbf{a}_{\hat{\omega}_p}) (\forall b_{\hat{\vartheta}_1} \in \mathbf{b}_{\hat{\vartheta}_1}) \cdots (\forall b_{\hat{\vartheta}_s} \in \mathbf{b}_{\hat{\vartheta}_s})$$
$$(\exists a_{\check{\omega}_1} \in \mathbf{a}_{\check{\omega}_1}) \cdots (\exists a_{\check{\omega}_q} \in \mathbf{a}_{\check{\omega}_q}) (\exists b_{\check{\vartheta}_1} \in \mathbf{b}_{\check{\vartheta}_1}) \cdots (\exists b_{\check{\vartheta}_t} \in \mathbf{b}_{\check{\vartheta}_t}) \quad (2.3)$$
$$(Ax = b)\}$$

or, which is equivalent, the set

$$\Sigma_{\alpha\beta}(\mathbf{A}, \mathbf{b}) := \{x \in \mathbb{R}^n \mid$$
$$(\forall \hat{A} \in \mathbf{A}^{\vee})(\forall \hat{b} \in \mathbf{b}^{\vee})(\exists \check{A} \in \mathbf{A}^{\exists})(\exists \check{b} \in \mathbf{b}^{\exists})((\hat{A} + \check{A})x = \hat{b} + \check{b})\}.$$

## 3. Outer Estimation Problem

The intersections of the AE-solution sets to interval linear systems with each orthant of the space $\mathbb{R}^n$ are easily proved to be convex polyhedral sets (see [15]). They are defined by systems of linear inequalities whose coefficients are the endpoints of the interval elements of the system (1.1). In principle, one could give a direct description of an AE-solution set by writing out the equations of all its bounding hyperplanes in each orthant, etc. But in general the complexity of such a process may grow not slower than the total number of orthants, i.e., exponentially with the dimension of the space $\mathbb{R}^n$. The direct explicit description of the solution sets becomes, as a result, extremely difficult, tedious, and practically even useless as the dimension of the system under consideration increases*.

On the other hand, a full description of the solution set usually is not even necessary in real-life situations. It suffices to change the exact solution set for some approximation (estimate) of it which is sufficient for practical purposes. For example, viability analysis and some system identification problems require *inner* estimation of the solution sets to interval equations, that is, computing simple subsets of the solution sets (see e.g. [21]). Alternatively, when analyzing the parametric sensitivity of a control system, one is often required to know guaranteed estimates of the state set within which our compensating control actions are able to hold the system in spite of the presence of uncontrolled perturbations. This is the case when *outer* estimates are needed, and the corresponding problem is usually formulated as follows:

> Find (quick and as sharp as possible) outer coordinate
> estimates of the solution set $\Sigma_{\alpha\beta}(\mathbf{A}, \mathbf{b})$ or, another way,
> evaluate $\inf\{x_k \mid x \in \Sigma_{\alpha\beta}(\mathbf{A}, \mathbf{b})\}$ from below and
> $\sup\{x_k \mid x \in \Sigma_{\alpha\beta}(\mathbf{A}, \mathbf{b})\}$ from above, $k = 1, \ldots, n$.
>
> $(3.1)$

In point of fact, the problem statement (3.1) prescribes seeking a box—rectangular parallelotope with the axis-aligned faces—that contains the solution set. The boxes are geometrical images of the interval vectors, so that we shall term a box enclosing the solution set as an *outer interval estimate* of this solution set. To sum up, it is convenient to reformulate the problem (3.1) in the following purely interval form:

> Find (quick and as sharp as possible) an outer
> interval estimate of the solution set $\Sigma_{\alpha\beta}(\mathbf{A}, \mathbf{b})$
> to a given interval linear system $\mathbf{A}x = \mathbf{b}$.

---

* Lakeyev managed to prove recently [9] that the complexity of recognition whether an AE-solution set to a given interval linear system is empty or not is NP-hard, that is, computationally intractable problem, provided that sufficiently many entries of $\mathbf{A}$ have the interval E-uncertainty.

The above problem is the main object under study in the present work. More precisely, we aim at developing a numerical technique for outer interval estimation of the generalized AE-solution sets to the interval systems (1.1). For simplicity, in the rest of the paper we consider only the interval linear systems $Ax = b$ with the square $n \times n$-matrices.

In the theory that we are presenting below, *Kaucher complete interval arithmetic* $\mathbb{IR}$ plays a crucial role. This arithmetic is a natural completion of the classical interval arithmetic $\mathbb{IR}$, so that $\mathbb{IR} \subset \mathbb{IR}$. The distinctive feature of the arithmetic $\mathbb{IR}$ is the presence of *improper* intervals $[\underline{x}, \overline{x}]$, $\underline{x} > \overline{x}$, apart from the ordinary *proper* intervals $[\underline{x}, \overline{x}]$ with $\underline{x} \leq \overline{x}$ forming the classical interval arithmetic. As a whole, the complete interval arithmetic has good algebraic and inclusion order properties, which facilitates easier symbolic manipulations, etc.

We remind that the *dualization* of an interval $\mathbf{v} \in \mathbb{IR}$ is

$$\text{dual } \mathbf{v} := [\overline{v}, \underline{v}],$$

i.e., reversing its endpoints. For interval vectors and matrices, the dualization operation is taken componentwise. *Modulus* (magnitude) of an interval $\mathbf{v} \in \mathbb{IR}$ is defined as

$$|\mathbf{v}| := \max\{|\underline{v}|, |\overline{v}|\}.$$

By "opp", we will denote taking the opposite element in the complete arithmetic $\mathbb{IR}$, while "$\ominus$" is the inverse operation to the addition:

$$\text{opp } \mathbf{v} := [-\underline{v}, -\overline{v}],$$

$$\mathbf{u} \ominus \mathbf{v} := \mathbf{u} + \text{opp } \mathbf{v} = [\underline{u} - \underline{v}, \overline{u} - \overline{v}].$$

The definition of the inclusion ordering on $\mathbb{IR}$ is as follows:

$$\mathbf{u} \subseteq \mathbf{v} \iff \underline{u} \geq \underline{v} \text{ and } \overline{u} \leq \overline{v}.$$

The detailed description of Kaucher complete interval arithmetic can be found e.g. in the original works [3]–[5], or in [14], [16].

## 4. Characterizations of AE-Solution Sets

For the generalized AE-solution sets to interval linear systems (1.1), the following analytic characterization is known [15], [17]:

THEOREM 4.1. *A point $x \in \mathbb{R}^n$ belongs to the solution set $\Sigma_{\alpha\beta}(\mathbf{A}, \mathbf{b})$ if and only if*

$$\mathbf{A}^{\forall} \cdot x - \mathbf{b}^{\forall} \subseteq \mathbf{b}^{\exists} - \mathbf{A}^{\exists} \cdot x, \tag{4.1}$$

*where all the operations and relations are those of the classical interval arithmetic.*

We introduce

DEFINITION 4.1. The interval matrix and interval vector

$$\mathbf{A}^c := \mathbf{A}^\forall + \text{dual } \mathbf{A}^\exists, \qquad \mathbf{b}^c := \text{dual } \mathbf{b}^\forall + \mathbf{b}^\exists$$

are called *characteristic* for the AE-solution set $\Sigma_{\alpha\beta}(\mathbf{A}, \mathbf{b})$ to the interval linear system (1.1) specified by the disjoint decomposition of $\mathbf{A}$ into $\mathbf{A}^\forall$ and $\mathbf{A}^\exists$, and of $\mathbf{b}$ into $\mathbf{b}^\forall$ and $\mathbf{b}^\exists$.

The new language Definition 4.1 suggests enables us to speak of a *solution set that corresponds to the characteristic matrix* $\mathbf{A}^c$ *and right-hand side vector* $\mathbf{b}^c$ (in the last Section). $\mathbf{A}^c$ and $\mathbf{b}^c$ actually express, in a concentrated form, both the types of interval uncertainty of all parameters and their intervals proper. In addition, the new concepts facilitate rewriting the result of Theorem 4.2 in a more concise form:

THEOREM 4.2. *A point* $x \in \mathbb{R}^n$ *belongs to the solution set* $\Sigma_{\alpha\beta}(\mathbf{A}, \mathbf{b})$ *if and only if*

$$\mathbf{A}^c \cdot x \subseteq \mathbf{b}^c \tag{4.2}$$

*in complete interval arithmetic.*

*Proof.* Notice that

$$\text{opp} \, (-\mathbf{v}) = \text{dual } \mathbf{v}$$

for any interval $\mathbf{v} \in \mathbb{IR}$. Therefore, adding (dual $\mathbf{b}^\forall$ + dual $(\mathbf{A}^\exists \cdot x)$) to both sides of (4.1) yields the following equivalent inclusion in the complete interval arithmetic

$$\mathbf{A}^\forall \cdot x + \text{dual } (\mathbf{A}^\exists \cdot x) \subseteq \text{dual } \mathbf{b}^\forall + \mathbf{b}^\exists. \tag{4.3}$$

Further, dual $(\mathbf{A}^\exists \cdot x) = (\text{dual } \mathbf{A}^\exists) \cdot x$, since $x$ is a point. So, (4.3) is equivalent to

$$\mathbf{A}^\forall \cdot x + (\text{dual } \mathbf{A}^\exists) \cdot x \subseteq \text{dual } \mathbf{b}^\forall + \mathbf{b}^\exists.$$

In the left-hand side, we can avail ourselves of the distributivity with respect to the point variable $x$, which results in

$$(\mathbf{A}^\forall + \text{dual } \mathbf{A}^\exists) \cdot x \subseteq \text{dual } \mathbf{b}^\forall + \mathbf{b}^\exists,$$

and that coincides with (4.2).                                                                 □

In our work, we will need a "fixed-point form characterization" of the AE-solution sets. To derive it, we add $(x \ominus \mathbf{A}^c x)$ to both sides of the inclusion (4.2), thus getting the equivalent relation

$$x \subseteq x + \text{opp} \, (\mathbf{A}^c x) + \mathbf{b}^c.$$

But opp $(\mathbf{A}^c x) = $ opp $(\mathbf{A}^c) x$ for the point $x$, we have therefore

$$x \subseteq x + (\text{opp } \mathbf{A}^c) x + \mathbf{b}^c,$$

Again, we can make use of the fact that $x$ is a point and factor it out in the right-hand side due to the distributivity. Overall,

$$x \in \Sigma_{\alpha\beta}(\mathbf{A}, \mathbf{b}) \qquad \Longleftrightarrow \qquad x \subseteq (I \ominus \mathbf{A}^c) x + \mathbf{b}^c.$$

It should be stressed that for $x \in \Sigma_{\alpha\beta}(\mathbf{A}, \mathbf{b}) \neq \emptyset$ the above proof implies the interval vector $(I \ominus \mathbf{A}^c) x + \mathbf{b}^c$ being proper.

To summarize, we get the following

THEOREM 4.3. *A point $x \in \mathbb{R}^n$ belongs to the solution set $\Sigma_{\alpha\beta}(\mathbf{A}, \mathbf{b})$ if and only if*

$$x \in (I \ominus \mathbf{A}^c) x + \mathbf{b}^c.$$

## 5. "Algebraic Approach" in Outer Estimation Problem

The problems of inner interval estimation of the generalized solution sets to interval systems of equations are known to be successfully solved by *algebraic approach* [13], [15], [17], a technique that changes the original estimation problem for a problem of computing an algebraic solution to an auxiliary equation in Kaucher complete interval arithmetic $\mathbb{IR}$. Recall

DEFINITION 5.1. An interval vector is called an *algebraic solution* to an interval equation (inequality, etc.) if substituting this vector into the equation and executing all interval operations according to the rules of the interval arithmetic result in an equality (inequality, etc.).

We are going to show how a similar approach (which we also shall call *algebraic*) may be applied to the problems of outer interval estimation of the AE-solution sets. An alternative technique that solves the same problem—generalized interval Gauss-Seidel iteration—has been presented in [18].

THEOREM 5.1. *Let an interval matrix $\mathbf{C} \in \mathbb{IR}^{n \times n}$ be such that the spectral radius $\rho(|\mathbf{C}|)$ of the matrix made up of the moduli of its entries is less than 1. Then, for any vector $\mathbf{d} \in \mathbb{IR}^n$, the algebraic solution to the interval linear system*

$$x = \mathbf{C}x + \mathbf{d} \tag{5.1}$$

*exists and is unique.*

*Proof.* In complete interval arithmetic $\mathbb{IR}$, the distance dist $(\cdot, \cdot)$ between the elements is known to be introduced as follows [5]:

$$\text{dist} (\mathbf{u}, \mathbf{v}) := \max\{|\underline{u} - \underline{v}|, |\overline{u} - \overline{v}|\} = |\mathbf{u} \ominus \mathbf{v}|.$$

It is worth noting as well that for any intervals $\mathbf{c}, \mathbf{u}, \mathbf{v} \in \mathbb{IR}$ the inequality

$$\text{dist}(\mathbf{cu}, \mathbf{cv}) \leq |\mathbf{c}| \cdot \text{dist}(\mathbf{u}, \mathbf{v})$$

is valid (see also [5]). This estimate holds true for the multidimensional case too if the distance between $\mathbf{u}, \mathbf{v} \in \mathbb{IR}^n$ is understood as the componentwise vector-valued metric (*pseudometric* according to the terminology by Collatz [2]). More precisely, for the interval vectors $\mathbf{u}, \mathbf{v}$ we define

$$\text{dist}(\mathbf{u}, \mathbf{v}) := \begin{pmatrix} \text{dist}(\mathbf{u}_1, \mathbf{v}_1) \\ \vdots \\ \text{dist}(\mathbf{u}_n, \mathbf{v}_n) \end{pmatrix} \in \mathbb{R}^n.$$

Then, for any interval matrix $C$ with the elements $\mathbf{c}_{ij} \in \mathbb{IR}$ and any interval vectors $\mathbf{u}, \mathbf{v}$ of the corresponding size, we have

$$\text{dist}(\mathbf{Cu}, \mathbf{Cv}) \leq |\mathbf{C}| \cdot \text{dist}(\mathbf{u}, \mathbf{v}). \tag{5.2}$$

To prove the inequality (5.2), let us remind that

$$\text{dist}(\mathbf{y} + \mathbf{z}, \mathbf{y}' + \mathbf{z}') \leq \text{dist}(\mathbf{y}, \mathbf{y}') + \text{dist}(\mathbf{z}, \mathbf{z}')$$

for any one-dimensional intervals $\mathbf{y}, \mathbf{y}', \mathbf{z}, \mathbf{z}' \in \mathbb{IR}$ (see [5]). We can therefore conclude that

$$\text{dist}((\mathbf{Cu})_i, (\mathbf{Cv})_i) = \text{dist}\left(\sum_{j=1}^{n} \mathbf{c}_{ij}\mathbf{u}_j, \sum_{j=1}^{n} \mathbf{c}_{ij}\mathbf{v}_j\right)$$

$$\leq \sum_{j=1}^{n} \text{dist}(\mathbf{c}_{ij}\mathbf{u}_j, \mathbf{c}_{ij}\mathbf{v}_j)$$

$$\leq \sum_{j=1}^{n} |\mathbf{c}_{ij}| \cdot \text{dist}(\mathbf{u}_j, \mathbf{v}_j)$$

for all $i = 1, 2, \ldots, n$, which proves the multidimensional estimate (5.2).

In the situation under study, for any $\mathbf{d} \in \mathbb{IR}^n$

$$\text{dist}(\mathbf{Cu} + \mathbf{d}, \mathbf{Cv} + \mathbf{d}) = \text{dist}(\mathbf{Cu}, \mathbf{Cv}) \leq |\mathbf{C}| \cdot \text{dist}(\mathbf{u}, \mathbf{v}).$$

If the spectral radius of the matrix $|\mathbf{C}|$ is less than 1, then we can apply the finite-dimensional version of Schröder's fixed-point theorem (see e.g. [1], [2], [11], [12]). Namely, the map $\mathbb{IR}^n \longrightarrow \mathbb{IR}^n$ which acts

$$\mathbf{x} \mapsto \mathbf{Cx} + \mathbf{d}$$

is a contraction with respect to the pseudometric "dist" and has thus a unique fixed-point that is an algebraic solution to the interval linear system (5.1). $\square$

THEOREM 5.2. *Let an AE-solution set* $\Sigma_{\alpha\beta}(\mathbf{A}, \mathbf{b})$ *of the interval linear system (1.1) be nonempty and* $\rho(|I \ominus \mathbf{A}^c|) < 1$. *Then the algebraic solution to the interval linear system*

$$x = (I \ominus \mathbf{A}^c) x + \mathbf{b}^c \qquad (5.3)$$

*(which exists and is unique by virtue of Theorem 5.1) is a proper interval vector enclosing the solution set* $\Sigma_{\alpha\beta}(\mathbf{A}, \mathbf{b})$.

*Proof.* Assume that $\mathbf{x}^*$ is an algebraic solution to the interval linear system (5.3). We are going to show that for any point $\tilde{x} \in \Sigma_{\alpha\beta}(\mathbf{A}, \mathbf{b})$ there holds $\tilde{x} \in \mathbf{x}^*$.

Due to Theorem 4.1, the membership $\tilde{x} \in \Sigma_{\alpha\beta}(\mathbf{A}, \mathbf{b})$ is equivalent to the inclusion

$$\tilde{x} \in (I \ominus \mathbf{A}^c) \tilde{x} + \mathbf{b}^c. \qquad (5.4)$$

Let us launch an iteration process in $\mathbb{IR}^n$ according to the following formulas:

$$\mathbf{x}^{(0)} := \tilde{x}, \qquad (5.5)$$

$$\mathbf{x}^{(k+1)} := (I \ominus \mathbf{A}^c) \mathbf{x}^{(k)} + \mathbf{b}^c. \qquad (5.6)$$

Using induction, it is fairly simple to prove that all the vectors generated by this process contain $\tilde{x}$. Indeed, for $\mathbf{x}^{(0)}$ it is true by construction. If $\tilde{x} \in \mathbf{x}^{(k)}$, then in view of (5.4) and inclusion monotonicity of the interval arithmetic operations in $\mathbb{IR}$ we arrive at

$$\tilde{x} \in (I \ominus \mathbf{A}^c) \tilde{x} + \mathbf{b}^c \subseteq (I \ominus \mathbf{A}^c) \mathbf{x}^{(k)} + \mathbf{b}^c = \mathbf{x}^{(k+1)}.$$

Therefore, $\tilde{x} \in \mathbf{x}^{(k)}$ for all integer $k$. In particular, the above means that all the interval vectors $\mathbf{x}^{(k)}$ must be *proper*.

Furthermore, the condition $\rho(|I \ominus \mathbf{A}^c|) < 1$ implies the convergence of the iteration process defined in the pseudometric space $\mathbb{IR}^n$ by the formulas (5.5)–(5.6) (see, e.g., [1], [2], [11], [12]). There is no difficulty realizing that the sequence $\mathbf{x}^{(k)}$, $k = 1, 2, \ldots$, converges to a fixed point of the map

$$x \mapsto (I \ominus \mathbf{A}^c) x + \mathbf{b}^c,$$

that is, to the unique algebraic solution $\mathbf{x}^*$ of the equation (5.3). Since the membership $\tilde{x} \in \mathbf{x}^{(k)}$ is equivalent to a system of $2n$ nonstrict inequalities, then it must hold in the limit as well,

$$\tilde{x} \in \lim_{k \to \infty} \mathbf{x}^{(k)} = \mathbf{x}^*,$$

while this limit interval $\mathbf{x}^*$ is proper too.                                   □

## 6. Implementation

We conclude the paper with some comments on practical implementation of the technique developed, i.e., on the methods for computing algebraic solutions to the main equation (5.3) and overall applicability of our approach. Notice that Theorems 5.1–5.2 give the necessary theoretical basis for constructing stationary iteration algorithms relying upon Schröder's contracting mapping theorem.

Another possibility is the subdifferential Newton method (see e.g. [14]) whose convergence is substantiated rigorously for the interval linear systems (5.3) with the matrices $A^c$ in which, along every row, the entries are either all proper or all improper. Empirically, it has been revealed that the method works well even for the general interval linear systems of the form (5.3), with arbitrarily mixed proper and improper entries in the matrix $A^c$ (although then the algorithm is no longer subdifferential, it is *quasidifferential Newton method*).

The key point of the applicability of our algebraic approach is the reduction of the original interval system (1.1) to the form (5.3) so that the requirement $\rho(|I \ominus A^c|) < 1$ is met. This cannot always be done.

In the classical problem of computing enclosures for the united solution set, one traditionally makes use of the so-called *preconditioning*—multiplying both sides of the system, from the left, by a point matrix. Such a transformation leads to widening of the united solution set, but a careful choice of the preconditioning point matrix improves the properties of the interval matrix of the system we thus obtain [6], [11]. Unfortunately, the above prescription fails when we turn to outer estimation of the generalized solution sets: they do not necessarily extend after the preconditioning, changing in a more complex way. Still, an outcome from our difficulty exists and it amounts to that we should precondition *the characteristic matrix and right-hand side vector* rather than the original interval system itself.

Let us turn to the analytical characterization of AE-solution sets that Theorem 4.2 gives:

$$x \in \Sigma_{\alpha\beta}(A, b) \qquad \Longleftrightarrow \qquad A^c x \subseteq b^c.$$

If $\Lambda$ is a point $n \times n$-matrix, then $A^c x \subseteq b^c$ implies the inclusion $\Lambda(A^c x) \subseteq \Lambda b^c$. The interval matrix product is known to be non-associative in the general case, but for point $\Lambda$ and $x$ there holds the equality $\Lambda(A^c x) = (\Lambda A^c) x$. Finally, we get

$$x \in \Sigma_{\alpha\beta}(A, b) \qquad \Longrightarrow \qquad (\Lambda A^c) x \subseteq \Lambda b^c. \qquad (6.1)$$

We can interpret $\Lambda A^c$ and $\Lambda b^c$ as characteristic matrix and right-hand side vector of another interval linear system, so the implication (6.1) establishes

THEOREM 6.1. *If $\Lambda$ is a point $n \times n$-matrix, then the AE-solution set corresponding to the characteristic matrix $A^c$ and right-hand side vector $b^c$ is contained in the AE-solution set corresponding to the characteristic matrix $\Lambda A^c$ and right-hand side vector $\Lambda b^c$.*

We can therefore replace our main problem (3.1) with the outer estimation of an AE-solution set defined by the new characteristic matrix $\Lambda A^c$ and right-hand side vector $\Lambda b^c$. If the interval matrix of the original system is not "too large," one may hope that a suitable choice of $\Lambda$ will cause the spectral radius $\rho(|I \ominus \Lambda A^c|)$

to become actually less than one. It is worth noting that, similar to the traditional case, taking $\Lambda$ as the inverse to the middle of $\mathbf{A}$ works reasonably well. Overall, we can consider the procedure summarized in Theorem 6.1 as a kind of *generalized preconditioning* of the interval linear system (1.1). Its detailed analysis is going to be presented in an expanded version of this short note.

## References

1. Alefeld, G. and Herzberger, J.: *Introduction to Interval Computations*, Academic Press, New York, 1983.
2. Collatz, L.: *Funktionalanalysis und Numerische Mathematik*, Springer-Verlag, Berlin—Höttingen—Heidelberg, 1964.
3. Gardeñes, E. and Trepat, A.: Fundamentals of SIGLA, an Interval Computing System over the Completed Set of Intervals, *Computing* **24** (1980), pp. 161–179.
4. Gardeñes, E., Trepat, A., and Mielgo, H.: Present Perspective of the SIGLA Interval System, *Freiburger Intervall-Berichte* (82/9) (1982), pp. 1–65.
5. Kaucher, E.: Interval Analysis in the Extended Interval Space $\mathbb{IR}$, *Computing Suppl.* **2** (1980), pp. 33–49.
6. Kearfott, R. B.: *Rigorous Global Search: Continuous Problems*, Kluwer Academic Publishers, Dordrecht, 1996.
7. Kelling, B.: Geometrische Untersuchungen zur eigenschränkte Lösungsmenge Intervallgleichungssysteme, *ZAMM* **74** (1994), pp. 625–628.
8. Kleene, S. C.: *Mathematical Logic*, John Wiley, New York, 1967.
9. Lakeyev, A. V.: Computational Complexity of Estimation of Generalized Solution Sets to Interval Linear Systems, in: *Proceedings of XI International Conference "Optimization Methods and Their Applications", Baikal, July 5–12, 1998 (section 4)*, Irkutsk, 1998, pp. 115–118.
10. Neumaier, A.: Tolerance Analysis with Interval Arithmetic, *Freiburger Intervall-Berichte* (86/9) (1986), pp. 5–19.
11. Neumaier, A.: *Interval Methods for Systems of Equations*, Cambridge University Press, Cambridge, 1990.
12. Ortega, J. M. and Rheinboldt, W. C.: *Iterative Solutions of Nonlinear Equations in Several Variables*, Academic Press, New York, 1970.
13. Shary, S. P.: A New Approach to the Analysis of Static Systems under Interval Uncertainty, in: Alefeld, G., Frommer, A., and Lang, B. (eds), *Scientific Computing and Validated Numerics*, Akademie Verlag, Berlin, 1996, pp. 118–132.
14. Shary, S. P.: Algebraic Approach in the "Outer Problem" for Interval Linear Equations, *Reliable Computing* **3** (1) (1997), pp. 103–135.
15. Shary, S. P.: Algebraic Approach to the Analysis of Linear Static Systems under Interval Uncertainty, *Izvestiya Akademii Nauk. Control Theory and Systems* (3) (1997), pp. 51–61 (in Russian).
16. Shary, S. P.: Algebraic Approach to the Interval Linear Static Identification, Tolerance and Control Problems, or One More Application of Kaucher Arithmetic, *Reliable Computing* **2** (1) (1996), pp. 3–33.
17. Shary, S. P.: Algebraic Solutions to Interval Linear Equations and Their Applications, in: Alefeld, G. and Herzberger, J. (eds), *Numerical Methods and Error Bounds*, Akademie Verlag, Berlin, 1996, pp. 224–233.
18. Shary, S. P.: Interval Gauss-Seidel Method for Generalized Solution Sets to Interval Linear Systems, in: *MISC'99—Workshop on Applications of Interval Analysis to Systems and Control, Girona, Spain, February 24–26, 1999*, Universitat de Girona, 1999, pp. 51–65.
19. Shary, S. P.: Solving the Linear Interval Tolerance Problem, *Mathematics and Computers in Simulation* **39** (1995), pp. 53–85.
20. Vatolin, A. A.: On Linear Programming Problems with Interval Coefficients, *J. Comp. Mathem. and Math. Phys.* **24** (1984), pp. 1629–1637 (in Russian).
21. Walter, E. and Pronzato, L.: *Identification of Parametric Models from Experimental Data*, Springer, Berlin-Heidelberg, 1997.

*Reliable Computing* **5**: 337–346, 1999.
© 1999 *Kluwer Academic Publishers.*

# A Real Polynomial Decision Algorithm Using Arbitrary-Precision Floating Point Arithmetic

ADAM STRZEBONSKI
*Wolfram Research Inc. and Jagiellonian University, 100 Trade Centre Drive, Champaign, IL 61820,
USA, e-mail: adams@wolfram.com*

(Received: 25 September 1998; accepted: 4 February 1999)

**Abstract.** We study the problem of deciding whether a system of real polynomial equations and inequalities has solutions, and if yes finding a sample solution. For polynomials with exact rational number coefficients the problem can be solved using a variant of the cylindrical algebraic decomposition (CAD) algorithm. We investigate how the CAD algorithm can be adapted to the situation when the coefficients are inexact, or, more precisely, *Mathematica* arbitrary-precision floating point numbers. We investigate what changes need to be made in algorithms used by CAD, and how reliable are the results we get.

## 1. Introduction

The problem of deciding whether a system of real polynomial equations and inequalities has solutions arises in many application areas, for example polynomial optimization, geometric theorem proving, computational geometry, or control theory (see [6]). The Cylindrical Algebraic Decomposition (CAD) algorithm (see [3]) can be used to solve this problem (see [2] for a collection of works on the subject). The CAD algorithm assumes that the input polynomials have exact rational number coefficients. In the following section we describe an algorithm based on CAD, which accepts input polynomials with *Mathematica* arbitrary-precision floating point number coefficients. Arbitrary-precision numbers are floating point numbers with unrestricted number of digits, which contain information about the number of digits considered accurate. The arithmetic automatically keeps track of precision (see [7] and [14]).

While a system of polynomial equations or inequalities with exact coefficients can either have solutions or not, allowing inexact coefficients results in a wider variety of possible answers. It may happen that some systems within the coefficients' error bounds have solutions and others do not. Also, we can have several kinds of inexact solutions varying in "strength" from a strong assertion that each tuple of numbers within the error bounds of the solution is a solution of all systems within the error bounds of coefficients, to some weaker notion of an inexact solution satisfying an equation or an inequality, like, for instance, a tuple of inexact numbers satisfies an equation if after substitution of the numbers for variables the difference of sides of the equation is less than some specified value. The possible results given by our algorithm are discussed in Section 3.

Finally, we present results of experiments with our implementation of the algorithm. We use randomly generated inequality systems, geometric inequalities from [11], and strong inequalities from [10]. We compare the performance of our algorithm and a similar algorithm based on CAD using exact computations.

## 2. The Algorithm

The input is a logical combination of polynomial equations and inequalities in variables $x_1, ..., x_n$, with arbitrary-precision floating-point number coefficients. We transform the system into the disjunctive normal form, i.e. an alternative of conjunctions of

$$p_k(x_1, ..., x_n) \, \rho \, 0$$

where $\rho$ is one of $<, \leq, =,$ or $\neq$. The main idea of CAD based decision algorithm is to construct a finite set $S$ of points in $\mathbb{R}^n$ such that if the input system has solutions one of the solutions must be in $S$. The algorithm consists of two phases.

### 2.1. PROJECTION

We construct a sequence $P_n, ..., P_1$, such that $P_i$ is a finite set of primitive (as polynomials in $x_i$), square free, and relatively prime polynomials in $x_1, ..., x_i$. $P_n$ is computed as the set of primitive, square free, and relatively prime factors of the input polynomials $p_k$. $P_{i-1}$ is obtained from $P_i$ using the projection operator. What is the projection operator depends on the input. In the general case we use the projection operator described in [8] and [9]. If during the sample point construction phase we see that the set of input polynomials was not well-oriented (as described in [8]), we start over with the projection operator described in [5]. When all input inequalities are strong we use the simpler projection operator described in [12].

The projection operators require computing of the resultants (and sometimes also the principal subresultant coefficients), and primitive, square free, and relatively prime factor sets for sets of polynomials. We compute g.c.d.'s using the Euclidean algorithm, removing the leading coefficients of the subsequent remainders if they can be zero within the error bounds. The cofactors are computed using the polynomial division algorithm. The resultants and subresultants are computed using the subresultant remainder sequence, again the leading coefficients which may be zero are removed. We do some sanity checks, for example if we divide polynomials that we know should divide exactly, we check whether the remainder is indeed zero (i.e. all coefficients contain zero within the error bounds). The g.c.d. and subresultants algorithms can therefore fail if one of the sanity checks fails (due to loss of precision during computations).

## 2.2. CONSTRUCTION OF SAMPLE POINTS

The set $S$ of sample points is constructed one coordinate at a time. Let $S_i$, for $1 \leq i \leq n$, denote the set of all $i$-tuples which are first $i$ coordinates of some point from $S$ (in particular $S_n = S$). For $S_1$ we choose all real roots of polynomials from $P_1$ and one point from each of the open intervals which remain after removing the roots from $\mathbb{R}$. Similarly, $S_{i+1}$ is constructed by taking, for each tuple $\{a_1, ..., a_i\} \in S_i$, tuples $\{a_1, ..., a_i, a_{i+1}\}$, with choices for $a_{i+1}$ being all real roots of polynomials $p(a_1, ..., a_i, x_{i+1})$, for $p \in P_{i+1}$, and one point from each of the open intervals which remain after removing the roots from $\mathbb{R}$. If all input inequalities are strong it suffices to take only the points from open intervals.

For a proof that so constructed set $S$ indeed has the required property see [5], [8], and [12]. Our algorithm does not construct the whole set $S$, but rather generates its elements one at a time, and checks if they satisfy the input system. It also tests plausibility of existence of solutions with a given $i$ first coordinates before constructing further coordinates, as suggested in [4].

To compute sample points we need to compute real roots of univariate polynomials with arbitrary-precision number coefficients. We use the following algorithm.

ALGORITHM RealRoots.

**Input:** Univariate polynomial $f$ with arbitrary-precision number coefficients.
**Output:** A list of arbitrary-precision numbers covering all real roots of $f$.

1. Compute $g = f / g.c.d.(f, f')$.

2. If the computation of $g$ fails due to loss of precision compute a bound on the absolute value of real roots of $f$, and return an inexact zero with accuracy low enough to cover all the numbers with absolute value smaller than the bound.

3. If the free term of $g$ may be zero within the error bounds, divide $g$ by its variable, and remember to include zero with an appropriate accuracy on the list of roots.

4. Fix precision of computations at somewhat more than the highest precision of a coefficient of $g$, and compute candidates for the positive and the negative roots of $g$ using the algorithm *PositiveRoots* described below.

5. Determine the correct precision of each root using the á posteriori validation method described in [13].

ALGORITHM PositiveRoots.

**Input:** An interval $int$, a pair of univariate square free polynomials $g$ and $h$ with precision $prec$ floating point number coefficients, and a homographic transformation $t$ of positive reals onto $int$, such that $h(x) = g(t(x))$.
**Output:** A list of precision $prec$ numbers approximating the roots of $g$ in $int$.

1. If all coefficients of $h$ are non-negative return an empty list. (Note that the roots of $g$ in $int$ correspond to the positive roots of $h$.)

2. If the free term of $h$ is negative, $g$ must have opposite signs at endpoints of $int$. Use a secant type method to compute an approximation of a root of $g$ in $int$. Deflate $g$ and $h$ and call *PositiveRoots* recursively.

3. Find a lower bound $lb$ for positive roots of $h$. Let $t'$ be the transformation defined by:

   a) if $lb < 1$, $t'(x) = x$,

   b) if $1 \le lb < 16$, $t'(x) = x + lb$,

   c) if $lb \ge 16$, $t'(x) = lb(x + 1)$.

4. Put $h = h \circ t'$, $t = t \circ t'$, and compute the corresponding new value of $int$.

5. Check if zero is a root of $h$. If yes deflate $g$ and $h$, and remember to add $t(0)$ to the root list.

6. Define $t'$ and $t''$ by: $t'(x) = x + 1$ and $t''(x) = 1/(x+1)$. Put $h_1 = h \circ t'$, $t_1 = t \circ t'$, $h_2 = h \circ t''$, and $t_2 = t \circ t''$.

7. Check if zero is a root of $h_1$. If yes deflate $g$, $h_1$ and $h_2$, and remember to add $t_1(0)$ to the root list.

8. Call *PositiveRoots* recursively with $g$, $h_1$ and $t_1$, and $g$, $h_2$ and $t_2$. The corresponding intervals are result of subdivision of $int$ with point $t(1)$.

The correctness of *PositiveRoots* follows from the correctness of the real root isolation algorithm described in [1].

When the input system is an alternative of conjunctions of equations and inequalities it is usually faster to look for a solution of each term of the alternative separately, unless all terms of the alternative contain inequalities with the same set of polynomials. Certainly when parallel computations are available it is faster to treat each term of an alternative separately.

## 3. Interpretation of Results

The algorithm presented here can give several types of answers with various degrees of "strength." Here we classify the possible answers we can get. Following the last remark of the previous section we assume for simplicity that the input system is a conjunction of equations and inequalities.

Let $a = \{a_1, ..., a_n\}$, $x = \{x_1, ..., x_n\}$, and let $p(x)$ be a polynomial with arbitrary-precision number coefficients. We call a point $a' \in \mathbb{R}^n$ in a *representative* of $a$ if all coordinates of $a'$ fall within the error bounds of $a$. Similarly, a polynomial $p' \in \mathbb{R}[x]$ is a *representative* of $p$ if all coefficients of $p'$ fall within the error bounds of coefficients of $p$.

We say that $a$ *satisfies*

1. a strong inequality $p(x) < 0$ iff after replacing $x$ with $a$ in $p(x)$ we get a negative arbitrary-precision number. A negative arbitrary-precision number represents

an interval with both endpoints negative. This means that for any representatives $a'$ of $a$, and $p'$ of $p$, $p'(a') < 0$;

2. a weak inequality $p(x) \leq 0$ iff after replacing $x$ with $a$ in $p(x)$ we get a negative or zero arbitrary-precision number. A zero arbitrary-precision number $u$ represents a symmetric interval around zero with radius $10^{-accuracy(u)}$. If $p(a)$ is a zero arbitrary-precision number there may be representatives $a'$ of $a$, and $p'$ of $p$, such that $p'(a') > 0$;

3. an equation $p(x) = 0$ iff after replacing $x$ with $a$ in $p(x)$ we get a zero arbitrary-precision number $u$. This tells us only that for any representatives $a'$ of $a$, and $p'$ of $p$, $p'(a') < 10^{-accuracy(u)}$. It does not guarantee the existence of representatives for which the equation $p'(a') = 0$ is satisfied.

If our algorithm finds a solution of a system of equations and inequalities the solution satisfies the equations and inequalities in the sense described above. Therefore if the input system consists only of inequalities, and we get a solution which satisfies the strong versions of all inequalities, all the representatives of the solution are solutions of all representatives of input inequalities, in particular all representatives of input inequalities are guaranteed to have solutions. Otherwise there may not exist even a single representative of the system which has solutions, however such a weak solution of equations may be all we need in practice (or all we can hope for if we have an inexact input), especially if substituting of the solution in the left hand sides of the equations gives zeroes with high accuracies.

In case when our algorithm finds no solutions the situation is more complicated. Consider the one-inequality system

$$(1.x^2 - 2.x + 1.)y^2 + 1. < 0$$

where the coefficients have some finite precision *prec*. There are representatives of the inequality which have solutions, however there are no arbitrary-precision numbers satisfying the inequality. Moreover, arbitrary-precision numbers satisfy the weak version of the inequality only if they make the left hand side a zero with a negative accuracy.

To prove that none of the representatives of the input system has solutions in addition to our algorithm not returning any solutions we need to check for the following two properties.

1. For any representatives of the input inequalities there must be a one-to-one correspondence between the projection polynomials computed with the representatives, and the projection polynomials computed with the arbitrary-precision number coefficients. Moreover, the polynomials computed with representatives should be representatives of the corresponding polynomials computed with the arbitrary-precision number coefficients.

2. If any polynomial used in our algorithm to compute sample points has multiple roots (or two polynomials have common roots), then for any representatives of

input inequalities, the corresponding polynomial would have roots with exactly the same multiplicities (or the corresponding pair of polynomials would have the same number of roots in common).

We know that the input system has the first property if none of the g.c.d.'s or contents computed in the projection phase comes out nonconstant (except possibly for the case when a monomial divides all terms of a polynomial), and if none of the (sub)resultants has an inexact zero leading coefficient. The second property is satisfied if all the roots of polynomials used to compute the first coordinate of sample points are different (i.e. no two roots have a common representative), and if a pair of polynomials used to compute the $j + 1$-st coordinate of sample points has common roots (or a polynomial has multiple roots), then the degree of the g.c.d. of the polynomials (or of the polynomial and its derivative) is equal to the number of zero leading principal subresultant coefficients (or subdiscriminants) of the corresponding $j + 1$-variate polynomials with respect to the $j + 1$-st variable, after substituting the first $j$ coordinates for the variables.

Our algorithm checks these properties (until it finds the first violation). In the example above if we project first with respect to $y$, the second property is not satisfied, which is detected because the first level of projection contains polynomial $1. x^2 - 2. x + 1.$ which has a double root. If we project first with respect to $x$, the first property is not satisfied because the discriminant of the input polynomial with respect to $x$ is $0. y^6 + 4. y^4$, so it has an inexact zero leading coefficient.

If one of the above properties is not satisfied and the algorithm does not find any solutions, the most we can hope for is to show that at least for some representatives of input inequalities the system has no solutions. This we can guarantee if none of the sample points generated by our algorithm is a *weak solution* i.e. solution satisfying the weak versions of the input inequalities. If we do find a weak solution all we can tell is that the system is "unlikely" to have solutions, especially if the weak solutions have high precision. The only way in which the representatives of the system, which have the same form of the projection as the computed inexact projection (i.e. their projection polynomials being representatives of the corresponding inexact projection polynomials), can have solutions is that the computed sample points have so low precision that they "cover" the solutions.

## 4. Examples

In this section we compare performance our decision algorithm for polynomial equation and inequality systems with arbitrary-precision number coefficients (PDAPC) with a CAD based decision algorithm for polynomial equation and inequality systems with rational number coefficients (PDRC). Both algorithms were implemented in the C kernel of *Mathematica*. The examples were run on a Pentium II, 233 MHz computer with 64 MB of RAM. The timings are in seconds. The systems of inequalities with arbitrary-precision number coefficients are equal to the corresponding systems with rational number coefficients, with the coeffi-

Table 1. Random equation and inequality systems.

| #var | degree | density | #eqn | #ineq | PDRC timing | PDAPC timing | prec | answer |
|------|--------|---------|------|-------|-------------|--------------|------|--------|
| 2 | 3 | 0.5 | 0 | 2 | 0.22 | 0.05 | 20 | yes1 |
| 2 | 5 | 0.5 | 0 | 2 | 2.88 | 0.44 | 20 | yes1 |
| 2 | 5 | 0.5 | 0 | 4 | 12.34 | 1. | 20 | yes1 |
| 2 | 5 | 1 | 0 | 6 | 127.1 | 4.33 | 20 | yes1 |
| 3 | 3 | 0.3 | 0 | 4 | 351.5 | 3.7 | 20 | yes1 |
| 4 | 2 | 0.3 | 0 | 6 | 207.7 | 1.76 | 20 | yes1 |
| 5 | 2 | 0.2 | 0 | 4 | 1.16 | 0.19 | 20 | yes1 |
| 2 | 3 | 0.5 | 0 | 10 | 39.94 | 11.33 | 20 | no1 |
| 2 | 3 | 0.5 | 1 | 2 | 3.22 | 0.47 | 20 | yes2 |
| 2 | 5 | 0.5 | 1 | 2 | 19.01 | 2.36 | 40 | yes2 |
| 2 | 3 | 0.3 | 3 | 0 | 23.26 | 0.31 | 20 | no1 |
| 3 | 3 | 0.2 | 2 | 2 | >1000 | 12.88 | 25 | yes2 |

cients computed to precision *prec*. The symbols in the column *answer* have the following meanings

- yes*1*: the input system contains no equations, and PDAPC found a solution which satisfies strong versions of all inequalities,
- *yes2*: PDAPC found a solution, but the above does not hold,
- *no1*: PDAPC proved that none of the representatives of the input system have solutions.
- *no2*: PDAPC proved that at least some of the representatives of the input system have no solutions.
- *no3*: PDAPC found no solutions, but none of the above holds.

The answers from PDRC in all cases agreed with answers from PDAPC.

## 4.1. RANDOM EXAMPLES

We have randomly generated examples with prescribed number of variables (#var), maximal total degree, density (ratio of the number of nonzero coefficients to the number of different monomials up to the given maximal total degree), number of equations (#eqn), and number of strong inequalities (#ineq). The coefficients were rational numbers with randomly selected 6-digit numerators and denominators. In case when PDAPC returns a solution we chose the precision of computations so that the solution has at least 16 digits of precision.

In all the random examples PDAPC was significantly faster than PDRC, and we always were able to choose precision large enough to avoid "weak" answers ("no2" or "no3").

Table 2. Geometric inequalities.

| example | #variables | degree | PDRC timing | PDAPC timing | prec | answer |
|---------|-----------|--------|-------------|--------------|------|--------|
| 1 | 2 | 3 | 1.41 | 0.57 | 35 | no3 |
| 2 | 3 | 3 | >1000 | 12.14 | 40 | no2 |
| 3 | 2 | 3 | 0.24 | 0.22 | 25 | no3 |
| 4 | 3 | 3 | 48.67 | 3.09 | 30 | no2 |
| 5 | 2 | 4 | 0.21 | 0.2 | 20 | no3 |
| 6 | 3 | 4 | >1000 | 11.16 | 90 | no2 |
| 7 | 2 | 6 | 3.28 | 0.93 | 20 | no3 |
| 8 | 3 | 6 | >1000 | 168.4 | 400 | no2 |
| 9 | 3/2 | 4/3 | 160. | 0.74 | 25 | no3 |
| 10 | 4/3 | 4/3 | >1000 | 11.16 | 35 | no2 |
| 11 | 3 | 4 | 3.04 | 1.53 | 20 | no3 |
| 12 | 4 | 4 | >1000 | 735.9 | 200 | no2 |

## 4.2. GEOMETRIC INEQUALITIES

Examples in this section have been chosen from [11]. In all examples $a$, $b$, and $c$ are sides of a triangle. All the original weak inequalities are homogeneous in $a$, $b$, and $c$, so one can set $a = 1$ without loss of generality, and we have done that in our computations. We have also tested a set of strong inequalities obtained from the original ones by adding 1 to the "greater or equal" side. Such inequalities are equivalent to the original inequalities, but they are no longer homogeneous. In all examples

$$TC = a \geq 0 \wedge b \geq 0 \wedge c \geq 0 \wedge a + b \geq c \wedge b + c \geq a \wedge c + a \geq b$$

denotes the triangle conditions. In all examples the algorithms checked that there are no solutions of the negation of the formula we want to prove. All of the original (odd-numbered) inequalities become identities if $a = b = c$, therefore we could not get from PDAPC answers any stronger than "no3" in odd-numbered examples (because there are weak solutions even in the exact coefficients case), and "no2" in even-numbered examples (because in any neighborhood of the negations of the formulas there are some solvable formulas). In all examples where a weak solution was found we used a precision high enough, so that the weak solution had precision (or accuracy if zero) of at least 16 digits.

1. $TC \Rightarrow a^3 + b^3 + c^3 \geq 3abc + 8(a - b)(b - c)(c - a)$.
2. $TC \Rightarrow a^3 + b^3 + c^3 + 1 > 3abc + 8(a - b)(b - c)(c - a)$.
3. $TC \Rightarrow (a + b + c)^3 \geq (2a + 2b - c)(2a + 2c - b)(2b + 2c - a)$.
4. $TC \Rightarrow (a + b + c)^3 + 1 > (2a + 2b - c)(2a + 2c - b)(2b + 2c - a)$.
5. $TC \Rightarrow a^4 + b^4 + c^4 + 5abc(a + b + c) \geq 2(ab + ac + bc)^2$.
6. $TC \Rightarrow a^4 + b^4 + c^4 + 5abc(a + b + c) + 1 > 2(ab + ac + bc)^2$.

Table 3. Strong inequalities.

| inequalities | degree | PDRC timing | PDAPC timing | prec | answer |
|---|---|---|---|---|---|
| $B1 \wedge B2$ | 2 | 0.11 | 0.36 | 20 | yes1 |
| $B1 \wedge B4$ | 2 | 0.1 | 0.27 | 20 | no2 |
| $B1 \wedge B2 \wedge B3$ | 2 | 0.93 | 1.52 | 20 | yes1 |
| $B1 \wedge B2 \wedge B4$ | 2 | 0.91 | 2.63 | 20 | no2 |
| $B1 \wedge C1$ | 2 | 0.11 | 0.34 | 20 | yes1 |
| $B1 \wedge C2$ | 2 | 0.17 | 0.46 | 20 | no3 |
| $T \wedge C1$ | 4 | 0.59 | 0.62 | 40 | yes1 |
| $T \wedge B2$ | 4 | 0.69 | 1.09 | 30 | yes1 |
| $HB1 \wedge HB2 \wedge HB3$ | 2 | 12.23 | 34.33 | 20 | no2 |
| $HT \wedge HB2 \wedge HB3$ | 4 | 9.63 | 32.85 | 80 | no2 |
| $T \wedge C1 \wedge B2$ | 4 | 2.52 | 2.1 | 20 | yes1 |
| $HT \wedge C1 \wedge HB2$ | 4 | 29.31 | 133.5 | 80 | no2 |

7. $TC \Rightarrow a^2b^4 + b^2c^4 + c^2a^4 \geq abc(a^3 + b^3 + c^3 + ab(b-a) + bc(c-b) + ca(a-c))$.

8. $TC \Rightarrow a^2b^4 + b^2c^4 + c^2a^4 + 1 > abc(a^3 + b^3 + c^3 + ab(b-a) + bc(c-b) + ca(a-c))$.

9. $TC \Rightarrow a^3 + b^3 + c^3 \geq 3abc + 2\sqrt{9 + 6\sqrt{3}}(a - b)(b - c)(c - a)$.

10. $TC \Rightarrow a^3 + b^3 + c^3 + 1 > 3abc + 2\sqrt{9 + 6\sqrt{3}}(a - b)(b - c)(c - a)$.

11. $TC \wedge k \geq 1 \Rightarrow 2k(a+b+c)^3 + 9(1+k)abc \geq (1+7k)(a+b+c)(ab+ac+bc)$.

12. $TC \wedge k \geq 1 \Rightarrow 2k(a+b+c)^3 + 9(1+k)abc + 1 > (1+7k)(a+b+c)(ab+ac+bc)$.

Note that, in examples 9 and 10, PDRC needs to use one more variable and one more 4-th degree equation than PDAPC, to represent the nested radical coefficient.

## 4.3. STRONG INEQUALITIES FROM [10]

These examples were chosen to demonstrate a version of CAD specialized for systems of strong inequalities, which used only sample points with rational number coordinates. Our implementation of PDRC also uses only sample points with rational number coordinates, and also uses a simpler version of projection, as described in [12]. Following the notation of [10] let us put.

$$B1 = x^2 + y^2 + z^2 < 1,$$

$$B2 = (x - 1)^2 + (y - 1)^2 + (z - 1)^2 < 1,$$

$$B3 = (x - 1)^2 + (y - 1)^2 + \left(z + \frac{1}{2}\right)^2 < 1,$$

$$B4 = \left(x - \frac{3}{2}\right)^2 + (y - 2)^2 + z^2 < 1,$$

$$C1 = x^2 + y^2 + z^2 + 2yz - 4y - 4z + 3 < 0 \wedge y - 1 < z \wedge z < y + 1,$$

$$C2 = x^2 + y^2 + z^2 + 2yz - 4y - 4z + 3 < 0 \wedge y + 1 < z \wedge z < y + 2,$$
$$T = z^4 + (2y^2 + 2x^2 + 6)z^2 + y^4 + 2x^2y^2 - 10y^2 + x^4 - 10x^2 + 9 < 0,$$
$$HB1 = B1 \wedge x + y + z < 0,$$
$$HB2 = B2 \wedge x + y + z > 3,$$
$$HB3 = B3 \wedge x + y + z < \frac{3}{2},$$
$$HT = T \wedge x + y < 0.$$

In these examples PDRC performs better than PDAPC.

## References

1. Akritas, A., Bocharov, A., and Strzebonski, A.: Implementation of Real Root Isolation Algorithms in Mathematica, in: *International Conference INTERVAL'94. Abstracts*, St.Petersburg, Russia, 1994, pp. 23–27.
2. Caviness, B. and Johnson, J., (eds): *Quantifier Elimination and Cylindrical Algebraic Decomposition*, Springer-Verlag, 1998.
3. Collins, G. E.: Quantifier Elimination for the Elementary Theory of Real Closed Fields by Cylindrical Algebraic Decomposition, *Lect. Notes Comput. Sci.* **33** (1975), pp. 134–183.
4. Collins, G. E. and Hong, H.: Partial Cylindrical Algebraic Decomposition for Quantifier Elimination, *J. Symbolic Comp.* **12** (1991), pp. 299–328.
5. Hong, H.: An Improvement of the Projection Operator in Cylindrical Algebraic Decomposition, in: *Proceedings of ISSAC*, 1990, pp. 261–264.
6. Jirstrand, M.: Constructive Methods for Inequality Constraints in Control, *Linkoping Studies in Science and Technology, Dissertations* **527** (1998).
7. Keiper, J. B. and Withoff, D.: Numerical Computation in Mathematica, in: *Course Notes, Mathematica Conference*, 1992.
8. McCallum, S.: An Improved Projection for Cylindrical Algebraic Decomposition, in: Caviness, B. and Johnson, J. (eds), *Quantifier Elimination and Cylindrical Algebraic Decomposition*, Springer-Verlag, 1998.
9. McCallum, S.: An Improved Projection for Cylindrical Algebraic Decomposition of Three Dimensional Space, *J. Symbolic Comp.* **5** (1988), pp. 141–161.
10. McCallum, S.: Using Cylindrical Algebraic Decomposition, *The Computer Journal* **36** (5) (1993), pp. 432–438.
11. Mitrinovic, D., Pecaric, J. E., and Volenec, V.: *Recent Advances in Geometric Inequalities*, Kluwer Academic Publishers, 1989.
12. Strzebonski, A.: An Algorithm for Systems of Strong Polynomial Inequalities, *The Mathematica Journal* **4** (4) (1994), pp. 74–77.
13. Strzebonski, A.: Computing in the Field of Complex Algebraic Numbers, *J. Symbolic Comp.* **24** (1997), pp. 647–656.
14. Wolfram, S.: *The Mathematica Book*, 3rd. Ed., 1996.

*Reliable Computing* **5**: 347–357, 1999.

# A Numerical Verification Method of Solutions for the Navier-Stokes Equations

YOSHITAKA WATANABE
*Computer Center, Kyushu University 33, Fukuoka 812-8581, Japan,*
*e-mail: watanabe@cc.kyushu-u.ac.jp*

and

NOBITO YAMAMOTO and MITSUHIRO T. NAKAO
*Graduate School of Mathematics, Kyushu University 33, Fukuoka 812-8581, Japan*

(Received: 9 November 1998; accepted: 15 January 1999)

**Abstract.** A numerical verification method of the solution for the stationary Navier-Stokes equations is described. This method is based on the infinite dimensional fixed point theorem using the Newton-like operator. We present a verification algorithm which generates automatically on a computer a set including the exact solution. Some numerical examples are also discussed.

## 1. Introduction

We proposed, in [12] and [10], a method to estimate the guaranteed a posteriori $H_0^1$ error bounds of the finite element solutions for the Stokes problem in mathematically rigorous sense. These papers also describe a method to derive the constructive $H_0^1$ a priori error estimates for the same problems based on the estimation of the largest eigenvalues for related matrices.

Furthermore, in [11], we clarified that an Aubin-Nitsche-like technique can also be applied to the constructive $L^2$ error estimates and establish the estimates both in a posteriori and a priori sense by using the results obtained in our previous works.

In this paper, we describe a numerical verification method of the solution for the stationary Navier-Stokes equations incorporating with a posteriori and a priori error estimates for the Stokes problem. This method is based on the method in [7], [8] for elliptic problems, but some essential extensions are necessary to deal with the convection term. Namely, special techniques are devised to overcome the difficulty from the low regularity caused by such term. We present a verification algorithm which automatically generates on a computer a set including the exact solution.

## 1.1. Navier-Stokes Equations

We consider the following stationary Navier-Stokes equations

$$\begin{cases} -v\Delta u + \nabla p = -(u \cdot \nabla)u + f & \text{in } \Omega, \\ \operatorname{div} u = 0 & \text{in } \Omega, \\ u = 0 & \text{on } \partial\Omega, \end{cases} \qquad (1.1)$$

where $\Omega$ is a convex polygonal domain in $\mathbb{R}^2$, $u = (u_1, u_2)^T$ the two-dimensional velocity field, $p$ a kinematic pressure field, $v > 0$ the viscosity constant and $f = (f_1, f_2)^T$ a pair of $L^2$ function on $\Omega$ which means a density of body forces per unit mass.

## 1.2. Some Function Spaces

We denote by $H^k(\Omega)$ the usual $k$-th order Sobolev space on $\Omega$, and define $(\cdot, \cdot)$ as the inner product in $L^2(\Omega)$ and put

$$\begin{aligned} H_0^1(\Omega) &\equiv \{v \in H^1(\Omega) ; v = 0 \text{ on } \partial\Omega\}, \\ L_0^2(\Omega) &\equiv \{v \in L^2(\Omega) ; (v, 1) = 0\}, \\ S &\equiv H_0^1(\Omega)^2 \times L_0^2(\Omega). \end{aligned}$$

The norm in $L^2(\Omega)$ and $H_0^1(\Omega)$ is denoted by $|q|_0 \equiv (q, q)^{1/2}$ and $|v|_1 \equiv |\nabla v|_0$, respectively. We also define $H^2(\Omega)$-seminorm $|\cdot|_2$ by

$$|u|_2 \equiv \left( \left| \frac{\partial^2 u}{\partial x^2} \right|_0^2 + 2 \left| \frac{\partial^2 u}{\partial x \partial y} \right|_0^2 + \left| \frac{\partial^2 u}{\partial y^2} \right|_0^2 \right)^{1/2}.$$

In what follows, since no confusion may arise, we will use the same notations for the corresponding norms and inner products in $L^2(\Omega)^2$ and $H_0^1(\Omega)^2$ as in $L^2(\Omega)$ and $H_0^1(\Omega)$, respectively. Finally, we define $H^{-1}(\Omega)^2$ as the dual space of $H_0^1(\Omega)^2$ and $\langle \cdot, \cdot \rangle$ as the duality pairing between $H^{-1}(\Omega)^2$ and $H_0^1(\Omega)^2$. The norm in $H^{-1}(\Omega)^2$ is denoted by $|u|_{-1}$.

## 2. Finite Element Approximation

We rewrite (1.1) in the weak form:

find $[u, p] \in H_0^1(\Omega)^2 \times L_0^2(\Omega)$ such that

$$\begin{cases} v(\nabla u, \nabla v) - (p, \operatorname{div} v) = -((u \cdot \nabla)u, v) + (f, v) & \forall v \in H_0^1(\Omega)^2, \\ (\operatorname{div} u, q) = 0 & \forall q \in L_0^2(\Omega). \end{cases} \qquad (2.1)$$

Note that, for $u \in H_0^1(\Omega)^2$, the term $(u \cdot \nabla)u$ does not belong to $L^2(\Omega)^2$ but $L^p(\Omega)^2$ $(p < 2)$ because of Sobolev's imbedding theorem (e.g. [2]).

Next, we introduce some finite element subspaces for the approximation of the velocity and pressure. Let $T_h$ be a family of triangulations of $\Omega \subset \mathbb{R}^2$, which consist of triangles or quadrilaterals dependent on a scale parameter $h > 0$. For $T_h$, we denote by $X_h \subset H_0^1(\Omega) \cap C(\overline{\Omega})$ and $Y_h \subset L_0^2(\Omega) \cap C(\overline{\Omega})$ the finite element subspaces for the approximation of each component of the velocity $u$ and the pressure $p$, respectively. Here, continuity assumptions of $X_h$ and $Y_h$ are necessary to obtain the guaranteed error bounds for the Stokes problem (see [10]).

We set $S_h \equiv X_h^2$. Furthermore, we assume, as the approximation property of $X_h$, that

$$\inf_{\xi \in X_h} |v - \xi|_1 \leq C_0 h |v|_2 \qquad \forall v \in H_0^1(\Omega) \cap H^2(\Omega),$$

where $C_0$ is a positive constant, independent of $v$ and $h$, which can be numerically determined. This assumption holds for many finite element subspaces (cf. [1]).

It is well-known, e.g. [2], that for each $\xi \in H^{-1}(\Omega)^2$, the weak form of the Stokes equation:

$$\nu(\nabla u, \nabla v) - (p, \operatorname{div} v) - (q, \operatorname{div} u) = \langle \xi, v \rangle \qquad \forall [v, q] \in S \tag{2.2}$$

has a unique solution $[u, p] \in S$. We suppose that for each $\xi \in H^{-1}(\Omega)^2$, there exists a unique solution $[u_h, p_h] \in S_h \times Y_h$ satisfying

$$\nu(\nabla u_h, \nabla v_h) - (p_h, \operatorname{div} v_h) - (q_h, \operatorname{div} u_h) = \langle \xi, v_h \rangle \qquad \forall [v_h, q_h] \in S_h \times Y_h. \tag{2.3}$$

The validity of this assumption can be checked by so-called discrete inf-sup condition on $S_h \times Y_h$ (cf. [2]). If $[u, p]$ is a solution of (2.2) and $[u_h, p_h]$ is a solution of (2.3), it can be easily seen that

$$|u - u_h|_1 \leq C_2 |\xi|_{-1}, \tag{2.4}$$

where $C_2 \equiv 2 / \nu$ is a positive constant. Moreover, for the case $\xi \in L^2(\Omega)^2$, using constructive a priori error estimates described in [10], a positive constant $C_1$, dependent on $\Omega$, $h$, $C_0$ and $\nu$ only, such that

$$|u - u_h|_1 \leq C_1 |\xi|_0, \tag{2.5}$$

can be computed.

A finite element solution $[u_h, p_h] \in S_h \times Y_h$ of problem (2.1) is defined by

$$\begin{cases} \nu(\nabla u_h, \nabla v_h) - (p_h, \operatorname{div} v_h) = -((u_h \cdot \nabla)u_h, v_h) + (f, v_h) & \forall v_h \in S_h, \\ -(\operatorname{div} u_h, q_h) = 0 & \forall q_h \in Y_h. \end{cases} \tag{2.6}$$

In actual calculation, we use the interval Newton method to enclose $[u_h, p_h]$ satisfying (2.6) in the small intervals. In what follows, we fix the spaces $X_h$, $S_h$, $Y_h$ and let $[u_h, p_h]$ denote an approximate solution of (2.1) satisfying (2.6). Using the component $u_h$ of this solution we consider the following Stokes problem

$$\begin{cases} -\nu \Delta \bar{u} + \nabla \bar{p} = -(u_h \cdot \nabla)u_h + f & \text{in } \Omega, \\ \operatorname{div} \bar{u} = 0 & \text{in } \Omega, \\ \bar{u} = 0 & \text{on } \partial\Omega. \end{cases} \tag{2.7}$$

By the assumption of uniqueness of the solution satisfying (2.3), the finite element solution $[\bar{u}, \bar{p}] \in S$ of (2.7) coincides with $[u_h, p_h]$. Consequently, setting $v_0 \in H_0^1(\Omega)^2$ and $p_0 \in L_0^2(\Omega)$ as

$$v_0 \equiv \bar{u} - u_h, \qquad p_0 \equiv \bar{p} - p_h,$$

we can compute the numerical estimates of $|v_0|_1$ and $|p_0|_0$ using a posteriori estimates for (2.7) because of $-(u_h \cdot \nabla)u_h + f \in L^2(\Omega)^2$ (cf. [10]–[12]). In what follows, $v_0$ is considered as an element in $H_0^1(\Omega)^2$ whose norm can be bounded, but explicit form is unknown. By (1.1) and (2.7) we have

$$\begin{cases} -\nu\Delta(u - \bar{u}) + \nabla(p - \bar{p}) = -(u \cdot \nabla)u + (u_h \cdot \nabla)u_h, \\ \operatorname{div}(u - \bar{u}) = 0. \end{cases}$$

Here, $w$ and $r$ are defined by

$$w \equiv u - \bar{u}, \qquad r \equiv p - \bar{p},$$

respectively. Then, $u = w + \bar{u} = w + v_0 + u_h$ implies the following residual form for the Navier-Stokes equation

$$\begin{cases} -\nu\Delta w + \nabla r = g(w) & \text{in } \Omega, \\ \operatorname{div} w = 0 & \text{in } \Omega, \\ w = 0 & \text{on } \partial\Omega, \end{cases} \tag{2.8}$$

where

$$g(w) \equiv -\big((u_h + v_0 + w) \cdot \nabla\big)(u_h + v_0 + w) + (u_h \cdot \nabla)u_h.$$

## 3. Fixed Point Formulation

First, note that, since the Stokes problem

$$\begin{cases} -\nu\Delta\hat{w} + \nabla\hat{r} = \xi & \text{in } \Omega, \\ \operatorname{div}\hat{w} = 0 & \text{in } \Omega, \\ \hat{w} = 0 & \text{on } \partial\Omega, \end{cases} \tag{3.1}$$

has a unique solution $[\hat{w}, \hat{r}] \in S$ for each $\xi \in H^{-1}(\Omega)^2$, denoting the solution $\hat{w}$ of (3.1) by $A\xi$, then, $A$ is a continuous linear operator from $H^{-1}(\Omega)^2$ to $H_0^1(\Omega)^2$. Thus, setting

$$F \equiv Ag,$$

(2.8) is rewritten as the fixed point problem in $H_0^1(\Omega)^2$:

$$w = Fw.$$

Concerning $F$, the following result is obtained in [6] (Chapter 5):

LEMMA 3.1. $F$ is a compact map from $H_0^1(\Omega)^2$ to $H_0^1(\Omega)^2$.

Next, from the assumption of $S_h$ and $Y_h$, for any $\xi \in H^{-1}(\Omega)^2$, there exists a unique $\bar{u}_h \in S_h$ satisfying (2.3). We denote this correspondence by $A_h : H^{-1}(\Omega)^2 \longrightarrow S_h$. We now define $S_h^*$ by

$$S_h^* \equiv \{v \in H_0^1(\Omega)^2 \mid v = (A - A_h)f, \ f \in H^{-1}(\Omega)^2\},$$

and $\overline{S_h^*}$ by the closure of $S_h^*$ with norm $|\cdot|_1$, and we introduce the product space $X$ by

$$X \equiv S_h \times \overline{S_h^*}$$

Then, $X$ is a Banach space with norm

$$\max\{|x_h|_1, |x^*|_1\}, \quad x = [x_h, x^*] \in X.$$

We define the linear map $P$ from $X$ to $H_0^1(\Omega)^2$ by

$$Px \equiv x_h + x^*, \quad x = [x_h, x^*] \in X,$$

and set

$$G \equiv g \circ P.$$

Then, the map $\tilde{F} : X \longrightarrow X$ defined by

$$\tilde{F}x \equiv [A_h Gx, (A - A_h)Gx]$$

is compact because of the compactness of the map $AG = F \circ P$. Thus, if we find a nonempty, bounded, convex and closed set $W \subset X$ such that $\tilde{F}W \subset W$, then there exists a fixed point $x$ of $\tilde{F}$ in $W$ by Schauder's fixed point theorem. Then, for this fixed point $x = [x_h, x^*] \in X$, $Px = x_h + x^* \in H_0^1(\Omega)$ is also a fixed point of $F$, namely, $Px$ is a solution of (2.1).

## 4. Newton-Like Method and Computer Algorithm

Now, we introduce the Newton-like method proposed in [7], [8]. First, we define the map $\hat{g}(w) : H_0^1(\Omega)^2 \longrightarrow H^{-1}(\Omega)^2$ by

$$\hat{g}(w) \equiv -(w \cdot \nabla)w,$$

and suppose that the restriction of the operator $P_1 - A_h \hat{g}'(u_h) : H_0^1(\Omega)^2 \longrightarrow S_h$ to $S_h$ has an inverse

$$[P_1 - A_h \hat{g}'(u_h)]_h^{-1} : S_h \longrightarrow S_h, \tag{4.1}$$

where $P_1$ is an $H_0^1$-projection from $H_0^1(\Omega)^2$ to $S_h$, and $\hat{g}'(u_h)$ denotes the Fréchet derivative of $\hat{g}$ at $u_h$. This assumption is equivalent to the invertibility of a matrix, which can be numerically checked in actual verified computations (e.g. [13]). Next, we define the Newton-like operator $N_h : X \longrightarrow S_h$ as

$$N_h x \equiv x_h - [P_1 - A_h \hat{g}'(u_h)]_h^{-1}(x_h - A_h Gx), \qquad x = [x_h, x^*],$$

and the compact operator $T : X \longrightarrow X$ by

$$Tx \equiv [N_h x, (A - A_h)Gx].$$

Then, under the invertibility assumption (4.1), two fixed point problems: $x = Tx$ and $x = \tilde{F}x$ are equivalent.

Now, for any $v_h \in S_h$, using real coefficients $\{a_i\}_{1 \le i \le 2n}$ and basis of $X_h$ : $\{\varphi_i\}_{1 \le i \le n}$, we represent $v_h$ as

$$v_h = \left( \sum_{i=1}^n a_i \varphi_i, \sum_{i=1}^n a_{n+i} \varphi_i \right)^T.$$

Then, we define $(v_h)_i$ by

$$(v_h)_i \equiv |a_i|, \qquad 1 \le i \le 2n.$$

Now, for any non-negative real vector $\{W_i\}_{1 \le i \le 2n+2}$, we define $W_h \subset S_h$ and $W^* \subset \overline{S_h^*}$ as

$$W_h \equiv \{w_h \in S_h; \ (w_h)_i \le W_i, \ 1 \le i \le 2n\},$$
$$W^* \equiv \{w \in \overline{S_h^*}; \ |w|_1 \le W_{2n+1} + W_{2n+2}\},$$

and the set $W \subset X$ as

$$W \equiv W_h \times W^*.$$

Then, we have the following computable verification condition.

THEOREM 4.1. *Let $W_h$, $W^*$ and $W$ be sets defined above. If the inclusions*

$$\begin{cases} N_h(W) \subset W_h, \\ (A - A_h)G(W) \subset W^* \end{cases} \tag{4.2}$$

*hold, then there exists a fixed point $x$ of $\tilde{F}$ in $W$.*

*Proof.* By the definition, $W$ is a non-empty, closed, convex and bounded set in $X$. And from (4.2), we have $N_h(W) \times (A - A_h)G(W) \subset W_h \times W^*$ in $X$. Then, by the compactness of operator $T$, we get

$$TW \subset W \quad \text{in } X.$$

Hence, from Schauder's fixed point theorem, we obtain the desired result.          □

Next, we propose a computer algorithm to construct the set $W$ which satisfies the verification condition (4.2). We use the similar iterative method with inflation to that in [7], [8], etc.

First, for $N = 0$, we take appropriate initial vector $W_i^{(0)}$ ($1 \le i \le 2n + 2$) and for $\{W_i^{(0)}\}_{1 \le i \le 2n+2}$, we define $W^{(0)} = W_h^{(0)} \times W^{*(0)}$. For $N \ge 1$, with a given $0 < \delta \ll 1$, we set

$$\overline{W}_i^{(N-1)} \equiv W_i^{(N-1)}(1 + \delta) \qquad 1 \le i \le 2n + 2,$$

and for $\{\overline{W}_i^{(N-1)}\}_{1 \le i \le 2n+2}$, define the $\delta$-*inflation* by $\overline{W}^{(N-1)} = \overline{W}_h^{(N-1)} \times \overline{W}^{*(N-1)}$.

Next, for the set $\overline{W}^{(N-1)}$, we construct the candidate set $W^{(N)} = W_h^{(N)} \times W^{*(N)}$ by

$$\begin{cases} W_h^{(N)} \equiv N_h \overline{W}^{(N-1)}, \\ W_{2n+1}^{(N)} \equiv C_1 \sup_{w \in \overline{W}^{(N-1)}} |G_1(w)|_0, \\ W_{2n+2}^{(N)} \equiv C_2 \sup_{w \in \overline{W}^{(N-1)}} |G_2(w)|_{-1}, \end{cases} \qquad (4.3)$$

where for each $w = [w_h, w^*] \in X$,

$$G_2(w) \equiv -((v_0 + w^*) \cdot \nabla)(v_0 + w^*) \in H^{-1}(\Omega)^2,$$
$$G_1(w) \equiv G(w) - G_2(w) \in L^2(\Omega)^2.$$

Here, $W_h^{(N)}$ is determined by the interval vector solution for the $2n$ dimensional linear system of equations with interval right-hand side (cf. [15], [16]). $W_{2n+1}^{(N)}$ corresponds to the a priori error estimates (2.5) for the finite element solution of the Stokes problem for the smooth part of $G(W)$, which is presented in [10]. On the other hand, $W_{2n+2}^{(N)}$ stands for the a priori estimates (2.4) for the solution of the Stokes problem with the less smooth ($H^{-1}$-element) right-hand side, which can be computed because of $G_2(w) \in L^1(\Omega)$ for all $w \in X$. For example, setting $\hat{w} = v_0 + w^*$, we have

$$(G_2(\hat{w}), v) \le |\hat{w}|_1 \|\hat{w}\|_{L^4} \|v\|_{L^4} \le C_{L^4}^2 |\hat{w}|_1^2 |v|_1 \quad \forall v \in H_0^1(\Omega)^2,$$

where $\| \cdot \|_{L^4}$ is the $L^4$-norm on $\Omega$ and $C_{L^4}$ a explicit constant in the Sobolev imbedding theorem (see [14]). Thus we obtain $\sup_{w \in \overline{W}^{(N-1)}} |G_2(w)|_{-1} \le C_{L^4}^2 |\hat{w}|_1^2$.
Here, in general, $|\hat{w}|_1^2 \ll 1$ because $\hat{w}$ is a residual part of the solution.

In the actual calculation on a computer, each quantity of (4.3) is computed in the over-estimated sense.

Now, we have the following verification condition in a computer.

THEOREM 4.2. *If, for a step $K$, we have*

$$W_i^{(K)} \le \overline{W}_i^{(K-1)}, \quad 1 \le i \le 2n + 2,$$

*then, in the set* $\overline{W}^{(K-1)} = \overline{W}_h^{(K-1)} \times \overline{W}^{*(K-1)} \subset X$ *constructed by* $\{\overline{W}_i^{(K-1)}\}_{1 \le i \le 2n+2}$, *there exists an element $x$ satisfying $x = Tx$.*

*Proof.* From Theorem 4.1, it is sufficient to check (4.2) holds for $\overline{W}^{(K-1)}$. Then, by the assumption and definition of the set $W^{(K)}$, we have

$$N_h \overline{W}^{(K-1)} \subset \overline{W}_h^{(K-1)}.$$

Next, for any $\psi \in (A - A_h)G(\overline{W}^{(K-1)})$, we can take $w \in \overline{W}^{(K-1)}$ such that

$$\psi = (A - A_h)G(w)$$
$$= (A - A_h)G_1(w) + (A - A_h)G_2(w).$$

Also, by virtue of (2.4) and (2.5), we get

$$|(A - A_h)G_1(w)|_1 \leq C_1 \sup_{w \in \overline{W}^{(K-1)}} |G_1(w)|_0 \leq \overline{W}_{2n+1}^{(K-1)},$$

$$|(A - A_h)G_2(w)|_1 \leq C_2 \sup_{w \in \overline{W}^{(K-1)}} |G_2(w)|_{-1} \leq \overline{W}_{2n+2}^{(K-1)}.$$

Hence, we obtain $\psi \in \overline{W^*}^{(K-1)}$, and thus

$$(A - A_h)G(\overline{W}^{(K-1)}) \subset \overline{W^*}^{(K-1)}$$

holds.                                                                $\square$

By virtue of the Newton-like operator $N_h$ and the constructive a priori error estimate (2.5), the above process should be successful as the parameter $h$ becomes small.

## 5. Numerical Examples

Let $\Omega$ be a rectangular domain in $\mathbb{R}^2$ such that $\Omega = (0, 1) \times (0, 1)$. Also let $\delta_x$ : $0 = x_0 < x_1 < \cdots < x_L = 1$ be a uniform partition in $x$ direction, and let $\delta_y$ be the same partition as $\delta_x$ for $y$ direction. We define the partition of $\Omega$ by $\delta \equiv \delta_x \otimes \delta_y$. $L$ denotes the number of partitions for the interval $(0, 1)$, i.e. $h = 1 / L$.

Further, we define the finite element subspace $X_h$ and $Y_h$ by $X_h \equiv \mathcal{M}_0^2(x) \otimes \mathcal{M}_0^2(y)$ where $\mathcal{M}_0^2(x)$, $\mathcal{M}_0^2(y)$ are sets of continuous piecewise quadratic polynomials on $(0, 1)$ under the above partition $\delta$ with homogeneous boundary condition, and set $Y_h \equiv \mathcal{M}_0^1(x) \otimes \mathcal{M}_0^1(y) \cap L_0^2(\Omega)$, where $\mathcal{M}_0^1(x)$, $\mathcal{M}_0^1(y)$ piecewise linear as well. By the result in [9], we can take the constant $C_0 = 1 / (2\pi)$.

We choose the vector function $f$ so that

$$u_1(x, y) = C \sin^2 \pi x \sin \pi y \cos \pi y,$$
$$u_2(x, y) = -C \sin^2 \pi y \sin \pi x \cos \pi x,$$
$$p(x, y) = -C^2 \cos 2\pi x \cos 2\pi y / 16$$

are the exact solutions of (1.1) for an arbitrary constant $C$. Figure 1 and Figure 2 show the pressure and velocity field on $\Omega$, respectively ($C = 1$).

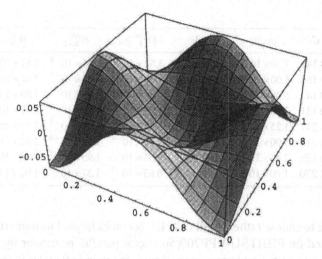

*Figure 1.* Pressure field $p$.

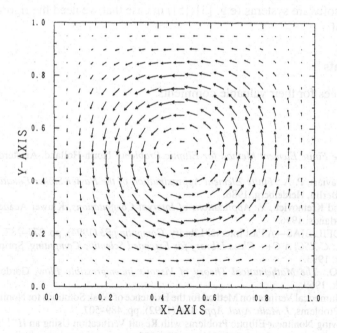

*Figure 2.* Velocity field $u = (u_1, u_2)_T$.

Table 1 shows verified values of $C$, $C_1$, $|v_0|_1$, $W_{2n+1}^{(K)}$, $W_{2n+2}^{(K)}$ and $\|W_h^{(K)}\|_\infty$, where $\|\cdot\|_\infty$ stands for the $L^\infty$-norm on $\Omega$. Here, $K$ is the iteration number satisfying the verification condition of Theorem 4.2.

Table 1. Verified values.

| $1/\nu$ | $L$ | $K$ | $C$ | $C_1$ | $\|u_h\|_\infty$ | $\|v_0\|_1$ | $\|W_h^{(K)}\|_\infty$ | $W_{2n+1}^{(K)}$ | $W_{2n+2}^{(K)}$ |
|---|---|---|---|---|---|---|---|---|---|
| 1 | 10 | 14 | 1/8 | 0.140 | $7.50\times10^{-2}$ | $2.61\times10^{-2}$ | $3.52\times10^{-3}$ | $2.44\times10^{-3}$ | $1.67\times10^{-4}$ |
| 2 | 10 | 12 | 1/10 | 0.165 | $5.00\times10^{-2}$ | $1.29\times10^{-2}$ | $1.33\times10^{-3}$ | $7.67\times10^{-4}$ | $7.67\times10^{-5}$ |
| 5 | 40 | 22 | 1/10 | 0.143 | $5.00\times10^{-2}$ | $3.55\times10^{-3}$ | $1.38\times10^{-3}$ | $7.59\times10^{-4}$ | $1.89\times10^{-5}$ |
| 10 | 40 | 23 | 1/20 | 0.131 | $2.38\times10^{-2}$ | $4.19\times10^{-2}$ | $4.26\times10^{-4}$ | $1.86\times10^{-4}$ | $1.47\times10^{-6}$ |
| 20 | 40 | 32 | 1/40 | 0.257 | $1.25\times10^{-2}$ | $5.81\times10^{-4}$ | $4.45\times10^{-4}$ | $1.96\times10^{-4}$ | $2.47\times10^{-6}$ |
| 40 | 40 | 27 | 1/100 | 0.511 | $5.00\times10^{-3}$ | $4.35\times10^{-4}$ | $2.90\times10^{-4}$ | $1.08\times10^{-4}$ | $2.42\times10^{-6}$ |
| 50 | 40 | 13 | 1/200 | 0.639 | $2.50\times10^{-3}$ | $2.68\times10^{-4}$ | $8.35\times10^{-5}$ | $1.95\times10^{-5}$ | $8.48\times10^{-7}$ |
| 100 | 40 | 15 | 1/400 | 1.270 | $1.00\times10^{-3}$ | $2.09\times10^{-4}$ | $6.63\times10^{-5}$ | $1.37\times10^{-5}$ | $1.02\times10^{-6}$ |

Up to now, we have to choose rather small $C$ as $1/\nu$ becomes large. The numerical examples are computed on FUJITSU VPP700/56 vector parallel processor by the usual computer arithmetic with double precision. Hence, the round off errors in these examples are neglected. However, from our experiences, the order of magnitude for the effect of round-off is under $10^{-10}$. Therefore, it is almost negligible compared with the truncation error which amounts to $10^{-3} \sim 10^{-2}$. Of course, we have to use those verification software systems (e.g. [3]–[5]) in case that we need the rigorous mathematical proof.

## Acknowledgements

We thank the referees for their valuable comments.

## References

1. Ciarlet, P. G.: *The Finite Element Method for Elliptic Problems*, North-Holland, Amsterdam, 1978.
2. Girault, V. and Raviart, P. A.: *Finite Element Approximation of the Navier-Stokes Equations*, Springer-Verlag, Berlin, Heidelberg, 1986.
3. Kearfott, R. B. and Kreinovich, V.: *Applications of Interval Computations*, Kluwer Academic Publishers, Netherlands, 1996.
4. Knüppel, O.: PROFIL/BIAS—A Fast Interval Library, *Computing* 53 (1994), pp. 277–287.
5. Kulisch, U. et. al.: *C-XSC, A C++ Class Library for Extended Scientific Computing*, Springer-Verlag, New York, 1993.
6. Ladyzhenskaya, O.: *The Mathematical Theory of Viscous Incompressible Flow*, Gordon & Breach, New York, 1969.
7. Nakao, M. T.: A Numerical Verification Method for the Existence of Weak Solutions for Nonlinear Boundary Value Problems, *J. Math. Anal. Appl.* 164 (1992), pp. 489–507.
8. Nakao M. T.: Solving Nonlinear Elliptic Problems with Result Verification Using an $H^{-1}$ Type Residual Iteration, *Computing, Suppl.* 9 (1993), pp. 161–173.
9. Nakao, M. T., Yamamoto, N., and Kimura, S.: On Best Constant in the Optimal Error Estimates for the $H_0^1$-projection into Piecewise Polynomial Spaces, *J. Approximation Theory* 93 (1998), pp. 491–500.
10. Nakao, M. T., Yamamoto, N., and Watanabe, Y.: A Posteriori and Constructive A Priori Error Bounds for Finite Element Solutions of the Stokes Equations, *J. Comput. Appl. Math.* 91 (1998), pp. 137–158.

11. Nakao, M. T., Yamamoto, N., and Watanabe, Y.: Constructive $L^2$ Error Estimates for Finite Element Solutions of the Stokes Equations, *Reliable Computing* **4** (2) (1998), pp. 115–124.
12. Nakao, M. T., Yamamoto, N., and Watanabe, Y.: Guaranteed Error Bounds for the Finite Element Solutions of the Stokes Problem, in: Alefeld, G., Frommer, A., and Lang, B. (eds): *Scientific Computing and Validated Numerics*, Proceedings of International Symposium on Scientific Computing, Computer Arithmetic and Validated Numerics SCAN-95, *Mathematical Research* **90** (1996), pp. 258–264.
13. Rump, S. M.: On the Solution of Interval Linear Systems, *Computing* **47** (1992), pp. 337–353.
14. Talenti, G.: Best Constant in Sobolev Inequality, *Ann. Math. Pure Appl.* **110** (1976), pp 353–372.
15. Watanabe, Y., and Nakao, M. T.: Numerical Verifications of Solutions for Nonlinear Elliptic Equations, *Japan J. Indust. Appl. Math.* **10** (1993) pp. 165–178.
16. Yamamoto, N.: A Numerical Verification Method for Solutions of Boundary Value Problems with Local Uniqueness by Banach's Fixed-point Theorem, *SIAM J. Numer. Anal.* **35** (1998) pp. 2004–2013.

*T. Csendes (ed.), Developments in Reliable Computing* 359–364.
© 1999 *Kluwer Academic Publishers.*

# Convex Sets of Full Rank Matrices

BARBARA KOŁODZIEJCZAK AND TOMASZ SZULC

{barbarak, tszulc}@math.amu.edu.pl

*Faculty of Mathematics and Computer Science, Adam Mickiewicz University, Matejki 48/49, 60-769 Poznań, Poland*

**Abstract.** Let $A = (a_{ij})$ and $B = (b_{ij})$ be $n$-by-$m$ real matrices. Using the notion of a block $P$-matrix [2] we give a necessary and sufficient condition for the set $r(A,B)$ $(c(A,B)$, resp.) of $n$-by-$m$ matrices whose rows (columns, resp.) are independent convex combinations of the rows (columns, resp.) of $A$ and $B$ to consist entirely of full row (column, resp.) rank matrices. This improves a result on the set $r(A,B)$ $(c(A,B)$, resp.) proven in [8]. Moreover, we also derive a sufficient condition for the interval of $A$ and $B$, i.e., for the set $i(A,B)$ of real $n$-by-$m$ matrices $(t_{ij}a_{ij} + (1 - t_{ij})b_{ij})$ with $t_{ij} \in [0,1]$ to have the abovementioned rank property.

**Keywords:** full row rank, full column rank, set of matrices

## 1. Introduction

Full rank matrices arise in a variety of applications. We restrict ourselves to note their role in the studies of solving underdetermined systems of linear algebraic equations [4] referring a reader to a survey of full rank oriented problems given in [6]. So, it seems to be natural to investigate the robustness of the property in question, i.e., to investigate if a full rank matrix $A = (a_{ij})$ does save the original property under unknown variations of its entries in a specified range. Among the results related to this problem we mention a full rank characterization of all convex combinations of $k$, $k \geq 2$, $n$-by-$m$ matrices $A_1, \ldots, A_k$ given in [6].

In this paper, continuing a research started in [8], we discuss the full rank property of some sets of matrices. Concerning the applicability of the proposed results a good example could be the question of establishing if a full row (column, resp.) rank matrix saves the original property under unknown variations of the entries of a row (column, resp.) of the matrix in a range of intervals. So, the paper can be viewed as a contribution to the problem of robustness of matrix properties.

## 2. Notation and definitions

By $diag(t_1, \ldots, t_s)$ we will denote the diagonal $s$-by-$s$ matrix with diagonal entries $t_1, \ldots, t_s$ and by $I_s$ and $\Theta_s$ the $s$-by-$s$ identity and null matrix, respectively ( we set $I_n = I$ and $\Theta_n = \Theta$). The $n$-by-$m$ matrix with all entries equal 1 will be denoted by $J$. For a rectangular matrix $X = (x_{ij})$ we denote its transpose by $X'$ and we introduce the absolute value of $X$ as the matrix $|X| = (|x_{ij}|)$; the spectral radius of a square $X$ will be denoted by $\rho(X)$.

DEFINITION 1. The Hadamard (entrywise) product of two $n$-by-$m$ matrices $Y = (y_{ij})$ and $Z = (z_{ij})$ is the $n$-by-$m$ matrix defined and denoted by $Y \circ Z = (y_{ij} z_{ij})$.

DEFINITION 2. [3]. Let

$$B = \begin{bmatrix} B_{11} & B_{12} \\ B_{21} & B_{22} \end{bmatrix}$$

be a partitioned matrix and $B_{22}$ its nonsingular (square) block. The Schur complement of $B_{22}$ in $B$ is then defined to be the matrix $B_{11} - B_{12} B_{22}^{-1} B_{21}$.

For real $n$-by-$m$ matrices $A = (a_{ij})$ and $B = (b_{ij})$ we define the following sets

$$r(A,B) := \{C : C = T_n A + (I - T_n)B, \ T_n = diag(t_1, \ldots, t_n), \ t_i \in [0,1]\},$$

$$c(A,B) := \{C : C = A T_m + B(I_m - T_m), \ T_m = diag(t_1, \ldots, t_m), \ t_i \in [0,1]\}$$

and

$$i(A,B) := \{C : C = T \circ A + (J - T) \circ B, \ T = (t_{ij}), \ t_{ij} \in [0,1]\}.$$

It is easy to see that $r(A,B)$, $c(A,B)$ and $i(A,B)$ are convex sets and that

$$r(A,B) \subseteq i(A,B) \quad \text{and} \quad c(A,B) \subseteq i(A,B).$$

Moreover, following [5], we notice that there always exist matrices $C_L, C_R \in i(A,B)$ such that $C_L \leq C \leq C_R$ (entrywise) for all $C \in i(A,B)$. So we can write $i(A,B) = [C_L, C_R]$ and call the set $i(A,B)$ an interval matrix. Finally, following the notation of [7], we can write

$$i(A,B) = [A_c - \Delta, A_c + \Delta] = \{C : A_c - \Delta \leq C \leq A_c + \Delta\}, \tag{1}$$

where $A_c = \frac{1}{2}(C_L + C_R)$ and $\Delta = \frac{1}{2}(C_R - C_L)$.

By $Q(s)$ we will denote a fixed partition of $Q = \{1, \ldots, q\}$ into $s$, $1 \leq s \leq q$, pairwise disjoint nonvoid subsets $Q_j$ of the cardinality $q_j$. $T^{(s)}$ is then the set of all diagonal $q$-by-$q$ matrices $T_q$ such that, for $i = 1, \ldots, s$, $T_q[Q_i] = t_i I_{q_i}$, where $t_i \in [0,1]$

and $T_q[Q_i]$ is the principal submatrix of $T_q$ with row and column indices in $Q_i$.

DEFINITION 3. [2]. Let $Q(s)$ be a partition of $Q$. A $q$-by-$q$ real matrix $W$ is called a block $P$-matrix with respect to the partition $Q(s)$, if for any $T_q \in \mathcal{T}^{(s)}$

$$\det(T_q W + (I_q - T_q)) \neq 0.$$

## 3. Results

We start with a full rank characterization of the set $r(A, B)$.

THEOREM 1 *The following are equivalent.*
*(i) $r(A, B)$ is of full row rank, i.e., each $C \in r(A, B)$ is of full row rank.*
*(ii) $B$ is of full row rank and the $2n$-by-$2n$ matrix*

$$\begin{bmatrix} AA' & AB' - AA' \\ \Theta & I \end{bmatrix} \begin{bmatrix} BA' & BB' - BA' \\ -I & I \end{bmatrix}^{-1} \tag{2}$$

*is a block $P$-matrix with respect to the partition $\{N_1, \ldots, N_n\}$ of $\{1, \ldots, 2n\}$, where $N_i = \{i, i+n\}$, $i = 1, \ldots, n$.*

**Proof:** We will proceed similarly as in the proof of Theorem 1 in [8].
(i) $\Rightarrow$ (ii): First we observe that, as $r(A, B)$ is of full row rank, the row rank property of $B$ is obvious. Now, from a well known rank property, we have that for all $T_n$'s with the diagonal entries in [0,1]

$$rank((T_n A + (I - T_n)B)(T_n A + (I - T_n)B)') = n.$$

So, for all $T_n$'s with the diagonal entries in [0,1]

$$\begin{aligned} (T_n A + (I - T_n)B)(T_n A + (I - T_n)B)' = \\ T_n AA' T_n + (I - T_n)BA' T_n + T_n AB'(I - T_n) + (I - T_n)BB'(I - T_n) \end{aligned} \tag{3}$$

is nonsingular. After slight manipulations saving the nonsingularity (3) becomes

$$T_n AA' + T_n (AB' - AA')(I - T_n) + (I - T_n)(BA' - BB')T_n + (I - T_n)BB'$$

which is the Schur complement of $I$ in

$$\begin{bmatrix} T_n AA' + (I - T_n)BA' & T_n(AB' - AA') + (I - T_n)(BB' - BA') \\ -(I - T_n) & I \end{bmatrix}. \tag{4}$$

Observe that we can write (4) as

$$\begin{bmatrix} T_n & \Theta \\ \Theta & T_n \end{bmatrix} \begin{bmatrix} AA' & AB' - AA' \\ \Theta & I \end{bmatrix} + \begin{bmatrix} I - T_n & \Theta \\ \Theta & I - T_n \end{bmatrix} \begin{bmatrix} BA' & BB' - BA' \\ -I & I \end{bmatrix}$$

and that, as $BB'$ is the Schur complement of $I$ in

$$\begin{bmatrix} BA' & BB' - BA' \\ -I & I \end{bmatrix}, \tag{5}$$

the matrix (5) is nonsingular. So, for all $T_n$'s with the diagonal entries in [0,1]

$$\begin{bmatrix} T_n & \Theta \\ \Theta & T_n \end{bmatrix} \begin{bmatrix} AA' & AB' - AA' \\ \Theta & I \end{bmatrix} \begin{bmatrix} BA' & BB' - BA' \\ -I & I \end{bmatrix}^{-1} + \begin{bmatrix} I - T_n & \Theta \\ \Theta & I - T_n \end{bmatrix}$$

is nonsingular and the implication in question follows by Definition 3.

(ii) $\Rightarrow$ (i): First we observe that the full row rank of $B$, and therefore the non-singularity of $BB'$, implies the nonsingularity of the matrix (5). Now, from the property of the matrix (2) after slight manipulations we get that, for all $T_n$'s with the diagonal entries in [0,1] the matrix (3) is nonsingular. So, from the earlier mentioned rank property

$$rank(T_n A + (I - T_n)B) = n$$

for all $T_n$'s with the diagonal entries in [0,1]. This shows that $r(A, B)$ is of full row rank and finishes the proof. ∎

THEOREM 2  *The following are equivalent.*
(i) *$c(A, B)$ is of full column rank, i.e., each $C \in c(A, B)$ is of full column rank.*
(ii) *$B$ is of full column rank and the 2m-by-2m matrix*

$$\begin{bmatrix} A'A & A'B - A'A \\ \Theta_m & I_m \end{bmatrix} \begin{bmatrix} B'A & B'B - B'A \\ -I_m & I_m \end{bmatrix}^{-1} \tag{6}$$

*is a block P-matrix with respect to the partition $\{M_1, \ldots, M_m\}$ of $\{1, \ldots, 2m\}$, where $M_i = \{i, i+m\}$, $i = 1, \ldots, m$.*

**Proof:**  The proof is entirely similar to the proof of Theorem 1 and is thus omitted. ∎

*Remark 1.* In (ii) of Theorem 1 (Theorem 2, resp.) one can assume that $A$ is of full row (column, resp.) rank and the matrix (2) ((6), resp.) is changed accordingly.

*Remark 2.* Concerning a practical verification of the property of the matrix (2) ((6), resp.) we mention that a real *P*-matrix, i.e., a matrix with all principal minors positive, is a block *P*-matrix with respect to any partition of the index set. Moreover, in [2] another sufficient condition and some necessary conditions for a real matrix to be a block *P*-matrix with respect to a given partition of the index set are given. Unfortunately, a numerically verifiable characterization of this notion remains unknown. What is more, an easily verifiable one cannot be expected at all. The reason is the following.

Set $m = n$ and $B = I$. Then it can readily be noted that the matrix (2) becomes the matrix $\begin{bmatrix} A & \Theta \\ I & A' \end{bmatrix}$. Next, the condition that this matrix is a block *P*-matrix with respect to the partition described in Theorem 1 converts into the following condition

$$\det\left(\begin{bmatrix} T_n & \Theta \\ \Theta & T_n \end{bmatrix}\begin{bmatrix} A & \Theta \\ I & A' \end{bmatrix} + \begin{bmatrix} I-T_n & \Theta \\ \Theta & I-T_n \end{bmatrix}\right) \neq 0,$$

for any $T_n = diag(t_1,...,t_n)$, $t_i \in [0,1]$ and this is obviously equivalent to $\det(T_n A + (I - T_n)) \neq 0$. The latter condition, as it is well known, is equivalent to the statement that $A$ is a *P*-matrix. Now, as shown by G.E.Coxson [1], the problem of determination of whether the given matrix is *P*-matrix is co-NP-complete.

Hence, the problem of determination for the given $A$ and $B$ whether the condition (ii) of Theorem 1 (or Theorem 2) holds for them, is co-NP-complete, is, in particular, NP-hard (even for the case of $m = n$, $B = I$). Consequently, sooner the polynomial algorithm, which solves this problem, does not exist.

*Remark 3.* Theorem 1 (Theorem 2, resp.) can be applied to establish if a full row (column, resp.) rank $n$-by-$m$ matrix will save the original property under unknown variations of some entries of its *s*th row (column, resp.), $1 \le s \le n$ ($1 \le s \le m$, resp.), in the given intervals. It should be noted that such a problem can be also solved by using the characterizations of the full rank property of all convex combinations of $n$-by-$m$ matrices $A_1,...,A_k$, $k \ge 2$, given in [6]. However, these characterizations refer to $k(k-1)n$-by-$k(k-1)n$ and $k(k-1)m$-by-$k(k-1)m$ matrices, respectively in the case of full row rank and full column rank, respectively and therefore, for $k > 2$, to larger matrices than (2) and (6), respectively.

We shall close the paper with two results on full rank property of the set $i(A,B)$. Using the notation introduced in Section 2 we have the following theorems.

THEOREM 3 *For the set $i(A,B)$ to be of full row rank, i.e., for each $C \in i(A,B)$ to be of full row rank, it suffices that $A_c$ is of full row rank and*

$$\rho(|I - A_c A_c'| + |A_c|\Delta') < 1.$$

**Proof:** Let $C$ be an arbitrary matrix from $i(A, B)$. Then, following [7, Section 5], we have

$$A_c C' = I - (I - A_c A_c' + A_c (A_c' - C'))$$

and since

$$\rho(I - A_c A_c' + A_c (A_c' - C')) \le \rho(|I - A_c A_c'| + |A_c||\Delta') < 1$$

it follows that $A_c C'$ is nonsingular. So, as $A_c C'$ is an $n$-by-$n$ matrix we get

$$rank(A_c C') = n$$

and from the property of the rank of the product of two matrices we obtain that $rank(C) = n$. So, the proof is completed. ∎

THEOREM 4 *For the set $i(A, B)$ to be of full column rank, i.e., for each $C \in i(A, B)$ to be of full column rank, it suffices that $A_c$ is of full column rank and*

$$\rho(|I_m - A_c' A_c| + |A_c'|\Delta) < 1.$$

**Proof:** The proof is entirely similar to the proof of Theorem 3 and is thus omitted. ∎

## Acknowledgment

The authors thank two anonymous referees for valuable comments and helpful suggestions on the original version of the paper.

## References

1. Coxson, G.E.: *The P-matrix problem is co-NP-complete*, Mathematical Programming **64** (1994), pp. 173-178.
2. Elsner, L. and Szulc, T.: *Block P-matrices*, Linear and Multilinear Algebra **44**(1998), pp. 1-12.
3. Fiedler, M.: *Special Matrices and Their Applications in Numerical Mathematics*, Nijhof, Dordrecht, 1986.
4. Golub, G.H. and Van Loan, C.F.: *Matrix Computations*, 3rd ed., Johns Hopkins University Press, Baltimore, 1996.
5. Johnson, C.R. and Tsatsomeros, M.J.: *Convex sets of nonsingular and P-matrices*, Linear and Multilinear Algebra **38**(1995), pp. 233-240.
6. Kołodziejczak, B. and Szulc, T.: *Convex combinations of matrices – full rank characterizations*, Linear Algebra Appl., to appear.
7. Rohn, J.: *Positive definiteness and stability of interval matrices*, SIAM J.Matrix Anal. Appl. **15**(1)(1994), pp. 175-184.
8. Szulc, T.: *Rank of convex combinations of matrices*, Reliable Computing **2**(2)(1996), pp. 181-185.

*T. Csendes (ed.), Developments in Reliable Computing 365–372.*
© 1999 *Kluwer Academic Publishers.*

# Multiaspect Interval Types

MICHAEL LERCH AND JÜRGEN WOLFF VON GUDENBERG

*Lehrstuhl Informatik II, Universität Würzburg, D-97074 Würzburg, Germany*
{lerch, wolff}@informatik.uni-wuerzburg.de

## 1. On the Way to a New Expression Concept

Until now, in nearly all programming languages expressions are based on operators and functions. The operands are evaluated and then the operator or function is called. The languages differ in provided standard operators and functions, supported data types and means for user defined types, functions and operators. Although most operators may be overloaded in modern languages, the evaluation semantics of an expression is not touched.

We are investigating new concepts for expression evaluation and treat expressions as a whole in order to increase accuracy or performance and give the user a facility to introduce an enhanced evaluation semantics. We think that optimizations like symbolic manipulation or rewriting expressions should not be exclusively in charge of the compiler, but may be influenced by the user.

In this paper we consider one aspect of this topic in more detail, we investigate how the enclosure of an interval expression may be sharpened. We begin with a discussion of algebraic properties of the mathematical and of the computer representable spaces which occur in scientific computing. A special emphasis is put on interval spaces. The algebraïc rules may be used in a symbolic preprocessing step.

We use multiaspect interval types for the computation of different enclosures of the same arithmetic expression $f$. We show how mutual intersection of intermediate results decreases overestimation of the range of $f$.

## 2. Algebraic Properties

### 2.1. Point Spaces

We start with a purely algebraic classification of point spaces with one operation with respect to the properties commutativity, associativity, cancellation rule, neutral element and unique inverse.

These are not independent, the cancellation rule is a consequence of associativity and the existence of inverse elements, e.g.

|  | Comm. | Assoc. | Canc. | Neutr. | Unique inv. |
|---|---|---|---|---|---|
| $(\mathbb{R}, +), (\mathbb{C}, +)$ | × | × | × | × | × |
| $(\mathbb{R}, *), (\mathbb{C}, *)$ | × | × | × | × | × |
| $(M\mathbb{R}, +), (M\mathbb{C}, +)$ | × | × | × | × | × |
| $(M\mathbb{R}, *) \ (M\mathbb{C}, *)$ |  | × |  | × |  |
| $(R, +), (C, +)$ | × |  |  | × | × |
| $(R, *), (C, *)$ | × |  |  | × |  |
| $(MR, +), (MC, +)$ | × |  |  | × | × |
| $(MR, *), (MC, *)$ |  |  |  | × |  |

Whereas for real or complex numbers all properties are fulfilled, multiplication for matrices is not commutative or cancellative nor do inverse elements exist.

Strictly speaking, none of the machine representable arithmetic spaces in the second part of the table bears any reasonable algebraic structure. From the algebraic point of view the missing associativity is annoying. If $R$ is a symmetric floating-point screen which supports denormalized numbers, like the IEEE 754 formats, unique additive inverses exist, except for the special cases ∞ or NaN. Despite that fact no cancellation law holds. Think of an addition of two small numbers to a very large one as a counter-example.

Combining the two operations $+, *$ together with the distributivity law, we obtain the well known field structure for scalar spaces and a non-commutative ring for matrices. The computer representable spaces again loose much of their algebraic structure and only fulfill the ringoid axioms [3] where the distributivity law only holds for the subset $\{-1, 0, 1\}$.

## 2.2. Set Spaces

For power set spaces and hence for theoretical interval spaces some of the properties for the corresponding point spaces are lost. Note that the cancellation law holds for addition of intervals but not for general unbounded sets. Inverse elements do not exist in traditional interval spaces, but in the space of directed intervals $\mathbb{D}$ [5].

The distributivity law is fulfilled only conditionally, in general we have subdistributivity. [8] or [9] give a complete characterization of the cases where equality holds. This can be exploited in an expression evaluation algorithm. A modified distributivity law is valid in $\mathbb{D}$ using inner operations. That opens up a range of further applications.

|  | Comm. | Assoc. | Canc. | Neutr. | Op. Closed |
|---|:---:|:---:|:---:|:---:|:---:|
| $(\mathbb{PR},+)$, $(\mathbb{PR},*)$ | × | × |  | × | × |
| $(\mathbb{PC},+)$, $(\mathbb{PC},*)$ | × | × |  | × | × |
| $(\mathbb{IR},+)$, $(\mathbb{IC}_R,+)$ | × | × | × | × | × |
| $(\mathbb{IR},*)$ | × | × |  | × | × |
| $(\mathbb{IC}_R,*)$ | × |  |  | × |  |
| $(\mathbb{IC}_C,+)$ | × | × | × | × | × |
| $(\mathbb{IC}_C,*)$ | × | × |  | × |  |
| $(\mathbb{IC}_S,+)$ | × | × |  | × | × |
| $(\mathbb{IC}_S,*)$ | × | × | × | × | × |

$\mathbb{IC}_R$, $\mathbb{IC}_C$ and $\mathbb{IC}_S$ denote the complex interval spaces with rectangles, circles [1] and sectors of circular rings [2] as interval elements.

As before, the corresponding matrix spaces keep the properties of their element spaces in case of addition and loose the commutativity and cancellation property in case of multiplication.

The machine representable spaces keep commutativity and the neutral element of the corresponding mathematical spaces. Associativity and cancellation law however is lost in general.

Note that, although directed rounding is used, subdistributivity does not hold in general, as can be proved by special counter-examples, e.g.:

$A = 8, B = C = [8.5, 8.6]$ in 2 digits decimal arithmetic yields

$A(B+C) = [130, 150] \supseteq [130, 140] = AB + AC$

An important property for a sharp enclosure of an expression is the measure of overestimation of the power set operations by the corresponding interval counterparts. If

$$\Diamond : \mathbb{PS} \to \mathbb{IS}$$

denotes the rounding from the power set to the interval space

$$(\text{RG}) \quad A \Diamond B = \Diamond(A \circ B)$$

defines the interval arithmetic in general. If, however,

$$(\text{OC}) \quad A \Diamond B = (A \circ B)$$

holds, the operation is closed in the interval space and a correct result may be obtained by an easily applicable formula.

As the table shows, different representations have different properties which can be exploited in applications. This leads us to the idea to combine various representations which we call aspects in one data type.

## 3.  Multiaspect Interval Data Types

### 3.1.  Data Types and Operations

Let us introduce some terms and notation.

A **multiaspect data type** is a collection of different representations — called **aspects** — of one item [11].

A **multiaspect interval type** is a multiaspect data type where the aspects are intervals.

A **multiaspect enclosure type** $(A_1 \times \ldots \times A_n)$ is a multiaspect interval type that represents a set of points $x$, i.e. each aspect $a_i \in A_i$ is a representation of an enclosure of $x$. The set $x$ may be empty, a single point or a bounded continuum. We can also consider discrete sets or arbitrary unconnected sets. Our enclosures, however, always are compact, connected sets.

The main idea with multiaspect enclosure types is to define arithmetic operations that yield enclosures better than every single aspect. Theoretically this can be achieved by intersection of all aspects. In practice it is not feasible to compute intersections of arbitrary aspects. Even for one aspect intersection is not closed usually. Similarly enclosures of arbitrary sets can hardly be defined.

Closed versions of arithmetic and set theoretic operations are defined for each aspect type $A_i$ separately. The intersection is denoted by $\cap_A$. By $[A](b)$ we denote the enclosure of the set $b \in B$ in the aspect type $A$.

To define arithmetic operations we introduce the normalization operator

$$\mathcal{N} : (A_1 \times \ldots \times A_n) \longrightarrow (A_1 \times \ldots \times A_n)$$

$$\mathcal{N} : (a_1, \ldots, a_n) \mapsto (\bigcap_{j=1}^{n}{}_{A_1} [A_1](a_j), \ldots, \bigcap_{j=1}^{n}{}_{A_n} [A_n](a_j))$$

Note that the set $x$ represented by the multiaspect type is not changed by the normalization. For better readability we restrict our further considerations to two aspect types $A$ and $B$.

For a 2-aspect type normalization reads

$$\mathcal{N}(a,b) := (a \cap_A [A](b), b \cap_B [B](a))$$

PROPOSITION 1   $\mathcal{N}$ *is idempotent:* $\mathcal{N}^2 \equiv \mathcal{N}$

Proof: The first component of $\mathcal{N}(\mathcal{N}(a,b))$ is $(a \cap_A [A](b)) \cap_A [A](b \cap_B [B](a))$. Since $b \cap_B [B](a) \subseteq B \subseteq [A](b) \subseteq a \cap_A [A](b)$, application of $[A]$ implies $[A](b \cap_B [B](a)) \subseteq a \cap_A [A](b)$, which completes the proof.

Apparently we need an enclosure function for each pair of aspects, in total $n(n-1)/2$. This number can be reduced to $n$ if we build a 'chain' of aspects

$$[A_{i+1}](a_i) \qquad i = 1, \ldots, n-1$$
$$[A_1](a_n)$$

Normalization produces an overestimation of the enclosures of the intersection but is easier to compute, formally:

**PROPOSITION 2** $[A](a \cap b) \subseteq a \cap_A [A](b)$

Proof: $(a \cap b) \subseteq [A]a \cap [A](b) \subseteq a \cap_A [A](b)$
enclosing in $[A]$ yields the assertion.

*Remark.* The proposition extends for types with more aspects. Intermediate intersections need not be enclosed as arbitrary sets.
$$[A](a \cap b \cap c) \subseteq a \cap [A](b \cap c) \subseteq a \cap_A [A](b) \cap_A [A](c)$$

**PROPOSITION 3** *A multiaspect enclosure consisting of one aspect and its enclosure* $(a, [B](a))$ *or* $([A](b), b)$ *is normalized.*

Proof: $a \subseteq [B](a) \Longrightarrow a \subseteq [A][B](a) \Longrightarrow a = a \cap [A][B](a)$

*Remark.* Normalization does not give the smallest enclosures in any aspect, even not for the available enclosures. $[A]((b) \cap [B](a)) \subseteq (a) \cap [B](a)$ is possible, e.g. let $a$ be circle containing a rectangle $b$.

*Remark.* The sharpness of the enclosure depends on the order of application.
$$[A][B](a)) \neq [B][A](a)$$

## Definition of Operations

Single argument constructors are provided, following proposition 3.

Arithmetic operations on multiaspect interval types are performed aspect-wise with subsequent normalization:

$$(a_1, b_1) \circ (a_2, b_2) := ((a_1 \circ a_2) \cap_A [A](b1 \circ b_2), (b_1 \circ b_2) \cap_B [B](a_1 \circ a_2))$$

Note that multiaspect enclosure types can be used to approximate arithmetic operations for arbitrary convex bodies [6].

### 3.2. Enclosure of Interval Expressions

Straightforward computation of an arithmetic expressions with multiaspect enclosure types is a very simple, transparent mechanism for an improvement of the range computation. Results are shown in section 6.

### 3.3. Multiaspectness in Algorithms

Some algorithms are tailored for the use of specific aspects, zero finding methods often use circles, whereas for the solution of liner systems rectangles may be more appropriate. A second idea for the use of multiaspectness is therefore to choose the appropriate aspect for each part of the algorithm. Of course, this can be realized by explicitly calling the single aspects. A fixed well known number of different aspects can also be combined in one data type and the proper aspects for the operations are selected dynamically by a global switch. Using the expression template facility [10, 4] this can be achieved without the overhead for the runtime test of the switch for an open number of aspects.

Multiaspect interval types may also be used in the inclusion test during the verification step in E–methods. If one aspect is completely contained in the interior of the previous value, the iteration can be stopped.

## 4. Conclusions, Rationale

Sharper enclosures of interval expressions may be obtained by symbolic manipulation and by the use of different aspects. Dynamic switching between different aspects may be helpful. So the requirements are the availability of the complete expression at runtime and an efficient evaluation and switching technique. In the next section we describe our implementation using arithmetic hierarchies and expression templates.

## 5. Our Implementation

We use C++ templates to provide an efficient mechanism for combining arbitrary aspect types:

```
template <class A, class B>
class MultiAspect;
```

This class template can be instantiated with any two types, e.g.:

```
typedef MultiAspect<Rectangle, Sector> RSInterval;
```

The arithmetic operations of the class `Multiaspect` are defined aspect-wise with a subsequent normalization:

```
MultiAspect<A,B> operator +(const MultiAspect<A,B>& x,
                            const MultiAspect<A,B>& y) {
    MultiAspect<A,B> result;

    result.aspect1 = x.aspect1 + y.aspect1;
    result.aspect2 = x.aspect2 + y.aspect2;

    A tempA = Enclosure<A,B>::enclose(result.aspect2);
    B tempB = Enclosure<A,B>::enclose(result.aspect1);

    result.aspect1.intersect(tempA);
    result.aspect2.intersect(tempB);

    return result;
}
```

Since nothing is known about the particular aspects a priori, the computation of the enclosures is delegated to a so-called *traits class* Enclosure. A traits class [7] is general class template which defines default behaviour for all possible instantiations. In our case this class is empty:

```
template <class A, class B>
class Enclosure { };
```

For the interesting cases *template specializations* have to be provided, e.g.:

```
template <>
class Enclosure<Rectangle, Sector>
{
public:
  static Rectangle enclose(const Sector& sect);
  static Sector enclose(const Rectangle& rect);
};
```

An important feature of all template techniques is runtime efficiency, because all instantiation and selection work is done at compile time.

Multiaspect data types with more than two aspects can be constructed by recursive instantiation of the MultiAspect template, e.g.:

```
typedef MultiAscpect<Sector, Circle> SCInterval;
typedef MultiAspect<Rectangle, SCInterval> RSCInterval;
```

To avoid the explosion of the number of specializations to be provided, we define a so-called *partial specialization* of Enclosure which covers all recursive instantiations:

```
template <class A, class B, class C>
class Enclosure<A, MultiAspect<B,C> >
{
public:
  static A enclose(const MultiAspect<B,C>& bc) {
    A temp1 = Enclosure<A,B>::enclose(bc.aspect1);
    A temp2 = Enclosure<A,C>::enclose(bc.aspect2);
    return temp1.intersect(temp2);
  }

  static MultiAspect<B,C> enclose(const A& a) {
    MultiAspect<B,C> result;
    result.aspect1 = Enclosure<A,B>::enclose(a);
    result.aspect2 = Enclosure<A,C>::enclose(a);
    return result;
  }
};
```

## 6. Experimental Results

For our tests we consider complex intervals, represented by the multiaspect data type RSInterval with rectangles and sectors of circular rings as aspects. We computed the range for various expression types (sums, products, polynomials, rational). Input intervals were generated randomly with different widths and position.

Each expression is evaluated for each aspect separately (single mode) and in multiaspect mode. Input data are rectangles or sectors — in this case the other aspect is generated by enclosure – or normalized multiaspect enclosures with non-empty point sets. For each aspect we measure the improvement by comparing the area of the results in single and multiaspect mode.

As expected, for sums the rectangular aspect could not be improved since addition is closed, whereas the sector aspect was improved by 74–85%, i.e. the area of the sector computed in multiaspect mode is only 15–26% of the area of the sector computed in single mode. Accordingly, for products sectors were optimal and rectangles could be improved by 66-90%. This holds for any type of input data.

For arbitrary arithmetical expressions improvements up to 90% were registered in nearly all cases for both aspects. We noticed a dependency of the distance from zero and of the order of involved operations. Sectors showed superior for input data close to zero, especially for expressions starting with multiplication.

Even better results could be obtained if input data were not normalized. Note that normalization is already a benefit of the multiaspect mode.

## 7. Acknowledgements

This work was partially supported by the Deutsche Forschungsgemeinschaft (DFG) under grant DFG-GZ 436 BUL 113/98/0.

## References

1. Alefeld, G. and Herzberger, J.: *Einführung in die Intervallrechnung*, BI, 1974.
2. Klatte, R. and Ullrich, Ch.: Complex sector arithmetic, *Computing* 24, 1980.
3. Kulisch, U. and Miranker, W.L.: *Computer Arithmetic in Theory and Practice*, Academic Press, 1981.
4. Lerch, M. and Wolff von Gudenberg, J: Expression Templates for Dot Product Expressions, to appear in *Reliable Computing* 5(1) (1999).
5. Markov, S.: On Directed Interval Arithmetic and its Applications, *JUCS* 1(7), 1995.
6. Markov, S.: On the Algebraic Properties of Convex Bodies and Some Applications, *preprint*, 1998.
7. Myers, N.C.: Traits: A New and Useful Template Technique, *C++ Report*, June 1995.
8. Ratschek, H.: Die Subistributivität der Intervallarithmetik, *ZAMM* 54, 1971.
9. Spaniol, O.: Die Distributivität in der Intervallarithmetik, *Computing* 5, 1970.
10. T. Veldhuizen: Expression Templates, *C++ Report* 7(5), June 1995.
11. Yakovlev, A.G.: Multiaspectness and Localization, *Interval Computations* No. 4, 1993.

*T. Csendes (ed.), Developments in Reliable Computing* 373–382.
© 1999 *Kluwer Academic Publishers.*                                    373

# MATLAB-Based Analysis of Roundoff Noise

REZSŐ DUNAY AND ISTVÁN KOLLÁR                    {dunay, kollar}@mit.bme.hu
*Department of Measurement and Information Systems, Technical University of Budapest, Budapest,
Műegyetem rakpart 9, Hungary, H-1521*

**Abstract.** All arithmetic operations can be decomposed into an infinitely accurate calculation and a
subsequent arithmetic rounding. The kind of rounding in use determines the properties of the whole
arithmetic system. The article describes a tool for the analysis of the effects of rounding errors during
arithmetic calculations in practical systems.

**Keywords:** arithmetic, rounding, quantization, finite word length, DSP.

## 1. Introduction

In cost-effective systems, or systems with low power consumption, even today low
bit-number, fixed-point arithmetic is used usually. The design of software for these
systems requires special considerations from the algorithmic point of view (e.g. the
optimization of low bit-number coefficients). In this paper we describe an affordable tool for the analysis of such systems. Since personal computers are available
to almost everyone, these were chosen as primary platform for our simulation environment.

The paper begins with enumeration and comparison of some terms, to define
the notation. The next part summarizes some special requirements of practical systems. Two sections deal with the problems of our common simulation environment
for numerical algorithm simulation, one of which is dedicated to the problems of
our MATLAB based implementation. The following section examine the applicability of our simulation environment for system analysis and verification. Finally
some conclusions are drawn.

## 2. Notation

Most of today's computers produce the result of all arithmetic operations by rounding the exact result to the destination accuracy. All such arithmetic operations can
be decomposed into an infinitely accurate calculation and a subsequent rounding
operation.

In engineering practice, quantities which may have only a finite number of possible values, are called *quantized* quantities. The unit which produces a quantized

value from its continuous input is usually referred to as a *quantizer*. Its operation is to round the input to the required accuracy, using the 'rules of rounding'. Thus we can state that all arithmetic operations can be simulated by an accurate operation and a subsequent quantizer.

It is straightforward to see that all fixed-point operations can be described by using a simple equidistant quantizer. There are some new results that analyze floating-point number systems too from this aspect [2].

## 3. Practical Considerations

There are many attributes of engineering work that originate in limited hardware or software resources, or in the nature of the problem to be solved. Two of these aspects will be mentioned here, which - especially together - have a great influence on the analysis of the arithmetical side of realizations.

*Feedback* plays a central role in many fields of engineering. It is also an inherent property of all iterative processes: they re-use the result of the previous iteration to calculate the new one. This is one of the most common approaches in engineering science, which can help to speed up systems, to decrease error levels, etc. In practice there are more feedback systems than purely feedforward ones. This is why feedback systems and their analysis is so important for engineers.

*Usually very large number of iterations* are necessary. Many digital signal processing systems - like process control units - work 24 hours a day, executing many thousands of steps in each second, processing the results of the previous steps together with the new incoming data. The systems are usually implemented on digital hardware with arithmetic often using no more than 16-bits.

All the examples, provided at the end of the article, illustrate the above two properties.

The hardware, engineers use not always conforms to the well-known IEEE standard [4]. Practical systems (using e.g. digital signal processors) use various rounding modes and a general simulation environment should be able to handle all of them. The most important quantization modes (or rounding rules) supported by our MATLAB toolbox are listed in Table 1.

Practical systems frequently use mixed accuracy data (e.g. different data types in programming languages, higher precision accumulator in DSP and common microprocessors), so a reasonable simulation environment should be able to simultaneously handle data types of different accuracy.

Though it is not a property of the system itself, but is greatly connected to its application, whether a 100% verified solution is needed or not. It is one of the major decisions in the design process and will be examined in more detail in section 6.

## 4. Simulation Environment

Our goal was to create an environment optimal for testing and analyzing roundoff effects. The simulation environment should provide a reliable means complying with the following requirements:

*Reproducibility:* arithmetic operations should produce a well-defined result. This is necessary to be able to reproduce and certify the experiments [5].

*Flexibility:* The user should be able to control all important aspects of all the operations to handle all kinds of different systems, see Table 1.

*Easy accessibility:* The user should be able to easily access the functionality, express the ideas as close as possible to the engineering notation.

The best solution would be to have a single environment capable of simulating the very same program using the native arithmetic of the host computer (this is fast), and using a user defined arithmetic (this is needed by engineers). Fortunately these options differ only in the underlying arithmetic (type of the data and the operations), but semantically (at the level of the user's program) they are the same, and

*Table 1.* The most important quantizer properties

| Property | Description | Values |
|----------|-------------|--------|
| Operation | Operation mode | quantize, idle, noise model |
| Type | Type of quantizer | fixed, floating-point |
| Precision | Number of 'mantissa' bits | 1..53 |
| FractBits | Number of fraction bits | 1..53 |
| Coding | Coding of the value | one's, two's compl., sign-magn. |
| RoundDir | Direction of rounding | trunc, round, floor, ceil, toinf |
| TreatHalf | Handling of .5 LSB values | floor, ceil, fix, toinf, toeven, toodd, rand |
| Overflow | Type of overflow | clip, modular, triang, inf, none |
| Emin, Emax | Exponent range | -1023..1022 |
| Underflow | Type of underflow | gradual, flush to zero |
| ExpCoding | Coding of the exponent | one's, two's compl., sign-magn., biased |
| ExpBias | Bias of the exponent | |
| LeadingBit | Leading bit present? | Hidden, present |
| UfCoding | Coding of underflow | exponent, leading zero |
| Data | Actual value | |
| PrecData | Accurate value | |

LSB: least significant bit

can effectively be implemented in any object oriented programming language. This means that once developed, the same user program could simulate the system using high precision IEEE arithmetic, but a low precision simulation is also possible (in theory a higher-than-native accuracy is also possible!). Our MATLAB toolbox is being written exactly for this purpose. We would like to give direct support for a wide range of Digital Signal Processors by creating quantizers that correspond to their registers, memory words and instructions.

## 5.  Implementation in MATLAB [3]

Engineers use MATLAB very frequently, as its numerical abilities are acceptable, while the description of algorithms is quite easy in its programming language.

MATLAB seems to be a perfect environment for our purpose. Starting with version 5.0 it is possible to use object oriented techniques to define our new data type for quantized data (quantized data class), instances of which can store quantized elements, together with the parameters of the used quantizer (e.g. the accuracy of the data itself is a property contained in the quantized variable itself). We defined only one single quantized data class, for efficiency reasons. The quantized versions of the basic operations belonging to our new data type are defined as required in MATLAB.

MATLAB itself is a homogenous system in the sense that practically only a single floating-point data type (usually the IEEE double precision) is applied. As stated before, in practical applications mixed accuracy data may be used. As MATLAB is an interpreted language (and we use an object oriented approach) the type (accuracy) of the result of a mixed precision operation is not defined by the operation itself, but it is only determined by the the accuracy of the operands. In our system the accuracy of the result is determined by the accuracy of the leftmost quantized-type input parameter. Mixed accuracy data leads to an important question concerning data conversions: a phenomenon called re-quantization should be avoided (the result of the re-quantization of a previously quantized data may not be equal to the data quantized by the second quantizer). The user should be given freedom in dealing with this problem, including string-to-number, number-to-string conversions.

In the current realization we use MATLAB's internal data type to store quantized data. The possibility of higher precision (longer mantissa) and larger dynamical range (longer exponent) are under investigation, but not yet supported.

## 6. Analysis and verification

In engineering practice analysis gives means for verification and validation. Its purpose is to prove - at least in some respect - the applicability of the designed system. It can basically be applied in two different ways:

*On-line verification:* the system is automatically checked during operation. It is based on the actual operational conditions and may need 100% reliability. In case of error (e.g. numerical inaccuracy or fault within the software) the necessary action (restart, user interaction) is taken. It is used only in mission-critical applications.

*Off-line validation:* the system behavior is analyzed off-line, checked for all expected operational conditions. 100% reliability is usually not needed and generally cannot be guaranteed.

From scientific point of view the above two methods are identical, but it can be an essential difference, whether the solution is 100% reliable ('validated solution') or not ('practical solution').

### 6.1. Validated solutions

The most clear verification process uses 100% validated results which in engineering terms is called 'worst case' design. Low bit-number, fixed-point arithmetic is used in many systems even today and usually they execute quite complex algorithms, briefly discussed in Section 3. The validation of these systems generally cannot be carried out with 100% reliability based on analytical methods (using System Theory) due to the lack of time, money or even *knowledge*. Because of this engineers need methods which can evaluate their algorithms in a (semi)automatic way.

Interval arithmetic provides an automatic way to calculate hard bounds, but it is automatic only when used in a 'naive' way. The examples at the end of the article demonstrate the well-known fact that the naive application of intervals usually leads to the exponential growth of the diameter of the intervals and in case of a low bit-number arithmetic the rate of the growth can be very high. This is mainly due to the fact that the naive approach does not take into account the structure of the modeled system.

Stable systems have an inherent property which causes injected errors to decay as time, i.e. iterations, go on. As quantization errors can be well modeled to be additive, in case of a stable system their effect should decrease as time passes. This 'forgetting' phenomenon is due to the above special property of all stable systems and is not taken into account during the naive application of interval arithmetic.

It would be very useful to develop algorithms that can automatically take into account the structural connections of the system and create a lot more optimistic result, than the naive approach [10]. An algorithm using the unification of interval arithmetic and the so called Bertram bound [11] is under investigation for linear systems.

### 6.2. *Practical solutions*

In practical applications the 'worst case' design is usually too expensive (e.g. dykes are not built high enough to beat off all flood, because it is cheaper to rebuild villages once every 100 years, than to construct higher dykes). In most cases designers use statistical methods based on 'confidence intervals' rather ('once every 100 years'). These methods do not guarantee limits to the results, but may provide intervals with reasonably high-level of confidence. To be able to produce useful results the meaning of reliability is reconsidered in a more practical way.

In the most frequently used, noise-model based confidence calculations the quantization - or roundoff - errors are modeled with independent additive noise (The 'amplitude' of this noise is of course connected to the width of the intervals in an interval arithmetic based solution, but the supposed independence and the additive nature of the added noise usually leads to more realistic results.).

Quantization theory provides also analytical ways to handle the effects of these independent noise sources (e.g. the effect of an Analog-Digital converter on the input of a system can be correctly analyzed). In complex practical systems, however simulation is used to calculate error levels on the output of the system. If we want to evaluate a statistical analysis on the output error, we have to execute many simulations on the same system, using the same input. The problem we have to face is that our deterministic system will always produce the same result for the same input. The solution to this problem may be the application of the Pseudo Quantization Noise. In this method instead of actually quantizing, a random value (noise with correctly designed properties) is added to the accurate result of the operations. Automatic replacement of actual quantizations with PQN is also built into our quantization toolbox.

This approach also has its limitations. The main problem is that the noise model is not always valid (like in the last example of section 7) and the independence of the separate noise sources cannot always be proven. There are techniques available (dithering) that aim at assuring the independence of noise components [8], [9].

## 7. Examples

The examples will demonstrate the simulation of some simple, but from engineering point of view interesting systems using low bit-number arithmetic. It is important to note here that the calculation of interval arithmetic on low bit-number systems needs our quantization toolbox, and cannot be done using any other interval arithmetic system.

The following examples are realizations of pre-designed Infinite Impulse Response (IIR) digital filters, implemented in a low bit-number arithmetic. The examples are simple, but they have a high relevance from engineering point of view. The simulations were carried out using our own quantization functions, interval calculations are also based on correctly implemented quantizers.

To avoid scaling problems, the simulations used a special floating-point arithmetic, having a 12 or 20 bits long mantissa, and a 4-bit exponent. The number format used - except for the bit numbers - corresponds to the IEEE standard (hidden leading bit, sign-magnitude coding, biased exponent). The one using a 12 bits long mantissa occupies 16 bits, so we could also call it a 'half-precision IEEE floating point number'.

In each example the following simulations were run:

*Low precision:* A simulation was carried out using the above special arithmetic, applying rounding, corresponding to the IEEE standard (round to even) [4]. The input signal and the coefficients were rounded to the lower precision using the IEEE round to even rule. The result is a dotted curve in all figures.

*Low precision interval arithmetic:* The quantization functions were used to calculate intervals. The input signal and the coefficients - like in the previous case - were rounded to the lower precision using the IEEE round to even rule. Intervals are shown as dashed lines (both the upper and the lower bound).

*IEEE double:* As a reference, the system was also simulated using IEEE double precision arithmetic. The 53 bits of the mantissa in these examples produce accurate enough result. In this case the input and the coefficients were not rounded. A continuous line is used on the graphs.

In all simulations the state variables of the systems were initialized to zero ($y(k) = x(k) = 0$, if $k < 0$).

In Fig. 1. the simulation result of the following two systems can be seen, using a 12 bits long mantissa:

$$y(k) = -x(k-1) + 0.999y(k-1), \tag{1}$$

$$y(k) = -x(k-2) + 1.3y(k-1) - 0.42y(k-2). \tag{2}$$

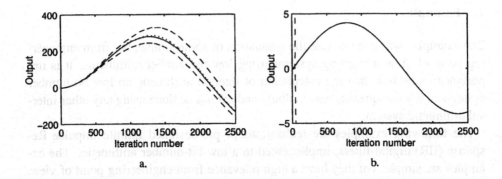

*Figure 1.* a. Results of system (1). b. Results of system (2). In both cases 12 bits long mantissa was used. Solid line: IEEE double precision; dotted line: low precision; dashed line: low precision interval arithmetic.

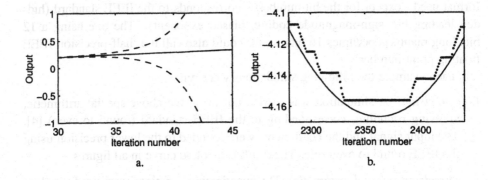

*Figure 2.* a. Exponential growth of the diameter of the intervals. b. The correlated error values. Magnified parts of Fig. 1b.

It can clearly be seen in the figures that the error of the 12 bits long IEEE rounding mode produces quite good results even after 2500 steps. The intervals get wide quite fast, and - as expected - in case of system (1) the diameter grows exponentially (magnified in Fig. 2a.). Fig. 2b. shows a magnified part of the simulation of system 2. In this picture the dots correspond to iteration steps of the 12-bit system. The graph well demonstrates that error values belonging to subsequent quantization steps are highly correlated.

Figure 3 shows a simulation example of an industry standard 1010 Hz Band rejection filter, declared in [1], which has 3 complex pole pairs and 3 complex zero pairs. Figure 3 shows the impulse response simulation of this system. Though the simulation used 20 mantissa bits, the intervals get unacceptable within 10 iteration steps.

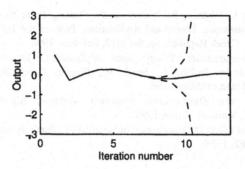

*Figure 3.* Simulation of the 1010 Hz Band rejection filter; impulse response; 20 bits long mantissa

## 8. Conclusions

Based on our MATLAB toolbox that is able to realize arbitrary quantizers it is possible to simulate models running on any arithmetic system. Our primary goal with the toolbox was to have an affordable tool for the statistical analysis of low bit-number systems, but we also want to use it for numerical validation leading to more realistic results than the 'naive' application of interval arithmetic.

The extension of the toolbox to directly support a wide range of Digital Signal Processors is under way. The directed versions of string-to-number and number-to-string conversion is also a task to be solved.

## References

1. IEEE Standard Equipment Requirements and Measurement Techniques. ANSI/IEEE Standard 734-1985, New York, NJ, Apr. 1985.
2. Kollár, I., "Statistical Theory of Quantization," Doctoral Thesis, Hungarian Academy of Sciences, Budapest, 1996, 416 p.
3. MATLAB User's Manual, Version 5.2, The MathWorks, Inc., Natick, MA, 1998.
4. IEEE Standard for Binary Floating-Point Arithmetic. ANSI/IEEE Standard 754-1985, New York, NJ, Aug. 1985.
5. Kahan, W., "The Baleful Effect of Computer Benchmarks upon Applied Mathematics, Physics and Chemistry", 1995.
   http://http.cs.berkeley.edu/~wkahan/ieee754status/baleful.ps
6. Hammer, I., Hocks, M., Kulisch, U., Ratz, D. "C++ Toolbox For Verified Computing", Springer-Verlag, 1995.
7. Rump, S. M., "From INTLAB to MATLAB", SCAN-98 IMACS/GAMM International Symposium on Scientific Computing, Computer Arithmetic and Validated Numerics, Budapest, Hungary, 22-25. Sep., 1998., Text Submitted to the special issue of Reliable Computing with the title "INTLAB Interval Laboratory".

8. Dunay, R., Kollár, I., Widrow, B., "Dithering for Floating Point Number Representation", Dithering in Measurement: Theory and Applications, Proc. of the 1st International On-Line Workshop, Prague, Czech Republic, pp. 9/1-9/12, Feb-Mar, 1998

9. Dithering in Measurement: Theory and Applications, 1st International On-Line Workshop, Edited by Holub, J., Smid, R., Prague, Czech Republic, Mar. 1998, http://measure.feld.cvut.cz/dithering98/

10. Ludyk G.,"CAE von Dynamischen Systemen, Analyse, Simulation, Entwurf von Regelungsstemen", Springer-Verlag, 1990.

11. Bertram, J. E., "The Effect of Quantization in Sampled-Feedback Systems", Trans. AIEE, Vol. 77, pt. 2, pp. 177–82, 1958.

*T. Csendes (ed.), Developments in Reliable Computing* 383–402.
© 1999 *Kluwer Academic Publishers.*

# SCAN-98 Collected Bibliography

GEORGE F. CORLISS                                                            georgec@mscs.mu.edu
*Department of Mathematics, Statistics, and Computer Science, Marquette University, P.O. Box 1881, Milwaukee, WI 53201–1881 USA*

**Abstract.** This is a bibliography of work related to interval arithmetic and validated computation. It represents the compilation of all works cited by the papers in this volume.

**Keywords:** Bibliography, validated computation.

This bibliography represents the common lists of references for all of the chapters in this papers in this volume. Each author compiled the references for his or her own paper. The separate bibliographies were merged into a single BibTex database. Because it includes all of the works cited by any paper in this volume, this bibliography includes many citations that are not directly related to interval analysis. We thank the authors of each article in this volume for their assistance in preparing this bibliography.

The electronic version of this bibliography contains fields not listed in the printed version including cross-references, ISBN, keywords, some abstracts, and electronic source (when available). The BibTex source is available from
`www.mscs.mu.edu/~georgec/Pubs/chapt.html#1999b`
at link [ scan98.bib ]. Corrections and updates are welcome and should be sent to the address above.

## References

1. J. P. ABBOTT AND R. P. BRENT, *Fast local convergence with single and multistep methods for nonlinear equations*, Austr. Math. Soc. (Series B), 19 (1997), pp. 173–199.
2. E. ADAMS, D. CORDES, AND R. J. LOHNER, *Enclosure of solutions of ordinary initial value problems and applications*, in Discretization in Differential Equations and Enclosures, E. Adams, R. Ansorge, C. Großmann, and H. G. Roos, eds., Akademie-Verlag, Berlin, 1987, pp. 9–28.
3. N. I. AHIESER, *Elements of the Elliptic Functions Theory*, Nauka, Moscow, 1970. In Russian.
4. A. AKRITAS, A. BOCHAROV, AND A. STRZEBONSKI, *Implementation of real root isolation algorithms in Mathematica*, in Conference Materials of Interval '94 (St. Petersburg, Russia, 1994), 1994, pp. 23–27.
5. G. ALEFELD AND J. HERZBERGER, *Einführung in die Intervallrechnung*, Springer-Verlag, Heidelberg, 1974.
6. ——, *Introduction to Interval Computations*, Academic Press, New York, 1983.
7. G. ALEFELD AND G. MEYER, *The Cholesky method for interval data*, Linear Algebra Appl., 194 (1993), pp. 161–182.

8. P. ALESSANDRI AND V. BERTHÉ, *Three distance theorems and combinatorics on words*, Research Report, Institut de Mathématiques de Luminy, Marseille, France, 1997.

9. E. ANDERSON, Z. BAI, C. BISCHOF, J. DEMMEL, J. DONGARRA, J. DU CROZ, A. GREENBAUM, S. HAMMARLING, A. MCKENNEY, S. OSTROUCHOV, AND D. C. SORENSEN, *LAPACK User's Guide, Release 2.0*, SIAM, Philadelphia, second ed., 1995.

10. ANSI/ISO, *Working paper for draft proposed international standard for information systems programming language C++*, Tech. Report ANSI X3J16/96-0225 ISO WG21/N1043, ANSI/ISO, 1996.

11. E. S. ARMSTRONG, *An extension of Bass' algorithm for stabilizing linear continuous constant systems*, IEEE Trans. Automat. Control, AC–20 (1975), pp. 153–154.

12. R. BARNHILL, G. FARIN, M. JORDAN, AND B. PIPER, *Surface/surface intersection*, Computer Aided Geometric Design, 4 (1987), pp. 3–16.

13. W. BARTH, R. LIEGER, AND M. SCHINDLER, *Ray tracing general parametric surfaces using interval arithmetic*, The Visual Computer, 10 (1994), pp. 363–371.

14. W. BARTH AND E. NUDING, *Optimale Lösung von Intervallgleichungssystemen*, Computing, 12 (1974), pp. 117–125.

15. H. BAUCH, K. JAHN, D. OELSCHLÄGEL, H. SÜSSE, AND V. WIEBIGKE, *Intervallmathematik Theorie und Anwendungen*, Teubner, Leipzig, 1987.

16. C. BENDSTEN AND O. STAUNING, *FADBAD, a flexible C++ package for automatic differentiation using the forward and backward methods*, Tech. Report 1996-x5-94, Department of Mathematical Modelling, Technical University of Denmark, Lyngby, Denmark, August 1996.

17. ——, *TADIFF, a flexible C++ package for automatic differentiation using Taylor series*, Tech. Report 1996-x5-94, Department of Mathematical Modelling, Technical University of Denmark, Lyngby, Denmark, April 1997.

18. F. BENHAMOU, F. GOUALARD, AND L. GRANVILLIERS, *Programming with the DecLIC language*, in Proceedings of the Second Workshop on Interval Constraints (October 1997, Port-Jefferson, NY), 1997.

19. F. BENHAMOU, D. MCALLESTER, AND P. VAN HENTENRYCK, *Clp(intervals) revisited*, in Proceedings of the 1994 International Symposium, M. Bruynooghe, ed., MIT Press, 1994, pp. 124–138.

20. F. BENHAMOU AND W. OLDER, *Applying interval arithmetic to real, integer and Boolean constraints*, Journal of Logic Programming, 32 (1997), pp. 1–24.

21. S. BERNER, *Ein paralleles Verfahren zur verifizierten globalen Optimierung*, PhD thesis, Bergische Universität GH Wuppertal, 1995.

22. J. E. BERTRAM, *The effect of quantization in sampled-feedback systems*, Trans. AIEE, 77 (1958), pp. 177–182.

23. M. BERZ AND G. HOFSTÄSTER, *Computation and application of Taylor polynomials with interval remainder bounds*, Reliable Computing, 4 (1998), pp. 83–97.

24. G. BIRKHOFF, *Lattice Theory*, AMS Colloquium Public., 25, AMS, New York, 1940.

25. G. BIRKHOFF AND R. S. VARGA, *Discretization errors for well-set Cauchy problems: I*, J. Math. and Phys., 44 (1965), pp. 1–23.

26. BLAGOVEST SENDOV AND G. BEER, *Hausdorff Approximations*, Kluwer, Dordrecht, Netherlands, 1990.

27. BLITZ++, *Homepage*. http://monet.uwaterloo.ca/blitz/, 1998.

28. J. E. BOBROW, *A direct minimization approach for obtaining the distance between convex polyhedra*, The International Journal of Robotics Research, 8 (1989), pp. 65–76.

29. K. BRAUNE, *Standard functions for real and complex point and interval arguments with dynamic accuracy*, Computing, Suppl., (1988), pp. 159–184.

30. R. P. BRENT, *Algorithm 524: MP, a FORTRAN multiple-precision arithmetic package*, ACM Trans. Math. Software, 4 (1978), pp. 71–81.

31. C. BREZINSKI AND M. R. ZAGLIA, eds., *Extrapolation Methods: Theory and Practice*, Studies in Computational Mathematics, No. 2, North-Holland, 1991.

32. ———, *A general extrapolation procedure revisited*, Advances Comput. Math., 2 (1994), pp. 461–477.

33. B. P. BUCKLES AND F. PETRY, *Genetic Algorithms*, IEEE Computer Society Press, 1992.

34. A. BUNSE–GERSTNER, R. BYERS, AND V. MEHRMANN, *Numerical methods for algebraic Riccati equations*, in Lecture Notes of the Workshop on "The Riccati Equation in Control, Systems, and Signals" (Como, Italy), S. Bittanti, ed., Pitagora Editrice, Bologna, 1989, pp. 107–115.

35. R. G. BURGER AND R. K. DYBVIG, *Printing floating point numbers quickly and accurately*, Sigplan Notices, 31 (1996), pp. 108–116.

36. B. F. CAVINESS AND J. R. JOHNSON, eds., *Quantifier Elimination and Cylindrical Algebraic Decomposition*, Texts and Monographs in Symbolic Computation, Springer-Verlag, Berlin, 1998.

37. L. CHEN, W. Q. HUANG, AND E. M. SONG, *A fast algorithm for computing the distance between two convex polyhedra*, Math. J. Chinese Univ., 16 (1994), pp. 345–359.

38. P. CHIN, R. M. CORLESS, AND G. F. CORLISS, *Globsol case study: Inexact greatest common denominators*, Technical Report, Marquette University Department of Mathematics, Statistics, and Computer Science, Milwaukee, Wisc., 1998.

39. ———, *Optimization strategies for the approximate GCD problem*, in Proceedings of the 1998 International Symposium on Symbolic and Algebraic Computation, O. Gloor, ed., New York, 1998, ACM Press, pp. 228–235.

40. D. CHIRIAEV, *Framework for reliable computing in Java(TM) programming language.* Preprint., 1998.

41. D. CHIRIAEV AND G. W. WALSTER, *Interval arithmetic specification.* www.mscs.mu.edu/~globsol/Papers/spec.ps, May 1998.

42. P. G. CIARLET, *The Finite Element Method for Elliptic Problems*, North-Holland, Amsterdam, 1978.

43. J. C. CLEARY, *Logical arithmetic*, Future Computing Systems, 2 (1987), pp. 125–149.

44. L. COLLATZ, *Funktionalanalysis und Numerische Mathematik*, Springer-Verlag, Berlin, 1964.

45. ———, *Differentialgleichungen*, 6. Aufl. Teubner, Stuttgart, 1981.

46. H. COLLAVIZZA, F. DELOBEL, AND M. RUEHER, *A note on partial consistencies over continuous domains solving techniques*, in Proceedings of the Fourth International Conference on Principles and Practice of Constraint Programming (CP '98), Lecture Notes in Computer Science, No. 1520, Berlin, 1998, Springer-Verlag.

47. G. E. COLLINS, *Quantifier elimination for the elementary theory of real closed fields by cylindrical algebraic decomposition*, in Automata Theory and Formal Languages, H. Brakhage, ed., Lecture Notes in Computer Science, No. 33, Springer-Verlag, Berlin, 1975, pp. 134–183.

48. G. E. COLLINS AND H. HONG, *Partial cylindrical algebraic decomposition for quantifier elimination*, J. Symbolic Comp., 12 (1991), pp. 299–328.

49. A. CONNELL AND R. M. CORLESS, *An experimental interval arithmetic package in Maple*, Interval Computations, 2 (1993), pp. 120–134.

50. R. M. CORLESS, *Cofactor iteration*, SIGSAM Bulletin: Communications in Computer Algebra, 30 (1996), pp. 34–38.

51. G. F. CORLISS, *Industrial applications of interval techniques*, in Computer Arithmetic and Self-Validating Numerical Methods, C. Ullrich, ed., Notes and Reports in Mathematics in Science and Engineering, No. 7, Academic Press, New York, 1990, pp. 73–90.

52. ——, *Comparing software packages for interval arithmetic*. Presented at SCAN '93, Vienna, 1993.

53. ——, *SCAN-98 collected bibliography*, in Developments in Reliable Computing, T. Csendes, ed., Kluwer, Dordrecht, Netherlands, 1999, pp. 383–402.

54. G. F. CORLISS AND R. B. KEARFOTT, *Rigorous global search: Industrial applications*, in Developments in Reliable Computing, T. Csendes, ed., Kluwer, Dordrecht, Netherlands, 1999, pp. 1–16.

55. G. F. CORLISS AND L. B. RALL, *Adaptive, self-validating quadrature*, SIAM J. Sci. Stat. Comput., 8 (1987), pp. 831–847.

56. ——, *Computing the range of derivatives*, in Computer Arithmetic, Scientific Computation, and Mathematical Modelling, E. W. Kaucher, S. M. Markov, and G. Mayer, eds., vol. 12 of IMACS Annals on Computing and Applied Mathematics, J. C. Baltzer, Basel, 1991, pp. 195–212.

57. G. F. CORLISS AND R. RIHM, *Validating an a priori enclosure using high-order Taylor series*, in Scientific Computing and Validated Numerics: Proceedings of the International Symposium on Scientific Computing, Computer Arithmetic and Validated Numerics - SCAN '95, G. Alefeld, A. Frommer, and B. Lang, eds., Akademie Verlag, Berlin, 1996, pp. 228–238.

58. G. E. COXSON, *The P-matrix problem is co-NP-complete*, Mathematical Programming, 64 (1994), pp. 173–178.

59. T. CSENDES, ed., *Developments in Reliable Computing*, Kluwer, Dordrecht, Netherlands, 1999.

60. J. DANIELS, S. RANJIT, R. B. KEARFOTT, AND G. F. CORLISS, *Globsol case study: Currency trading (Swiss Bank Corp.)*, Technical Report, Department of Mathematics, Statistics and Computer Science, Marquette University, Milwaukee, Wisc., 1998.

61. G. DARBOUX, *Sur les dèveloppements en série des fonctions d'une seule variable*, J. des Mathématique pures et appl., (1876), pp. 291–312. 3ème série, t. II.

62. E. DAVIS, *Constraint propagation with interval labels*, Journal of Artificial Intelligence, 32 (1987), pp. 281–331.

63. P. J. DAVIS AND P. RABINOWITZ, *Methods of Numerical Integration*, Academic Press, New York, 2nd ed., 1984.

64. J. P. DELAHAYE, *Optimalité du procédé $\delta^2$ d'Aitken pour l'accélération de la convergence linéaire*, RAIRO, Anal. Numér., 15 (1981), pp. 321–330.

65. J. P. DELAHAYE AND B. GERMAIN-BONNE, *Résultats négatifs en accélération de la convergence*, Numer. Math., 35 (1980), pp. 443–457.

66. N. S. DIMITROVA AND S. M. MARKOV, *On the interval-arithmetic presentation of the range of a class of monotone functions of many variables*, in Computer Arithmetic, Scientific Computation and Mathematical Modelling, E. W. Kaucher, S. M. Markov, and G. Mayer, eds., J. C. Baltzer, Basel, 1991, pp. 213–228.

67. ——, *A validated Newton type method for nonlinear equations*, Interval Computation, 2 (1994), pp. 27–51.

68. ——, *Verified computation of fast decreasing polynomials*, in Developments in Reliable Computing, T. Csendes, ed., Kluwer, Dordrecht, Netherlands, 1999, pp. 229–240. Also Reliable Computing 5/3 (1999), pp. 229–240.

69. N. S. DIMITROVA, S. M. MARKOV, AND E. POPOVA, *Extended interval arithmetics: New results and applications*, in Computer Arithmetic and Enclosure Methods, L. Atanassova and J. Herzberger, eds., Amsterdam, 1992, North-Holland, pp. 225–232.

70. T. A. DIVERIO, *Uso efetivo da matemática intervalar em supercomputadores vetoriais*, in Porto Alegre: CPGCC da UFRGS, 1995, p. 291. Tese de doutorado.

71. T. A. DIVERIO AND ET AL., *LIBAVI.A Biblioteca de rotinas intervalares – Manual de utiliza-cao*, in Porto Alegre: CPGCC da UFRGS, 1995, p. 350.

72. T. A. DIVERIO, U. A. L. FERNANDES, AND D. M. CLAUDIO, *Errors in vector processing and the libavi.a library*, Reliable Computing, 2 (1996), pp. 103–110.

73. T. A. DIVERIO, P. O. A. NAVAUX, D. M. CLAUDIO, C. A. HÖLBIG, R. L. SAGULA, AND U. A. L. FERNANDES, *High performance with high accuracy laboratory*, Revista de Informática Teórica e Aplicada, RITA, Porto Alegre: Informática da UFRGS, 3 (1997), pp. 35–54.

74. D. DOBKIN, J. HERSHBERGER, D. KIRKPATRICK, AND S. SURI, *Computing the intersection – depth of polyhedra*, Algorithmica, 9 (1993), pp. 528–533.

75. D. DOBKIN AND D. KIRKPATRICK, *Determining the separation of preprocessed polyhedra — A unified approach*, in Automata, Languages and Programming (Coventry, 1990), Lecture Notes in Computer Science, No. 443, Springer-Verlag, New York, 1990, pp. 400–413.

76. J. J. DONGARRA, *Performance of various computers using standard linear equations software.* www.netlib.org/benchmark/performance.ps, 1997.

77. J. J. DONGARRA, J. J. DU CROZ, I. S. DUFF, AND S. J. HAMMARLING, *A set of level 3 basic linear algebra subprograms*, ACM Trans. Math. Software, 16 (1990), pp. 1–17.

78. J. J. DONGARRA, J. J. DU CROZ, S. J. HAMMARLING, AND R. J. HANSEN, *An extended set of Fortran basic linear algebra subprograms*, ACM Trans. Math. Software, 14 (1990), pp. 1–17.

79. R. DUNAY AND I. KOLLÁR, *MATLAB-based analysis of roundoff noise*, in Developments in Reliable Computing, T. Csendes, ed., Kluwer, Dordrecht, Netherlands, 1999, pp. 373–382.

80. R. DUNAY, I. KOLLÁR, AND B. WIDROW, *Dithering for floating point number representation*, in 1st International On-Line Workshop (Prague, Czech Republic, Mar. 1998), J. Holub and R. Smid, eds., 1998. http://measure.feld.cvut.cz/dithering98/.

81. E. DYLLONG, W. LUTHER, AND W. OTTEN, *An accurate distance-calculation algorithm for convex polyhedra*, in Developments in Reliable Computing, T. Csendes, ed., Kluwer, Dordrecht, Netherlands, 1999, pp. 241–254. Also Reliable Computing 5/3 (1999), pp. 241–254.

82. H. EDELSBRUNNER, *Algorithms in Combinatorial Geometry*, Monographs on Theoretical Computer Science, No. 10, Springer-Verlag, Berlin, 1987.

83. EGCS, *egcs project homepage.* http://egcs.cygnus.com, 1998.

84. B. L. EHLE, *On Padé approximations to the exponential function and A-stable methods for the numerical solution of initial value problems*, SIAM J. Math. Anal., 4 (1973), pp. 671–680.

85. P. EIJGENRAAM, *The Solution of Initial Value Problems Using Interval Arithmetic*, Mathematical Centre Tracts, No. 144, Stichting Mathematisch Centrum, Amsterdam, 1981.

86. L. ELSNER AND T. SZULC, *Block p-matrices*, Linear and Multilinear Algebra, 44 (1998), pp. 1–12.

87. J. S. ELY, *The VPI software package for variable precision interval arithmetic*, Interval Computations, 2 (1993), pp. 135–153.

88. W. H. ENRIGHT, T. E. HULL, AND B. LINDBERG, *Comparing numerical methods for stiff systems of ODEs*, BIT, 15 (1975), pp. 10–48.

89. M. D. ERCEGOVAC, T. LANG, J.-M. MULLER, AND A. TISSERAND, *Reciprocation, square root, inverse square root, and some elementary functions using small multipliers*, Technical Report RR97-47, LIP, École Normale Supérieure de Lyon, November 1997.

90. A. FACIUS, *The need for higher precision in solving large sparse linear systems*, in Developments in Reliable Computing, T. Csendes, ed., Kluwer, Dordrecht, Netherlands, 1999, pp. 17–30.

91. B. FALTINGS, *Arc-consistency for continuous variables*, Artificial Intelligence, 65 (1991), pp. 363–376.

92. P. M. FARMWALD, *High bandwidth evaluation of elementary functions*, in Proceedings of the 5th IEEE Symposium on Computer Arithmetic, K. S. Trivedi and D. E. Atkins, eds., Los Alamitos, CA, 1981, IEEE Computer Society Press.

93. X. FENG, R. YANG, Y. YAN, Y. ZHU, G. F. CORLISS, AND R. B. KEARFOTT, *Globsol case study: Parameter optimization for the eddy current compensation of MRI coils (General Electric Medical)*, Technical Report, Department of Mathematics, Statistics and Computer Science, Marquette University, Milwaukee, Wisc., 1998.

94. A. FIAT, R. KARP, M. LUBY, L. MCGEOCH, D. SLEATOR, AND N. E. YOUNG, *Competitive paging algorithms*, J. Algorithms, 12 (1991), pp. 685–699.

95. M. FIEDLER, *Special Matrices and Their Applications in Numerical Mathematics*, Nijhof, Dordrecht, 1986.

96. G. FREILING AND G. JANK, *Non-symmetric matrix Riccati equations*, J. Analysis Appl., 14 (1995), pp. 259–284.

97. E. C. FREUDER, *Synthesizing constraint expressions*, Communications of the ACM, 21 (1978), pp. 958–966.

98. F. FRITZ, *Development of a finite element analysis program based on interval arithmetic*, master's thesis, Marquette University Department of Mathematics, Statistics, and Computer Science, Milwaukee, Wisc., May 1999.

99. F. FRITZ, P. THALACKER, G. F. CORLISS, AND R. B. KEARFOTT, *Globsol User Guide*, Technical Report, Department of Mathematics, Statistics and Computer Science, Marquette University, Milwaukee, Wisc., 1998. www.mscs.mu.edu/~globsol/User_Guide.

100. A. FROMMER, *Lösung linearer Gleichungssysteme auf Parallelrechnern*, Vieweg-Verlag, Wiesbaden, 1990.

101. A. FROMMER AND P. MAASS, *Fast CG-based methods for Thikonov-Phillips regularization*, SIAM J. Sci. Comp., (to appear).

102. A. FROMMER AND A. WEINBERG, *Verified error bounds for linear systems through the Lanczos process*, in Developments in Reliable Computing, T. Csendes, ed., Kluwer, Dordrecht, Netherlands, 1999, pp. 255–268. Also Reliable Computing 5/3 (1999), pp. 255–268.

103. E. GARDEÑES AND A. TREPAT, *Fundamentals of SIGLA, an interval computing system over the completed set of intervals*, Computing, 24 (1980), pp. 161–179.

104. E. GARDEÑES, A. TREPAT, AND H. MIELGO, *Present perspective of the SIGLA interval system*, Freiburger Intervall-Berichte, 82 (1982), pp. 1–65.

105. D. M. GAY, *Correctly rounded binary-decimal and decimal-binary conversions*, Numerical Analysis Manuscript 90–10, AT&T Bell Laboratories, November 1990.

106. E. G. GILBERT, D. W. JOHNSON, AND S. S. KEERTHI, *A fast procedure for computing the distance between complex objects in three-dimensional space*, IEEE Journal of Robotics and Automation, 4 (1988), pp. 193–203.

107. V. GIRAULT AND P. A. RAVIART, *Finite Element Approximation of the Navier-Stokes Equations*, Springer-Verlag, Berlin, 1986.

108. G. GOLUB AND G. MEURANT, *Matrices, moments and quadrature*, in Numerical Analysis 1993 (Dundee 1993), Pitman Res. Notes Math., No. 302, Longman Sci. Tech., Harlow, 1994.

109. ———, *Matrices, moments and quadrature II. How to compute the error in iterative methods*, BIT, 37 (1997), pp. 687–705.

110. G. GOLUB AND J. H. WELSCH, *Calculation of Gauss quadrature rules*, Math. Comp., 23 (1969), pp. 221–230.

111. G. H. GOLUB AND C. F. VAN LOAN, *Matrix Computations*, Johns Hopkins University Press, Baltimore, 3rd ed., 1996.

112. S. GOTTSCHALK, *Separating axis theorem*, Technical Report TR96-024, Department of Computer Science, UNC Chapel Hill, 1996.

113. L. GRANVILLIERS, *On the combination of box-consistency and hull-consistency*, in Proceedings of the Workshop ECAI on Non Binary Constraints (Brighton, Aug., 1998), 1998.

114. J. H. GRAY AND L. B. RALL, *INTE: A UNIVAC 1108/1110 program for numerical integration with rigorous error estimation*, MRC Technical Summary Report No. 1428, Mathematics Research Center, University of Wisconsin - Madison, 1975.

115. A. GREENBAUM, *Iterative Methods for Solving Linear Systems*, Frontiers in Applied Mathematics, No. 17, SIAM, Philadelphia, 1997.

116. R. GREGORY AND D. KARNEY, *A Collection of Matrices for Testing Computational Algorithms*, Wiley, New York, 1969.

117. A. GRIEWANK, *ODE solving via automatic differentiation and rational prediction*, in Numerical Analysis 1995, D. F. Griffiths and G. A. Watson, eds., Pitman Research Notes in Mathematics Series, No. 344, Addison-Wesley Longman, 1996.

118. A. GRIEWANK, G. F. CORLISS, P. HENNEBERGER, G. KIRLINGER, F. A. POTRA, AND H. J. STETTER, *High-order stiff ODE solvers via automatic differentiation and rational prediction*, in Numerical Analysis and Its Applications, Lecture Notes in Computer Science, No. 1196, Springer, Berlin, 1997, pp. 114–125.

119. A. GRIEWANK, D. JUEDES, AND J. UTKE, *ADOL–C, a package for the automatic differentiation of algorithms written in C/C++*, ACM Trans. Math. Software, 22 (1996), pp. 131–167.

120. A. GRIEWANK, J. UTKE, AND A. WALTHER, *Evaluating higher derivative tensors by forward propagation of univariate Taylor series*, Preprint IOKOMO-09-97t, TU Dresden, Inst. of Scientific Computing, 1997. Revised June 1998.

121. E. HAIRER, S. P. NØRSETT, AND G. WANNER, *Solving Ordinary Differential Equations I. Nonstiff Problems*, Computational Mechanics, No. 8, Springer-Verlag, Berlin, 2nd revised ed., 1993.

122. R. HAMMER, M. HOCKS, U. KULISCH, AND D. RATZ, *Numerical Toolbox for Verified Computing I: With Algorithms and Pascal-XSC Programs*, Springer-Verlag, Berlin, 1994.

123. ———, *C++ Toolbox for Verified Computing I: Basic Numerical Problems: Theory, Algorithms, and Programs*, Springer-Verlag, Berlin, 1995.

124. E. R. HANSEN, *Global Optimization Using Interval Analysis*, Marcel Dekker, New York, 1992.

125. P. HANSEN, *Regularization tool, a MATLAB package for the analysis and solution of discrete ill-posed problems; version 2.0 for MATLAB 4.0*, Technical Report 92–03, UNIC, 1992.

126. H. HASSLER AND N. TAKAGI, *Function evaluation by table look-up and addition*, in Proceedings of the 12th IEEE Symposium on Computer Arithmetic (Bath, UK, July 1995), S. Knowles and W. McAllister, eds., Los Alamitos, CA, 1995, IEEE Computer Society Press.

127. G. L. HAVILAND AND A. A. TUSZYNSKI, *A CORDIC arithmetic processor chip*, IEEE Trans. on Computers, C-29 (1980), pp. 68–79.

128. G. HEINDL, *Zur Einschließung der Werte von Peanofunktionalen*, Z. angew. Math. Mech., 75 (1995), pp. 637–638.

129. ———, *A representation of the interval hull of a tolerance polyhedron describing inclusions of function values and slopes*, in Developments in Reliable Computing, T. Csendes, ed., Kluwer, Dordrecht, Netherlands, 1999, pp. 269–278. Also Reliable Computing 5/3 (1999), pp. 269–278.

130. F. D. HÉLÈNE COLLAVIZZA AND M. RUEHER, *Comparing partial consistencies*, in Developments in Reliable Computing, T. Csendes, ed., Kluwer, Dordrecht, Netherlands, 1999, pp. 213–228. Also Reliable Computing 5/3 (1999), pp. 213–228.

131. P. HENRICI, *Circular Arithmetic and the Determination of Polynomial Zeros*, Lecture Notes, No. 228, Springer, Heidelberg, 1971.

132. C. HERMITE, *Extrait d'une lettre de M. Ch. Hermite à M. Borchardt sur la formule d'interpolation de Lagrange*, J. de Crelle, 84 (1878), p. 70. Oeuvres, tome III, p. 432–443.

133. M. R. HESTENES AND E. STIEFEL, *Methods of conjugate gradients for solving linear systems*, Journal of Research of the National Bureau of Standards, 49 (1952), pp. 409–436.

134. W. HOFSCHUSTER AND W. KRÄMER, *A fast public domain interval library in ANSI C*, in Proceedings of the 15th IMACS World Congress on Scientific Computation, Modelling and Applied Mathematics, Vol. 2, A. Sydow, ed., 1997, pp. 395–400.

135. O. HOLZMANN, B. LANG, AND H. SCHÜTT, *Newton's constant of gravitation and verified numerical quadrature*, Reliable Computing, 2 (1996), pp. 229–239.

136. H. HONG, *An improvement of the projection operator in cylindrical algebraic decomposition*, in Proceedings of ISSAC 1990, 1990, pp. 261–264.

137. H. HONG AND V. STAHL, *Safe starting regions by fixed points and tightening*, Computing, 53 (1994), pp. 323–335.

138. J. HORMIGO, J. VILLALBA, AND E. L. ZAPATA, *A hardware approach to interval arithmetic for sine and cosine functions*, in Developments in Reliable Computing, T. Csendes, ed., Kluwer, Dordrecht, Netherlands, 1999, pp. 31–42.

139. R. HORST AND P. M. PARDALOS, *Handbook of Global Optimization*, Kluwer, Dordrecht, Netherlands, 1995.

140. E. G. HOUGHTON, R. F. EMNETT, J. D. FACTOR, AND C. L. SABHARWAL, *Implementation of a divide-and-conquer method for intersection of parametric surfaces*, Computer Aided Geometric Design, 2 (1985), pp. 173–183.

141. Y. H. HU, *The quantization effects of the CORDIC algorithm.*, IEEE Trans. on Signal Processing, 40 (1992), pp. 834–844.

142. E. HUBER, *Intersecting general parametric surfaces using bounding volumes*, in Tenth Canadian Conference on Computational Geometry - CCCG '98, 1998, pp. 52–54.

143. E. HUBER AND W. BARTH, *Surface-to-surface intersection with complete and guaranteed results*, in Developments in Reliable Computing, T. Csendes, ed., Kluwer, Dordrecht, Netherlands, 1999, pp. 189–202.

144. D. HUSUNG, *ABACUS — Programmierwerkzeug mit hochgenauer Arithmetik für Algorithmen mit verifizierten Ergebnissen*, diplomarbeit, Universität Karlsruhe, 1988.

145. ———, *Precompiler for Scientific Computation (TPX)*, Technical Report 91.1, Inst. f. Informatik III, TU Hamburg-Harburg, 1989.

146. E. HYVÖNEN, *Constraint reasoning based on interval arithmetic: The tolerance propagation approach*, Artificial Intelligence, 58 (1992), pp. 71–112.

147. IBM, *High-Accuracy Arithmetic Subroutine Library (ACRITH); General Information Manual, 3rd Edition, GC 33–6163–02; Program Description and User's Guide, SC 33–6164–02*, IBM Deutschland GmbH, 3rd ed., 1986.

148. IEEE, *IEEE standard equipment requirements and measurement techniques (IEEE/ANSI 734-1985)*, Technical Report, IEEE, 1985.

149. M. IRI, *Automatic differentiation*, Bull. Jap. Soc. Industrial and Applied Mathematics, 3 (1993).

150. ———, *Guaranteed accuracy and fast automatic differentiation*, KITE Journal of Electronics Engineering, 4 (1993), pp. 34–40.

151. ———, *The role of automatic differentiation in nonlinear analysis and high-quality computation*, Technical Report TRISE 96-05, Dept. of Information and System Engineering, Faculty of Science and Engineering, Chuo University, Tokyo, Japan, 1996.

152. K. G. IVANOV AND V. TOTIK, *Fast decreasing polynomials*, Constructive Approx., 6 (1990), pp. 1–20.

153. C. JANSSON, *Interval linear systems with symmetric matrices, skew-symmetric matrices and dependencies in right hand side*, Computing, 46 (1991), pp. 265–274.

154. ———, *A global optimization method using interval arithmetic*, IMACS Annals of Computing and Applied Mathematics, (1992).

155. ———, *On self-validating methods for optimization problems*, in Topics in Validated Computations: Proceedings of the IMACS-GAMM International Workshop on Validated Computations (Oldenburg, 30 Aug. - 3 Sept., 1993), J. Herzberger, ed., Studies in Computational Mathematics, No. 5, Amsterdam, 1994, North-Holland / Elsevier, pp. 381–439.

156. ———, *Construction of convex lower and concave upper bound functions*, Bericht 98.1, Forschungsschwerpunktes Informations- und Kommunikationstechnik der Technischen Universität Hamburg-Harburg, März 1998.

157. M. JIRSTRAND, *Constructive Methods for Inequality Constraints in Control*, PhD thesis, Linkoping Studies in Science and Technology, 1998. Dissertations, No. 527.

158. C. R. JOHNSON AND M. J. TSATSOMEROS, *Convex sets of nonsingular and P-matrices*, Linear and Multilinear Algebra, 38 (1995), pp. 233–240.

159. W. KAHAN, *The baleful effect of computer benchmarks upon applied mathematics, physics and chemistry*. http://http.cs.berkeley.edu/~wkahan/ieee754status/baleful.ps, 1998.

160. E. W. KAUCHER, *Algebraische Erweiterungen der Intervallrechnung unter Erhaltung der Ordnungs- und Verbandstrukturen*, Computing Suppl., 1 (1977), pp. 65–79.

161. ———, *Interval analysis in the extended interval space IR*, Computing, Suppl., 2 (1980), pp. 33–49.

162. R. B. KEARFOTT, *A Fortran 90 environment for research and prototyping of enclosure algorithms for nonlinear equations and global optimization*, ACM Trans. Math. Software, 21 (1995), pp. 63–78.

163. ———, *A review of techniques in the verified solution of constrained global optimization problems*, in Applications of Interval Computations, R. B. Kearfott and V. Kreinovich, eds., Kluwer, Dordrecht, Netherlands, 1996, pp. 23–59.

164. ———, *Rigorous Global Search: Continuous Problems*, Kluwer, Dordrecht, Netherlands, 1996.

165. ———, *A specific proposal for interval arithmetic in Fortran*. http://interval.usl.edu/F90/f96-pro.asc, 1996.

166. R. B. KEARFOTT, M. DAWANDE, K. S. DU, AND C. HU, *INTLIB: A portable Fortran 77 interval elementary function library*, Interval Computing, 3 (1992), pp. 96–105.

167. ———, *Algorithm 737: INTLIB: A portable Fortran 77 interval standard library*, ACM Trans. Math. Software, 20 (1994), pp. 447–459.

168. R. B. KEARFOTT AND K. S. DU, *The cluster problem in multivariate global optimization*, J. of Global Optimization, 5 (1994), pp. 253–265.

169. R. B. KEARFOTT AND V. KREINOVICH, *Applications of Interval Computations*, Kluwer, Dordrecht, Netherlands, 1996.

170. R. B. KEARFOTT AND M. NOVOA, *Algorithm 681: INTBIS: A portable interval Newton / bisection package*, ACM Trans. Math. Software, 16 (1990), pp. 152–157.

171. G. KEDEM, *Automatic differentiation of computer programs*, ACM Trans. Math. Software, 6 (1980), pp. 150–165.

172. J. B. KEIPER AND D. WITHOFF, *Numerical computation in Mathematica*, in Course Notes, Mathematica Conference, 1992.

173. B. KELLING, *Geometrische Untersuchungen zur eigenschränkte Lösungsmenge Intervallgleichungssysteme*, ZAMM, 74 (1994), pp. 625–628.

174. R. KELNHOFFER, *Applications of Interval Methods to Parametric Set Estimation from Bounded Error Data*, PhD thesis, Marquette University Department of Electrical and Computer Engineering, 1997.

175. R. KLATTE, U. KULISCH, M. NEAGA, D. RATZ, AND C. ULLRICH, *PASCAL-XSC, Sprachbeschreibung mit Beispielen*, Springer-Verlag, Berlin, 1991.

176. ———, *PASCAL-XSC, Language Reference with Examples*, Springer-Verlag, Berlin, 1992.

177. R. KLATTE, U. KULISCH, A. WIETHOFF, C. LAWO, AND M. RAUCH, *C–XSC, A C++ Class Library for Extended Scientific Computing*, Springer-Verlag, Berlin, 1993.

178. R. KLATTE AND C. P. ULLRICH, *Complex sector arithmetic*, Computing, 24 (1980).

179. S. C. KLEENE, *Mathematical Logic*, Wiley, New York, 1967.

180. O. KNÜPPEL, *BIAS – basic interval arithmetic subroutines*, Technical Report 93.3, Inst. f. Informatik III, TU Hamburg–Harburg, July 1993.

181. ———, *PROFIL/BIAS — A fast interval library*, Computing, 53 (1994), pp. 277–287.

182. ———, *PROFIL/BIAS and extensions, version 2.0*, Technical Report, Inst. f. Informatik III, TU Hamburg–Harburg, 1998.

183. L. V. KOLEV, *Interval Methods for Circuit Analysis*, World Scientific, 1993.

184. I. KOLLÁR, *Statistical theory of quantization*, Doct. Sci. Thesis, Hungarian Academy of Sciences, Budapest, 1996.

185. B. KOŁODZIEJCZAK AND T. SZULC, *Convex sets of full rank matrices*, in Developments in Reliable Computing, T. Csendes, ed., Kluwer, Dordrecht, Netherlands, 1999, pp. 359–364.

186. ———, *Convex combinations of matrices – Full rank characterizations*, Linear Algebra Appl., (to appear).

187. K. KOTA AND J. R. CAVALLARO, *Numerical accuracy and hardware tradeoffs for CORDIC arithmetic for special purpose processors*, IEEE Trans. on Computers, 42 (1993), pp. 769–779.

188. W. KRATZ AND E. STICKEL, *Numerical solution of matrix polynomial equations by Newton's method*, IMA J. Numer. Anal., 7 (1987), pp. 355–369.

189. R. KRAWCZYK, *Newton-Algorithmen zur Bestimmung von Nullstellen mit Fehlerschranken*, Computing, 4 (1969), pp. 187–201.

190. R. KRIER, *Komplexe Kreisarithmetik*, PhD thesis, Universität Karlsruhe, 1973.

191. F. KRÜCKEBERG, *Ordinary differential equations*, in Topics in Interval Analysis, E. R. Hansen, ed., Clarendon Press, Oxford, 1969, pp. 91–97.

192. D. KUCK AND B. KUHN, *Kuck Associates Homepage*. www.kai.com, 1998.

193. W. KÜHN, *Rigorously computed orbits of dynamical systems without the wrapping effect*, Computing, 61 (1998), pp. 47–67.

194. ———, *Towards an optimal control of the wrapping effect*, in Developments in Reliable Computing, T. Csendes, ed., Kluwer, Dordrecht, Netherlands, 1999, pp. 43–52.

195. U. W. KULISCH, ed., *Wissenschaftliches Rechnen mit Ergebnisverifikation — Eine Einführung*, Akademie Verlag, Berlin, and Vieweg Verlag, Wiesbaden, 1989.

196. U. W. KULISCH AND W. L. MIRANKER, *Computer Arithmetic in Theory and Practice*, Academic Press, New York, 1981.

197. ———, *A New Approach to Scientific Computation*, Academic Press, New York, 1983.

198. ———, *The arithmetic of the digital computer: A new approach*, SIAM Review, 28 (1986), pp. 1–40.

199. L. V. KUPRIYANOVA, *Inner estimation of the united solution set of interval linear algebraic system*, Reliable Computing, 1 (1995), pp. 15–31.

200. O. LADYZHENSKAYA, *The Mathematical Theory of Viscous Incompressible Flow*, Gordon & Breach, 1969.

201. A. V. LAKEYEV, *Linear algebraic equations in Kaucher arithmetic*, Reliable Computing, Supplement (Extended Abstracts of APIC '95: International Workshop on Applications of Interval Computations, El Paso, TX, Feb. 23–25, 1995), (1995), pp. 130–133.

202. ——, *Computational complexity of estimation of the generalized solution sets to interval linear systems*, in Proceedings of XI International Conference "Optimization Methods and Their Applications"(Baikal, July 5–12, 1998), Irkutsk, 1998, pp. 115–118.

203. ——, *On existence and uniqueness of solutions of linear algebraic equations in Kaucher's interval arithmetic*, in Developments in Reliable Computing, T. Csendes, ed., Kluwer, Dordrecht, Netherlands, 1999, pp. 53–66.

204. J. D. LAMBERT, *Computational Methods in Ordinary Differential Equations*, Wiley, 1977.

205. C. LÁNCZOS, *An iteration method for the solution of the eigenvalue problem of linear differential and integral operators*, J. Res. Nat. Bur. Standards, 45 (1950), pp. 255–282.

206. B. LANG, *A comparison of subdivision strategies for verified multi-dimensional Gaussian quadrature*, in Developments in Reliable Computing, T. Csendes, ed., Kluwer, Dordrecht, Netherlands, 1999, pp. 67–76.

207. D. LASSER, *Kurven und Flächenverschneidungsmethoden*, in Geometrische Verfahren der Graphischen Datenverarbeitung, J. L. Encarnação, ed., Springer-Verlag, Berlin, 1990, pp. 61–87.

208. C. LAWO, *C-XSC, A programming environment for verified scientific computing and numerical data processing*, in Scientific Computing with Automatic Result Verification, E. Adams and U. Kulisch, eds., Academic Press, Orlando, Fla., 1992, pp. 71–86.

209. C. L. LAWSON, R. J. HANSON, D. KINCAID, AND F. T. KROGH, *Algorithm 539: Basic linear algebra subprograms for Fortran usage*, ACM Trans. Math. Software, 5 (1979), pp. 308–325.

210. Y. LEBBAH AND O. LHOMME, *Acceleration methods for numeric CSPs*, in Proceedings of the Fifteenth National Conference on Artificial Intelligence (AAAI-98), Madison, WI, USA, July 26-30 1998, MIT Press, pp. 19–24.

211. ——, *Prediction by extrapolation for interval tightening methods*, in Developments in Reliable Computing, T. Csendes, ed., Kluwer, Dordrecht, Netherlands, 1999, pp. 159–166.

212. J. H. M. LEE AND M. H. VAN EMDEN, *Interval computation as deduction in CHIP*, Journal of Logic Programming, 16 (1993), pp. 255–276.

213. V. LEFÈVRE, *An algorithm that computes a lower bound on the distance between a segment and $Z^2$*, Research Report 97-18, Laboratoire de l'Informatique du Parallélisme, Lyon, France, June 1997.

214. ——, *An algorithm that computes a lower bound on the distance between a segment and $Z^2$*, in Developments in Reliable Computing, T. Csendes, ed., Kluwer, Dordrecht, Netherlands, 1999, pp. 203–212.

215. V. LEFÈVRE, J.-M. MULLER, AND A. TISSERAND, *Towards correctly rounded transcendentals*, in Proceedings of the 13th IEEE Symposium on Computer Arithmetic, Asilomar, USA, 1997, IEEE Computer Society Press, Los Alamitos, CA, 1997.

216. M. LERCH, *Expression concepts in scientific computing*, in Developments in Reliable Computing, T. Csendes, ed., Kluwer, Dordrecht, Netherlands, 1999, pp. 119–132.

217. M. LERCH AND J. WOLFF VON GUDENBERG, *Expression templates for dot product expressions*, Reliable Computing, 5 (1999), pp. 69–80.

218. ——, *Multiaspect interval types*, in Developments in Reliable Computing, T. Csendes, ed., Kluwer, Dordrecht, Netherlands, 1999, pp. 365–372.

219. O. LHOMME, *Consistency techniques for numeric CSPs*, in Proceedings of International Joint Conference on Artificial Intelligence (IJCAI-93), R. Bajcsy, ed., Chambery, France, Aug. 1993, Morgan Kaufmann, pp. 232–238.

220. O. LHOMME AND M. RUEHER, *Application des techniques CSP au raisonnement sur les intervalles*, Revue d'intelligence Artificielle (Dunod), 11 (1997), pp. 283–312.

221. R. J. LOHNER, *A verified solver for linear systems with band structure*. To be included in *Toolbox for Verified Computing II*.

222. ——, *Enclosing the solutions of ordinary initial and boundary value problems*, in Computer Arithmetic: Scientific Computation and Programming Languages, E. W. Kaucher, U. W.' Kulisch, and C. Ullrich, eds., Wiley-Teubner Series in Computer Science, Stuttgart, 1987, pp. 255–286.

223. ——, *Einschließung der Lösung gewöhnlicher Anfangs– und Randwertaufgaben und Anwendungen*, PhD thesis, Universität Karlsruhe, 1988.

224. G. LUDYK, *CAE von Dynamischen Systemen, Analyse, Simulation, Entwurf von Regelungsstemen*, Springer-Verlag, Berlin, 1990.

225. W. LUTHER AND W. OTTEN, *The complex arithmetic geometric mean and multiple–precision matrix functions*, in Scientific Computing and Validated Numerics: Proceedings of the International Symposium on Scientific Computing, Computer Arithmetic and Validated Numerics - SCAN '95, G. Alefeld, A. Frommer, and B. Lang, eds., Berlin, 1996, Akademie Verlag, pp. 52–58.

226. ——, *Verified calculation of the solution of algebraic Riccati equation*, in Developments in Reliable Computing, T. Csendes, ed., Kluwer, Dordrecht, Netherlands, 1999, pp. 105–118.

227. A. MACKWORTH, *Consistency in networks of relations*, Journal of Artificial Intelligence, 8 (1977), pp. 99–118.

228. S. M. MARKOV, *Interval differential equations*, in Interval Mathematics 1980, K. Nickel, ed., Academic Press, New York, 1980, pp. 145–164.

229. ——, *Some applications of extended interval arithmetic to interval iterations*, Computing Suppl., 2 (1980), pp. 69–84.

230. ——, *On the presentation of ranges of monotone functions using interval arithmetic*, Interval Computations, 4 (1992), pp. 19–31.

231. ——, *On directed interval arithmetic and its applications*, JUCS, 1 (1995), pp. 510–521.

232. ——, *On the algebraic properties of convex bodies and some applications*. Preprint, 1998.

233. ——, *An iterative method for algebraic solution to interval equations*, Applied Numerical Mathematics, 30 (1999), pp. 225–239.

234. ——, *On the algebraic properties of convex bodies and some applications*, J. Convex Analysis, (submitted).

235. S. M. MARKOV AND R. ALT, *On the relation between stochastic and interval arithmetic*. Preprint, 1998.

236. S. M. MARKOV AND K. OKUMURA, *The contribution of T. Sunaga to interval analysis and reliable computing*, in Developments in Reliable Computing, T. Csendes, ed., Kluwer, Dordrecht, Netherlands, 1999, pp. 167–188.

237. MATHWORKS, *MATLAB User's Manual, Version 5.2*, The MathWorks, Inc., Natick, Mass., 1998.

238. O. MAYER, *Algebraische und metrische Strukturen in der Intervallrechnung und einige Anwendungen*, Computing, 5 (1970), pp. 144–162.

239. S. MCCALLUM, *An improved projection for cylindrical algebraic decomposition*, in Quantifier Elimination and Cylindrical Algebraic Decomposition, B. Caviness and J. Johnson, eds., Springer-Verlag, Berlin, 1988.

240. ——, *An improved projection for cylindrical algebraic decomposition of three dimensional space*, J. Symbolic Comp., 5 (1988), pp. 141–161.

241. ——, *Solving polynomial strict inequalities using cylindrical algebraic decomposition*, The Computer Journal, 36 (1993), pp. 432–438.

242. C. C. MCGEOCH, *How to stay competitive*, Amer. Math. Monthly, 101 (1994), pp. 897–901.

243. V. MEHRMANN AND E. TAN, *Defect correction methods for the solution of algebraic Riccati equations*, IEEE Trans. Automat. Control, AC–33 (1988), pp. 695–698.

244. M. MILANESE, J. NORTON, H. PIET-LAHANIER, AND E. WATER, eds., *Bounding Approaches to System Identification*, Plenum Press, London, 1996.

245. D. S. MITRINOVIĆ, J. E. PEČARIĆ, AND V. VOLENEC, *Recent Advances in Geometric Inequalities*, Kluwer, Dordrecht, Netherlands, 1989.

246. M. B. MONAGAN, K. O. GEDDES, K. M. HEAL, G. LABAHN, AND S. M. VORKOETTER, *Maple V Programming Guide*, Springer-Verlag, Berlin, 1998.

247. U. MONTANARI, *Networks of constraints: Fundamental properties and applications to picture processing*, Information Science, 7 (1974), pp. 95–132.

248. R. E. MOORE, *Interval Arithmetic and Automatic Error Analysis in Digital Computing*, PhD thesis, Department of Computer Science, Stanford University, 1962.

249. ———, *The automatic analysis and control of error in digital computation based on the use of interval numbers*, in Error in Digital Computation, Vol. I, L. B. Rall, ed., Wiley, New York, 1965, pp. 61–130.

250. ———, *Automatic local coordinate transformations to reduce the growth of error bounds in interval computation of solutions of ordinary differential equations*, in Error in Digital Computation, Vol. II, L. B. Rall, ed., Wiley, New York, 1965, pp. 103–140.

251. ———, *Interval Analysis*, Prentice Hall, Englewood Cliffs, N.J., 1966. German translation: Intervallanalyse, translated by D. Pfaffenzeller, R. Oldenburg, München, 1968.

252. ———, *A test for existence of solutions for non-linear systems*, SIAM J. Numer. Anal., 4 (1977), pp. 611–615.

253. ———, *Methods and Applications of Interval Analysis*, SIAM, Philadelphia, Penn., 1979.

254. ———, *Reliability in Computing: The Role of Interval Methods in Scientific Computations*, Academic Press, 1988.

255. J.-M. MULLER, *Elementary Functions: Algorithms and Implementations*, Birkhaüser, Boston, 1997.

256. ———, *A few results on table-based methods*, in Developments in Reliable Computing, T. Csendes, ed., Kluwer, Dordrecht, Netherlands, 1999, pp. 279–288. Also Reliable Computing 5/3 (1999), pp. 279–288.

257. K. MUSCH AND G. SCHUMACHER, *Interval analysis for embedded systems*, in Developments in Reliable Computing, T. Csendes, ed., Kluwer, Dordrecht, Netherlands, 1999, pp. 149–158.

258. N. C. MYERS, *Traits: A new and useful template technique*, C++ Report, (1995).

259. M. T. NAKAO, *A numerical verification method for the existence of weak solutions for nonlinear boundary value problems*, J. Math. Anal. Appl., 164 (1992), pp. 489–507.

260. ———, *Solving nonlinear elliptic problems with result verification using an $H^{-1}$ type residual iteration*, Computing, Suppl., 9 (1993), pp. 161–173.

261. M. T. NAKAO, N. YAMAMOTO, AND S. KIMURA, *On best constant in the optimal error estimates for the $H_0^1$-projection into piecewise polynomial spaces*, J. Approximation Theory, (to appear).

262. M. T. NAKAO, N. YAMAMOTO, AND Y. WATANABE, *Guaranteed error bounds for the finite element solutions of the Stokes problem*, in Scientific Computing and Validated Numerics: Proceedings of the International Symposium on Scientific Computing, Computer Arithmetic and Validated Numerics - SCAN '95, G. Alefeld, A. Frommer, and B. Lang, eds., Berlin, 1996, Akademie Verlag, pp. 258–264.

263. ———, *Constructive $L^2$ error estimates for finite element solutions of the Stokes equations*, Reliable Computing, 4 (1998), pp. 115–124.

264. ———, *A posteriori and constructive a priori error bounds for finite element solutions of the Stokes equations*, J. Comput. Appl. Math., 91 (1998), pp. 137–158.

265. N. S. NEDIALKOV, *Computing Rigorous Bounds on the Solution of an Initial Value Problem for an Ordinary Differential Equation*, PhD thesis, Department of Computer Science, University of Toronto, Toronto, Canada, M5S 3G4, February 1999.

266. N. S. NEDIALKOV AND K. R. JACKSON, *An interval Hermite-Obreschkoff method for computing rigorous bounds on the solution of an initial value problem for an ordinary differential equation*, in Developments in Reliable Computing, T. Csendes, ed., Kluwer, Dordrecht, Netherlands, 1999, pp. 289–310. Also Reliable Computing 5/3 (1999), pp. 289–310.

267. N. S. NEDIALKOV, K. R. JACKSON, AND G. F. CORLISS, *Validated solutions of initial value problems for ordinary differential equations*, Appl. Math. Comp., (to appear).

268. A. NEUMAIER, *A distributive interval arithmetic*, Freiburger Intervall-Berichte, 82 (1982), pp. 31–38. Inst. f. Angewandte Mathematik, U. Freiburg i. Br.

269. ——, *Tolerance analysis with interval arithmetic*, Freiburger Intervall-Berichte, 86 (1986), pp. 5–19.

270. ——, *Interval Methods for Systems of Equations*, Cambridge University Press, Cambridge, 1990.

271. A. NEUMAIER, S. DALLWIG, AND H. SCHICHL, *GLOPT – A program for constrained global optimization*, in Developments in Global Optimization, I. Bomze et al., ed., Kluwer, Dordrecht, 1997, pp. 19–36.

272. NIST, *Matrix market*. http://math.nist.gov/matrixmarket/, 1998.

273. N. OBRESCHKOFF, *Neue Quadraturformeln*, Abh. Preuss. Akad. Wiss. Math. Nat. Kl., 4 (1940).

274. ——, *Sur le quadrature mecaniques*, Spisanie Bulgar. Akad. Nauk. (Journal of the Bulgarian Academy of Sciences), 65 (1942), pp. 191–289.

275. K. OKUMURA, *An application of interval operations to electric network analysis*, Bull. of the Japan Society for Industrial and Applied Mathematics, 3 (1993), pp. 15–27. In Japanese.

276. ——, *Recent topic of circuit analysis — an application of interval arithmetic*, Journal of System Control, 40 (1996), pp. 393–400. In Japanese.

277. K. OKUMURA, S. SAEKI, AND A. KISHIMA, *On an improvement of an algorithm using interval analysis for solution of nonlinear circuit equations*, Trans. of IECEJ, J69-A (1986), pp. 489–496. In Japanese.

278. W. J. OLDER AND A. VELINO, *Extending Prolog with constraint arithmetic on real intervals*, in Proc. of IEEE Canadian conference on Electrical and Computer Engineering, New York, 1990, IEEE Computer Society Press.

279. ——, *Constraint arithmetic on real intervals*, in Constraint Logic Programming: Selected Research, F. Benhamou and A. Colmerauer, eds., MIT Press, 1993, pp. 175–196.

280. J. M. ORTEGA AND W. C. RHEINBOLDT, *Iterative Solution of Nonlinear Equations in Several Variables*, Academic Press, New York, 1970.

281. C. C. PAIGE, *The Computation of Eigenvalues and Eigenvectors of Very Large Sparse Matrices*, PhD thesis, University of London, 1971.

282. B. PARLETT, *The Symmetric Eigenvalue Problem*, Prentice-Hall, Englewood Cliffs, NJ, 1980.

283. M. PETROVICH, *Calculation with numerical intervals*, Beograd, (1932). In Serbian, Serbian title: *Racunanje sa brojnim razmacima*.

284. L. S. PONTRJAGIN, *Topological Groups*, Princeton Univ. Press, Princeton, 1946.

285. E. POPOVA AND C. ULLRICH, *Embedding directed intervals in Mathematica*, Revista de Informatica Teorica e Applicada, 3 (1996), pp. 99–115.

286. J. E. POTTER, *Matrix quadratic solutions*, J. SIAM Appl. Math., 14 (1966), pp. 496–501.

287. D. PRIEST, *Fast table-driven algorithms for interval elementary functions*, in Proceedings of the 13th Symposium on Computer Arithmetic (Pacific Grove, CA, July 6–9, 1997), IEEE Computer Society Press, New York, 1997, pp. 168–174.

288. PROLOGIA, *PrologIV Constraints Inside*, Parc technologique de Luminy - Case 919 13288, Marseille cedex 09 (France), 1996.

289. S. QUINLAN, *Efficient distance computation between non-convex objects*, in IEEE Intern. Conf. on Robotics and Automation, IEEE, 1994, pp. 3324–3329.

290. L. B. RALL, *Automatic Differentiation: Techniques and Applications*, Lecture Notes in Computer Science, No. 120, Springer-Verlag, Berlin, 1981.

291. A. RALSTON AND P. RABINOWITZ, *A First Course in Numerical Analysis*, McGraw-Hill, New York, 2nd ed., 1978.

292. S. RANJIT, *Risk management of currency portfolios*, master's thesis, Marquette University Department of Economics, Milwaukee, Wisc., April 1998.

293. H. RATSCHEK, *Die subdistributivität der intervallarithmetik*, ZAMM, 54 (1971).

294. H. RATSCHEK AND J. ROKNE, *Computer Methods for the Range of Functions*, Ellis Horwood / Prentice-Hall, Chichester, 1984.

295. H. RATSCHEK AND G. SCHRÖDER, *Representation of semigroups as systems of compact convex sets*, Proc. Amer. Math. Soc., 65 (1977), pp. 24–28.

296. D. RATZ, *An inclusion algorithm for global optimization in a portable PASCAL-XSC implementation*, in Computer Arithmetic and Enclosure Methods, L. Atanassova and J. Herzberger, eds., Amsterdam, 1992, North-Holland, pp. 329–338.

297. ———, *Box-splitting strategies for the interval Gauss–Seidel step in a global optimization method*, Computing, 53 (1994), pp. 337–354.

298. D. RATZ AND T. CSENDES, *On the selection of subdivision directions in interval branch-and-bound methods for global optimization*, Journal of Global Optimization, 7 (1995), pp. 183–207.

299. P. L. RICHMAN, *Automatic error analysis for determining precision*, Comm. ACM, 15 (1972), pp. 813–817.

300. R. RIHM, *On a class of enclosure methods for initial value problems*, Computing, 53 (1994), pp. 369–377.

301. J. ROHN, *Systems of linear interval equations*, Linear Algebra and Its Applications, 126 (1989), pp. 39–78.

302. ———, *NP-hardness results for linear algebraic problems with interval data*, in Topics in Validated Computations: Proceedings of the IMACS-GAMM International Workshop on Validated Computations (Oldenburg, 30 Aug. - 3 Sept., 1993), J. Herzberger, ed., Studies in Computational Mathematics, No. 5, Amsterdam, 1994, North-Holland / Elsevier, pp. 463–471.

303. ———, *Positive definiteness and stability of interval matrices*, SIAM J. Matrix Anal. Appl., 15 (1994), pp. 175–184.

304. M. RUEHER AND C. SOLNON, *Concurrent cooperating solvers within the reals*, Reliable Computing, 3 (1997), pp. 325–333.

305. S. M. RUMP, *Kleine Fehlerschranken bei Matrixproblemen*, PhD thesis, Universität Karlsruhe, 1980.

306. ———, *Improved iteration schemes for the validation algorithms for dense and sparse nonlinear systems*, Computing, 57 (1992), pp. 77–84.

307. ———, *On the solution of interval linear systems*, Computing, 47 (1992), pp. 337–353.

308. ———, *Validated solution of large linear systems*, in Computing Supplementum 9, Validation Numerics, R. Albrecht, G. Alefeld, and H. J. Stetter, eds., Springer-Verlag, Berlin, 1993, pp. 191–212.

309. ———, *Verification methods for dense and sparse systems of equations*, in Topics in Validated Computations: Proceedings of the IMACS-GAMM International Workshop on Validated Computations (Oldenburg, 30 Aug. - 3 Sept., 1993), J. Herzberger, ed., Studies in Computational Mathematics, No. 5, North-Holland / Elsevier, Amsterdam, 1994, pp. 63–136.

310. ———, *Fast and parallel interval arithmetic*, BIT, 39 (1999), pp. 539–560.

311. ———, *INTLAB — INTerval LABoratory*, in Developments in Reliable Computing, T. Csendes, ed., Kluwer, Dordrecht, Netherlands, 1999, pp. 77–105.

312. Y. SAAD, *Iterative Methods for Sparse Linear Systems*, PWS, Boston, 1996.

313. R. L. SAGULA, *Avaliacao de Desempenho de Bibliotecas Intervalares*, PhD thesis, Porto Alegre; II da UFRGS, 1997.

314. R. L. SAGULA, T. DIVERIO, AND J. NETTO, *Performance evaluation technique STU and libavi library*, in Developments in Reliable Computing, T. Csendes, ed., Kluwer, Dordrecht, Netherlands, 1999, pp. 131–140.

315. R. L. SAGULA AND T. A. DIVERIO, *Interval software performance*. Presented at the International Symposium on Scientific Computing, Computer Arithmetic and Validated Numerics, SCAN '97 (Lyon, France, Sept. 10–12, 1997), 1997.

316. D. SARMA AND D. W. MATULA, *Faithful bipartite ROM reciprocal tables*, in Proceedings of the 12th IEEE Symposium on Computer Arithmetic (Bath, UK, July 1995), S. Knowles and W. McAllister, eds., Los Alamitos, CA, 1995, IEEE Computer Society Press, pp. 17–28.

317. Y. SATO, M. HIRATA, T. MARUYAMA, AND Y. ARITA, *Efficient collision detection using fast distance-calculation algorithms for convex and non-convex objects*, in Proc. IEEE Intern. Conf. on Robotics and Automation (Minneapolis, Minnesota 1996), IEEE, 1996, pp. 771–778.

318. M. J. SCHULTE AND J. STINE, *Accurate function approximation by symmetric table lookup and addition*, in Proceedings of ASAP '97, Los Alamitos, CA, 1997, IEEE Computer Society Press.

319. ———, *Symmetric bipartite tables for accurate function approximation*, in Proceedings of the 13th IEEE Symposium on Computer Arithmetic, T. Lang, J.-M. Muller, and N. Takagi, eds., Los Alamitos, CA, 1997, IEEE Computer Society Press, pp. 175–183.

320. M. J. SCHULTE AND E. E. SWARTZLANDER, JR., *Variable-precision, interval arithmetic coprocessors*, Reliable Computing, 2 (1996), pp. 47–62.

321. M. J. SCHULTE, V. A. ZELOV, A. AKKAS, AND J. C. BURLEY, *The interval-enhanced GNU Fortran compiler*, in Developments in Reliable Computing, T. Csendes, ed., Kluwer, Dordrecht, Netherlands, 1999, pp. 311–322. Also Reliable Computing 5/3 (1999), pp. 311–322.

322. M. J. SCHULTE, V. A. ZELOV, G. W. WALSTER, AND D. CHIRIAEV, *Single-number interval I/O*, in Developments in Reliable Computing, T. Csendes, ed., Kluwer, Dordrecht, Netherlands, 1999, pp. 141–148.

323. G. SCHUMACHER, *Genauigkeitsfragen bei algebraisch-numerischen Algorithmen auf Skalar- und Vektorrechnern*, PhD thesis, Universität Karlsruhe, 1989.

324. ———, *Computer aided numerical analysis*, (to appear, 2000).

325. G. SCHUMACHER AND K. MUSCH, *Specification of a test case generator for ranges of data*, Deliverable T3/3/2 ESPRIT Project No. 23920, OMI/SAFE, 1998.

326. L. SCHWARTZ, *Theorie des distributions, I*, Hermann, Paris, 1950.

327. E. SCHWARZ, *High-Radix Algorithms for High-Order Arithmetic Operations*, PhD thesis, Dept. of Electrical Engineering, Stanford University, 1992.

328. C. E. SHANNON, *The Mathematical Theory of Communication*, University of Illinois Press, Urbana, 1949.

329. S. P. SHARY, *Solving the linear interval tolerance problem*, Mathematics and Computers in Simulation, 39 (1995), pp. 53–85.

330. ———, *Algebraic approach to the interval linear static identification, tolerance and control problems, or one more application of Kaucher arithmetic*, Reliable Computing, 2 (1996), pp. 3–33.

331. ———, *Algebraic solutions to interval linear equations and their applications*, in Numerical Methods and Error Bounds, G. Alefeld and J. Herzberger, eds., Akademie Verlag, Berlin, 1996, pp. 224–233.

332. ———, *A new approach to the analysis of static systems under interval uncertainty*, in Scientific Computing and Validated Numerics: Proceedings of the International Symposium on Scientific Computing, Computer Arithmetic and Validated Numerics - SCAN '95, G. Alefeld, A. Frommer, and B. Lang, eds., Akademie Verlag, Berlin, 1996, pp. 118–132.

333. ———, *Algebraic approach in the "outer problem" for interval linear equations*, Reliable Computing, 3 (1997), pp. 103–135.

334. ———, *Algebraic approach to the analysis of linear static systems under interval uncertainty*, Izvestiya Akademii Nauk. Control Theory and Systems, 3 (1997), pp. 51–61. In Russian.

335. ———, *Interval Gauss-Seidel method for generalized solution sets to interval linear systems*, in MISC '99 — Workshop on Applications of Interval Analysis to Systems and Control (Girona, Spain, February 24-26, 1999), Universidad de Girona, 1999, pp. 51–65.

336. ———, *Outer estimation of generalized solution sets to interval linear systems*, in Developments in Reliable Computing, T. Csendes, ed., Kluwer, Dordrecht, Netherlands, 1999, pp. 323–336. Also Reliable Computing 5/3 (1999), pp. 323–336.

337. M. A. SHAYMAN, *Geometry of the algebraic Riccati equation, part I and part II*, SIAM J. Control and Optimization, 21 (1983), pp. 375–394, 395–409.

338. SIEMENS, *ARITHMOS, Benutzerhandbuch*, bibl.-nr. u 2900-i-z87-1, Siemens AG,, 1986.

339. V. SIMA, *Algorithms for Linear-Quadratic Optimization*, Marcel Dekker, New York, 1996.

340. H. SIMON, *The Lánczos Algorithm for Solving Symmetric Linear Systems*, PhD thesis, University of California, Berkley, 1982.

341. S. SKELBOE, *Computation of rational functions*, BIT, 14 (1974), pp. 87–95.

342. V. T. SÓS, *On the distribution mod 1 of the sequence nα*, Ann. Univ. Sci. Budapest, Eötvös Sect. Math., 1 (1958), pp. 127–134.

343. O. SPANIOL, *Die distributivität in der intervallarithmetik*, Computing, 5 (1970).

344. O. STAUNING, *Automatic Validation of Numerical Solutions*, PhD thesis, Department of Mathematical Modelling, Technical University of Denmark, Lyngby, Denmark, October 1997. Also appeared at Technical Report IMM-PHD-1997-36.

345. STEVENSON, D., CHAIRMAN, FLOATING-POINT WORKING GROUP, MICROPROCESSOR STANDARDS SUBCOMMITTEE, *IEEE standard for binary floating point arithmetic (IEEE/ANSI 754-1985)*, tech. report, IEEE, 1985.

346. J. E. STINE AND M. J. SCHULTE, *Hardware support for interval arithmetic*, in Proceedings of 8th Great Lakes Symposium on VLSI, February 1998, pp. 208–213.

347. U. STORCK, *Verified calculation of the nodes and weights for Gaussian quadrature formulas*, Interval Computations, 4 (1993), pp. 114–124.

348. ———, *Verifizierte Berechnung mehrfach geschachtelter Integrale der Gaskinetik*, PhD thesis, Universität Karlsruhe, 1995.

349. A. H. STROUD, *Approximate Calculation of Multiple Integrals*, Prentice-Hall, New York, 1971.

350. A. STRZEBOŃSKI, *An algorithm for systems of strong polynomial inequalities*, The Mathematica Journal, 4 (1994), pp. 74–77.

351. ———, *Computing in the field of complex algebraic numbers*, J. Symbolic Comp., 24 (1997), pp. 647–656.

352. ———, *A real polynomial decision algorithm using arbitrary-precision floating point arithmetic*, in Developments in Reliable Computing, T. Csendes, ed., Kluwer, Dordrecht, Netherlands, 1999, pp. 337–346. Also Reliable Computing 5/3 (1999), pp. 337–346.

353. T. SUNAGA, *Geometry of numerals*, master's thesis, University of Tokyo, February 1956.

354. ———, *A basic theory of communication*, Memoirs, 2 (1958), pp. 426–443.

355. ———, *Theory of an interval algebra and its application to numerical analysis*, RAAG Memoirs, 2 (1958), pp. 29–46, 547–564.

356. ——, *Algebra of Analysis and Synthesis of Automata*, PhD thesis, University of Tokyo, Tokyo, February 1961. In Japanese.

357. ——, *Differential decreasing speed using small number of differences of teeth of gears*, Trans. of Japan Society of Mechanical Engineers, 39 (1973), pp. 3209–3216. In Japanese.

358. ——, *Design and Planning for Production*, Corona Publ., Tokyo, 1979. In Japanese.

359. J. SURÁNYI, *Über die Anordnung der Vielfachen einer reellen Zahl mod 1*, Ann. Univ. Sci. Budapest, Eötvös Sect. Math., 1 (1958), pp. 107–111.

360. S. SWIERCZKOWSKI, *On successive settings of an arc on the circumference of a circle*, Fundamenta Math., 46 (1958), pp. 187–189.

361. T. SZULC, *Rank of convex combinations of matrices*, Reliable Computing, 2 (1996), pp. 181–185.

362. N. TAKAGI, T. ASADA, AND S. YAJIMA, *Redundant CORDIC methods with a constant scale factor for sine and cosine computation*, IEEE Transactions on Computers, 40 (1991), pp. 989–995.

363. G. TALENTI, *Best constant in Sobolev inequality*, Ann. Math. Pure Appl., 110 (1976), pp. 353–372.

364. P. T. P. TANG, *Table-driven implementation of the exponential function in IEEE floating-point arithmetic*, ACM Trans. Math. Software, 15 (1989), pp. 144–157.

365. ——, *Table-driven implementation of the logarithm function in IEEE floating-point arithmetic*, ACM Trans. Math. Software, 16 (1990), pp. 378–400.

366. ——, *Table lookup algorithms for elementary functions and their error analysis*, in Proceedings of the 10th IEEE Symposium on Computer Arithmetic (Grenoble, France, June 1991), P. Kornerup and D. W. Matula, eds., Los Alamitos, CA, 1991, IEEE Computer Society Press, pp. 232–236.

367. ——, *Table-driven implementation of the expm1 function in IEEE floating-point arithmetic*, ACM Trans. Math. Software, 18 (1992), pp. 211–222.

368. P. J. THALACKER, K. JULIEN, P. G. TOUMANOFF, J. P. DANIELS, G. F. CORLISS, AND R. B. KEARFOTT, *Globsol case study: Portfolio management (Banc One)*, Technical Report, Department of Mathematics, Statistics and Computer Science, Marquette University, Milwaukee, Wisc., 1998.

369. M. TIENARI, *On the control of floating-point mantissa length in iterative computations*, in Proceedings of the International Computing Symposium, A. Günter et al., ed., North Holland, 1974, pp. 315–322.

370. E. TSANG, *Foundations of Constraint Satisfaction*, Computation in Cognitive Science, Academic Press, 1993.

371. C. ULLRICH, *Scientific programming language concepts*, ZAMM, 76 (1996), pp. 57–60.

372. P. VAN HENTENRYCK, D. MCALLESTER, AND D. KAPUR, *Solving polynomial systems using a branch and prune approach*, SIAM Journal on Numerical Analysis, 34 (1997), pp. 797–827.

373. P. VAN HENTENRYCK, L. MICHEL, AND Y. DEVILLE, *Numerica: A Modeling Language for Global Optimization*, MIT Press, Cambridge, Mass., 1997.

374. P. VAN HENTENRYCK AND J.-F. PUGET, *A constraint satisfaction approach to a circuit design problem*, Journal of Global Optimization, 13 (1998), pp. 75–93.

375. R. VAN IWAARDEN, *An Improved Unconstrained Global Optimization Algorithm*, PhD thesis, University of Colorado at Denver, Denver, Colorado, 1996.

376. ——, *IADOL-C*. Available from the author., 1997.

377. R. S. VARGA, *On higher order stable implicit methods for solving parabolic differential equations*, J. Math. and Phys., 40 (1961), pp. 220–231.

378. A. A. VATOLIN, *On linear programming problems with interval coefficients*, J. Comp. Mathem. and Math. Phys., 24 (1984), pp. 1629–1637. In Russian.

379. T. VELDHUIZEN, *Expression templates*, C++ Report, 7 (1995).

380. T. VELDHUIZEN AND K. PONNAMBALAM, *Linear algebra with C++ template metaprograms*, Dr. Dobb's Journal of Software Tools, 21 (1996), pp. 38–44.

381. J. E. VOLDER, *The CORDIC computing technique*, IRE Transactions on Electronic Computers, EC-8 (1959), pp. 330–334. Reprinted in Earl E. Swartzlander, Jr., Computer Arithmetic, vol. 1, IEEE Computer Society Press, 1990, 14–17.

382. J. VON NEUMAN AND H. H. GOLDSTINE, *Numerical inverting of matrices of high order*, Bull. AMS, 53 (1947), pp. 1021–1099.

383. G. W. WALSTER, *M77 reference manual: Minnesota FORTRAN 1977 standards version*, Technical Report, University Computer Center, University of Minnesota, Minneapolis, Minn., 1983.

384. ———, *Philosophy and practicality of interval arithmetic*, in Reliability in Computing: The Role of Interval Methods in Scientific Computing, R. E. Moore, ed., Academic Press, Boston, 1988, pp. 309–323.

385. ———, *Stimulating hardware and software support for interval arithmetic*, in Applications of Interval Computations, R. B. Kearfott and V. Kreinovich, eds., Kluwer, Dordrecht, Netherlands, 1996, pp. 405–416.

386. ———, *The extended real interval system*. www.mscs.mu.edu/~globsol/Papers/extended_intervals.ps, April 1998.

387. G. W. WALSTER AND E. R. HANSEN, *Interval algebra, composite functions and dependence in compilers*. Submitted to Reliable Computing. Available at www.mscs.mu.edu/~globsol/Papers/composite.ps, 1998.

388. E. WALTER AND L. PRONZATO, *Identification of Parametric Models from Experimental Data*, Communications and Control Engineering, Springer-Verlag, Berlin, 1997.

389. W. V. WALTER, *ACRITH-XSC: A Fortran-like language for verified scientific computing*, in Scientific Computing with Automatic Result Verification, E. Adams and U. Kulisch, eds., Academic Press, Boston, 1993, pp. 45–70.

390. J. S. WALTHER, *A unified algorithm for elementary functions*, in Proc. Spring Joint Computer Conf., 1971, pp. 379–385.

391. S. WANG, V. PIURI, AND E. E. SWARTZLANDER, JR., *Hybrid CORDIC algorithms*, IEEE Trans. on Computers, 46 (1997), pp. 1202–1207.

392. Z. WANG, P. TONELLATO, G. F. CORLISS, AND R. B. KEARFOTT, *Globsol case study: Gene prediction (Genome Therapeutics)*, Technical Report, Department of Mathematics, Statistics and Computer Science, Marquette University, Milwaukee, Wisc., 1998.

393. G. WANNER, *On the integration of stiff differential equations*, Technical Report, Université de Genéve, Section de Mathematique, Geneva, October 1976.

394. ———, *On the integration of stiff differential equations*, in Proceedings of the Colloquium on Numerical Analysis, Internat. Ser. Numer. Math., No. 37, Basel, 1977, Birkhäuser, pp. 209–226.

395. M. WARMUS, *Calculus of approximations*, Bull. Acad. Polon. Sci., Cl. III, IV (1956), pp. 253–259.

396. ———, *Approximations and inequalities in the calculus of approximations. Classification of approximate numbers*, Bull. Acad. Polon. Sci., Ser. math. astr. et phys., IX (1961), pp. 241–245.

397. Y. WATANABE AND M. T. NAKAO, *Numerical verifications of solutions for nonlinear elliptic equations*, Japan J. Indust. Appl. Math., 10 (1993), pp. 165–178.

398. Y. WATANABE, N. YAMAMOTO, AND M. T. NAKAO, *A numerical verification method of solutions for the Navier-Stokes equations*, in Developments in Reliable Computing, T. Csendes, ed., Kluwer, Dordrecht, Netherlands, 1999, pp. 347–358. Also Reliable Computing 5/3 (1999), pp. 347–358.

399. WATERLOO MAPLE, *Maple V Learning Guide*, Springer-Verlag, Berlin, 1998.

400. K. WERNER, *Verifizierte Berechnung der inversen Weierstraß-Funktion und der elliptischen Funktion von Jacobi in beliebigen Maschinenarithmetiken*, PhD thesis, Universität Duisburg, Duisburg, 1996.

401. J. C. WILLEMS, *Least squares stationary optimal control and the algebraic Riccati equation*, IEEE Trans. Autom. Contr., 16 (1971), pp. 621–634.

402. J. WOLFF VON GUDENBERG, *Hardware support for interval arithmetic*, in Scientific Computing and Validated Numerics: Proceedings of the International Symposium on Scientific Computing, Computer Arithmetic and Validated Numerics - SCAN '95, G. Alefeld, A. Frommer, and B. Lang, eds., Berlin, 1996, Akademie Verlag, pp. 32–37.

403. S. WOLFRAM, *The Mathematica Book*, Cambridge Univ. Press, Cambridge, 3rd. ed., 1996.

404. W. F. WONG AND E. GOTO, *Fast hardware-based algorithms for elementary function computations using rectangular multipliers*, IEEE Transactions on Computers, 43 (1994), pp. 278–294.

405. ———, *Fast evaluation of the elementary functions in single precision*, IEEE Transactions on Computers, 44 (1995), pp. 453–457.

406. A. G. YAKOVLEV, *Multiaspectness and localization*, Interval Computations, 4 (1993), pp. 195–209.

407. N. YAMAMOTO, *A numerical verification method for solutions of boundary value problems with local uniqueness by Banach's fixed-point theorem*, SIAM J. Numer. Anal., 35 (1998), pp. 2004–2013.

408. W. YANG AND G. CORLISS, *Bibliography of computational differentiation*, in Computational Differentiation: Techniques, Applications, and Tools, M. Berz, C. H. Bischof, G. F. Corliss, and A. Griewank, eds., SIAM, Philadelphia, Penn., 1996, pp. 393–418.

409. R. YOUNG, *The algebra of many-valued quantities*, Math. Annalen, 104 (1932), pp. 260–290.

410. V. A. ZELOV AND M. J. SCHULTE, *Implementing and testing interval I/O in the GNU Fortran compiler*, Technical Report, Lehigh University, December 1997.

411. V. A. ZELOV, G. W. WALSTER, D. CHIRIAEV, AND M. J. SCHULTE, *Java(TM) programming environment for accurate numerical computing*. presented at the *International Workshop on Modern Software Tools for Scientific Computing*, Oslo, Norway, September 1998.

412. J. ZEMKE, *b4m - BIAS for Matlab*, Technical Report, Inst. f. Informatik III, Technische Universität Hamburg-Harburg, 1998.

413. G. M. ZIEGLER, *Lectures on Polytopes*, Graduate Texts in Mathematics, No. 152, Springer-Verlag, Berlin, 1995.

414. R. ZURMÜHL AND S. FALK, *Matrizen und ihre Anwendungen*, Springer-Verlag, Berlin, 1984.

415. V. S. ZYUZIN, *On a way of representation of the interval numbers*. Presented at SCAN '98 (Budapest, Sept. 22-25, 1998), 1998.